A GUIDEBOOK TO MECHANISM IN ORGANIC CHEMISTRY

A guidebook to mechanism in organic chemistry

Peter Sykes M.Sc., Ph.D.
Fellow of Christ's College, Cambridge

Longman
London and New York

LONGMAN GROUP LIMITED
Longman House, Burnt Mill, Harlow, Essex, UK

Published in the United States of America
by Longman Inc., New York

First published 1961
Second impression 1962
Third impression 1962
Fourth impression 1963
Fifth impression 1963
Second edition 1965
Second impression 1966
Third impression 1967
Fourth impression 1969
Third edition 1970
Second impression 1971
Fourth edition 1975
Second impression 1977
Third impression 1978
Fourth impression 1980
Fifth edition 1981
Second impression 1982

Translations into
German, 1964, 1976, 1982
Japanese, and Spanish, 1964
French, 1966
Italian, 1967
Portuguese, 1969

British Library Cataloguing in Publication Data

Sykes, Peter
    A guidebook to mechanism in organic chemistry.—
    5th ed.
    1. Chemistry, Physical organic
    2. Chemical reactions
    I. Title
    547′.1′39   QD476   80-40776
    ISBN 0-582-44121-8

Filmset in Northern Ireland at The Universities
Press (Belfast) Ltd. and printed in Singapore by
Selector Printing Co (Pte) Ltd.

For Joyce

Contents

Foreword

Fifty years ago the student taking up organic chemistry—and I speak from experience—was almost certain to be referred to one or other of a few textbooks generally known by the name of their authors—e.g. Holleman, Bernthsen, Schmidt, Karrer and Gattermann. On these texts successive generations of chemists were nurtured, and not in one country alone, for they were translated into several languages. These, the household names of fifty years ago, have for the most part gone. In past times of course the total number of books available was rather small and it is only in the last quarter of a century that we have seen a veritable flood of organic chemical textbooks pouring into book-sellers' lists. The increase in the number of texts may be in part due to the rise in student numbers but the primary reason for it is the revolu-tionary impact of mechanistic studies on our approach to organic chemistry at the elementary level. With the plethora of books avail-able, however, it is now much more difficult for an author to become a household name wherever the subject is taught. Yet this has indeed happened to Dr. Peter Sykes through his *Guidebook to Mechanism in Organic Chemistry*.

In the Foreword which I was privileged to write for the First Edition in 1961 I described not only my own view of what was happen-ing in organic chemistry but also the type of approach to teaching it which was favoured by Dr. Sykes. Having known and watched him over many years first as student, then as colleague, and always as friend, I was confident that he had written an excellent book which, in my view at least, would add new interest to the study of organic chemistry. But its success has far exceeded even my high expectations and in its later editions it has been revised and refined without ever losing the cutting edge of the original.

The present volume continues the tradition. Once again the recent literature has been combed for new examples the better to exemplify principles of reactions. Of particular interest is an admirable chapter dealing with reactions controlled by orbital symmetry. Until I read it I was not convinced that this very important new development in the theory of organic reactions could be simply yet usefully communicated to students at an elementary level. To have succeeded in doing so only underlines further Dr. Sykes' gifts as a teacher and writer and I am sure that this new edition of the Guidebook will more than equal the success of its predecessors.

Cambridge   TODD

Preface to fifth edition

When preparations were being made for the last edition (4th, 1975), the intention was to incorporate a number of relatively small changes in content, to redesign several reaction schemes, and to incorporate more cogent examples as and when this proved possible. In the event the outcome was very different: the book was almost entirely rewritten, and a new chapter—on reactions controlled by orbital symmetry—was added.

This time no such complete rewriting was envisaged, nor was it found to be necessary: indeed, had it proved to be necessary then something must have been sadly wrong with the previous radical revision! A number of changes have been made, however, including the exclusion of pK_b as a measure of the strength of bases, and its replacement by pK_a to allow the use of a single continuous scale for measuring the strength of acids <u>and</u> bases.

The major change this time has been the addition of a new chapter on linear free energy relationships—describing the attempts to relate structure and reactivity on a quantitative basis. This development has been made in response to suggestions from a number of quarters, but only after considerable investigation and trial to ensure that the topic could be treated simply, yet adequately, at the existing level of sophistication of the 'Guidebook': I now indeed believe that it can.

Once again considerable care has been taken to resist any needless increase either in the size or in the intellectual demand of the book. Coupled with this has been a constant striving to keep explanations as simple and as clear as possible. It is therefore hoped that this new edition will continue to find favour among sixth-form students, as well as with those in universities, polytechnics, technical colleges, colleges of education, and colleges of further education.

I should, as always, like to thank the many readers who have so kindly taken the trouble to draw my attention to errors in previous editions, and who have also made numerous suggestions for improvement; wherever possible these have been incorporated. It is a source of great pleasure that readers are so helpful in this respect, and I should warmly appreciate receiving further such suggestions for the improvement of this new edition.

Finally, acknowledgement is made to the copyright holders, where relevant, for kind permission to reprint figures in Chapter 13 as follows: the American Chemical Society for Fig. 13.1 (Hammett, L. P. and Pfluger, H. L., *J. Amer. Chem. Soc.*, 1933, **55**, 4083), Fig. 13.2 (Hammett, L. P., and Pfluger H. L., *J. Amer. Chem. Soc.*, 1933, **55,** 4086), Fig. 13.3 (Hammett, L. P., *Chem. Rev.*, 1935, **17**, 131)‚ Fig. 13.4 (Taft, R. W. and Lewis, I. C., *J. Amer. Chem. Soc.*, 1958, **80**, 2437), Fig. 13.5 (Brown, H. C. and Okamoto, Y., *J. Amer. Chem. Soc.*, 1957, **79,** 1915), Fig. 13.6 (Brown, H. C., Schleyer, P. von R. *et al.*, *J. Amer. Chem. Soc.*, 1970, **92,** 5244), Fig. 13.8 (Hart, H. and Sedor, F. A., *J. Amer. Chem. Soc.*, 1967, **89,** 2344); the Chemical Society and Professor J. A. Leisten for Fig. 13.7 (Leisten, J. A. and Kershaw, D. N., *Proc. Chem. Soc.*, 1960, 84).

Cambridge
September 1980

PETER SYKES

1

Structure, reactivity, and mechanism

The chief advantage of a mechanistic approach, to the vast array of
disparate information that makes up organic chemistry, is the way in
which a relatively small number of guiding principles can be used, not
only to explain and interrelate existing facts, but to forecast the outcome of changing the conditions under which already known reactions
are carried out, and to foretell the products that may be expected
from new ones. It is the business of this chapter to outline some of
these guiding principles, and to show how they work. As it is the
compounds of carbon with which we shall be dealing, something
must be said about the way in which carbon atoms can form bonds
with other atoms, especially with other carbon atoms.

1.1   ATOMIC ORBITALS

The carbon atom has, outside its nucleus, six electrons which, on the
Bohr theory of atomic structure, were believed to be arranged in
orbits at increasing distance from the nucleus. These orbits corres-

ponded to gradually increasing levels of energy, that of lowest energy, the 1s, accommodating two electrons, the next, the 2s, also accommodating two electrons, and the remaining two electrons of the carbon atom going into the 2p level, which is actually capable of accommodating a total of six electrons.

The Heisenberg indeterminacy principle, and the wave-mechanical view of the electron, have made it necessary to do away with anything so precisely defined as actual orbits. Instead the wave-like electrons are now symbolised by *wave functions*, ψ, and the precise, classical orbitals of Bohr are superseded by three-dimensional *atomic orbitals* of differing energy level. The size, shape and orientation of these atomic orbitals—regions in which there is the greatest probability of finding an electron corresponding to a particular, quantised energy level—are each delineated by a wave function, ψ_A, ψ_B, ψ_C, etc. The orbitals are indeed rather like three-dimensional electronic contour maps, in which ψ^2 determines the relative probability of finding an electron at a particular point in the orbital.

The relative *size* of atomic orbitals, which is found to increase as their energy level rises, is defined by the principal quantum number, n, their *shape* and *spatial orientation* (with respect to the nucleus and each other) by the subsidiary quantum numbers, l and m, respectively.* Electrons in orbitals also have a further designation in terms of the *spin* quantum number, which can have the values $+\frac{1}{2}$ or $-\frac{1}{2}$. One limitation that theory imposes on such orbitals is that each may accommodate not more than two electrons, these electrons being distinguished from each other by having opposed (*paired*) spins.† This follows from the Pauli exclusion principle, which states that no two electrons in any atom may have exactly the same set of quantum numbers.

It can be shown, from wave-mechanical calculations, that the 1s orbital (quantum numbers $n = 1$, $l = 0$, $m = 0$, corresponding to the classical K shell) is spherically symmetrical about the nucleus of the atom, and that the 2s orbital (quantum numbers $n = 2$, $l = 0$, $m = 0$) is similarly spherically symmetrical, but at a greater distance from the nucleus; there is a region between the two latter orbitals where the probability of finding an electron approaches zero (a *spherical nodal surface*).

As yet, this marks no radical departure from the classical picture of orbitals, but with the 2p level (the continuation of the L shell) a difference becomes apparent. Theory now requires the existence of *three 2p orbitals* (quantum numbers $n = 2$, $l = 1$, with $m = +1$, 0, and

* n can have values of $1, 2, 3, \ldots, l$ values of $0, 1, 2, \ldots, n - 1$, and m values of $0, \pm 1, \pm 2, \ldots, \pm l$. We shall normally be concerned only with l values of 0 and 1, the corresponding orbitals being referred to (from spectroscopic terminology) as s and p orbitals, respectively, e.g. 1s, 2s, 2p orbitals, etc.

† One electron with spin quantum number $+\frac{1}{2}$, the other $-\frac{1}{2}$.

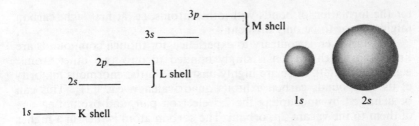

−1, respectively), all of the same shape and energy level (orbitals having the same energy level are described as *degenerate*), but differing from each other in their spatial orientation. They are in fact arranged mutually at right-angles along notional x, y and z axes and, therefore, designated as $2p_x$, $2p_y$ and $2p_z$, respectively. Further, these three $2p$ orbitals are found not to be spherically symmetrical, like the $1s$ and $2s$, but 'dumb-bell' shaped with a plane, in which there is zero probability of finding an electron (*nodal plane*), passing through the nucleus (at right-angles to the x, y and z axes, respectively), and so separating the two halves of each dumb-bell:

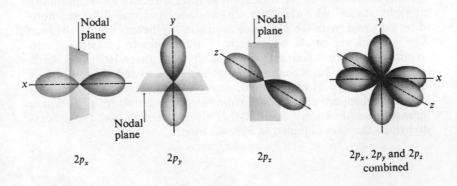

The six electrons of the carbon atom are then accommodated in atomic orbitals of increasing energy level until all are assigned (the *aufbau*, or build-up, principle). Thus two electrons, with paired spins, will go into the $1s$ orbital, a further two into the $2s$ orbital, but at the $2p$ level the remaining two electrons could be accommodated either in the same, e.g. $2p_x$, or different, e.g. $2p_x$ and $2p_y$, orbitals. Hund's rule, which states that two electrons will avoid occupying the same orbital so long as there are other energetically equivalent, i.e. degenerate, orbitals still empty, will apply, and the *electron configuration* of carbon will thus be $1s^2 2s^2 2p_x^1 2p_y^1$, with the $2p_z$ orbital remaining unoccupied. This represents the *ground state* of the free carbon atom in which only *two* unpaired electrons (in the $2p_x$ and $2p_y$ orbitals) are available for the formation of bonds with other atoms, i.e. at first sight carbon might appear to be only divalent.

This, however, is contrary to experience, for though compounds are known in which carbon is singly bonded to only two other atoms, e.g. CCl_2 (p. 260), these are highly unstable: in the enormous majority of its compounds carbon exhibits quadrivalency, e.g. CH_4. This can be achieved by uncoupling the $2s^2$ electron pair, and promoting one of them to the vacant $2p_z$ orbital. The carbon atom is then in a higher energy (*excited*) state, $1s^2 2s^1 2p_x^1 2p_y^1 2p_z^1$, but as it now has *four* unpaired electrons it is able to form *four*, rather than only *two*, bonds with other atoms or groups. The large amount of energy produced by forming these two extra bonds considerably outweighs that required [≈ 406 kJ (97 kcal) mol^{-1}] for the initial $2s^2$ uncoupling, and $2s \rightarrow 2p$ promotion.

1.2 HYBRIDISATION

A carbon atom combining with four other atoms clearly does not use the one $2s$ and the three $2p$ atomic orbitals that would now be available, for this would lead to the formation of three directed bonds, mutually at right angles (with the three $2p$ orbitals), and one different, non-directed bond (with the spherical $2s$ orbital). Whereas in fact, the four C—H bonds in, for example, methane are known to be identical and symmetrically (tetrahedrally) disposed at an angle of 109° 28′ to each other. This may be accounted for on the basis of redeploying the $2s$ and the three $2p$ atomic orbitals so as to yield four new (identical) orbitals, which are capable of forming stronger bonds (*cf.* p. 5). These new orbitals are known as sp^3 *hybrid* atomic orbitals, and the process by which they are obtained as *hybridisation*:

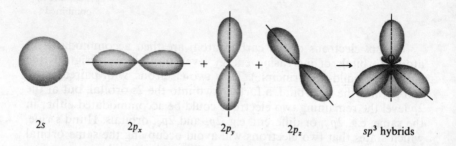

$2s$ $2p_x$ $2p_y$ $2p_z$ sp^3 hybrids

It should, however, be firmly emphasised, despite the diagram above, that hybridisation is a mathematical device in calculation and not a physical reality.

Similar, but different, redeployment is envisaged when a carbon atom combines with three other atoms, e.g. in ethene (ethylene) (p. 8): three sp^2 hybrid atomic orbitals disposed at 120° to each other in the same plane (*plane trigonal hybridisation*) are then employed. Finally, when carbon combines with two other atoms, e.g. in ethyne (acetylene) (p. 9): two sp^1 hybrid atomic orbitals disposed at 180° to each other (*digonal hybridisation*) are employed. In each case the *s* orbital is always involved as it is the one of lowest energy level.

These are all valid ways of deploying one 2*s* and three 2*p* atomic orbitals—in the case of sp^2 hybridisation there will be one unhybridised *p* orbital also available (p. 8), and in the case of sp^1 hybridisation there will be two (p. 10). Other, equally valid, modes of hybridisation are also possible in which the hybrid orbitals are not necessarily identical with each other. Hybridisation takes place so that the atom concerned can form as strong bonds as possible, and so that the other atoms thus bonded (and the electron pairs constituting the bonds) are as far apart from each other as possible, i.e. so that the total intrinsic energy of the resultant compound is at a minimum.

1.3 BONDING IN CARBON COMPOUNDS

Bond formation between two atoms is then envisaged as the progressive overlapping of the atomic orbitals of the two participating atoms, the greater the possible overlapping (the *overlap integral*), the stronger the bond so formed. The relative overlapping powers of atomic orbitals have been calculated as follows:

$$s = 1{\cdot}00$$
$$p = 1{\cdot}72$$
$$sp^1 = 1{\cdot}93$$
$$sp^2 = 1{\cdot}99$$
$$sp^3 = 2{\cdot}00$$

It will thus be apparent why the use of hybrid orbitals, e.g. sp^3 hybrid orbitals in the combination of one carbon and four hydrogen atoms to form methane, results in the formation of stronger bonds.

When the atoms have come sufficiently close together, it can be shown that their two atomic orbitals are replaced by two *molecular orbitals*, one having less energy and the other more than the energies of the two separate atomic orbitals. These two new molecular orbitals spread over both atoms and either may contain the two electrons (Fig. 1.1):

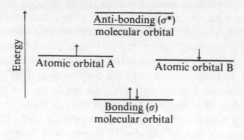

Fig. 1.1

The molecular orbital of reduced energy is called the *bonding orbital*, and its occupancy results in the formation of a stable bond between the two atoms. In the above case, the pair of electrons constituting the bond tend to be concentrated between the two positively charged atomic nuclei, which can thus be thought of as being held together by the negative charge between them. The molecular orbital of increased energy is called the *anti-bonding orbital*; this corresponds to a state in which the inter-nuclear space remains largely empty of electrons, and thus results in repulsion between the two positively charged atomic nuclei. The anti-bonding orbital remains empty in the ground state of the molecule, and need not here be further considered in the formation of stable bonds between atoms.

If overlap of the two atomic orbitals has taken place along their major axes, the resultant bonding molecular orbital is referred to as a σ orbital,* and the bond formed as a σ bond. The σ molecular orbital, and the electrons occupying it, are found to be *localised* symmetrically about the internuclear axis of the atoms that are bonded to each other. Thus on combining with hydrogen, the four hybrid sp^3 atomic orbitals of carbon overlap with the $1s$ atomic orbitals of four hydrogen atoms to form four identical, strong σ bonds, making angles of 109° 28' with each other (the regular tetrahedral angle), in methane. A similar, exactly regular, tetrahedral structure will result with, for example, CCl_4, but where the atoms bonded to carbon are not all the same, the spatial arrangement may depart slightly from the exactly symmetrical while remaining tetrahedral.

1.3.1 Carbon–carbon single bonds

The combination of two carbon atoms, for example in ethane, results from the axial overlap of two sp^3 atomic orbitals, one from each

---

* The anti-bonding molecular orbital is referred to as a σ^* orbital.

carbon atom, to form a strong σ bond between them. The carbon–carbon bond length in saturated compounds is found to be pretty constant—0·154 nm (1·54 Å). This refers, however, to a carbon–carbon single bond between sp^3 hybridised carbons. A similar single bond between two sp^2 hybridised carbons, $=CH-CH=$, is found on average to be about 0·147 nm (1·47 Å) in length, and one between two sp^1 hybridised carbons, $\equiv C-C\equiv$, about 0·138 nm (1·38 Å). This is not really surprising, for an s orbital and any electrons in it are held closer to, and more tightly by, the nucleus than is a p orbital and any electrons in it. The same effect will be observed with hybrid orbitals as their s component increases, and for two carbon atoms bonded to each other the nuclei are drawn inexorably closer together on going from

$$sp^3\text{--}sp^3 \longrightarrow sp^2\text{--}sp^2 \longrightarrow sp^1\text{--}sp^1.$$

We have not, however, defined a unique structure for ethane; the σ bond joining the two carbon atoms is symmetrical about a line joining the two nuclei, and, theoretically, an infinite variety of different structures is still possible, defined by the position of the hydrogens on one carbon atom relative to the position of those on the other. The two extremes, of all the possible species, are known as the *eclipsed* and *staggered* forms:

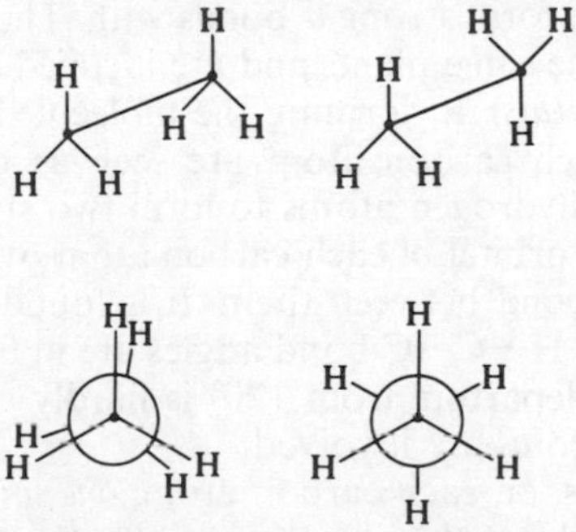

Eclipsed   Staggered

The above quasi three-dimensional representations are known as 'sawhorse' and Newman projections, respectively. The eclipsed and staggered forms, and the infinite variety of possible structures lying between them as extremes, are known as *conformations* of the ethane molecule; conformations being defined as different arrangements of the same group of atoms that can be converted into one another without the breaking of any bonds.

The staggered conformation is likely to be the more stable of the two for hydrogen atoms on the adjacent carbons are as far apart from each other as they can get [0·310 nm (3·1 Å)], and any so-called 'non-bonded' interaction between them is thus at a minimum; whereas in the eclipsed conformation they are suffering the maximum of crowding [0·230 nm (2·3 Å), slightly less than the sum of their van der Waals radii]. The long cherished principle of free rotation about a carbon–carbon single bond is not contravened, however, as it has been shown that the eclipsed and staggered conformations differ by only ≈ 12 kJ (3 kcal) mol^{-1} in energy content at 25°, and this is small enough to allow their ready interconversion through the agency of ordinary thermal motions at room temperature—the rotation frequency at 25° being $\approx 10^{12}$ sec^{-1}. That such crowding *can* lead to a real restriction of rotation about a carbon–carbon single bond has been confirmed by the isolation of two forms of $CHBr_2CHBr_2$, though admittedly only at low temperatures where collisions between molecules do not provide enough energy to effect the interconversion.

1.3.2   Carbon–carbon double bonds

In ethene each carbon atom is bonded to only *three* other atoms, two hydrogens and one carbon. Strong σ bonds are formed with these three atoms by the use of *three* orbitals derived by hybridising the 2s and, this time, *two* only of the carbon atom's 2p atomic orbitals—an atom will normally only mobilise as many hybrid orbitals as it has atoms or groups to form strong σ bonds with. The resultant sp^2 hybrid orbitals all lie in the same plane, and are inclined at 120° to each other (*plane trigonal orbitals*). In forming the molecule of ethene, two of the sp^2 orbitals of each carbon atom are seen as overlapping with the 1s orbitals of two hydrogen atoms to form two strong σ C—H bonds, while the third sp^2 orbital of each carbon atom overlap axially to form a strong σ C—C bond between them. It is found experimentally that the H—C—H and H—C—C bond angles are in fact 116·7° and 121·6°, respectively. The departure from 120° is hardly surprising seeing that different trios of atoms are involved.

This then leaves, on each carbon atom, *one unhybridised* 2p atomic orbital at right angles to the plane containing the carbon and hydrogen atoms. These two 2p atomic orbitals are parallel to each other and can themselves overlap, resulting in the formation of a bonding molecular orbital spreading over both carbon atoms and situated above and below the plane (i.e. it has a node in the plane of the molecule) containing the two carbon and four hydrogen atoms ($\diagdown$ indicates bonds to atoms lying *behind* the plane of the paper, and $\blacktriangleleft$ bonds to those lying in *front* of it):

This new bonding molecular orbital is known as a π orbital,* and the electrons that occupy it as π electrons. The new π bond that is thus formed has the effect of drawing the carbon atoms closer together, and the C=C distance in ethene is found to be 0·133 nm (1·33 Å), compared with a C—C distance of 0·154 nm (1·54 Å) in ethane. The *lateral* overlap of the p atomic orbitals that occurs in forming a π bond is less effective than the axial overlap that occurs in forming a σ bond, and the former is thus weaker than the latter. This is reflected in the fact that the energy of a carbon–carbon double bond, though more than that of a single bond is, indeed, less than twice as much. Thus the C—C bond energy in ethane is 347 kJ (83 kcal) mol^{-1}, while that of C=C in ethene is only 598 kJ (143 kcal) mol^{-1}.

The lateral overlap of the two $2p$ atomic orbitals, and hence the strength of the π bond, will clearly be at a maximum when the two carbon and four hydrogen atoms are exactly coplanar, for it is only in this position that the p atomic orbitals are exactly parallel to each other, and thus capable of the maximum overlapping. Any disturbance of this coplanar state, by twisting about the σ bond joining the two carbon atoms, would lead to reduction in π overlapping, and hence a decrease in the strength of the π bond: it will thus be resisted. A theoretical justification is thus provided for the long observed resistance to rotation about a carbon–carbon double bond. The distribution of the π electrons in two lobes, above and below the plane of the molecule, and extending beyond the carbon–carbon bond axis, means that a region of negative charge is effectively waiting there to welcome any electron-seeking reagents (e.g. oxidising agents); so that it comes as no surprise to realise that the characteristic reactions of a carbon–carbon double bond are predominantly with such reagents (*cf.* p. 175). Here the classical picture of a double bond has been replaced by an alternative, in which the two bonds joining the carbon atoms, far from being identical, are considered to be different in nature, strength and position.

1.3.3 Carbon–carbon triple bonds

In ethyne each carbon atom is bonded to only *two* other atoms, one hydrogen and one carbon. Strong σ bonds are formed with these two atoms by the use of *two* hybrid orbitals derived by hybridising the $2s$ and, this time, *one* only of the carbon atom's $2p$ atomic orbitals. The resultant *digonal sp*^{1} hybrid orbitals are co-linear. Thus, in forming the molecule of ethyne, these hybrid orbitals are used to form strong σ bonds between each carbon atom and one hydrogen atom, and between the two carbon atoms themselves, resulting in a linear molecule

---

* An anti-bonding, π*, molecular orbital is also formed (*cf.* p. 12).

having *two* unhybridised 2*p* atomic orbitals, at right angles to each other, on each of the two carbon atoms. The atomic orbitals on one carbon atom are parallel to those on the other, and can thus overlap with each other resulting in the formation of *two* π bonds in planes at right angles to each other:

$$H—C{\equiv}C—H \quad \rightarrow \quad H—C—C—H \quad \equiv \quad H—C—C—H$$

The ethyne molecule is thus effectively sheathed in a cylinder of negative charge. The C≡C bond energy is 812 kJ (194 kcal) mol^{-1}, so that the increment due to the third bond is less than that occurring on going from a single to a double bond. The C≡C bond distance is 0·120 nm (1·20 Å) so that the carbon atoms have been drawn still further together, but here again the decrement on going C=C → C≡C is smaller than that on going C—C → C=C.

1.3.4 Carbon–oxygen and carbon–nitrogen bonds

An oxygen atom has the electron configuration $1s^2 2s^2 2p_x^2 2p_y^1 2p_z^1$, and it too, on combining with other atoms, can be looked upon as utilising hybrid orbitals so as to form the strongest possible bonds. Thus on combining with the carbon atoms of two methyl groups, to form methoxymethane (dimethyl ether), CH_3—$\ddot{O}$—CH_3, the oxygen atom could use four sp^3 hybrid orbitals: two to form σ bonds by overlap with an sp^3 orbital of each of the two carbon atoms, and the other two to accommodate its two lone pairs of electrons. The C—O—C bond angle is found to be 110°, the C—O bond length, 0·142 nm (1·42 Å), and the bond energy, 360 kJ (86 kcal) mol^{-1}.

An oxygen atom can also form a double bond to carbon; thus in propanone (acetone), $Me_2C{=}\ddot{O}:$, the oxygen atom could use three sp^2 hybrid orbitals: one to form a σ bond by overlap with an sp^2 orbital of the carbon atom, and the other two to accommodate the two lone pairs of electrons. This leaves an unhybridised *p* orbital on both oxygen and carbon, and these can overlap with each other laterally (*cf.* C=C, p. 9) to form a π bond:

The C—C—O bond angle is found to be $\approx 120°$, the C=O bond length, $0\cdot122$ nm ($1\cdot22$ Å), and the bond energy, 750 kJ (179 kcal) mol^{-1}. The fact that this is very slightly greater than twice the C—O bond energy, whereas the C=C bond energy is markedly less than twice that of C—C, may be due in part to the fact that the lone pairs on oxygen are further apart, and so more stable, in C=O than in C—O; there being no equivalent circumstance with carbon. The fact that carbon–oxygen, unlike carbon–carbon, bonds are polar linkages also plays a part.

A nitrogen atom, with the electron configuration $1s^2 2s^2 2p_x^1 2p_y^1 2p_z^1$, can also be looked upon as using hybrid orbitals in forming single, C—N, double, C=N, and triple, C≡N, bonds with carbon. In each case one such orbital is used to accommodate the nitrogen lone pair of electrons; in double and triple bond formation one and two π bonds, respectively, are also formed by lateral overlap of the unhybridised p orbitals on nitrogen and carbon. Average bond lengths and bond energies are single, $0\cdot147$ nm ($1\cdot47$ Å) and 305 kJ (73 kcal) mol^{-1}, double, $0\cdot129$ nm ($1\cdot29$ Å) and 616 kJ (147 kcal) mol^{-1}, and triple, $0\cdot116$ nm ($1\cdot16$ Å) and 893 kJ (213 kcal) mol^{-1}.

1.3.5 Conjugation

When we come to consider molecules that contain more than one multiple bond, e.g. dienes with two C=C bonds, it is found that compounds in which the bonds are *conjugated* (alternating multiple and single; 1) are slightly more stable than those in which they are *isolated* (2):

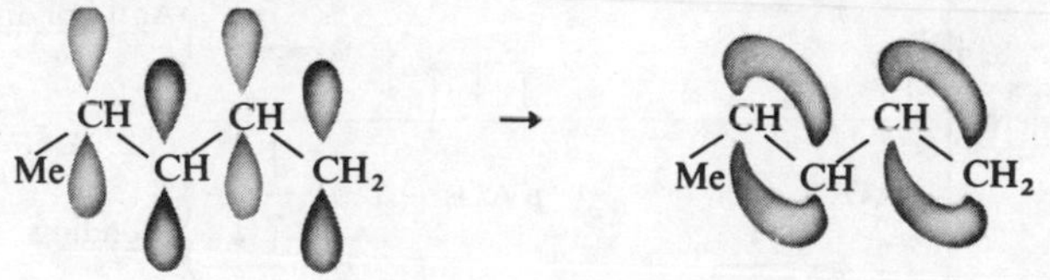

(1a) (1b)

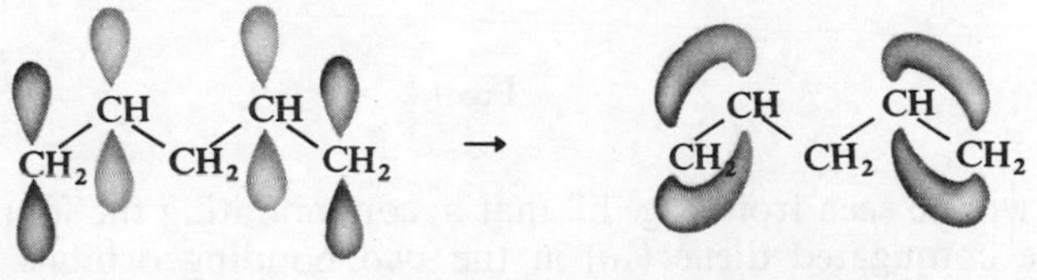

(2a) (2b)

This greater thermodynamic stability (lower energy content) of conjugated molecules is revealed in (1) having a lower heat of combustion, and a lower heat of hydrogenation than (2); and also in the general observation that isolated double bonds can often be made to migrate quite readily so that they become conjugated:

$$MeCH{=}CH{-}CH_2{-}\underset{\underset{Me}{|}}{C}{=}O \xrightarrow[\text{catalyst}]{\text{Base}} MeCH_2{-}CH{=}CH{-}\underset{\underset{Me}{|}}{C}{=}O$$

Conjugation is not of course confined to carbon–carbon multiple bonds.

With both (1*a*) and (2*a*) above, lateral overlap of the *p* atomic orbitals on adjacent carbon atoms could lead to the formation of two localised π bonds as shown, and the compounds would thus be expected to resemble ethene, only twice as it were! This is indeed found to be the case with (2), but (1) is found to behave differently in terms of its slightly greater stability (referred to above), in spectroscopic behaviour (see below), and in undergoing addition reactions more readily than does an isolated diene (p. 191). On looking more closely, however, it is seen that with (1*a*), but not with (2*a*), lateral overlap could take place between <u>all four</u> *p* atomic orbitals on adjacent carbon atoms. Such overlap will result in the formation of four molecular orbitals (Fig. 1.2), two bonding (ψ_1 and ψ_2) and two antibonding (ψ_3 and ψ_4)—the overlap of *n* atomic orbitals always gives rise to *n* molecular orbitals):

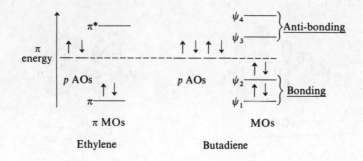

Fig. 1.2

It will be seen from Fig. 1.2 that accommodating the four electrons of the conjugated diene (1*a*) in the two bonding orbitals as shown leads to a lower total energy for the compound than—by analogy with ethene—accommodating them in two localised π bonds. The electrons are said to be *delocalised*, as they are now held in common by the whole of the conjugated system rather than being localised over two carbon atoms in π bonds, as in ethene or in (1*b*). Accommodation of the four electrons in the bonding molecular orbitals ψ_1 and ψ_2 results in electron distribution in a charge cloud as in (3):

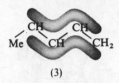

(3)

For such delocalisation to occur the four *p* atomic orbitals in (1*a*) would have to be essentially parallel, and this would clearly impose considerable restrictions on rotation about the C_2—C_3 bond in (3), which is indeed observed in practice as highly preferred conformations. It might also be expected that the π electron density between C_2 and C_3 would result in this bond having some double bond character, e.g. in its being shorter than a C—C single bond. The observed bond length is indeed short—0·147 nm (1·47 Å)—though no shorter than might be expected for a single bond between sp^2 hybridised carbon atoms (*cf.* p. 9). The stabilisation energy of a simple conjugated diene, compared with the corresponding isolated one, is relatively small—*ca.* 17 kJ (4 kcal) mol^{-1}—and even this cannot be ascribed wholly to electron delocalisation: the state of hybridisation of the carbon atoms involved, and the consequent differing strengths of the σ bonds between them, must also be taken into account.

Delocalisation is, however, much involved in stabilising the excited states of dienes, and of polyenes in general, i.e. in lowering the energy level of their excited states. The effect of this is to reduce the energy gap between ground and excited states of conjugated molecules, as compared with those containing isolated double bonds, and this energy gap is progressively lessened as the extent of conjugation increases. This means that the amount of energy required to effect the promotion of an electron, from ground to excited state, decreases with increasing conjugation, i.e. the wavelength at which the necessary radiation is absorbed increases. Simple dienes absorb in the ultraviolet region, but as the extent of conjugation increases the absorption gradually moves towards the visible range, i.e. the compound becomes coloured. This is illustrated by the series of $\alpha\omega$-diphenylpolyenes below:

$C_6H_5(CH{=}CH)_nC_6H_5$	*Colour*
$n = 1$	colourless
$n = 2$–4	yellow
$n = 5$	orange
$n = 8$	red

1.3.6   Benzene and aromaticity

One of the major problems of elementary organic chemistry is the detailed structure of benzene. The known planar structure of the molecule implies sp^2 hybridisation with p atomic orbitals, at right angles to the plane of the nucleus, on each of the six carbon atoms (4):

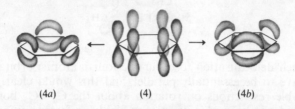

(4a)   (4)   (4b)

Overlapping could, of course, take place, 1,2; 3,4; 5,6; *or* 1,6; 5,4; 3,2, leading to formulations corresponding to the Kekulé structures (4a and 4b); but, as an alternative, all six adjacent p orbitals could overlap, as with conjugated dienes (p. 12), resulting in the formation of six molecular orbitals, three bonding ($\psi_1 \rightarrow \psi_3$) and three anti-bonding ($\psi_4 \rightarrow \psi_6$), with energy levels as represented below (Fig. 1.3):

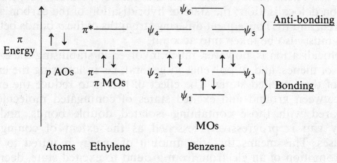

Fig. 1.3

The bonding MO of lowest energy (ψ_1) is cyclic and embraces all six carbon atoms, i.e. is delocalised. It has a nodal plane in the plane of the ring, so that there are two annular lobes, one above and one below the plane of the ring, of which only the upper one (looking down from above) is shown in (5a): two electrons are thus accommodated. The two further bonding MOs (ψ_2 and ψ_3), of equal energy (degenerate), also encompass all six carbon atoms, but each has a further nodal plane, at right angles to the plane of the ring, in addition to the one in the plane of the ring. Each MO thus has four lobes of which only the upper pair (looking down from above) are shown in (5b) and (5c):

each of these two MOs accommodates two more electrons—thus making six in all:

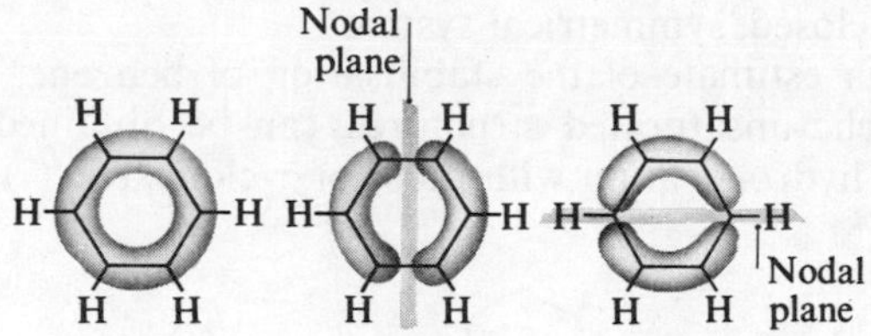

(5a)   (5b)   (5c)

The net result is an annular electron cloud, above and below the plane of the ring (6):

(6)

The influence of this cloud of negative charge on the type of reagents that will attack benzene is discussed below (p. 130).

Support for the above view is provided by the observation that all the carbon–carbon bond lengths in benzene are exactly the same,* 0·140 nm (1·40 Å), i.e. benzene is a regular hexagon with bond lengths somewhere in between the normal values for a single [0·154 nm (1·54 Å)] and a double [0·133 nm (1·33 Å)] bond. This regularity may be emphasised by avoiding writing Kekulé structures for benzene, as these are clearly an inadequate representation, and using instead:

There remains, however, the question of the much remarked thermodynamic stability of benzene. Part of this no doubt arises from the disposition of the three plane trigonal σ bonds about each carbon at their optimum angle of 120° (the regular hexagonal angle), but a larger part stems from the use of cyclic, delocalised molecular orbitals to accommodate the six residual electrons; this is a considerably more stable (lower energy) arrangement than accommodating the electrons

* As also are all the C—H bond lengths at 0·108 nm (1·08 Å).

in three localised π molecular orbitals, as is apparent from Fig. 1.3 (p. 14). The much greater stabilisation in benzene than in, for example, conjugated dienes (*cf.* p. 13) presumably stems from benzene being a cyclic, i.e. closed, symmetrical system.

A rough estimate of the stabilisation of benzene, compared with simple cyclic unsaturated structures, can be obtained by comparing its heat of hydrogenation with those of cyclohexene (7) and cyclohexa-1,3-diene (8):

$$+ \; H_2 \; \rightarrow \qquad \Delta H = -120 \, \text{kJ} \, (-28 \cdot 6 \, \text{kcal}) \, \text{mol}^{-1}$$

(7)

$$+ \; 2H_2 \; \rightarrow \qquad \Delta H = -232 \, \text{kJ} \, (-55 \cdot 6 \, \text{kcal}) \, \text{mol}^{-1}$$

(8)

$$+ \; 3H_2 \; \rightarrow \qquad \Delta H = -208 \, \text{kJ} \, (-49 \cdot 8 \, \text{kcal}) \, \text{mol}^{-1}$$

The heat of hydrogenation of the cyclic diene (8) is very nearly twice that of cyclohexene (7), and the heat of hydrogenation of the three double bonds in a Kekulé structure might thus be expected to be of the order of $3 \times -120 \, \text{kJ} \, (-28 \cdot 6 \, \text{kcal}) \, \text{mol}^{-1} = -360 \, \text{kJ} \, (-85 \cdot 8 \, \text{kcal}) \, \text{mol}^{-1}$; but when 'real' benzene is hydrogenated only $-208 \, \text{kJ} \, (-49 \cdot 8 \, \text{kcal}) \, \text{mol}^{-1}$ are evolved. 'Real' benzene is thus thermodynamically more stable than the hypothetical 'cyclohexatriene' by $151 \, \text{kJ} \, (36 \, \text{kcal}) \, \text{mol}^{-1}$; this compares with only $\approx 17 \, \text{kJ} \, (4 \, \text{kcal}) \, \text{mol}^{-1}$ by which a conjugated diene is stabilised, with respect to its analogue in which there is no interaction between the electrons of the double bonds.

In marked contrast to benzene above, the heat of hydrogenation of cyclooctatetraene (9) to cyclooctane (10) is $-410 \, \text{kJ} \, (98 \, \text{kcal}) \, \text{mol}^{-1}$, while that of cyclooctene (11) is $-96 \, \text{kJ} \, (23 \, \text{kcal}) \, \text{mol}^{-1}$:

$$\xrightarrow{\;4H_2\;} \qquad \xleftarrow{\;H_2\;}$$

(9)   (10)   (11)

$$\Delta H = -410 \, \text{kJ} \, (-98 \, \text{kcal}) \, \text{mol}^{-1} \qquad\qquad \Delta H = -96 \, \text{kJ} \, (-23 \, \text{kcal}) \, \text{mol}^{-1}$$

The difference between ΔH for (9) and $4 \times \Delta H$ for (11) is thus *minus* $26 \, \text{kJ} \, (-6 \, \text{kcal}) \, \text{mol}^{-1}$: cyclooctatetraene, unlike benzene, exhibits no characteristic stabilisation when compared with the relevant hypothetical cyclic polyene (it is in fact slightly *de*stabilised), i.e. it is not aromatic. This lack of aromatic character is, on reflection, not really surprising for cyclic *p* orbital overlap, as with benzene, would require (9) to be flat with a consequent C—C—C bond angle of 135°, resulting in very considerable ring strain for an array of sp^2 hybridised carbons (preferred angle 120°). Such strain can be relieved by puckering of the ring, but only at the expense of sacrificing the possibility of overall *p* orbital overlap. That such puckering occurs can be seen from X-ray crystallographic measurements, which show cyclooctatetraene to have the 'tub' structure (9*a*) with alternating double [0·133 nm (1·33 Å)], and single [0·146 nm (1·46 Å)], carbon–carbon bonds:

(9*a*)

Conditions necessary for cyclic polyenes to possess aromatic character are referred to below.

The amount by which benzene is stabilised compared with the hypothetical 'cyclohexatriene' should properly be called its *stabilisation energy*, it is, however, often called its *delocalisation energy*, which immediately begs the question as to how much of the stabilisation is actually due to delocalisation of the 6π electrons in benzene. The term *resonance energy*, though still widely used, is highly unsatisfactory on semantic grounds as it immediately conjures up visions of rapid oscillation between one structure and another, e.g. the Kekulé structures, thus entirely misrepresenting the real state of affairs (*cf.* p. 19).

The requirements necessary for the occurrence of aromatic stabilisation, and character, in cyclic polyenes appear to be: (*a*) that the molecule should be flat (to allow of cyclic overlap of *p* orbitals); and (*b*) that all the bonding orbitals should be completely filled. This latter condition is fulfilled in cyclic systems with $4n + 2\pi$ electrons* (*Hückel's rule*), and the arrangement that occurs by far the most commonly in aromatic compounds is when $n = 1$, i.e. that with 6π electrons. 10π electrons ($n = 2$) are present in naphthalene [12, stabilisation energy, 255 kJ (61 kcal) mol^{-1}], and 14π electrons ($n = 3$) in anthracene (13) and phenanthrene (14)—stabilisation energies, 352 and 380 kJ (84 and 91 kcal) mol^{-1}, respectively:

(12)	(13)	(14)
10πe(*n* = 2)	14πe(*n* = 3)	14πe(*n* = 3)

* Significantly, cyclooctatetraene with 8π electrons (4*n*, *n* = 2) has already been shown not to be aromatic (p. 16).

Though these substances are not monocyclic like benzene—and Hückel's rule should not, strictly, apply to them—the introduction of the transannular bond, that makes them bi- and tricyclic, respectively, seems to cause relatively little perturbation, so far as delocalisation of the π electrons over the cyclic group of ten or fourteen carbon atoms is concerned.

Quasi-aromatic structures are also known in which the stabilised cyclic species is an ion, e.g. the cycloheptatrienyl (tropylium) cation (15, *cf.* p. 105), the cyclopentadienyl anion (16, *cf.* p. 268), both of which have 6πe (*n* = 1), and even more surprisingly the cyclopropenyl cation (17, *cf.* p. 105) which has 2πe (*n* = 0):

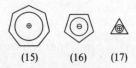

(15)   (16)   (17)

Further, the ring structure need not be purely carbocyclic, and pyridine (18, *cf.* p. 162), for example, with a nitrogen atom in the ring and 6πe (*n* = 1), is as highly stabilised as benzene:

(18)

A useful experimental criterion of aromatic character, in addition to those already mentioned, arises from the position of the signal from hydrogen atoms attached to the ring carbons in the compound's nuclear magnetic resonance (n.m.r.) spectrum.* The position of the n.m.r. signal from a hydrogen atom depends on the nature, i.e. local environment, of the carbon (or other atom) to which it is attached. Thus the proton signal of cyclooctatetraene is seen at τ4·32, and such a position is typical of protons in a non-aromatic cyclic polyene, while the proton signal of benzene is seen at τ2·8, which is found to be typical of aromatic compounds in general.

1.3.7 Conditions necessary for delocalisation

The difficulty in finding a satisfactory representation for the carbon–carbon bonding in benzene brings home to us the fact that our normal way of writing bonds between atoms as single, double or triple, involving

---

* A useful simple account of the use of n.m.r. (and other) spectra in an organic context may be found in: Williams, D. H. and Fleming, I. *Spectroscopic Methods in Organic Chemistry*, McGraw-Hill, 3rd Edition, 1980

two, four and six electrons, respectively, is clearly inadequate: some bonds involve other, even fractional, numbers of electrons. This is seen very clearly in the ethanoate (acetate) anion (19),

$$CH_3C \underset{O^\ominus}{\overset{O}{\diagdown}}$$

(19)

where, in flat contradiction of the above formula, X-ray crystallography shows that the two oxygen atoms are indistinguishable from each other, the two carbon–oxygen bond distances being the same, i.e. involving the same number of electrons.

These difficulties have led to the convention of representing molecules that cannot adequately be written as a single classical structure by a combination of two or more classical structures, the so-called *canonical structures*, linked by a double-headed arrow. The way in which one of these structures can be related to another often being indicated by curved arrows, the <u>tail</u> of the curved arrow indicating where an electron pair moves <u>from</u> and the <u>head</u> of the arrow where is moves to* :

$$CH_3C \overset{O}{\underset{O^\ominus}{\diagdown}} \;\;\leftrightarrow\;\; CH_3C \overset{O^\ominus}{\underset{O}{\diagdown}} \;\;\equiv\;\; \left[CH_3C \overset{O}{\underset{O}{\diagdown}} \right]^\ominus$$

(19a)   (19b)   (19ab)

It cannot be too firmly emphasized, however, that the ethanoate anion does not have two possible, and alternative, structures which are rapidly interconvertible, but a <u>single, real</u> structure (19ab)—sometimes referred to as a hybrid—for which the classical (canonical) structures (19a) and (19b) are less exact, limiting approximations.

A certain number of limitations must be borne in mind, however, when considering delocalisation and its representation through two or more classical structures as above. Broadly speaking, the more canonical structures that can be written for a compound, the greater the delocalisation of electrons, and the more stable the compound will be. These structures must not vary too widely from each other in energy content, however, or those of higher energy will contribute so little to the hybrid as to make their contribution virtually irrelevant. The stabilising effect is particularly marked when the structures have the same energy content, as with (19a) and (19b) above. Structures

* We shall, however, subsequently write canonical structures, e.g. (19a) and (19b), linked by a double arrow, but <u>without</u> any curved arrows. These will be used only to indicate a <u>real</u> movement of electron pairs, i.e. during actual bond-forming/bond-breaking in the course of a reaction.

involving separation of charge (*cf.* p. 24) may be written but, other things being equal, these are usually of higher energy content than those in which such separation has not taken place, and hence contribute correspondingly less to the hybrid. The structures written must all contain the same number of paired electrons, and the constituent atoms must all occupy essentially the same positions relative to each other in each canonical structure. If delocalisation is to be significant, all atoms attached to unsaturated centres must lie in the same plane or nearly so; examples where delocalisation, with consequent stabilisation, is actually prevented by steric factors are discussed subsequently (p. 26).

1.4 THE BREAKING AND FORMING OF BONDS

A covalent bond between two atoms can be broken in essentially the following ways:

$$R \cdot + \cdot X$$
$$R : X \rightarrow R : ^{\ominus} + X^{\oplus}$$
$$R^{\oplus} + : X^{\ominus}$$

In the first case each atom separates with one electron, leading to the formation of highly reactive entities called radicals, owing their reactivity to their unpaired electron; this is referred to as *homolytic fission* of the bond. Alternatively, one atom may hold on to both electrons, leaving none for the other, the result in the above case being a negative and a positive ion, respectively. Where R and X are not identical, the fission can, of course, take place in either of two ways, as shown above, depending on whether R or X retains the electron pair. Either of these processes is referred to as *heterolytic fission*, the result being the formation of an *ion pair*. Formation of a covalent bond can take place by the reversal of any of these processes, and also, of course, by the attack of first-formed radicals or ions on other species:

$$R \cdot + Br-Br \rightarrow R-Br + Br \cdot \qquad \text{(p. 314)}$$

$$R^{\oplus} + H_2O \rightarrow R-OH + H^{\oplus} \qquad \text{(p. 106)}$$

Such radicals or ion pairs are formed transiently as reactive intermediates in a very wide variety of organic reactions, as will be shown below. Reactions involving radicals tend to occur in the gas phase and in solution in non-polar solvents, and to be catalysed by light and by the addition of other radicals (p. 291). Reactions involving ionic intermediates take place more readily in solution in polar solvents, because of the greater ease of separation of charge therein, and very often because of the stabilisation of the resultant ion pairs through solvation. Many of these ionic intermediates can be considered as carrying their charge on a carbon atom, though the ion is often stabilised by delocalisation of the charge, to a greater or lesser extent, over other carbon atoms, or atoms of different elements:

$$CH_2=CH-CH_2-\overset{..}{\underset{..}{O}}H \;\overset{H^{\oplus}}{\rightleftharpoons}\; CH_2=CH-CH_2-\overset{\oplus}{O}H_2 \;\overset{-H_2O}{\rightleftharpoons}\; \left[\begin{array}{c} CH_2=CH-\overset{\oplus}{C}H_2 \\ \updownarrow \\ \overset{\oplus}{C}H_2-CH=CH_2 \end{array}\right]$$

$$CH_3-\overset{\overset{\displaystyle O}{\|}}{C}-CH_3 \;\overset{\ominus OH}{\rightleftharpoons}\; \left[\begin{array}{c} \overset{\overset{\displaystyle O}{\|}}{CH_3-C-\overset{\ominus}{C}H_2} \\ \updownarrow \\ \overset{\overset{\displaystyle O^{\ominus}}{|}}{CH_3-C=CH_2} \end{array}\right] + H_2O$$

When a positive charge is carried on carbon the entity is known as a *carbonium ion*, and when a negative charge, a *carbanion*. Though such ions be formed only transiently and be present only in minute concentration, they are nevertheless often of paramount importance in controlling the reactions in which they participate.

These three types, radicals, carbonium ions and carbanions, by no means exhaust the possibilities of transient intermediates in which carbon is the active centre; others include the electron-deficient species *carbenes*, R_2C: (p. 260), and also *benzynes* (p. 171).

1.5 FACTORS INFLUENCING ELECTRON-AVAILABILITY

In the light of what has been said above, any factors that influence the relative availability of electrons (the *electron density*) in particular bonds, or at particular atoms, in a compound might be expected to affect very considerably its reactivity towards a particular reagent: a position of high electron availability will be attacked with difficulty if at all by, for example, $^{\ominus}OH$, whereas a position of low electron availability is likely to be attacked with ease, and *vice versa* with a positively charged reagent. A number of such factors have been recognised.

1.5.1 Inductive effects

In a covalent single bond between unlike atoms, the electron pair forming the σ bond is never shared absolutely equally between the two atoms; it tends to be attracted a little more towards the more electronegative atom of the two. Thus in an alkyl chloride (20), the electron density tends to be greater nearer chlorine than carbon, as the former is the more electronegative; this is generally represented as in (20*a*) or (20*b*):

$$\overset{\delta+}{\underset{/}{\overset{\backslash}{C}}}\!-\!\overset{\delta-}{Cl} \qquad \overset{\backslash}{\underset{/}{C}}\!\rightarrow\!Cl$$

$$(20a) \qquad\qquad (20b)$$

If the carbon atom bonded to chlorine is itself attached to further carbon atoms, the effect can be transmitted further:

$$C\!-\!C\!-\!C\!\rightarrow\!C\!\twoheadrightarrow\!Cl$$

$$4\quad 3\quad 2\quad 1$$

The effect of the chlorine atom's partial appropriation of the electrons of the carbon–chlorine bond is to leave C_1 slightly electron-deficient; this it seeks to rectify by, in turn, appropriating slightly more than its share of the electrons of the σ bond joining it to C_2, and so on down the chain. The effect of C_1 on C_2 is less than the effect of Cl on C_1, however, and the transmission quickly dies away in a saturated chain, usually being too small to be noticeable beyond C_2.

Most atoms and groups attached to carbon exert such *inductive effects* in the same direction as chlorine, i.e. they are electron-withdrawing, owing to their being more electronegative than carbon, the major exception being alkyl groups which are electron-donating.[*] Though the effect is quantitatively rather small, it is responsible for the increase in basicity that results when one of the hydrogen atoms of ammonia is replaced by an alkyl group (p. 65), and, in part at any rate, for the readier substitution of the aromatic nucleus in methylbenzene than in benzene itself (p. 152).

All inductive effects are permanent polarisations in the ground state of a molecule, and are therefore manifested in its physical properties, for example, its dipole moment.

1.5.2   Mesomeric (conjugative) effects

These are essentially electron redistributions that can take place in unsaturated, and especially in conjugated, systems *via* their π orbitals. An example is the carbonyl group (p. 200), whose properties are not accounted for entirely satisfactorily by the classical formulation (21*a*),

---

[*] The metal atoms in, for example, lithium alkyls and Grignard reagents, both of which compounds are largely covalent, are also electron-donating, leading to negatively polarised carbon atoms in each case: $R\!\leftarrow\!Li$ and $R\!\leftarrow\!MgHal$ (*cf*. p. 217).

nor by the extreme dipole (21*b*) obtainable by shift of the π electrons:

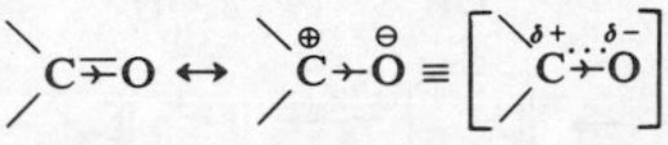

$$\text{>C}\!\overline{\overline{\mp}}\!\text{O} \leftrightarrow \text{>}\overset{\oplus}{\text{C}}\!\rightarrow\!\overset{\ominus}{\text{O}} \equiv \left[\text{>}\overset{\delta+}{\text{C}}\!\overset{\delta-}{\overline{\div}}\!\text{O} \right]$$

(21*a*)   (21*b*)   (21*ab*)

The actual structure is somewhere in between, i.e. (21*ab*) a hybrid of which (21*a*) and (21*b*) are the canonical forms. There will also be an inductive effect, as shown in (21*ab*) but this will be much smaller than the mesomeric effect as σ electrons are much less readily shifted than π electrons.

If the C=O group is conjugated with C=C, the above polarisation can be transmitted further *via* the π electrons, e.g. (22):

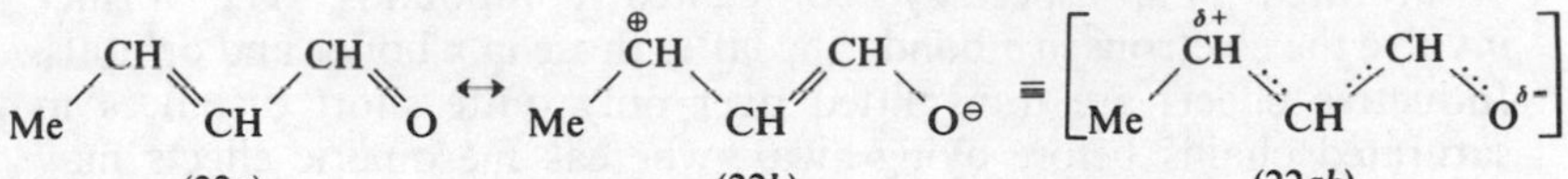

(22*a*)   (22*b*)   (22*ab*)

Delocalisation takes place (*cf.* 1,3-dienes, p. 13), so that an electron-deficient atom results at C_3, as well as at C_1 as in a simple carbonyl compound. The difference between this transmission *via* a conjugated system, and the inductive effect in a saturated system, is that here the effect suffers much less diminution by its transmission, and the polarity alternates.

The stabilisation that can result by delocalisation of a positive or negative charge in an ion, *via* its π orbitals, can be a potent feature in making the formation of the ion possible in the first place (*cf.* p. 54). It is, for instance, the stabilisation of the phenoxide anion (23), by delocalisation of its charge *via* the delocalised π orbitals of the nucleus, that is largely responsible for the acidity of phenol (*cf.* p. 55):

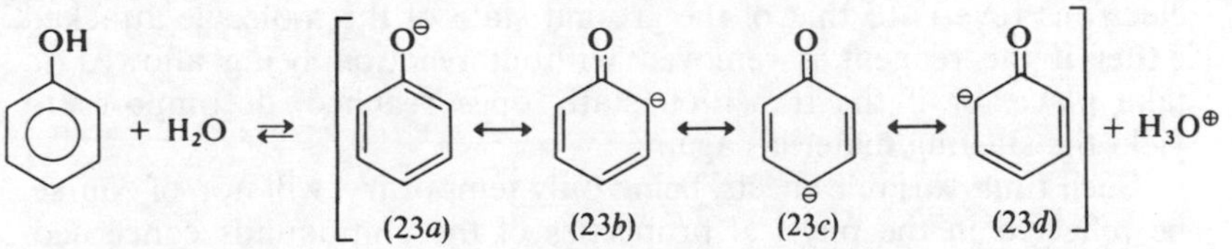

(23*a*)   (23*b*)   (23*c*)   (23*d*)

An apparently similar delocalisation can take place in undissociated phenol (24) itself, involving an unshared electron pair on the oxygen atom,

$$\left[\quad \underset{(24a)}{:OH} \quad \longleftrightarrow \quad \underset{(24b)}{\overset{\oplus}{OH}} \quad \longleftrightarrow \quad \underset{(24c)}{\overset{\oplus}{OH}} \quad \longleftrightarrow \quad \underset{(24d)}{\overset{\oplus}{OH}} \quad \right]$$

but this involves separation of charge, and will thus be correspondingly less effective than the stabilisation of the phenoxide ion which does not.

Mesomeric, like inductive, effects are permanent polarisations in the ground state of a molecule, and are therefore manifested in the physical properties of the compounds in which they occur. The essential difference between inductive and mesomeric effects is that the former occur essentially in saturated groups or compounds, the latter in unsaturated and, especially, conjugated compounds. The former involve the electrons in σ bonds, the latter those in π bonds and orbitals. Inductive effects are transmitted over only quite short distances in saturated chains before dying away, whereas mesomeric effects may be transmitted from one end to the other of quite large molecules provided that conjugation (i.e. delocalised π orbitals) is present, through which they can proceed.

1.5.3 Time-variable effects

Some workers have sought to distinguish between effects such as the two considered above, which are permanent polarisations manifested in the ground state of a molecule, and changes in electron distribution that may result either on the close approach of a reagent or, more especially, in the transition state (p. 38) that may result from its initial attack. The time-variable factors, by analogy with the permanent effects discussed above, have been named the *inductomeric* and *electromeric* effects, respectively. Any such effects can be looked upon as *polarisabilities* rather than as polarisations, for the distribution of electrons reverts to that of the ground state of the molecule attacked either if the reagent is removed without reaction being allowed to take place, or if the transition state, once reached, decomposes to yield the starting materials again.

Such time-variable effects, being only temporary, will not, of course, be reflected in the physical properties of the compounds concerned. It has often proved impossible to distinguish experimentally between permanent and time-variable effects, but it cannot be too greatly emphasised that, despite the difficulties in distinguishing what proportions of a given effect are due to permanent and what to time-variable factors, the actual close approach of a reagent may have a profound effect in enhancing reactivity in a reactant molecule, and so in promoting reaction.

1.5.4 Hyperconjugation

The relative magnitude of the inductive effect of alkyl groups is normally found to follow the order,

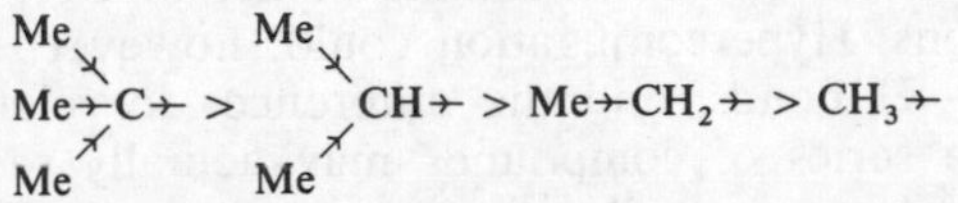

$$\begin{array}{cc} \text{Me} & \text{Me} \\ \diagdown & \diagdown \\ \text{Me} \rightarrow \text{C} \rightarrow > & \text{CH} \rightarrow > \text{Me} \rightarrow \text{CH}_2 \rightarrow > \text{CH}_3 \rightarrow \\ \diagup & \diagup \\ \text{Me} & \text{Me} \end{array}$$

as would be expected. When, however, the alkyl groups are attached to an unsaturated system, e.g. a double bond or a benzene nucleus, this order is found to be disturbed, and in the case of some conjugated systems actually reversed. It thus appears that alkyl groups are capable, in these circumstances, of giving rise to electron release by a mechanism different from the inductive effect. This has been explained as proceeding by an extension of the conjugative or mesomeric effect, delocalisation taking place in the following way:

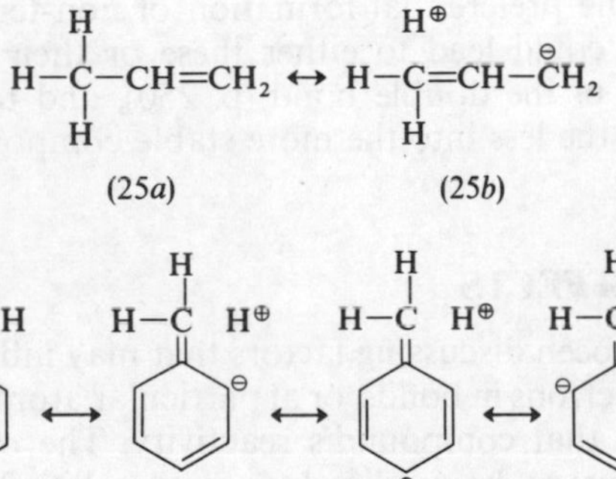

$$\begin{array}{ccc} & \text{H} & & \text{H}^{\oplus} \\ & | & & | \\ \text{H}-\text{C}-\text{CH}=\text{CH}_2 & \leftrightarrow & \text{H}-\text{C}=\text{CH}-\overset{\ominus}{\text{CH}}_2 \\ & | & & | \\ & \text{H} & & \text{H} \\ & (25a) & & (25b) \end{array}$$

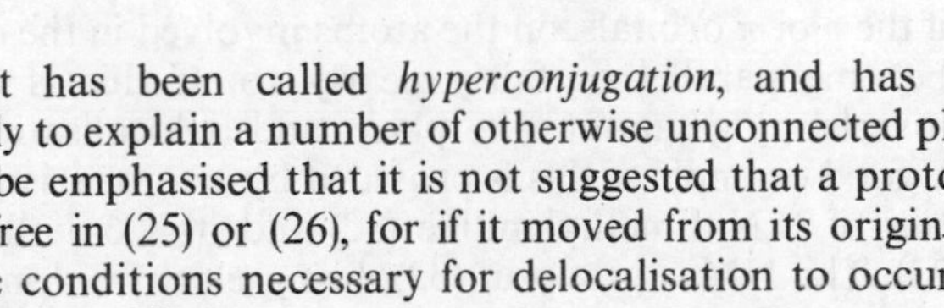

$$\begin{array}{cccc} \text{H} & \text{H} & \text{H} & \text{H} \\ | & | & | & | \\ \text{H}-\text{C}-\text{H} & \text{H}-\text{C} \; \text{H}^{\oplus} & \text{H}-\text{C} \; \text{H}^{\oplus} & \text{H}-\text{C} \; \text{H}^{\oplus} \\ & & & \\ (26a) & (26b) & (26c) & (26d) \end{array}$$

This effect has been called *hyperconjugation*, and has been used successfully to explain a number of otherwise unconnected phenomena. It should be emphasised that it is not suggested that a proton actually becomes free in (25) or (26), for if it moved from its original position one of the conditions necessary for delocalisation to occur would be controverted (p. 20).

Reversal of the expected (inductive) order of electron-donation to $CH_3 > MeCH_2 > Me_2CH > Me_3C$ could be explained on the basis of hyperconjugation being dependent on the presence of hydrogen on the carbon atoms α- to the unsaturated system. This is clearly at a maximum with CH_3 (25) and non-existent with Me_3C (29),

$$
\begin{array}{cccc}
\overset{\displaystyle H}{\underset{\displaystyle H}{H-C-CH=CH_2}} & \overset{\displaystyle Me}{\underset{\displaystyle H}{H-C-CH=CH_2}} & \overset{\displaystyle Me}{\underset{\displaystyle H}{Me-C-CH=CH_2}} & \overset{\displaystyle Me}{\underset{\displaystyle Me}{Me-C-CH=CH_2}} \\
(25) & (27) & (28) & (29)
\end{array}
$$

hence the increased electron-donating ability of CH_3 groups under these conditions. Hyperconjugation could, however, involve C—C as well as C—H bonds, and the differences in relative reactivity observed in a series of compounds may actually result from the operation of solvent, as well as hyperconjugative, effects.

Hyperconjugation has also been invoked to account for the greater thermodynamic stability of alkenes in which the double bond is not terminal, e.g. (30), compared with isomeric compounds in which it is, e.g. (31): in (30) there are <u>nine</u> 'hyperconjugable' α-hydrogen atoms, compared with only <u>five</u> in (31):

$$
\begin{array}{cc}
\overset{\displaystyle CH_3}{CH_3-C=CH-CH_3} & \overset{\displaystyle CH_3}{MeCH_2-C=CH_2} \\
(30) & (31)
\end{array}
$$

This results in the preferential formation of non-terminal alkenes, in reactions which could lead to either these or their terminal isomers on introduction of the double bond (p. 250), and to the fairly ready isomerisation of the less into the more stable compound.

1.6   STERIC EFFECTS

We have to date been discussing factors that may influence the relative availability of electrons in bonds, or at particular atoms, in a compound, and hence affect that compound's reactivity. The operation of these factors may, however, be modified or even nullified by the influence of steric factors; thus effective delocalisation *via* π orbitals can only take place if the p or π orbitals, on the atoms involved in the delocalisation, can become parallel or fairly nearly so. If this is prevented, significant overlapping cannot take place, and delocalisation may be inhibited. A good example of this is provided by a comparison between the behaviour of N,N-dimethylaniline (32) and its 2,6-dialkyl derivatives, e.g. (33). The NMe_2 group in (32), being electron-donating (due to the unshared electron pair on nitrogen interacting with the delocalised π orbitals of the nucleus), activates the nucleus towards attack by the diazonium cation $PhN_2^{\oplus}$, i.e. towards azo-coupling, leading to substitution at the *p*-position (*cf.* p. 152):

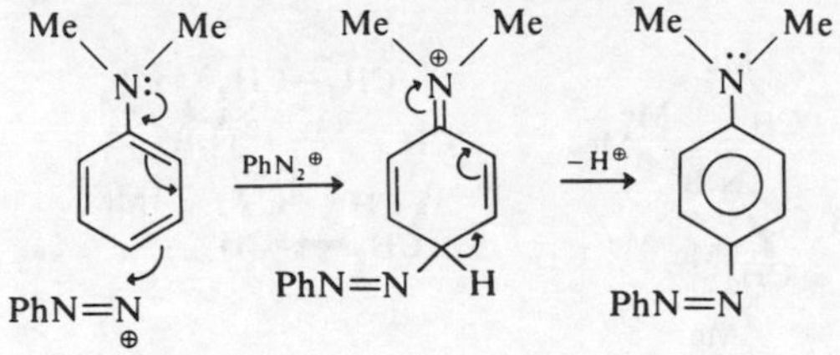

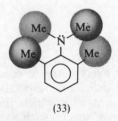

(32)

The 2,6-dimethyl derivative (33) does not couple under these conditions, however, despite the fact that the methyl groups that have been introduced are much too far away for their bulk to interfere directly with attack at the *p*-position. The failure to couple at this position is, in fact, due to the two methyl groups, in the *o*-positions to the NMe_2, interfering sterically with the two methyl groups attached to nitrogen, and so preventing these lying in the same plane as the benzene nucleus. This means that the *p* orbitals on nitrogen, and on the ring carbon atom to which it is attached, are prevented from becoming parallel to each other, and their overlapping is thus inhibited. Electronic interaction with the nucleus is thus largely prevented, and transfer of charge, as in (32), does not take place (*cf.* p. 70):

(33)

The most common steric effect, however, is the classical *steric hindrance*, in which it is apparently the sheer bulk of groups that is influencing the reactivity of a site in a compound directly: by impeding approach of a reagent to the reacting centre, and by introducing crowding in the transition state (*cf.* p. 38), and not by promoting or inhibiting electron-availability. This has been investigated closely in connection with the stability of the complexes formed by trimethyl-boron with a wide variety of amines. Thus the complex (34) formed with triethylamine dissociates extremely readily, whereas the complex (35) with quinuclidine, which can be looked upon as having three ethyl groups on nitrogen that are 'held back' from assuming a con-formation that would interfere sterically with attack on the nitrogen atom, is very stable:

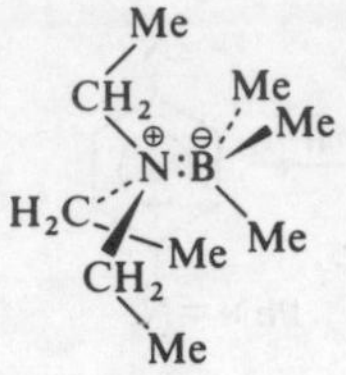

(34) 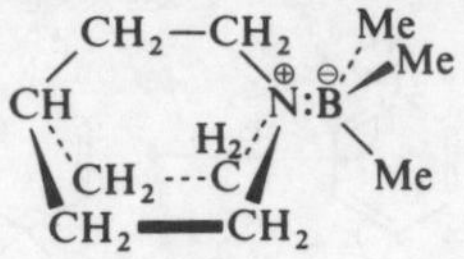   (35)

That this difference is not due to differing electron availability at the nitrogen atom in the two cases is confirmed by the fact that the two amines differ very little in their strengths as bases (*cf.* p. 72): the uptake of a proton constituting very much less of a steric obstacle than the uptake of the relatively bulky BMe_3. Esterification and ester hydrolysis are other reactions particularly susceptible to steric inhibition (*cf.* p. 237).

It should be emphasised that such steric inhibition is only an extreme case, and any factors which disturb or inhibit a particular orientation of the reactants with respect to each other, short of preventing their close approach, can also profoundly affect the rate of reactions: a state of affairs that is often encountered in reactions in biological systems.

1.7 REAGENT TYPES

Reference has already been made to electron-donating and electron-withdrawing groups, their effect being to render a site in a molecule electron-rich or electron-deficient, respectively. This will clearly influence the type of reagent with which the compound will most readily react. An electron-rich species such as phenoxide anion (36)

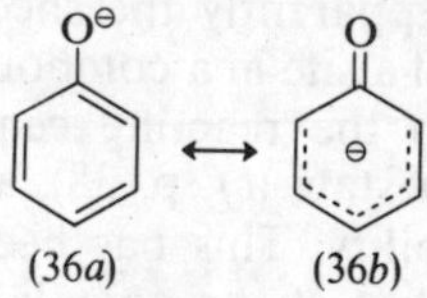

(36a)   (36b)

will tend to be most readily attacked by positively charged cations such as $C_6H_5N_2^{\oplus}$, a diazonium cation (p. 145), or by other species which, though not actually cations, possess an atom or centre that is electron-deficient; for example, the sulphur atom of sulphur trioxide (37) in sulphonation (p. 139):

$$\overset{\delta-}{O} = \overset{\delta-}{\underset{\underset{\delta-}{O}}{\overset{\overset{\delta-}{O}}{S}}} {}^{+++}$$

(37)

Such reagents, because they tend to attack the substrate at a position (or positions) of high electron density, are referred to as *electrophilic* reagents or *electrophiles*.

Conversely, an electron-deficient centre, such as the carbon atom in chloromethane (38)

$$\overset{\delta+}{H_3C} \rightarrow \overset{\delta-}{Cl}$$

(38)

will tend to be attacked most readily by (negatively charged) anions such as $^{\ominus}OH$, $^{\ominus}CN$, etc., or by other species which, though not actually anions, possess an atom or centre which is electron-rich; for example, the nitrogen atom in ammonia or amines, H_3N: or R_3N:. Such reagents, because they tend to attack the substrate at a position (or positions) of low electron density, i.e. where the atomic nucleus is short of its normal complement of orbital electrons, are referred to as *nucleophilic* reagents or *nucleophiles*.

It must be emphasised that only a *slightly* unsymmetrical distribution of electrons is required for a reaction's course to be dominated: the presence of a full-blown charge on a reactant certainly helps, but is far from being essential. Indeed the requisite unsymmetrical charge distribution may be induced by the mutual polarisation of reagent and substrate on their close approach, as when bromine adds to ethene (p. 177).

This electrophile/nucleophile dichotomy can be looked upon as a special case of the acid/base idea. The classical definition of acids and bases is that the former are proton-donors, and the latter proton-acceptors. This was made more general by Lewis, who defined acids as compounds prepared to accept electron pairs, and bases as substances that could provide such pairs. This would include a number of compounds not previously thought of as acids and bases, e.g. boron trifluoride (39),

$$F_3B + :NMe_3 \rightleftarrows F_3\overset{\ominus}{B}:\overset{\oplus}{N}Me_3$$

(39)   (40)

which acts as an acid by accepting the electron pair on nitrogen in trimethylamine to form the complex (40), and is therefore referred to as a *Lewis acid*. Electrophiles and nucleophiles in organic reactions can be looked upon essentially as acceptors and donors, respectively, of electron pairs, from and to other atoms—most frequently *carbon*. Electrophiles and nucleophiles also, of course, bear a relationship to oxidising and reducing agents, for the former can be looked upon as electron-acceptors and the latter as electron-donors. A number of the more common electrophiles and nucleophiles are listed below:

Electrophiles:

$H^{\oplus}, H_3O^{\oplus}, {}^{\oplus}NO_2, {}^{\oplus}NO, PhN_2{}^{\oplus}, R_3C^{\oplus}$

$\overset{*}{S}O_3, \overset{*}{C}O_2, \overset{*}{B}F_3, \overset{*}{A}lCl_3, \overset{*}{I}Cl, Br_2, O_3$

Nucleophiles:

$H^{\ominus}, BH_4{}^{\ominus}, H\overset{*}{S}O_3{}^{\ominus}, HO^{\ominus}, RO^{\ominus}, RS^{\ominus}, {}^{\ominus}CN, RCO_2{}^{\ominus}, RC{\equiv}C^{\ominus}, {}^{\ominus}CH(CO_2Et)_2$

$O:, N:, S:, \overset{*}{R}MgBr, \overset{*}{R}Li$ (41)

Where a reagent is starred, the star indicates the atom that accepts electrons from, or donates electrons to, the substrate as the case may be. No clear distinction can necessarily be made between what constitutes a reagent and what a substrate, for though ${}^{\oplus}NO_2$, ${}^{\ominus}OH$, etc., are normally thought of as reagents, the carbanion (41) could, at will, be either reagent or substrate, when reacted with, for example, an alkyl halide. The reaction of the former on the latter is a nucleophilic attack, while that of the latter on the former would be looked upon as an electrophilic attack; but no matter from which reactant's standpoint a reaction is viewed, its essential nature is not for a moment in doubt.

It should be remembered that reactions involving radicals as the reactive entities are also known. These are much less susceptible to variations in electron density in the substrate than are reactions involving polar intermediates, but they are greatly affected by the addition of small traces of substances that either liberate or remove radicals. They are considered in detail below (p. 304).

1.8   REACTION TYPES

There are essentially four general types of reaction which organic compounds can undergo:

    (*a*) Displacement (substitution)
    (*b*) Addition
    (*c*) Elimination
    (*d*) Rearrangement

In (*a*) it is displacement from carbon that is normally referred to, but the atom displaced can be either hydrogen or another atom or group.

In electrophilic substitution it is often hydrogen that is displaced, classical aromatic substitution (p. 130) being a good example:

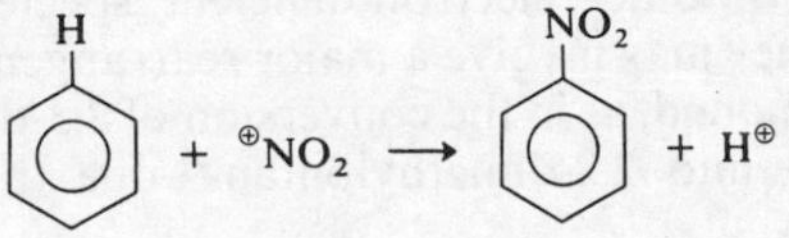

$$\text{(benzene with H)} + {}^{\oplus}NO_2 \longrightarrow \text{(benzene with } NO_2\text{)} + H^{\oplus}$$

In nucleophilic substitution it is often an atom other than hydrogen that is displaced (p. 77),

$$NC^{\ominus} + R-Br \longrightarrow NC-R + Br^{\ominus}$$

but nucleophilic displacement of hydrogen is also known (p. 164). Radical-induced displacement is also known, for example the halogenation of alkanes, though it should be emphasised that this last is not *direct* displacement on carbon (*cf.* p. 314).

Addition reactions, too, can be electrophilic, nucleophilic or radical in character, depending on the type of species that initiates the process. Addition to simple carbon–carbon double bonds is normally either electrophile-, or radical-, induced; e.g. addition of HBr,

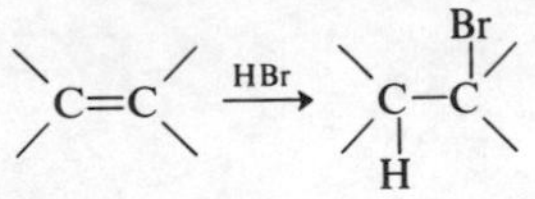

$$\text{C=C} \xrightarrow{HBr} \text{C--C (with Br, H)}$$

which can be initiated by the attack of either $H^{\oplus}$ (p. 181) or $Br\cdot$ (p. 308) on the double bond. By contrast, the addition reactions exhibited by the carbon–oxygen double bond, in simple aldehydes and ketones, are usually nucleophilic in character (p. 201). An example is the base-catalysed formation of cyanohydrins in liquid HCN:

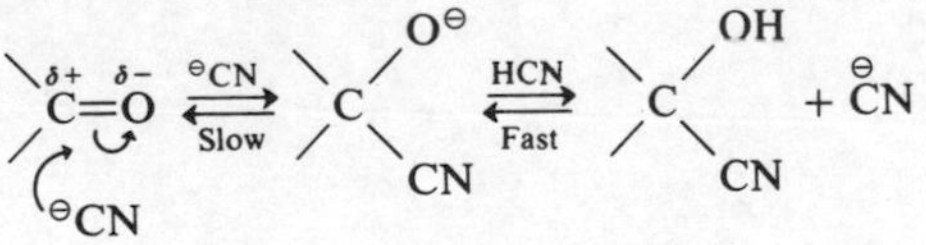

$$\underset{{}^{\ominus}CN}{\overset{\delta+ \;\; \delta-}{C=O}} \underset{Slow}{\overset{{}^{\ominus}CN}{\rightleftharpoons}} \underset{CN}{\overset{O^{\ominus}}{C}} \underset{Fast}{\overset{HCN}{\rightleftharpoons}} \underset{CN}{\overset{OH}{C}} + {}^{\ominus}CN$$

Elimination reactions are, of course, essentially the reversal of addition reactions; the most common type is the loss of hydrogen and another atom or group from adjacent carbon atoms to yield alkenes (p. 240):

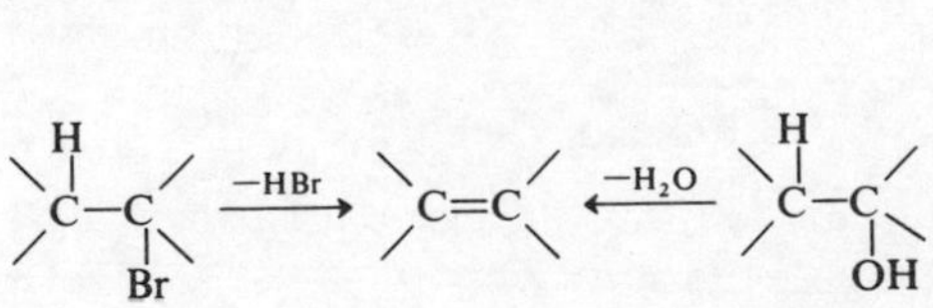

$$\underset{Br}{\overset{H}{C-C}} \xrightarrow{-HBr} C=C \xleftarrow{-H_2O} \underset{OH}{\overset{H}{C-C}}$$

Rearrangements may also proceed *via* intermediates that are essentially cations, anions, or radicals, though those involving carbonium ions, or other electron-deficient species, are by far the most common. They may involve a major rearrangement of the carbon skeleton of a compound, as in the conversion of 2,3-dimethylbutan-2,3-diol (pinacol, 42) into 2,2-dimethylbutan-3-one (pinacolone, 43, *cf.* p. 112):

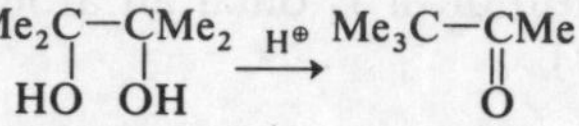

$$\underset{\underset{\text{HO \quad OH}}{|\qquad|}}{\text{Me}_2\text{C}-\text{CMe}_2} \xrightarrow{\text{H}^\oplus} \underset{\underset{\text{O}}{||}}{\text{Me}_3\text{C}-\text{CMe}}$$

(42)   (43)

The actual rearrangement step in such changes is often followed by a further displacement, addition or elimination, before a stable end-product is obtained.

2
Energetics, kinetics, and the investigation of mechanism

We have now listed a number of electronic and steric factors that can influence the reactivity of a compound in a given situation, and also the types of reagent that might be expected to attack particular centres in such a compound especially readily. We have as yet, however, had little to say directly about how these electronic and steric factors, varying from one structure to another, actually operate in energetic and kinetic terms to influence the course and rate of a reaction. These considerations are of major importance, not least for the light they might be expected to throw on the detailed pathway by which a reaction proceeds.

2.1 ENERGETICS OF REACTION

In considering that conversion of starting materials into products which constitutes an organic reaction, one of the things that we particularly want to know is '*how far* will the reaction go over towards products?' Systems tend to move toward their most stable state, so we might expect that the more stable the products are, compared with the starting materials, the further over in the former's favour any equilibrium between them might be expected to lie, i.e. the larger $\Delta_{stability}$ is in the diagram (Fig. 2.1) below, the greater the expected conversion into products:

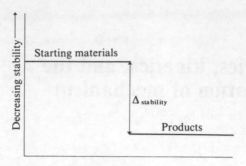

Fig. 2.1

However, it quickly becomes apparent that the simple energy change that occurs on going from starting materials to products, and that may readily be measured as the heat of reaction, ΔH^*, is not an adequate measure of the difference in stability between them, for there is often found to be no correlation between ΔH and the equilibrium constant for the reaction, K. Highly exothermic reactions are known with only small equilibrium constants (little conversion of starting materials into products), and some reactions with large equilibrium constants are known that are actually endothermic (enthalpy of products higher than that of starting materials): clearly some factor in addition to enthalpy must be concerned in the relative stability of chemical species.

That this should be so is a corollary of the Second Law of Thermodynamics which is concerned essentially with probabilities, and with the tendency for ordered systems to become disordered: a measure of the degree of disorder of a system being provided by its *entropy*, S. In seeking their most stable condition, systems tend towards *minimum* energy (actually enthalpy, H) and *maximum* entropy (disorder or randomness), a measure of their relative stability must thus embrace a compromise between H and S, and is provided by the *Gibb's free energy*, G, which is defined by,

$$G = H - TS$$

where T is the absolute temperature. The free energy change during a reaction, at a particular temperature, is thus given by,

$$\Delta G = \Delta H - T\Delta S$$

* H is a measure of the heat content, or *enthalpy*, of a compound, and ΔH is preceded by a minus sign if the products have a lower heat content than the starting materials: when there is such a *decrease* in enthalpy the reaction is *exothermic*.

and it is found that the change in free energy in going from starting materials to products, $\Delta G^{\ominus}$ ($\Delta G^{\ominus}$ refers to the change under standard conditions: at unit activity; less exactly at unit, i.e. molar, concentration), is related to the equilibrium constant, K, for the change by the relation,

$$-\Delta G^{\ominus} = 2 \cdot 303 RT \log K$$

i.e. the larger the *decrease* in free energy (hence, *minus* $\Delta G^{\ominus}$) on going from starting materials to products, the larger the value of K, and the further over the equilibrium lies in favour of the products. The situation of minimum free energy thus corresponds to the position of equilibrium for starting materials/products. For a reaction in which there is *no* free energy change ($\Delta G^{\ominus} = 0$) $K = 1$, which corresponds to 50% conversion of starting materials into products. Increasing *positive* values of $\Delta G^{\ominus}$ imply rapidly *decreasing* fractional values of K (the relationship is a logarithmic one), corresponding to extremely little conversion into products, while increasing *negative* values of $\Delta G^{\ominus}$ imply correspondingly rapidly *increasing* values of K. Thus a $\Delta G^{\ominus}$ of -42 kJ (-10 kcal) mol^{-1} corresponds to an equilibrium constant of $\approx 10^7$, and essentially complete conversion into products. A knowledge of the standard free energies of starting materials and of products, which have been measured for a larger number of organic compounds, thus enables us to predict the expected extent of the conversion of the former into the latter.

The ΔH factor for the change can be equated with the difference in energies between the bonds in the starting materials and the bonds in the products, and an approximate value of ΔH for a reaction can often be predicted from tables of standard bond energies: which is hardly unexpected, as it is from ΔH data that the average bond energies were compiled in the first place!

The entropy factor cannot be explained quite so readily, but effectively it relates to the number of possible ways in which their total, aggregate energy may be shared out among an assembly of molecules, and also the number of ways in which an individual molecule's quanta of energy may be shared out for translational, rotational, and vibrational purposes, of which the translational is likely to be by far the largest in magnitude. Thus for a reaction in which there is an *increase* in the number of molecular species on going from starting materials to products,

$$A \rightleftarrows B + C$$

there is likely to be a sizeable *increase* in entropy because of the gain in translational freedom. The $-T\Delta S$ term may then be large enough to outweigh the $+\Delta H$ term of an endothermic reaction, thus leading to a negative value for ΔG, and an equilibrium that lies well over in favour of products. If the reaction is exothermic anyway (ΔH negative),

ΔG will of course be even more negative, and the equilibrium constant, K, correspondingly larger still. Where the number of participating species *decreases* on going from starting materials to products there is likely to be a *decrease* in entropy (ΔS negative); hence,

$$\Delta G = \Delta H - (-)T\Delta S$$

and unless the reaction is sufficiently exothermic (ΔH negative and large enough) to counterbalance this, ΔG will be positive, and the equilibrium thus well over in favour of starting materials.

Cyclisation reactions may also be attended by a decrease in entropy,

$$CH_3(CH_2)_3CH{=}CH_2 \rightleftharpoons \begin{array}{c} H_2 \\ C \\ H_2C \diagup \quad \diagdown CH_2 \\ | \qquad | \\ H_2C \diagdown \quad \diagup CH_2 \\ C \\ H_2 \end{array}$$

for though there is no necessary change in translational entropy, a constraint is imposed on rotation about the carbon–carbon single bonds: this is essentially free in the open chain starting material, but is greatly restricted in the cyclic product. This rotational entropy term is, however, smaller in size than the translational entropy term involved in reactions where the number of participating species decreases on forming products. A fact that is reflected in the preference for *intra-* rather than *inter*-molecular hydrogen bonding in 1,2-diols:

$$\begin{array}{ccc} \underset{\textstyle intra\text{-molecular}}{\begin{array}{c} H \quad\; H \\ O \quad O \\ \diagdown | \quad | \diagup \\ {>}C{-}C{<} \end{array}} & & \underset{\textstyle inter\text{-molecular}}{\begin{array}{c} H \qquad\; H{\cdots} \qquad H \quad H \\ O \quad O \qquad O \quad O \\ \diagdown | \quad | \diagup \quad \diagdown | \quad | \diagup \\ {>}C{-}C{<} \quad {>}C{-}C{<} \end{array}} \end{array}$$

It should not be overlooked that the entropy term involves temperature ($T\Delta S$) while the enthalpy (ΔH) term does not, and their relative contributions to the free energy change may be markedly different for the same reaction carried out at widely differing temperatures.

2.2 KINETICS OF REACTION

Though a negative value for $\Delta G^{\ominus}$ is a necessary condition for a reaction to take place at all under a given set of conditions, further information is still needed as the $-\Delta G^{\ominus}$ value tells us nothing about *how fast* the starting materials are converted into products. Thus for the oxidation of cellulose,

$$(C_6H_{10}O_5)_n + 6nO_2 \rightleftharpoons 6nCO_2 + 5nH_2O$$

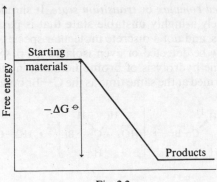

Fig. 2.2

$\Delta G^{\ominus}$ is negative and large in magnitude, so that the equilibrium lies essentially completely over in favour of CO_2 and H_2O; but a newspaper (very largely cellulose) can be read in the air (or even in an oxygen tent!) for long periods of time without it noticeably fading away to gaseous products: the *rate* of the conversion is extremely slow under these conditions despite the very large $-\Delta G^{\ominus}$, though it is, of course, speeded up at higher temperatures. The conversion of starting materials into products, despite a negative $\Delta G^{\ominus}$, is rarely if ever a mere run down-hill (Fig. 2.2), there is generally a barrier to be overcome *en route* (Fig. 2.3):

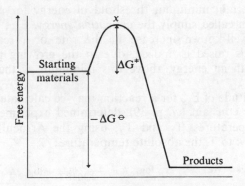

Fig. 2.3

2.2.1   Reaction rate and free energy of activation

The position x in the *energy profile* above (Fig. 2.3) corresponds to the least stable configuration through which the starting materials pass during their conversion into products, and is generally referred to as an *activated complex* or *transition state*. It should be emphasised that this is merely a highly unstable state that is passed through in a dynamic process, and <u>not</u> a discrete molecular species, an intermediate, that can actually be detected or even isolated (*cf.* p. 48). An example is (1) in the alkaline hydrolysis of bromomethane, in which the HO—C bond is being formed at the same time as the C—Br bond is being broken,

$$HO^{\ominus} + \begin{array}{c} H \\ | \\ C-Br \\ | \\ H \end{array} \rightarrow \left[HO\cdots\overset{\delta-}{C}\cdots\overset{\delta-}{Br} \right]^{\ddagger} \rightarrow HO-C \begin{array}{c} H \\ | \\ | \\ H \end{array} + Br^{\ominus}$$

(1)

and the three hydrogen atoms attached to carbon are passing through a configuration in which they all lie in one plane (at right-angles to the plane of the paper). This reaction is discussed in detail below (p. 77).

The height of the barrier in (Fig. 2.3), $\Delta G^{\ddagger}$, is called the *free energy of activation* for the reaction (the higher it is the slower the reaction), and can be considered as being made up of enthalpy ($\Delta H^{\ddagger}$) and entropy ($T\Delta S^{\ddagger}$) terms:

$$\Delta G^{\ddagger} = \Delta H^{\ddagger} - T\Delta S^{\ddagger}$$

$\Delta H^{\ddagger}$ (the *enthalpy of activation*) corresponds to the energy necessary to effect the stretching or even breaking of bonds that is an essential prerequisite for reaction to take place (e.g. stretching of the C—Br bond in 1). Thus reacting molecules must bring with them to any collision a certain minimum threshold of energy for reaction to be possible (often called simply the *activation energy*, E_{act}, but related to $\Delta H^{\ddagger}$); the well-known increase in the rate of a reaction as the temperature is raised is, indeed, due to the growing proportion of molecules with an energy above this minimum as the temperature rises.

The magnitude of E_{act} for a reaction may be calculated from values of k, the rate constant (*cf.* p. 39), determined experimentally at two different temperatures, T_1 and T_2, using the Arrhenius expression which relates k to T, the absolute temperature:

$$k = Ae^{-E/RT} \quad \text{or} \quad \log_{10} k = -\frac{E_{act}}{2.303RT} + \log_{10} A$$

Where R is the gas constant (8.32 Joules mol^{-1} deg^{-1}), and A is a constant for the reaction—independent of temperature—that is related to the proportion of the total number of collisions between reactant molecules that result in successful conversion into products. The value

* The symbol $\ddagger$ will often be applied to a structure to indicate that it is intended as an attempted representation of a transition state (T.S.).

of F_{act} may then be obtained graphically by plotting values of $\log_{10} k$ against $1/T$, or by conversion of the above equation into,

$$\log_{10} k_1/k_2 = -\frac{E_{act}}{2 \cdot 303R}\left[\frac{1}{T_1} - \frac{1}{T_2}\right]$$

and subsequent calculation.

The $\Delta S^{\neq}$ term (the *entropy of activation*) again relates to randomness. It is a measure of the change in degree of organisation, or ordering, of both the reacting molecules themselves and of the distribution of energy within them, on going from starting materials to the transition state; $\Delta S^{\neq}$ is related to the A factor in the Arrhenius equation above. If formation of the transition state requires the imposition of a high degree of organisation in the way the reactant molecules must approach each other, and also of the concentration of their energy in particular linkages so as to allow of their ultimate breakage, then the attainment of the transition state is attended by a sizeable decrease in entropy (randomness), and the possibility of its formation is correspondingly low.

2.2.2 Kinetics and the rate-limiting step

Experimentally, the measurement of reaction rates consists in investigating the rate at which starting materials disappear and/or products appear at a particular (constant) temperature, and seeking to relate this to the concentration of one, or all, of the reactants. The reaction may be monitored by a variety of methods, e.g. directly by the removal of aliquots followed by their titrimetric determination, or indirectly by observation of colorimetric, conductimetric, spectroscopic, etc., changes. Whatever method is used the crucial step normally involves matching the crude kinetic data against variable possible functions of concentration, either graphically or by calculation, until a reasonable fit is obtained. Thus for the reaction,

$$CH_3Br + {}^{\ominus}OH \rightarrow CH_3OH + Br^{\ominus}$$

it comes as no surprise to find a rate equation,

$$\text{Rate} = k[CH_3Br][{}^{\ominus}OH]$$

where k is known as the *rate constant* for the reaction. The reaction is said to be *second order* overall; *first order* with respect to CH_3Br, and *first order* with respect to ${}^{\ominus}OH$.

Such coincidence of stoichiometry and rate law is fairly uncommon, the former is commonly no guide at all to the latter, which can only be obtained by experiment. Thus for the base catalysed bromination of propanone,

$$CH_3COCH_3 + Br_2 \xrightarrow{\ominus OH} CH_3COCH_2Br + HBr$$

we find the rate equation,

$$Rate = k[CH_3COCH_3][^{\ominus}OH]$$

i.e. bromine does not appear, though $[^{\ominus}OH]$ does (*cf.* p. 287). Clearly bromine must be involved at some stage in the overall reaction as it is incorporated into the final product, but it patently cannot be involved in the step whose rate we are actually measuring. The overall reaction must thus involve at least two steps: one in which bromine is not involved (whose rate we are measuring), and one in which it is. In fact, very few organic reactions are one step processes as depicted in Fig. 2.3. This is obvious enough in an extreme example such as the formation of hexamine,

$$6CH_2O + 4NH_3 \rightarrow C_6H_6N_4 + 6H_2O$$

where the chance of the <u>simultaneous</u> collision of *six* molecules of CH_2O and *four* of NH_3 in a *ten*-body collision is effectively non-existent. But even where the stoichiometry is less extreme, reactions are normally composite, consisting of a number of successive steps (usually two-body collisions) of which we are actually measuring the slowest, and thus *rate-limiting*, one—the kinetic 'bottleneck' on the production line converting starting materials into products:

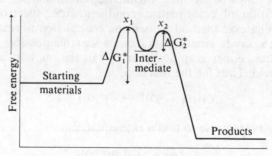

Fig. 2.4

In (Fig. 2.4) starting materials are being converted *via* transition state x_1 into an intermediate, which then decomposes into products *via* a second transition state x_2. As depicted above the formation of the intermediate *via* x_1, is the more energy-demanding ($\Delta G_1^{\ddagger} > \Delta G_2^{\ddagger}$)

of the two steps, and hence will be the slower, i.e. the step whose rate our kinetic experiments will actually be measuring. It is followed by a fast (less energy demanding), non rate-limiting conversion of the intermediate into products. The above bromination of propanone, can, under certain conditions, be said to follow an idealised pattern corresponding to (Fig. 2.4), in which slow, rate-limiting removal of proton by base results in the formation of the carbanion intermediate (2), which then undergoes rapid, non rate-limiting attack by Br_2 to yield bromopropanone and bromide ion as the products:

$$HO^{\ominus} \overset{\frown}{} H$$
$$\underset{CH_2COCH_3}{|} \xrightarrow[\text{slow}]{\ominus OH} \overset{\frown}{} \underset{\underset{(2)}{\ominus}CH_2COCH_3}{Br-Br} \xrightarrow[\text{fast}]{Br_2} \underset{CH_2COCH_3}{\overset{Br + Br^{\ominus}}{|}}$$
$$+$$
$$H_2O$$

It should be emphasised that though this explanation is a reasonable deduction from the experimentally established rate equation, the latter cannot be claimed to *prove* the former. Our experimentally determined rate equation will give us information about the species that are involved up to and including the rate-limiting step of a reaction: the rate equation does indeed specify the *composition* but not, other than by inference, the *structure* of the transition state for the rate-limiting step. It gives no direct information about intermediates nor, except by default as it were, about the species that are involved in rapid, non rate-limiting processes beyond this rate-limiting step.

In considering the effect that a change of conditions, e.g. of solvent or in the structure of the starting material, might be expected to have on the rate of a reaction, we need to know what effect such changes will have on the stability (free energy level) of the transition state: any factors which serve to stabilise it will lead to its more rapid formation, and the opposite will also apply. It is seldom possible to obtain such detailed information about these high-energy transition states: the best we can commonly do is to take the relevant intermediates as models for them, and see what effect such changes might be expected to have on these. Such a model is not unreasonable; the transiently formed intermediate in Fig. 2.4 closely resembles, in terms of free energy level, the transition state that precedes it, and might be expected to resemble it in structure as well. Certainly such an intermediate is normally likely to be a better model for the transition state than the starting material would be. Thus σ complexes (Wheland intermediates) in aromatic electrophilic substitutions are used as models for the transition states that are their immediate precursors (p. 150).

The effect of a catalyst is to increase the rate at which a reaction will take place; this is done by making available an alternative path of less energetic demand, often through the formation of a new, and more stable (lower energy), intermediate:

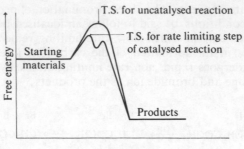

Fig. 2.5

Thus the rate of hydration of an alkene, directly with water,

$$\diagdown C = C \diagup + H_2O \rightarrow \diagdown \underset{H}{\overset{OH}{C} - C} \diagup$$

is often extremely slow, but it can be greatly speeded up by the presence of an acid catalyst, which effects initial protonation of the alkene to a carbonium ion intermediate. This is then followed by easy and rapid attack on the positively charged carbonium ion by a water molecule acting as a nucleophile, and finally by liberation of a proton which is able to function again as a catalyst (p. 184):

$$\underset{H_2O:}{\diagdown C = C \diagup} \overset{H^{\oplus}}{\rightleftharpoons} \diagdown \overset{\oplus}{C} - \underset{H}{C} \diagup \overset{H_2O}{\rightleftharpoons} \diagdown \underset{H}{\overset{\overset{\oplus}{O}H_2}{C} - C} \diagup \overset{-H^{\oplus}}{\rightleftharpoons} \diagdown \underset{H}{\overset{OH}{C} - C} \diagup$$

The details of acid/base catalysis are discussed subsequently (p. 73).

2.2.3 Kinetic versus thermodynamic control

Where a starting material may be converted into two or more alternative products, e.g. in electrophilic attack on an aromatic species that already carries a substituent (p. 149), the proportions in which the alternative products are formed is often determined by their relative rate of formation: the faster a product is formed the more of it there will be in the final product mixture; this is known as *kinetic control*. This is not always what is observed however, for if one or more of the alternative reactions is reversible, or if the products are readily inter-convertible directly under the conditions of the reaction, the composition of the final product mixture may be dictated not by the relative rates of formation of the different products, but by their relative thermodynamic stabilities in the reaction system: we are then seeing *thermodynamic* or *equilibrium control*. Thus the nitration of methylbenzene is found to be kinetically controlled, whereas the Friedel–Crafts alkylation of the same species is often thermodynamically controlled (p. 160). The form of control that operates may also be influenced by the reaction condition, thus the sulphonation of naphthalene with concentrated H_2SO_4 at 80° is essentially kinetically controlled, whereas at 160° it is thermodynamically controlled (p. 162).

2.3   INVESTIGATION OF REACTION MECHANISMS*

It is seldom, if ever, possible to provide complete and entire information, structural, energetic, and stereochemical, about the pathway that is traversed by any chemical reaction: no reaction mechanism can ever be proved to be correct! Sufficient data can nevertheless usually be gathered to show that one or more theoretically possible mechanisms are just not compatible with the experimental results, and/or to demonstrate that of several remaining alternatives one is a good deal more likely than the others.

2.3.1   The nature of the products

Perhaps the most fundamental information about a reaction is provided by establishing the structure of the products that are formed during its course, and relating this information to the structure of the starting material. Where, as is often the case with organic reactions, more than one product is obtained then it is usually an advantage to know also the relative proportions in which the products are obtained, e.g. in establishing, among other things, whether kinetic or thermodynamic control is operating (*cf*. p. 42). In the past this had to be done laboriously—and often imprecisely—by manual isolation of the products, but may now often be achieved much more easily and exactly by gas/liquid chromatography or, indirectly, by spectroscopic methods.

The importance of establishing the correct structure of the reaction product is best illustrated by the confusion that can result when this has been assumed, wrongly, as self-evident, or established erroneously. Thus the yellow triphenylmethyl radical (3, *cf*. p. 292), obtained from the action of metals on triphenylmethyl halides in 1900, readily forms

---

* This topic is studied specifically, and in detail, in the author's *The Search for Organic Reaction Pathways*, Longman, 1972.

a colourless dimer (m.w. = 486) which was—reasonably enough—assumed to have the structure of hexaphenylethane (4). Only after nearly 70 years (in 1968) has it been shown [its n.m.r. spectrum (*cf.* p. 18) exhibits one non-aromatic **H**] that the dimer actually has the structure (5):

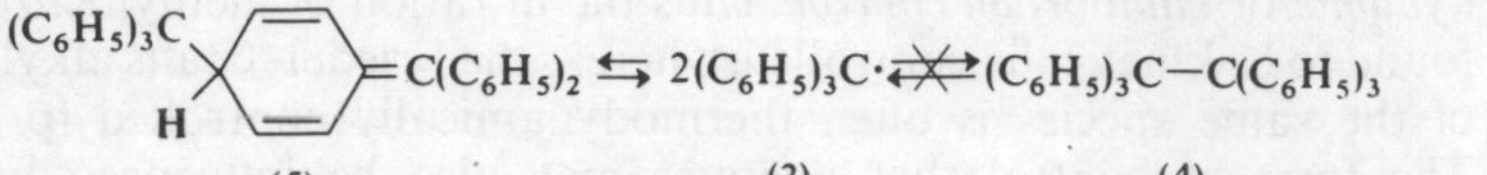

$$(C_6H_5)_3C \underset{H}{\diagdown} \hspace{-0.5em} \diagup\hspace{-1em}\big\langle\hspace{-0.3em}\bigcirc\hspace{-0.3em}\big\rangle\hspace{-1em}= C(C_6H_5)_2 \rightleftharpoons 2(C_6H_5)_3C \cdot \rightleftharpoons (C_6H_5)_3C{-}C(C_6H_5)_3$$

$$\quad\quad (5) \quad\quad\quad\quad\quad (3) \quad\quad\quad (4)$$

At which point numerous small details of the behaviour of (3) and of its dimer, that had previously appeared anomalous, promptly became understandable.

Information about the products of a reaction can be particularly informative when one of them is quite unexpected. Thus the reaction of chloro-4-methylbenzene (*p*-chlorotoluene, 6) with amide ion, $^{\ominus}NH_2$, in liquid ammonia (p. 170) is found to lead not only to the expected 4-methylphenylamine (*p*-toluidine, 7), but also to the quite unexpected 3-methylphenylamine (*m*-toluidine, 8), which is in fact the major product:

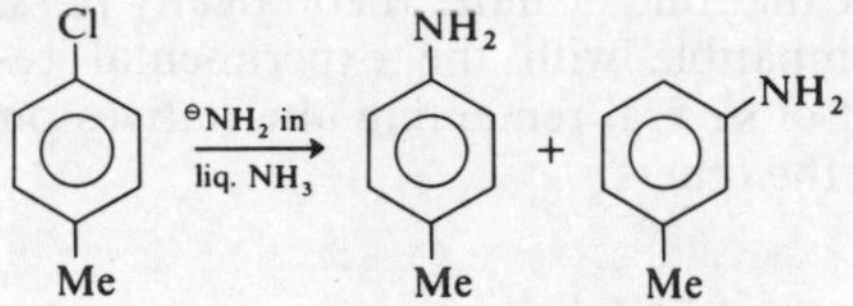

$$\quad\quad (6) \quad\quad\quad\quad (7) \quad\quad\quad\quad (8)$$

Expected   Unexpected

The latter clearly cannot be obtained from (6) by a simple substitution process, and either must be formed from (6) *via* a different pathway than (7), or if the two products are formed through some common intermediate then clearly (7) cannot be formed by a direct substitution either.

2.3.2 Kinetic data

The largest body of information about reaction pathways has come—and still does come— from kinetic studies as we shall see, but the interpretation of kinetic data in mechanistic terms (*cf.* p. 39) is not always quite as simple as might at first sight be supposed. Thus the effective reacting species, whose concentration really determines the reaction rate, may differ from the species that was put into the reaction mixture to start with, and whose changing concentration we are actually seeking to measure. Thus in aromatic nitration the effective attacking species is usually $^{\oplus}NO_2$ (p. 133), but it is HNO_3 that we put into the reaction mixture, and whose changing concentration we are measuring; the relationship between the two may well be complex and so, therefore, may be the relation between the rate of reaction and $[HNO_3]$. Despite the fact that the essential reaction is a simple one, it may not be easy to deduce this from the quantities that we can readily measure.

Then again, if the hydrolysis in aqueous solution of the alkyl halide, RHal, is found to follow the rate equation,

$$\text{Rate} = k_1[\text{RHal}]$$

it is not necessarily safe to conclude that the rate-determining step does not involve the participation of water, simply on the grounds that $[H_2O]$ does not appear in the rate equation; for if water is being used as the solvent it will be present in very large excess, and its concentration would remain virtually unchanged whether or not it actually participated in the rate-limiting stage. The point could perhaps be settled by carrying out the hydrolysis in another solvent, e.g. HCO_2H, and by using a much smaller concentration of water as a potential nucleophile. The hydrolysis may then be found to follow the rate equation,

$$\text{Rate} = k_2[\text{RHal}][\text{H}_2\text{O}]$$

but the actual mechanism of hydrolysis could well have changed on altering the solvent, so that we are not, of necessity, any the wiser about what actually went on in the original aqueous solution.

The vast majority of organic reactions are carried out in solution, and quite small changes in the solvent used can have the profoundest effects on reaction rates and mechanisms. Particularly is this so when polar intermediates, for example carbonium ions or carbanions as constituents of ion pairs, are involved, for such species normally carry an envelope of solvent molecules about with them. This greatly affects their stability (and their ease of formation), and is strongly influenced by the composition and nature of the solvent employed, particularly its polarity and ion-solvating capabilities. By contrast, reactions that involve radicals (p. 291) are much less influenced by the nature of the solvent (unless this is itself capable of reacting with radicals), but are greatly influenced by the addition of radical sources (e.g. peroxides) or radical absorbers (e.g. quinones), or by light which may initiate reaction through the production of radicals by photochemical activation, e.g. $Br_2 \xrightarrow{hv} Br\cdot \; \cdot Br$.

A reaction that is found, on kinetic investigation, to proceed unexpectedly faster or slower than the apparently similar reactions, under comparable conditions, of compounds of related structure suggests the operation of a different, or modified, pathway from the general one that might otherwise have been assumed for the series. Thus the observed rates of hydrolysis of the chloromethanes with strong bases are found, under comparable conditions, to vary as follows,

$$CH_3Cl \gg CH_2Cl_2 \ll CHCl_3 \gg CCl_4$$

clearly suggesting that trichloromethane undergoes hydrolysis in a different manner from the other compounds (*cf.* p. 260).

2.3.3 The use of isotopes

It is often a matter of some concern to know whether a particular bond has, or has not, been broken in a step up to and including the rate-limiting step of a reaction: simple kinetic data cannot tell us this, and further refinements have to be resorted to. If, for example, the bond concerned is C—H, the question may be settled by comparing the rates of reaction, under the same conditions, of the compound in which we are interested, and its exact analogue in which this bond has been replaced by a C—D linkage. The two bonds will have the same chemical nature as isotopes of the same element are involved, but their vibration frequencies, and hence their dissociation energies, will be slightly different because atoms of different *mass* are involved: the greater the mass, the stronger the bond. This difference in bond strength will, of course, be reflected in different rates of breaking of the two bonds under comparable conditions: the stronger C—D bond being broken more slowly than the C—H bond; quantum-mechanical calculation suggests a maximum rate difference, k_H/k_D, of ≈ 7 at 25°.

Thus in the oxidation

$$Ph_2C\overset{\displaystyle OH}{\underset{\displaystyle H}{\diagup}}\quad\xrightarrow[\ominus OH]{MnO_4^{\ominus}}\quad Ph_2C{=}O$$

it is found that Ph_2CHOH is oxidised 6·7 times as rapidly as Ph_2CDOH; the reaction is said to exhibit a *primary kinetic isotope effect*, and breaking of the C—H bond must clearly be involved in the rate-limiting step of the reaction. By contrast benzene, C_6H_6, and hexadeuterobenzene, C_6D_6, are found to undergo nitration at essentially the same rate, and C—H bond-breaking, that must occur at some stage in the overall process,

$$\text{H}-\hspace-6pt\bigcirc \;+\; {}^{\oplus}NO_2 \;\longrightarrow\; \text{NO}_2-\hspace-6pt\bigcirc \;+\; H^{\oplus}$$

thus cannot be involved in the rate-limiting step (*cf.* p. 134).

Primary kinetic isotope effects are also known involving pairs of isotopes other than hydrogen/deuterium, but as the relative mass difference must needs be smaller their maximum values will be correspondingly smaller. Thus the following have been observed:

$$HO^\ominus + {}^{(14)}_{12}CH_3 - I \rightarrow HO - {}^{(14)}_{12}CH_3 + I^\ominus \qquad \frac{k_{^{12}C}}{k_{^{14}C}} = 1.09\,(25°)$$

$$PhCH_2 - {}^{(37)}_{35}Cl + H_2O \rightarrow PhCH_2 - OH + H^\oplus\,{}^{(37)}_{35}Cl^\ominus \qquad \frac{k_{^{35}Cl}}{k_{^{37}Cl}} = 1.0076\,(25°)$$

It should be emphasised that primary kinetic isotope effects are observed experimentally with values intermediate between the maximum calculated value and unity (i.e. *no* isotope effect): these too can be useful, as they may supply important information about the breaking of particular bonds in the transition state.

Isotopes can also be used to solve mechanistic problems that are non-kinetic. Thus the aqueous hydrolysis of esters to yield an acid and an alcohol could, in theory, proceed by cleavage at (*a*) alkyl/oxygen fission, or (*b*) acyl/oxygen fission:

$$
\begin{array}{c}
(a) \qquad (a)\searrow RC\overset{\displaystyle O}{\overset{\|}{}} - O - H + H^{18}O - R' \\[2ex]
R\overset{\displaystyle O}{\overset{\|}{C}} + O + R' \xrightarrow{H_2{}^{18}O} \\[2ex]
(b) \qquad (b)\nearrow \quad RC\overset{\displaystyle O}{\overset{\|}{}} - {}^{18}OH + H - OR'
\end{array}
$$

If the reaction is carried out in water enriched in the heavier oxygen isotope ^{18}O, (*a*) will lead to an alcohol which is ^{18}O enriched and an acid which is not, while (*b*) will lead to an ^{18}O enriched acid but a normal alcohol. Most simple esters are in fact found to yield an ^{18}O enriched acid indicating that hydrolysis, under these conditions, proceeds *via* (*b*) acyl/oxygen fission (p. 234). It should of course be emphasised that these results are only valid provided that neither acid nor alcohol, once formed, can itself exchange its oxygen with water enriched in ^{18}O, as has indeed been shown to be the case.

Heavy water, D_2O, has often been used in a rather similar way. Thus in the Cannizzaro reaction of benzaldehyde (p. 212),

$$PhC\overset{\displaystyle O}{\overset{\|}{}}-H + PhC\overset{\displaystyle O}{\overset{\|}{}}-H \xrightarrow[H_2O]{\ominus OH} PhC\overset{\displaystyle O}{\overset{\|}{}}-O^\ominus + PhC\overset{\displaystyle OH}{\underset{H}{\overset{|}{}}}-H$$

$$(9)$$

the question arises of whether the second hydrogen atom that becomes attached to carbon, in the molecule of phenylmethanol (benzyl alcohol,

9) that is formed, comes from the solvent (H_2O) or from a second molecule of benzaldehyde. Carrying out the reaction in D_2O is found to lead to the formation of no PhCHDOH, thus demonstrating that the second hydrogen atom could not have come from water, and must therefore have been provided by direct transfer from a second molecule of benzaldehyde.

A wide range of other isotopic labels, e.g. ^3H (or T), ^{13}C, ^{14}C, ^{15}N, ^{32}P, ^{35}S, ^{37}Cl, ^{131}I, etc., have also been used to provide important mechanistic information; the carbon isotopes are particularly useful as this element occurs in all organic compounds. The major difficulties with isotopic labelling studies are: (a) selective synthesis so that the isotopic label is introduced only into a known position (or positions) in the test compound; (b) selective degradation of the reaction product to establish the site of the label; and (c) monitoring label content by radioactive (e.g. ^3H, ^{14}C, ^{32}P, ^{35}S, etc.), mass spectroscopic (e.g. D, ^{13}C, ^{15}N, ^{18}O, etc.) or other measurements.

2.3.4   The study of intermediates

Among the most concrete evidence obtainable about the mechanism of a reaction is that provided by the actual isolation of one or more intermediates from the reaction mixture. Thus in the Hofmann reaction (p. 121), by which amides are converted into amines,

$$\underset{\text{O}}{\overset{\text{O}}{\underset{\|}{RC}}}-NH_2 \xrightarrow[\ominus OH]{Br_2} RNH_2$$

it is, with care, possible to isolate the N-bromoamide, RCONHBr, its anion, $RCONBr^\ominus$, and an isocyanate, RNCO; thus going some considerable way to elucidate the overall mechanism of the reaction. It is of course necessary to establish beyond all doubt that any species isolated really is an intermediate—and not merely an alternative product—by showing that it may be converted, under the normal reaction conditions, into the usual reaction products at a rate at least as fast as the overall reaction under the same conditions. It is also important to establish that the species isolated really is on the direct reaction pathway, and not merely in equilibrium with the true intermediate.

It is much more common not to be able to isolate any intermediates at all, but this does not necessarily mean that none are formed, merely that they may be too labile or transient to permit of their isolation. Their occurrence may then often be inferred from physical, particularly spectroscopic, measurements made on the system. Thus in the formation of oximes from a number of carbonyl compounds by reaction

with hydroxylamine (p. 215),

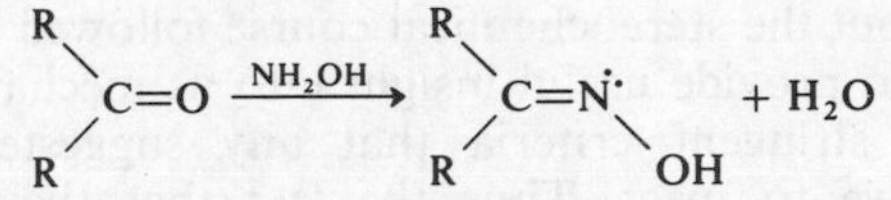

$$\begin{array}{c} R \\ \diagdown \\ \diagup \\ R \end{array} C{=}O \xrightarrow{\ NH_2OH\ } \begin{array}{c} R \\ \diagdown \\ \diagup \\ R \end{array} C{=}\overset{\cdot}{N} \begin{array}{c} \\ \\ OH \end{array} + H_2O$$

the infra-red absorption band characteristic of C=O in the starting material disappears rapidly, and may have gone completely before the band characteristic of C=N in the product even begins to appear. Clearly an intermediate must be formed, and further evidence suggests that it is the *carbinolamine* (10),

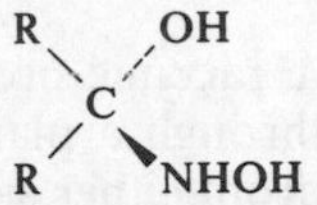

$$\begin{array}{c} R \diagdown \quad \diagup OH \\ C \\ R \diagup \quad \diagdown NHOH \end{array}$$

(10)

which forms rapidly and then breaks down only slowly to yield the products, the oxine and water.

Where we have reason to suspect the involvement of a particular species as a labile intermediate in the course of a reaction, it may be possible to confirm our suspicions by introducing into the reaction mixture, with malice aforethought, a reactive species which we should expect our postulated intermediate to react with particularly readily. It may then be possible to divert the labile intermediate from the main reaction pathway—to *trap* it—and to isolate a stable species into which it has been unequivocally incorporated. Thus in the hydrolysis of trichloromethane with strong bases (*cf.* p. 46), the highly electron-deficient *dichlorocarbene*, CCl$_2$, which has been suggested as a labile intermediate (p. 260), was 'trapped' by introducing into the reaction mixture the electron-rich species *cis* but-2-ene (11), and then isolating the resultant stable cyclopropane derivative (12), whose formation can hardly be accounted for in any other way:

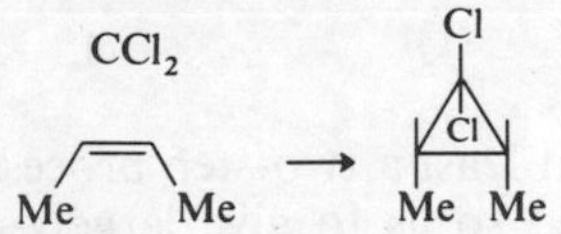

$$CCl_2$$

Me   Me   Me Me

(11)   (12)

The successful study of intermediates not only provides one or more signposts which help define the detailed pathway traversed by a reaction, the intermediates themselves may also provide inferential evidence about the transition states for which they are often taken as models (*cf.* p. 41).

2.3.5   Stereochemical criteria

Information about the stereochemical course followed by a particular reaction can also provide useful insight into its mechanism, and may well introduce stringent criteria that any suggested mechanistic scheme will have to meet. Thus the fact that the base-catalysed bromination of an optically active ketone such as (13)

$$\text{PhCO}\overset{*}{\text{C}}\text{HMeEt} \xrightarrow[\ominus\text{OH}]{\text{Br}_2} \text{PhCO}\overset{*}{\text{C}}\text{BrMeEt}$$
$$(+) \qquad\qquad\qquad (\pm)$$

(13)

leads to an optically inactive racemic product (p. 288), indicates that the reaction must proceed through a planar intermediate, which can undergo attack equally well from either side leading to equal amounts of the two mirror-image forms of the product. Then again, the fact that cyclopentene (14) adds on bromine under polar conditions to yield the *trans* dibromide (15) only, indicates that the mechanism of

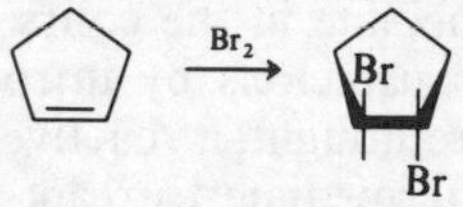

(14)   (15)

the reaction cannot simply be direct, one-step addition of the bromine molecule to the double bond, for this must lead to the *cis* dibromide (16):

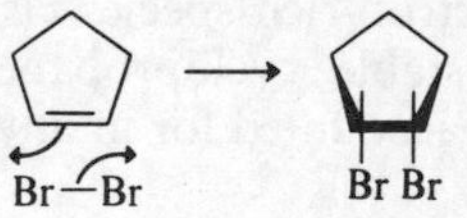

(14)   (16)

The addition must be at least a two-step process (*cf.* p. 176). Reactions like this, which proceed so as to give largely—or even wholly—one stereoisomer out of the two alternatives possible, are said to be *stereoselective*.

Then again, many elimination reactions are found to occur much more readily in that member of a pair of geometrical isomerides in which the atoms or groups to be eliminated are TRANS to each other, than in the isomer in which they are CIS (p. 249). As is seen in the relative ease of elimination from *anti* and *syn* aldoxime acetates to yield the same cyanide:

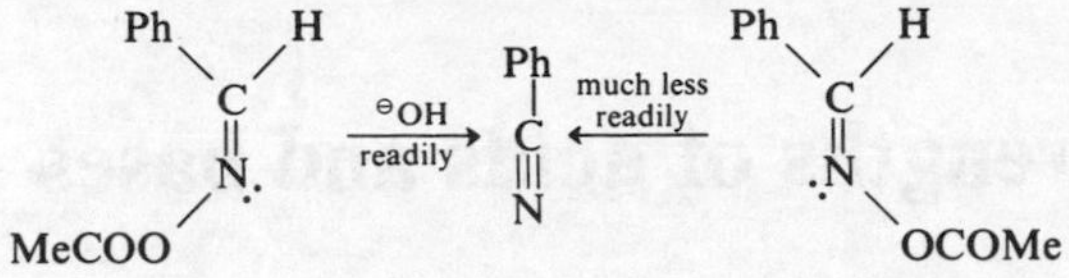

$$
\underset{Anti}{\overset{\displaystyle \underset{MeCOO}{\overset{Ph}{\diagdown}}\overset{H}{\diagup}}{C}}
\quad \overset{\ominus OH}{\underset{readily}{\longrightarrow}} \quad
\overset{Ph}{\underset{N}{\overset{|}{\underset{\parallel}{C}}}}
\quad \overset{much\ less}{\underset{readily}{\longleftarrow}} \quad
\underset{Syn}{\overset{Ph\diagdown \diagup H}{C}}
$$

This clearly sets limitations to which any mechanism advanced for the reaction will have to conform, and gives the lie to that prime tenet of 'lasso chemistry': groups are eliminated most readily when closest together:

This clearly sets limitations to which any mechanism advanced for the reaction will have to conform, and gives the lie to that prime tenet of 'lasso chemistry': groups are eliminated most readily when closest together:

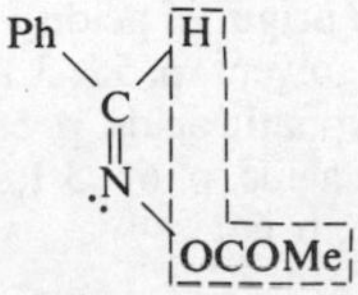

The degree of success with which a suggested mechanism can be said to delineate the course of a particular reaction is not determined solely by its ability to account for the known facts; the acid test is how successful it is at forecasting a change in rate, or even in the nature of the products formed, when the conditions under which the reaction is carried out, or the structure of the starting material, are changed. Some of the suggested mechanisms we shall encounter measure up to these criteria better than do others, but the overall success of a mechanistic approach to organic reactions is demonstrated by the way in which the application of a few relatively simple guiding principles can bring light and order to bear on a vast mass of disparate information about equilibria, reaction rates, and the relative reactivity of organic compounds. We shall now go on to consider some simple examples of this.

3

The strengths of acids and bases

Modern electronic theories of organic chemistry have been highly
successful in a wide variety of fields in correlating behaviour with
structure, not least in accounting for the relative strengths of organic
acids and bases. According to the definition of Arrhenius, acids are
compounds that yield hydrogen ions, $H^\oplus$, in solution while bases
yield hydroxide ions, $^\ominus OH$. Such definitions are reasonably adequate
if reactions in water only are to be considered, but the acid/base
relationship has proved so useful in practice that the concepts of both
acids and bases have become considerably more generalised. Thus
Brønsted defined acids as substances that would give up protons,
i.e. *proton donors*, while bases were *proton acceptors*. The first ionisation of sulphuric acid in aqueous solution is then looked upon as:

$$H_2SO_4 + H_2O \rightleftarrows H_3O^\oplus + HSO_4{}^\ominus$$

| *Acid* | *Base* | *Con-jugate acid* | *Con-jugate base* |

Here water is acting as a base by accepting a proton, and is thereby
converted into its so-called *conjugate acid*, $H_3O^\oplus$, while the acid,
H_2SO_4, by donating a proton is converted into its *conjugate base*,
$HSO_4{}^\ominus$.

The more generalised picture provided by Lewis, who defined acids as molecules or ions capable of coordinating with unshared electron pairs, and bases as molecules or ions which have such unshared electron pairs available for coordination, has already been referred to (p. 29). Lewis acids include such species as boron trifluoride (1) which reacts with trimethylamine to form a solid salt (m.p. 128°):

$$Me_3N\!:\!\downarrow BF_3 \rightleftarrows Me_3\overset{\oplus}{N}\!:\!\overset{\ominus}{BF_3}$$

(1)

Other common examples are aluminium chloride, tin (IV) chloride, zinc chloride, etc. We shall, at this point, be concerned essentially with proton acids, and the effect of structure on the strength of a number of organic acids and bases will now be considered in turn. Compounds in which it is a C—H bond that is ionised will be considered subsequently (p. 264), however.

3.1 ACIDS

3.1.1 pK_a

The strength of an acid, HA, in water, i.e. the extent to which it is dissociated, may be determined by considering the equilibrium:

$$H_2O: + HA \rightleftarrows H_3O^{\oplus} + A^{\ominus}$$

Then the equilibrium constant, in water, is given by:

$$K_a \approx \frac{[H_3O^{\oplus}][A^{\ominus}]}{[HA]}$$

The [H_2O] term is incorporated into K_a because water is present in such excess that its concentration does not change significantly. It should be emphasised that K_a, the *acidity constant* of the acid in water, is only approximate (as above) if concentrations are used instead of the more correct activities; it is a reasonable assumption, however, provided the solution is fairly dilute. The acidity constant is influenced by the composition of the solvent in which the acid is dissolved (see below) and by other factors, but it does, nevertheless, serve as a useful guide to comparative acid strength. In order to avoid writing negative powers of 10, K_a is generally converted into pK_a (p$K_a = -\log_{10} K_a$); thus while K_a for ethanoic(acetic) acid in water at 25° is $1{\cdot}79 \times 10^{-5}$, p$K_a = 4{\cdot}76$. The smaller the numerical value of pK_a, the stronger the acid to which it refers.

Very weak acids, those with pK_a greater than ≈ 16, will not be detectable as acids at all in water, as the [$H_3O^{\oplus}$] they will produce therein will be less than that produced by the autolysis of water itself:

$$H_2O + H_2O \rightleftarrows H_3O^{\oplus} + {}^{\ominus}OH$$

Their relative acidities (pK_as) thus cannot be measured in water at all. Further, when acids are sufficiently strong (low enough pK_a), they will all be essentially fully ionised in water, and will thus all appear to be of the same strength, e.g. HCl, HNO_3, $HClO_4$, etc. This is known as the *levelling effect* of water.

The range of comparative pK_a measurement can, however, be extended by, in the first case, providing a stronger, and the latter case a weaker, base than H_2O as solvent. By carrying out measurements in a range of solvents of increasing basicity (and by using an acid that is near the bottom limit of the acidity range in one solvent and near the top of the range in the next as a common reference in each case) it is possible to carry determination of acid strengths on down to acids as weak as methane ($pK_a \approx 43$).

3.1.2 The origin of acidity in organic compounds

Among the factors that may influence the acidity of an organic compound, HA, are:

(*a*) The strength of the H—A bond.
(*b*) The electronegativity of A.
(*c*) Factors stabilising $A^{\ominus}$ compared with HA.
(*d*) The nature of the solvent.

Of these (*a*) is not normally found to be a limiting factor, but the effect of (*b*) is reflected in the fact that the pK_a of methanol, CH_3O—H, is ≈ 16 while that of methane, CH_3—H, is ≈ 43, oxygen being considerably more electronegative than carbon. By contrast the pK_a of methanoic (formic) acid is 3·77. This is in part due to the electron-withdrawing carbonyl group enhancing the electron affinity of the oxygen atom to which the incipient proton is attached, but much more important is (*c*): the stabilisation possible in the resultant methanoate anion compared with the undissociated methanoic acid molecule:

$$\left[\begin{array}{c} HC \diagup\!\!\!\diagdown \begin{array}{l} O \\ \ddot{O}{-}H \end{array} \\ \updownarrow \\ HC \diagup\!\!\!\diagdown \begin{array}{l} O^{\ominus} \\ \underset{\oplus}{O}{-}H \end{array} \end{array} \right] + H_2O \rightleftharpoons \left[\begin{array}{c} HC \diagup\!\!\!\diagdown \begin{array}{l} O \\ O^{\ominus} \end{array} \\ \updownarrow \\ HC \diagup\!\!\!\diagdown \begin{array}{l} O^{\ominus} \\ O \end{array} \end{array} \right] + H_3O^{\oplus}$$

There is extremely effective delocalisation, with consequent stabilisation, in the methanoate anion involving as it does two canonical structures of identical energy, and though delocalisation can take place in the methanoic acid molecule also, this involves separation of charge and will consequently be much less effective as a stabilising influence (*cf.* p. 20). The effect of this differential stabilisation is somewhat to discourage the recombination of proton with the methanoate anion, the equilibrium is to this extent displaced to the right, and methanoic acid is, by organic standards, a moderately strong acid.

With alcohols there is no such factor stabilising the alkoxide anion $RO^\ominus$, relative to the alcohol itself, and alcohols are thus very much less acidic than carboxylic acids. With phenols, however, there is again the possibility of relative stabilisation of the anion (2), by delocalisation of its negative charge through interaction with the π orbitals of the aromatic nucleus:

(2a)　　　(2b)　　　(2c)　　　(2d)

Delocalisation also occurs in the undissociated phenol molecule (*cf.* p. 23) but, involving charge separation, this is less effective than in the anion (2), thus leading to some reluctance on the part of the latter to recombine with a proton. Phenols are indeed found to be stronger acids than alcohols (the pK_a of phenol itself is 9·95) but considerably weaker than carboxylic acids. This is due to the fact that delocalisation of the negative charge in the carboxylate anion involves structures of identical energy content (see above), and of the centres involved two are highly electronegative oxygen atoms; whereas in the phenoxide anion (2) the structures involving negative charge on the nuclear carbon atoms are likely to be of higher energy content than the one in which it is on oxygen and, in addition, of the centres here involved only one is a highly electronegative oxygen atom. The relative stabilisation of the anion, with respect to the undissociated molecule, is thus likely to be less effective with a phenol than with a carboxylic acid, leading to the lower relative acidity of the former.

3.1.3 The influence of the solvent

Despite the above discussion on the influence of internal structural features on a compound's acidity, the real determining role is often exerted by the solvent, and this is particularly the case when, as commonly, the solvent is water.

Water has the initial disadvantage as a ionising solvent for organic compounds that some of them are insufficiently soluble in their unionised form to dissolve in it in the first place. That limitation apart, water is a singularly effective ionising solvent on account (*a*) of its high dielectric constant ($\epsilon = 80$), and (*b*) of its ion-solvating ability. The first property achieves an effect because the higher the dielectric constant ('polarity') of a solvent the lower the electrostatic energy of any pairs of ions present in it; hence the more readily such ions are formed, the more stable they are in solution, and the less ready are they to recombine with each other.

Ions in solution strongly polarise nearby solvent molecules, thereby collecting a solvation envelope of solvent molecules around them: the greater the extent to which this can take place, the greater the stabilisation of the ion, which is in effect stabilising itself by spreading or delocalising its charge. The peculiar effectiveness of water, as an ion-solvating medium, arises from the fact that H_2O is extremely readily polarised, and also relatively small in size; because of this it can solvate, and thereby stabilise, both cations and anions. The effect is particularly marked with anions for powerful 'hydrogen-bonded' type solvation can occur (see below). Similar H-bonded type solvation cannot in general occur with cations, but in the particular case of acids, the initial cation, $H^\oplus$, can also solvate through hydrogen bonding with the solvent water molecules:

$$H-Y + nH_2O \rightleftharpoons H-\overset{\oplus}{O}\Big\langle\begin{matrix} H & HO \\ \\ H & HO-H \end{matrix}\Big\rangle^{\!\!\!/} \;+\; H\cdots\overset{..}{\underset{..}{Y}}{}^{\ominus}\cdots H \begin{matrix} H-OH \\ \\ OH \end{matrix}$$

Alcohols, just so long as they are not too bulky, e.g. MeOH, share something of water's abilities and, for example, HCl is found to be a strong acid in methanol also. It should not, however, be forgotten that the prime requirement of the solvent is that it should be capable of functioning as a base: the weaker the base, the smaller the dissociation of the acid. Thus we find that in, for example, methylbenzene (toluene) HCl occurs as such, i.e. it is almost wholly undissociated.

3.1.4 Simple aliphatic acids

The replacement of the non-hydroxylic hydrogen atom of methanoic acid by an alkyl group might be expected to produce a weaker acid, as the electron-donating inductive effect of the alkyl group would reduce the residual electron affinity of the oxygen atom carrying the incipient proton, and so reduce the strength of the acid. In the alkyl-substituted anion the increased electron availability on oxygen would serve to promote its recombination with proton, as compared with the methanoate anion/methanoic acid system:

$$\left[Me \rightarrow C \begin{matrix} \nearrow O \\ \nwarrow O \end{matrix} \right]^{\ominus} \qquad \left[H - C \begin{matrix} \nearrow O \\ \nwarrow O \end{matrix} \right]^{\ominus}$$

We should thus expect the equilibrium to be shifted to the left compared with that for methanoic acid/methanoate anion, and it is in fact found that the pK_a of ethanoic acid is 4·76, compared with 3·77 for methanoic acid. However, the overall change in structure effected in so small a molecule as methanoic acid on replacement of H by CH$_3$ makes it doubtful whether so simple an argument is really valid; it could well be that the relative solvation possibilities in the two cases are markedly affected by the considerably different shapes of, as well as by the relative charge distribution in, the two small molecules.

It is important to remember that the value of the acidity constant, K_a, of an acid is related to the standard free energy change for the ionisation, $\Delta G^{\ominus}$, by the relation

$$-\Delta G^{\ominus} = 2\cdot303RT \log K_a$$

and that $\Delta G^{\ominus}$ includes both enthalpy and entropy terms:

$$\Delta G^{\ominus} = \Delta H^{\ominus} - T\Delta S^{\ominus}$$

Thus it is found for the ionisation of ethanoic acid in water at 25° ($K_a = 1\cdot79 \times 10^{-5}$) that $\Delta G^{\ominus} = 27\cdot2$ kJ (6·5 kcal), $\Delta H^{\ominus} = -0\cdot5$ kJ (0·13 kcal), and $\Delta S^{\ominus} = -92$ J (22 cal) deg^{-1} [i.e. $T\Delta S^{\ominus} = -27\cdot6$ kJ (6·6 kcal)]; while for methanoic acid ($K_a = 17\cdot6 \times 10^{-5}$) the corresponding figures are: $\Delta G^{\ominus} = 21$ kJ (5·1 kcal), $\Delta H^{\ominus} = -0\cdot3$ kJ (0·07 kcal), and $\Delta S^{\ominus} = -74$ J (18 cal) deg^{-1} [i.e. $T\Delta S^{\ominus} = -21\cdot3$ kJ (5·17 kcal)]. The surprisingly small $\Delta H^{\ominus}$ values almost certainly arise from the fact that the energy required for dissociation of the O—H bond in the undissociated carboxylic acids is cancelled out by that evolved in solvating the resultant ions.

The differing $\Delta G^{\ominus}$s, and hence the differing K_as, for the two acids thus result from the different values of the two entropy ($\Delta S^{\ominus}$) terms. There are two species on each side of the equilibrium and differences in translational entropy on dissociation will thus be small. However, the two species are neutral molecules on one side of the equilibrium and ions on the other. The main feature that contributes to $\Delta S^{\ominus}$ is thus the solvation sheaths of water molecules that surround RCO$_2{}^{\ominus}$ and H$_3$O$^{\oplus}$, and the consequent restriction, in terms of increased orderliness, that is thereby imposed on the solvent water molecules; the increase in orderliness not being quite so great as might have been expected as there is already a good deal of orderliness in liquid water itself. The difference in strength between methanoic and ethanoic acids thus does indeed relate to the differential solvation of their anions, as was suggested above.

Further substitution of alkyl groups in ethanoic acid has much less effect than this first introduction and, being now essentially a second-order effect, the influence on acid strength is not always regular, steric and other influences playing a part; pK_a values are observed as follows:

$$Me_2CHCO_2H \qquad Me_3CCO_2H$$
$$4\cdot86 \qquad\qquad 5\cdot05$$

$$CH_3CO_2H \quad MeCH_2CO_2H$$
$$4\cdot76 \qquad\quad 4\cdot88$$

$$Me(CH_2)_2CO_2H \quad Me(CH_2)_3CO_2H$$
$$4\cdot82 \qquad\qquad 4\cdot86$$

If there is a doubly bonded carbon atom adjacent to the carboxyl group the acid strength is increased. Thus propenoic (acrylic) acid, $CH_2{=}CHCO_2H$, has a pK_a of $4\cdot25$ compared with $4\cdot88$ for the saturated analogue, propanoic acid. This is due to the fact that the unsaturated α-carbon atom is sp^2 hybridised, which means that electrons are drawn closer to the carbon nucleus than in a saturated, sp^3 hybridised atom due to the rather larger s contribution in the sp^2 hybrid. The result is that sp^2 hybridised carbon atoms are less electron-donating that saturated sp^3 hybridised ones, and so propenoic acid though still weaker than methanoic acid is stronger than propanoic. The effect is much more marked with the sp^1 hybridised carbon atom of a triple bond, thus the pK_a of propynoic (propiolic) acid, $HC{\equiv}CCO_2H$, is $1\cdot84$. An analogous situation occurs with the hydrogen atoms of ethene and ethyne; those of the former are little more acidic than the hydrogens in ethane, whereas those of ethyne are sufficiently acidic to be readily replaceable by a number of metals (*cf.* p. 266).

3.1.5 Substituted aliphatic acids

The effect of introducing electron-withdrawing substituents into simple aliphatic acids is more marked. Thus halogen, with an inductive effect acting in the opposite direction to alkyl, might be expected to increase the strength of an acid so substituted, and this is indeed observed as pK_a values show:

$$CH_3{\rightarrow}CO_2H \qquad F{\leftarrow}CH_2{\leftarrow}CO_2H$$
$$4\cdot76 \qquad\qquad\quad 2\cdot66$$

$$Cl{\leftarrow}CH_2{\leftarrow}CO_2H \qquad \overset{\displaystyle Cl}{\underset{\displaystyle Cl}{{\searrow}\atop{\nearrow}}}CH{\leftarrow}CO_2H$$
$$2\cdot86 \qquad\qquad\qquad\qquad 1\cdot29$$

$$Br{\leftarrow}CH_2{\leftarrow}CO_2H$$
$$2\cdot90$$

$$I{\leftarrow}CH_2{\leftarrow}CO_2H \qquad Cl{\leftarrow}\overset{\displaystyle Cl}{\underset{\displaystyle Cl}{C}}{\leftarrow}CO_2H$$
$$3\cdot16 \qquad\qquad\qquad 0\cdot65$$

The relative effect of the different halogens is in the expected order, fluorine being the most electronegative (electron-withdrawing) and producing a hundredfold increase in strength of fluoroethanoic acid as compared with ethanoic acid itself. The effect is very much greater than that produced, in the opposite direction, by the introduction of an alkyl group, and the introduction of further halogens still produces large increases in acid strength: trichloroethanoic is thus a very strong acid.

Here again it is important to remember that K_a (and hence pK_a) is related to $\Delta G^\ominus$ for the ionisation, and that $\Delta G^\ominus$ includes both $\Delta H^\ominus$ and $\Delta S^\ominus$ terms. In this series of halogen-substituted ethanoic acids $\Delta H^\ominus$ is found to differ little from one compound to another, the observed change in $\Delta G^\ominus$ along the series being due largely to variation in $\Delta S^\ominus$. This arises from the substituent halogen atom effecting delocalisation of the negative charge over the whole of the anion,

$$\left[F{\leftarrow}CH_2{\leftarrow}C \underset{O}{\overset{O}{\Big\backslash\!\!\!/}} \right]^\ominus \qquad \left[CH_3{\rightarrow}C \underset{O}{\overset{O}{\Big\backslash\!\!\!/}} \right]^\ominus$$

the latter thus imposes correspondingly less powerful restriction on the water molecules surrounding it than does the unsubstituted ethanoate anion whose charge is largely concentrated, being confined substantially to $CO_2{}^\ominus$. There is therefore a smaller decrease in entropy on ionisation of the halogen-substituted ethanoic acids than with ethanoic acid itself. This is particularly pronounced with CF_3CO_2H for whose ionisation $\Delta G^\ominus = 1{\cdot}3$ kJ ($0{\cdot}3$ kcal) compared with $27{\cdot}2$ kJ ($6{\cdot}5$ kcal) for CH_3CO_2H, while $\Delta H^\ominus$ differs very little from one acid to the other.

The introduction of a halogen atom further away from the carboxyl group than the adjacent α-position has much less influence. Its inductive effect quickly dies away down a saturated chain, with the result that the negative charge becomes progressively less spread, i.e. more concentrated, in the carboxylate anion. The acid thus increasingly resembles the corresponding simple aliphatic acid itself, as the following pK_a values show:

$$\begin{array}{cc}
\text{MeCH}_2\text{CH}_2\text{CO}_2\text{H} & \overset{\text{Cl}}{\underset{|}{\text{MeCH}_2\text{CHCO}_2\text{H}}} \\
4{\cdot}82 & 2{\cdot}84 \\[2ex]
\overset{\text{Cl}}{\underset{|}{\text{CH}_2}}\text{CH}_2\text{CH}_2\text{CO}_2\text{H} & \overset{\text{Cl}}{\underset{|}{\text{MeCHCH}_2\text{CO}_2\text{H}}} \\
4{\cdot}52 & 4{\cdot}06
\end{array}$$

Other electron-withdrawing groups, e.g. $R_3N^\oplus$, CN, NO_2, SO_2R, CO, CO_2R increase the strength of simple aliphatic acids, as also do hydroxyl and methoxyl groups. The unshared electrons on the oxygen atoms of the last two groups are not able to exert a mesomeric effect, in the opposite direction to their inductive effect, owing to the intervening saturated carbon atoms. These effects are seen in the pK_a values:

$$O_2N \leftarrow CH_2 \leftarrow CO_2H \qquad EtO_2C \leftarrow CH_2 \leftarrow CO_2H$$
$$1.68 \qquad\qquad 3.35$$

$$\overset{\oplus}{Me_3N} \leftarrow CH_2 \leftarrow CO_2H \qquad MeCO \leftarrow CH_2 \leftarrow CO_2H$$
$$1.83 \qquad\qquad 3.58$$

$$NC \leftarrow CH_2 \leftarrow CO_2H \qquad Me\overset{..}{O} \leftarrow CH_2 \leftarrow CO_2H$$
$$2.47 \qquad\qquad 3.53$$

$$H\overset{..}{O} \leftarrow CH_2 \leftarrow CO_2H$$
$$3.83$$

3.1.6   Phenols

Analogous effects can be observed with substituted phenols, the presence of electron-withdrawing groups in the nucleus increasing their acidity. In the case of a nitro substituent, the inductive effect would be expected to fall off with distance on going o- $\rightarrow$ m- $\rightarrow$ p-nitrophenol, but there would also be an electron-withdrawing mesomeric effect when the nitro group is in the o- or p-, but not in the m-position; and this too would promote ionisation by stabilisation (though delocalisation) of the resultant anion. We might therefore expect o- and p-nitrophenols to be more acidic than the m-compound which is, in fact, found to be the case. Introduction of further NO_2 groups promotes acidity markedly, thus 2,4,6-trinitrophenol (picric acid) is found to be a very strong acid:

	pK_a
C_6H_5OH	9.9
$o\text{-}O_2NC_6H_4OH$	7.2
$m\text{-}O_2NC_6H_4OH$	8.35
$p\text{-}O_2NC_6H_4OH$	7.14
$2,4\text{-}(O_2N)_2C_6H_3OH$	4.01
$2,4,6\text{-}(O_2N)_3C_6H_2OH$	1.02

Here again $\Delta H^{\ominus}$ is found to vary only very slightly between o-, m- and p- nitrophenols, the differing $\Delta G^{\ominus}$ values observed for the three arising from differences in the $T\Delta S^{\ominus}$ terms, i.e. from variations in the solvation patterns of the three anions, due to the differing distribution of negative charge in them.

The effect of introducing electron-donating alkyl groups into the benzene nucleus is found to be small:

	pK_a
C_6H_5OH	9.95
$o\text{-}MeC_6H_4OH$	10.28
$m\text{-}MeC_6H_4OH$	10.08
$p\text{-}MeC_6H_4OH$	10.19

The resulting substituted phenols are very slightly weaker acids, but the effect is marginal and irregular, indicating that the effect of such substituents in destabilising the phenoxide ion, by disturbing the interaction of its negative charge with the delocalised π orbitals of the aromatic nucleus, is small, as might have been expected.

3.1.7 Aromatic carboxylic acids

Benzoic acid, with a pK_a of 4·20, is a stronger acid than its saturated analogue cyclohexane carboxylic acid ($pK_a = 4·87$); suggesting that a phenyl group, like a double bond, is here less electron-donating—compared with a saturated carbon atom—towards the carboxyl group, due to the sp^2 hybridised carbon atom to which the carboxyl group is attached (*cf.* p. 58). The introduction of alkyl groups into the benzene nucleus has very little effect on the strength of benzoic acid (*cf.* similar introduction in aliphatic acids),

	pK_a
$C_6H_5CO_2H$	4·20
$m\text{-MeC}_6H_4CO_2H$	4·24
$p\text{-MeC}_6H_4CO_2H$	4·34

but electron-withdrawing groups increase its strength, the effect, as with the phenols, being most pronounced when they are in the *o*- and *p*-positions:

	pK_a
$C_6H_5CO_2H$	4·20
$o\text{-O}_2NC_6H_4CO_2H$	2·17
$m\text{-O}_2NC_6H_4CO_2H$	3·45
$p\text{-O}_2NC_6H_4CO_2H$	3·43
$3,5\text{-(O}_2N)_2C_6H_3CO_2H$	2·83

The particularly marked effect with *o*-NO$_2$ may be due to the very short distance over which the powerful inductive effect is operating, but more direct interaction between the neighbouring groups themselves cannot be ruled out.

The presence of groups such as OH, OMe, or halogen having an electron-withdrawing inductive effect, but an electron-donating mesomeric effect when in the *o*- and *p*-positions, may, however, cause the *p*-substituted acids to be weaker than the *m*- and, on occasion, weaker even than the unsubstituted acid itself, e.g. *p*-hydroxybenzoic acid:

	pK_a of $XC_6H_4CO_2H$				
	H	Cl	Br	OMe	OH
o-	4·20	2·94	2·85	4·09	2·98
m-	4·20	3·83	3·81	4·09	4·08
p-	4·20	3·99	4·00	4·47	4·58

It will be noticed that this compensating effect becomes more pronounced in going $Cl \approx Br \rightarrow OH$, i.e. in increasing order of readiness with which the atom attached to the nucleus will part with its electron pairs.

It is important to emphasise, however, that here—as in the cases above—it is probably the effect of differing charge distributions in the anions on their patterns of solvation, i.e. on the $T\Delta S^{\ominus}$ term relating to the degree of ordering induced locally in the assembly of solvent molecules, that is responsible for the observed differences in pK_a.

The behaviour of *o*-substituted acids is, as seen above, often anomalous. Their strength is sometimes found to be considerably greater than expected due to direct interaction between the adjacent groups. Thus intramolecular hydrogen bonding (*cf.* p. 36) stabilises the anion (4) from *o*-hydroxybenzoic(salicyclic) acid (3) by delocalising its charge, an advantage not shared by its *m*- and *p*-isomers, nor by *o*-methoxybenzoic acid:

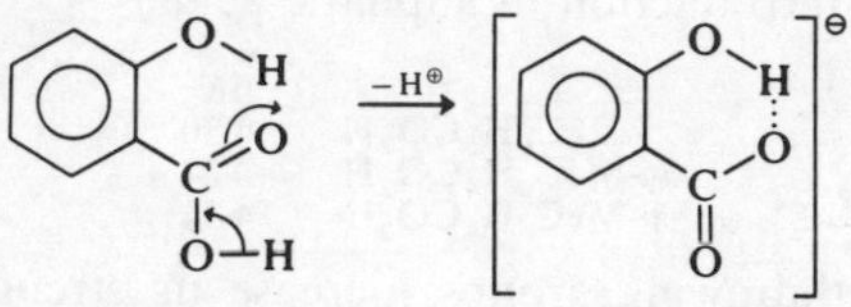

(3)   (4)

Intramolecular hydrogen bonding can, of course, operate in the undissociated acid as well as in the anion, but it is likely to be considerably more effective in the latter than in the former—with consequent relative stabilisation—because the negative charge on oxygen in the anion will lead to stronger hydrogen bonding. The effect is even more pronounced where hydrogen bonding can occur with hydroxyl groups in both *o*-positions, and 2,6-dihydroxybenzoic acid is found to have $pK_a = 1.30$.

3.1.8   Dicarboxylic acids

As the carboxyl group itself has an electron-withdrawing inductive effect, the presence of a second such group in an acid might be expected to make it stronger, as shown by the following pK_a values:

HCO_2H	HO_2CCO_2H
3·77	1·23
CH_3CO_2H	$HO_2CCH_2CO_2H$
4·76	2·83
$CH_3CH_2CO_2H$	$HO_2CCH_2CH_2CO_2H$
4·88	4·19
$C_6H_5CO_2H$	$HO_2CC_6H_4CO_2H$
4·17	*o*-2·98
	m-3·46
	p-3·51

The effect is very pronounced, but falls off sharply as soon as the carboxyl groups are separated by more than one saturated carbon atom. *Cis*-butenedioic(maleic) acid (5, $pK_a^1 = 1.92$) is a much stronger acid than *trans*-butenedioic(fumaric) acid (6, $pK_a^1 = 3.02$), due to the intramolecular hydrogen bonding that can take place with the former, but not with the latter, leading to relative stabilisation of the *cis* (maleate, 7) mono-anion (*cf. o*-hydroxybenzoic acid above):

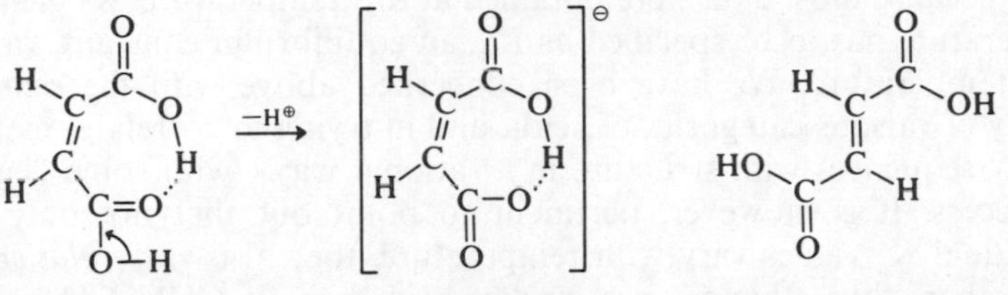

(5) (7) (6)

The second dissociation of *trans*-butenedioic acid ($pK_a^2 = 4.38$) occurs more readily than that of the *cis*-acid ($pK_a^2 = 6.23$), however, because of the greater difficulty in removing a proton from the negatively charged cyclic system in the anion (7) derived from the latter. Ethane-dioic(oxalic), propane-1,3-dioic(malonic) and butane-1,4-dioic(succinic) acids are each weaker in their second dissociations than methanoic, ethanoic and propanoic acids, respectively. This is because the second proton has to be removed from a negatively charged species containing an electron-donating substituent, i.e. $CO_2^{\ominus}$, which might be expected to destabilise the anion with respect to the undissociated acid, as compared with the unsubstituted system:

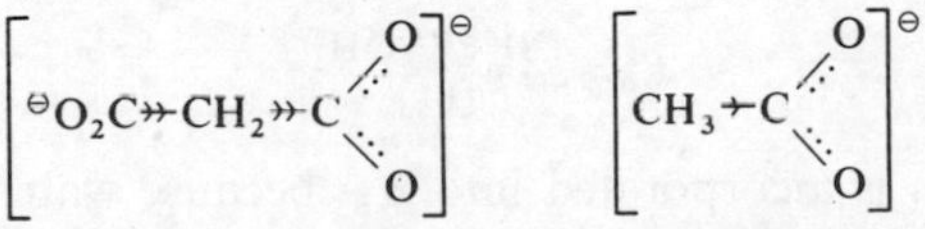

It is interesting, however, that the $\Delta H^{\ominus}$ term for the first dissociation (K^1) of butane-1,4-dioic acid is in fact +ve, i.e. the reaction is endothermic, while $\Delta H^{\ominus}$ for the second dissociation (K^2) is −ve, i.e. the reaction is exothermic (despite having to remove a proton from a negatively charged species): exactly the opposite of what might have been expected! The values of K^1 and K^2 are thus again determined by the entropy component of $\Delta G^{\ominus}$ in each case, reflecting the less constraint imposed on surrounding solvent water molecules by the singly charged anion (K^1), in which the negative charge is to some extent localised, than by the doubly charged species (K^2) in which it is largely concentrated (*cf.* the effect in haloethanoate *v.* ethanoate anions, p. 59).

3.1.9 pK_a and temperature

We have already seen (p. 55) that the K_a, and hence pK_a, value for an acid is <u>not</u> an intrinsic attribute of the species itself, because it varies from one solvent to another: the value depending on the overall system of which the acid is a constituent. Values are normally quoted for aqueous solution, unless otherwise specified, because most data are available for that solvent. Most values are also quoted as at 25°, again because most data were obtained at this temperature. A constant temperature has to be specified as K_a, an equilibrium constant, varies with temperature. We have been concerned above with the *relative* acidity of various categories of acids, and in trying to correlate relative acidity sequences with structure in a rational way—with some degree of success. It is, however, pertinent to point out that not only do individual K_a values vary with temperature, they also vary *relative* to each other: thus ethanoic is a weaker acid than Et_2CHCO_2H below 30°, but a stronger acid above that temperature. Such reversals of relative acidity with change of temperature are found to be fairly common; it thus behoves us not to split too many fine hairs about correlating relative acidity with structure at 25°!

3.2 BASES

3.2.1 pK_b, p$K_{BH^\oplus}$ and pK_a

The strength of a base, B:, in water, may be determined by considering the equilibrium:

$$B: + HOH \rightleftarrows BH^\oplus + {}^\ominus OH$$

The equilibrium constant in water, K_b, is then given by:

$$K_b \approx \frac{[BH^\oplus][^\ominus OH]}{[B:]}$$

The $[H_2O]$ term is incorporated into K_b, because water is present in such excess that its concentration does not change significantly; here again, concentrations can commonly be used instead of the more correct activities provided the solution is reasonably dilute.

It is, however, now more usual to describe the strength of bases also in terms of K_a and pK_a, thereby establishing a single continuous scale for both acids <u>and</u> bases. To make this possible we use, as our reference reaction for bases, the equilibrium

$$BH^\oplus + H_2O \rightleftarrows B: + H_3O^\oplus$$

for which we can then write,

$$K_a \approx \frac{[B:][H_3O^\oplus]}{[BH^\oplus]}$$

where K_a (and pK_a) is a measure of the acid strength of the conjugate acid, $BH^{\oplus}$, of the base, B:. This measure of the readiness with which $BH^{\oplus}$ will part with a proton is, conversely, a measure of the lack of readiness with which the base, B:, will accept one: the stronger $BH^{\oplus}$ is as an acid, the weaker B: will be as a base. Thus the smaller the numerical value of pK_a for $BH^{\oplus}$, the weaker B: is as a base. When using pK to quote the strength of a base, B:, $pK_{BH^{\oplus}}$ should actually be specified but it has become common—though incorrect—to write it simply as pK_a.

Taking as an example $NH_4^{\oplus}$, with a pK_a value of 9·25,

$$NH_4^{\oplus} + H_2O \rightleftarrows NH_3 + H_3O^{\oplus}$$

it is found that $\Delta G^{\ominus} = 52·7\ kJ\ (12·6\ kcal)$, $\Delta H^{\ominus} = 51·9\ kJ\ (12·4\ kcal)$, and $\Delta S^{\ominus} = -2·9\ J\ (0·7\ cal)\ deg^{-1}$ [i.e. $T\Delta S^{\ominus} = -0·8\ kJ\ (0·2\ kcal)$] at 25°. Thus the position of the above equilibrium is effectively determined by $\Delta H^{\ominus}$, the effect of $\Delta S^{\ominus}$ being all but negligible: a result that is in marked contrast to the behaviour of many acids as we have seen above (p. 57). The reason for the small effect of $\Delta S^{\ominus}$ is that here there is one charged species (a positive ion) on each side of the equilibrium, and these ions have closely comparable effects in restricting the solvent water molecules that surround them, so that their entropies of solvation tend to cancel each other out.

3.2.2   Aliphatic bases

As increasing strength in nitrogenous bases is related to the readiness with which they are prepared to take up protons and, therefore, to the availability of the unshared electron pair on nitrogen, we might expect to see an increase in basic strength on going: $NH_3 \rightarrow RNH_2 \rightarrow R_2NH \rightarrow R_3N$, due to the increasing inductive effect of successive alkyl groups making the nitrogen atom more negative. An actual series of amines was found to have related pK_a values as follows, however:

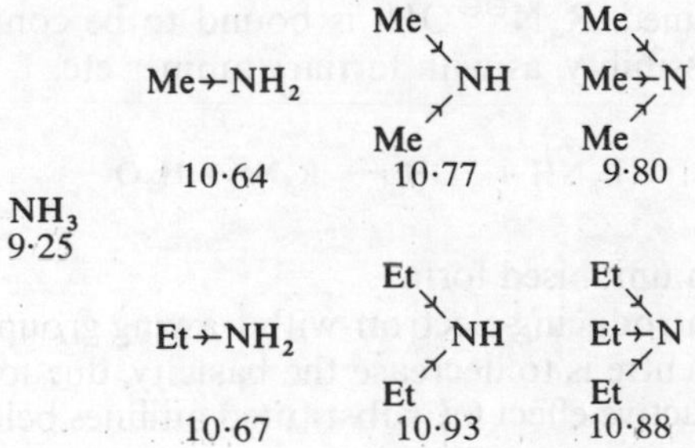

It will be seen that the introduction of an alkyl group into ammonia increases the basic strength markedly as expected. The introduction of a second alkyl group further increases the basic strength, but the net effect of introducing the second alkyl group is very much less marked than with the first. The introduction of a third alkyl group to yield a tertiary amine, however, actually decreases the basic strength in both the series quoted. This is due to the fact that the basic strength of an amine in water is determined not only by electron-availability on the nitrogen atom, but also by the extent to which the cation, formed by uptake of a proton, can undergo solvation, and so become stabilised. The more hydrogen atoms attached to nitrogen in the cation, the greater the possibilities of powerful solvation *via* hydrogen bonding between these and water:

$$R_2\overset{\oplus}{N}\begin{array}{c}H\cdots:OH_2\\ \diagup \\ \diagdown \\ H\cdots:OH_2\end{array} \; > \; R_3\overset{\oplus}{N}-H\cdots:OH_2$$

Thus on going along the series, $NH_3 \rightarrow RNH_2 \rightarrow R_2NH \rightarrow R_3N$, the inductive effect will tend to *increase* the basicity, but progressively less stabilisation of the cation by hydration will occur, which will tend to *decrease* the basicity. The *net* effect of introducing successive alkyl groups thus becomes progressively smaller, and an actual changeover takes place on going from a secondary to a tertiary amine. If this is the real explanation, no such changeover should be observed if measurements of basicity are made in a solvent in which hydrogen-bonding cannot take place; it has, indeed, been found that in chlorobenzene the order of basicity of the butylamines is

$$BuNH_2 < Bu_2NH < Bu_3N$$

though their related pK_a values in water are 10·61, 11·28 and 9·87.

Tetraalkylammonium salts, e.g. $R_4N^{\oplus}I^{\ominus}$, are known, on treatment with moist silver oxide, AgOH, to yield basic solutions comparable in strength with the mineral alkalis. This is readily understandable for the base so obtained, $R_4N^{\oplus \ominus}OH$, is bound to be completely ionised as there is no possibility, as with tertiary amines etc.,

$$R_3\overset{\oplus}{N}H + {}^{\ominus}OH \rightarrow R_3N: + H_2O$$

of reverting to an unionised form.

The effect of introducing electron-withdrawing groups, e.g. Cl, NO_2, close to a basic centre is to decrease the basicity, due to their electron-withdrawing inductive effect (*cf.* substituted anilines below, p. 69); thus

the amine

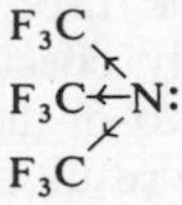

$$F_3C$$
$$F_3C \rightarrow N:$$
$$F_3C$$

is found to be virtually non-basic, due to the three powerfully electron-withdrawing CF_3 groups.

The change is also pronounced with $C=O$, for not only is the nitrogen atom, with its electron pair, bonded to an electron-withdrawing group through an sp^2 hybridised carbon atom (*cf.* p. 58), but an electron-withdrawing mesomeric effect can also operate:

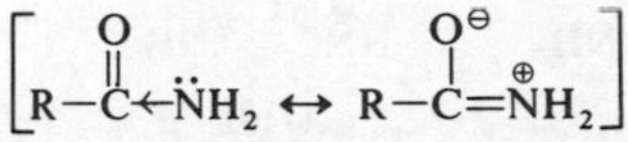

$$\left[R-\overset{\overset{\textstyle O}{\|}}{C}\leftarrow\overset{..}{N}H_2 \leftrightarrow R-\overset{\overset{\textstyle O^{\ominus}}{|}}{C}=\overset{\oplus}{N}H_2 \right]$$

Thus amides are found to be only very weakly basic in water [pK_a for ethanamide(acetamide) $= -0\cdot5$], and if two $C=O$ groups are present the resultant imides, far from being basic, are often sufficiently acidic to form alkali metal salts, e.g. benzene-1,2-dicarboximide (phthalimide, 8):

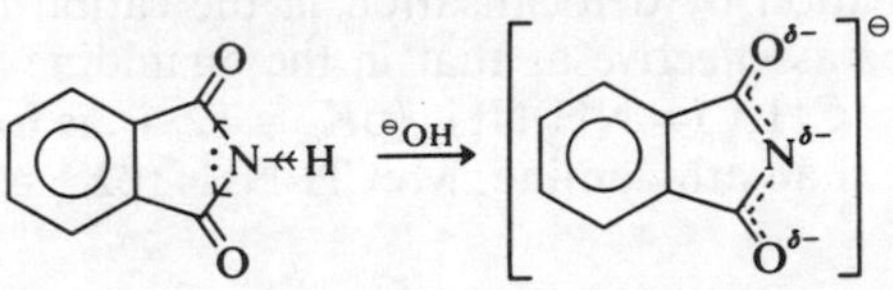

(8)

The effect of delocalisation in *increasing* the basic strength of an amine is seen in guanidine, $HN=C(NH_2)_2$ (9), which, with the exception of the tetraalkylammonium hydroxides above, is among the strongest organic nitrogenous bases known, with a related pK_a of $\approx13\cdot6$. Both the neutral molecule, and the cation, $H_2\overset{\oplus}{N}=C(NH_2)_2$ (10), resulting from its protonation, are stabilised by delocalisation;

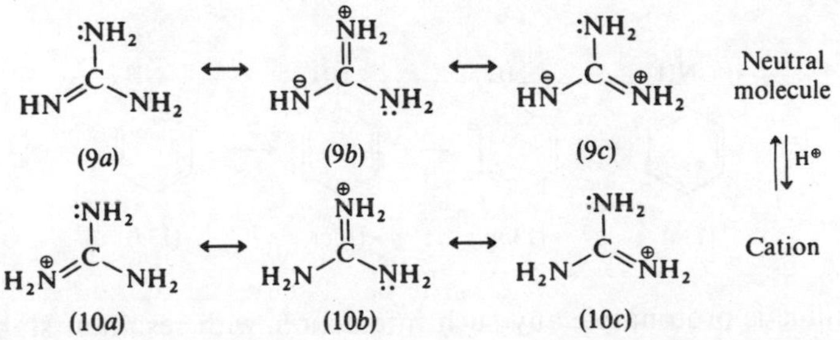

but in the cation the positive charge is spread symmetrically by contribution to the hybrid of three exactly equivalent structures of equal energy. No comparably effective delocalisation occurs in the neutral molecule (in which two of the contributing structures involve separation of charge), with the result that the cation is greatly stabilised with respect to it, thus making protonation 'energetically profitable' and guanidine an extremely strong base.

A somewhat analogous situation occurs with the amidines, $RC(=NH)NH_2$ (11):

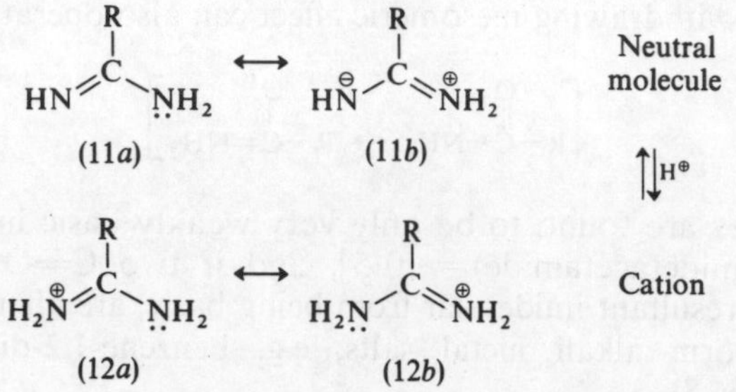

(11a)   (11b)   Neutral molecule

(12a)   (12b)   Cation

While stabilisation by delocalisation in the cation (12) would not be expected to be as effective as that in the guanidine cation (10) above, ethanamidine, $CH_3C(=NH)NH_2$ ($pK_a = 12.4$), is found to be a much stronger base than ethylamine, $MeCH_2NH_2$ ($pK_a = 10.67$).

3.2.3   Aromatic bases

The exact reverse of the above is seen with aniline (13), which is a very weak base ($pK_a = 4.62$) compared with ammonia ($pK_a = 9.25$) or methylamine ($pK_a = 10.64$). In aniline the nitrogen atom is again bonded to an sp^2 hybridised carbon atom but, more significantly, the unshared electron pair on nitrogen can interact with the delocalised π orbitals of the nucleus:

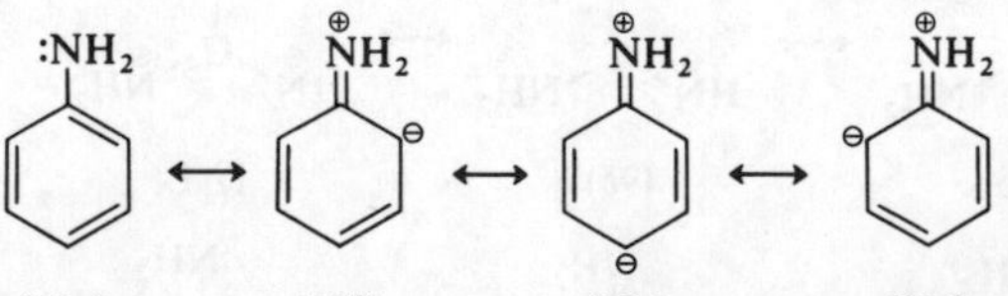

(13a)   (13b)   (13c)   (13d)

If aniline is protonated, any such interaction, with resultant stabilisation, in the anilinium cation (14) is prohibited, as the electron pair on

N is no longer available:

$$H\overset{\oplus}{\underset{}{:}}\overset{\oplus}{N}H_2$$

(14)

The aniline molecule is thus stabilised with respect to the anilinium cation, and it is therefore 'energetically *un*profitable' for aniline to take up a proton; it thus functions as a base with the utmost reluctance ($pK_a = 4\cdot62$, compared with cyclohexylamine, $pK_a = 10\cdot68$). The base-weakening effect is naturally more pronounced when further phenyl groups are introduced on the nitrogen atom; thus diphenylamine, Ph_2NH, is an extremely weak base ($pK_a = 0\cdot8$), while triphenylamine, Ph_3N, is by ordinary standards not basic at all.

The major influence of mesomeric stabilisation—of the aniline molecule (13) with respect to the anilinium cation (14)—in determining aniline's strength as a base is borne out by the small, and irregular, effect on its pK_a of introducing methyl groups on the nitrogen atom, or into the benzene nucleus. These groups would not be expected to influence markedly the interaction of the nitrogen atom's lone pair with the delocalised π orbital system of the benzene nucleus (*cf.* the small effect produced by introducing alkyl groups into the benzene nucleus of phenol, p. 60). Thus the substituted anilines are found to have related pK_a values:

$C_6H_5NH_2$	C_6H_5NHMe	$C_6H_5NMe_2$	$MeC_6H_4NH_2$
4·62	4·40	4·38	*o*- 4·38
			m- 4·67
			p- 5·00

The small, base-strengthening inductive effect that alkyl groups usually exert is not large enough to influence the stabilisation of the aniline molecule to any significant extent, and may well be modified by steric and solvation considerations. A group with a more powerful (electron-withdrawing) inductive effect, e.g. NO_2, is found to have rather more influence. Electron-withdrawal is intensified when the nitro group is in the *o*- or *p*-position, for the interaction of the unshared pair of the amino nitrogen with the delocalised π orbital system of the benzene nucleus is then enhanced. The neutral molecule is thus stabilised even further with respect to the cation, resulting in further weakening as a base. Thus the nitro-anilines are found to have related pK_a values:

PhNH_2	NO_2C_6H_4NH_2
4·62	*o*- −0·28
	m- 2·45
	p- 0·98

The extra base-weakening effect, when the substituent is in the *o*-position, is due in part to the short distance over which its inductive effect is operating, and also to direct interaction, both steric and by hydrogen bonding, with the NH_2 group (*cf.* the case of *o*-substituted benzoic acids, p. 62). *o*-Nitroaniline is such a weak base that its salts are largely hydrolysed in aqueous solution, while 2,4-dinitroaniline is insoluble in aqueous acids, and 2,4,6,-trinitroaniline resembles an amide; it is indeed called picramide and readily undergoes hydrolysis to picric acid (2,4,6-trinitrophenol).

With substituents such as OH and OMe that have unshared electron pairs, an electron-donating, i.e. base-strengthening, mesomeric effect can be exerted from the *o*- and *p*-, but not from the *m*-position, with the result that the *p*-substituted aniline is a stronger base than the corresponding *m*-compound. The *m*-compound is a weaker base than aniline itself, due to the electron-withdrawing inductive effect exerted by the oxygen atom in each case. As so often, the effect of the *o*-substituent remains somewhat anomalous, due to direct interaction with the NH_2 group by both steric and polar effects. The substituted anilines are found to have related pK_a values as follows:

$PhNH_2$	$HOC_6H_4NH_2$	$MeOC_6H_4NH_2$
4·62	*o*- 4·72	*o*- 4·49
	m- 4·17	*m*- 4·20
	p- 5·50	*p*- 5·29

An interesting case is provided by 2,4,6-trinitro-N,N-dimethylaniline (15) and 2,4,6-trinitroaniline (16), where the former is found to be about forty thousand times stronger as a base than the latter (by contrast N,N-dimethylaniline and aniline itself are of almost the same basic strength). This is due to the fact that the NMe_2 group is sufficiently large to interfere sterically with the very large NO_2 groups in both *o*-positions. Rotation about ring-carbon to nitrogen bonds allows the O atoms of NO_2 and the Me groups of NMe_2 to move out of each other's way, but the *p* orbitals on the N atoms are now no longer parallel to the *p* orbitals of the ring-carbon atoms. As a consequence, mesomeric shift of the unshared electron pair on $\overset{.}{N}Me_2$ to the oxygen atoms of the NO_2 groups, *via* the *p* orbitals of the ring-carbon atoms (*cf.* p. 69), is inhibited, and the expected base-weakening—by mesomeric electron-withdrawal—does not take place (*cf.* p. 27). The base-weakening influence of the three nitro groups in (15) is thus confined essentially to their inductive effects:

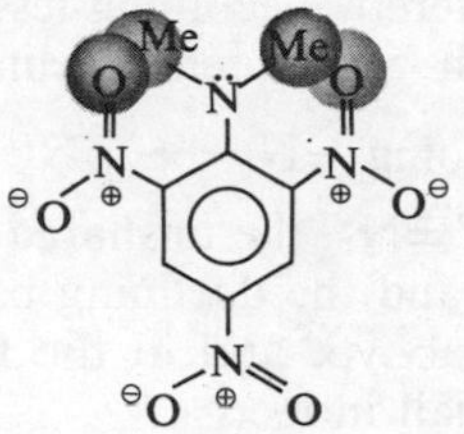

(15)

In 2,4,6-trinitroaniline (16), however, the NH_2 group is sufficiently small for no such limitation to be imposed; hydrogen-bonding between the oxygen atoms of the o-NO_2 groups and the hydrogen atoms of the NH_2 group may indeed help to hold these groups in the required, planar, orientation. The p orbitals may thus assume a parallel orientation, and the strength of (16) as a base is enormously reduced by the powerful electron-withdrawing mesomeric effect of the three NO_2 groups:

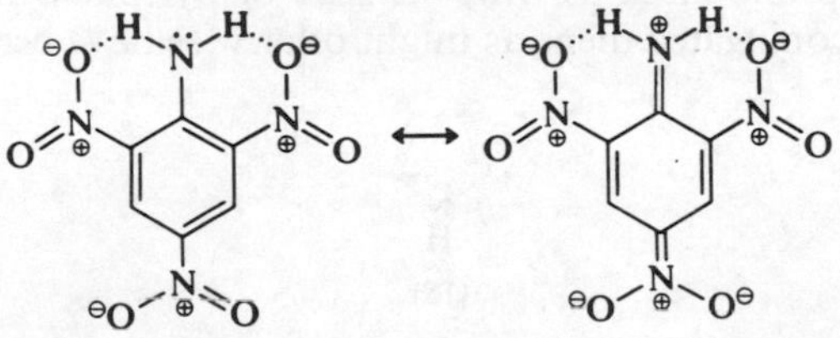

(16a)   (16b)

3.2.4 Heterocyclic bases

Pyridine, , is an aromatic compound (*cf.* p. 18), the N atom is sp^2 hybridised, and contributes one electron to the $6\pi e$ ($4n + 2$, $n = 1$) system; this leaves a lone pair of electrons available on nitrogen (accommodated in an sp^2 hybrid orbital), and pyridine is thus found to be basic ($pK_a = 5\cdot04$). It is, however, a very much weaker base than aliphatic tertiary amines (e.g. $Et_3\ddot{N}$, $pK_a = 10\cdot88$), and this weakness is found to be characteristic of bases in which the nitrogen atom is multiply bonded. This is due to the fact that as the nitrogen atom becomes progressively more multiply bonded, its lone-pair of electrons is accommodated in an orbital that has progressively more *s* character. The electron pair is thus drawn closer to the nitrogen nucleus, and held more tightly by it, thereby becoming less available for forming a bond with a proton, with consequent decline in the basicity of the compound (*cf.* p. 58). On going $\diagdown$N: $\longrightarrow$ $\diagdown$N: $\longrightarrow$ $\equiv$N: in, for example, $R_3N:$ $\longrightarrow C_5H_5N:$ $\longrightarrow RC\equiv N:$, the unshared pairs are in sp^3, sp^2 and sp^1 orbitals, respectively, and the declining basicity is reflected in the two pK_a values quoted above, and in the fact that the basicity of alkyl cyanides is very small indeed.

With quinuclidine (17), however, the unshared electron pair is

(17)

again in an sp^3 orbital and its related pK_a (10·58) is found to be very little different from that of triethylamine (10·88).

Pyrrole (18) is found to exhibit some aromatic character (though this is not so pronounced as with benzene or pyridine), and does not behave like a conjugated diene as might otherwise have been expected:

$$\underset{\substack{\text{N} \\ \text{H}}}{\diagup\diagdown\ddot{}}$$

(18)

For such aromaticity to be achieved, six π electrons $(4n + 2, n = 1)$ from the ring atoms must fill the three bonding molecular orbitals (*cf.* p. 17). This necessitates the contribution of two electrons by the nitrogen atom and, though the resultant electron cloud will be deformed towards nitrogen because of the more electronegative nature of that atom as compared with the four carbons, nitrogen's electron pair will not be readily available for taking up a proton (18*a*):

$$\underset{\substack{\text{N} \\ \text{H}}}{\bigcirc} \overset{H^\oplus}{\rightleftharpoons} \underset{\substack{\text{N} \\ \text{H}}}{\overset{\oplus}{\diagdown}}\overset{H}{\underset{H}{}}$$

(18*a*)   (19)

Protonation, if forced upon pyrrole, is found to take place not on nitrogen but on the α-carbon atom (19). This occurs because incorporation of the nitrogen atom's lone pair of electrons into the aromatic $6\pi e$ system leaves the N atom positively polarised; protons tend to be repelled by it, and are thus taken up by the adjacent α-carbon atom.

The basicity situation rather resembles that already encountered with aniline (p. 68) in that the cation (19) is destabilised with respect to the neutral molecule (18*a*). The effect is much more pronounced with pyrrole, however, for to function as a base it has to lose all aromatic character, and consequent stabilisation: this is reflected in its related pK_a (0·4) compared with aniline's of 4·62, i.e. pyrrole is a very weak base indeed. It can in fact function as an acid, albeit a very weak one, in that the H atom of the NH group may be removed by strong bases, e.g. $^{\ominus}HN_2$, the resultant anion (20) being stabilised by delocalisation (*cf.* the cyclopentadienyl anion, p. 268):

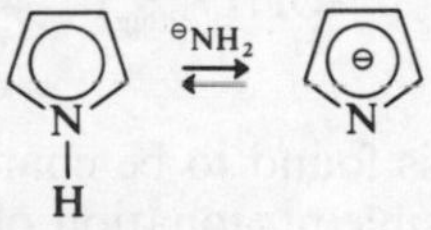

$$\underset{(18a)}{\overset{\displaystyle \underset{H}{\overset{|}{N}}}{\bigcirc}} \quad \overset{\ominus NH_2}{\underset{\rightleftharpoons}{}} \quad \underset{(20)}{\overset{\displaystyle \underset{N}{\ominus}}{\bigcirc}}$$

No such considerations can, of course, apply to the fully reduced pyrrole, pyrrolidine (21),

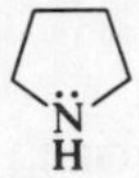

$$\underset{(21)}{\overset{\displaystyle \underset{\underset{H}{|}}{\overset{\displaystyle\ddot{N}}{}}}{\bigcirc}}$$

which is found to have a related pK_a of 11·27, closely resembling that of diethylamine (10·93).

3.3. ACID/BASE CATALYSIS

Catalysis in homogeneous solution has already been referred to (p. 41) as operating by making available an alternative reaction path of lower energetic demand, often *via* a new and more stable (lower energy) intermediate: by far the most common, and important, catalysts in organic chemistry are acids and bases.

3.3.1 Specific and general acid catalysis

The simplest case is that in which the reaction rate is found to be $\propto$ pH, rising as the pH falls, i.e. $\propto [H_3O^{\oplus}]$ if water is the solvent. A common example (*cf.* p. 207) is the hydrolysis of simple acetals, e.g. $MeCH(OEt)_2$, where it is found that:

$$\text{Rate} = k[H_3O^{\oplus}][MeCH(OEt)_2]$$

This is known as *specific acid catalysis*, specific in that $H_3O^\oplus$ is the only acidic species that catalyses the reaction: the reaction rate is found to be unaffected by the addition of other potential proton donors (acids) such as $NH_4^\oplus$, provided that $[H_3O^\oplus]$, i.e. pH, is not changed, indirectly, by their addition. The mechanism of the above acetal hydrolysis is believed to be,

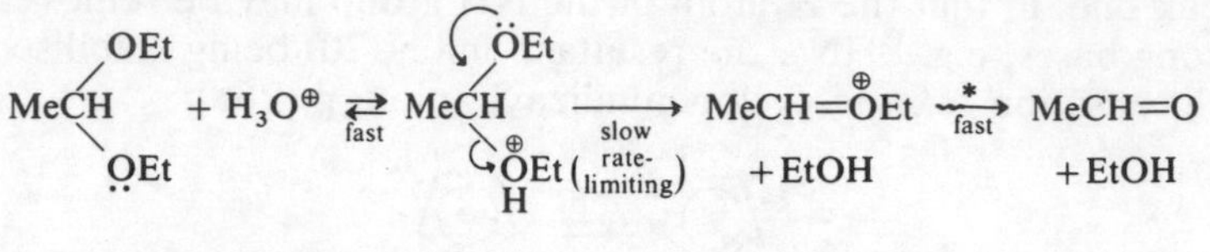

and specific acid catalysis is found to be characteristic of reactions in which there is rapid, reversible protonation of the substrate *before* the slow, rate-limiting step.

Reactions are also known which are catalysed not only by $H_3O^\oplus$, but by other acids in the system as well; e.g. in the hydrolysis of orthoesters such as $MeC(OEt)_3$ in the presence of an acid, HA, where it is found that:

$$\text{Rate} = k_{H_3O^\oplus}[H_3O^\oplus][MeC(OEt)_3] + k_{HA}[HA][MeC(OEt)_3]$$

This is known as *general acid catalysis*, general because the catalysis is by proton donors in general, and not by $H_3O^\oplus$ alone. General acid catalysis often only becomes important at higher pHs, e.g. $\approx$ pH 7 when $[H_3O^\oplus] \approx 10^{-7}$, while [HA] may be 1–2, molar; general acid catalysis will still occur at lower pHs, but may then be masked by the greater contribution by $H_3O^\oplus$. The above orthoester hydrolysis is believed to proceed (only HA is shown here but $H_3O^\oplus$ will do the same thing),

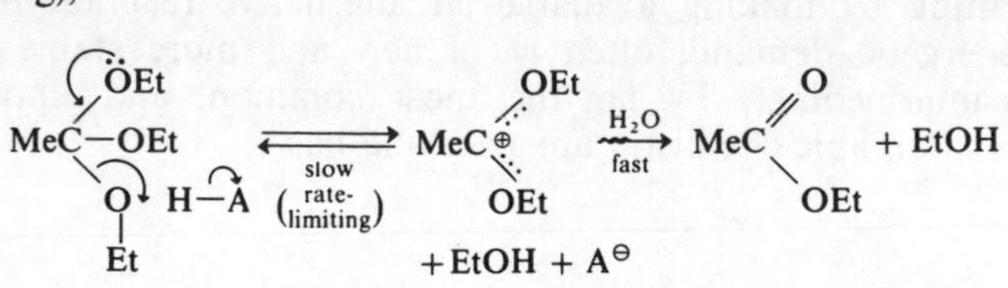

and general acid catalysis is characteristic of reactions in which protonation of the substrate is slow, i.e. rate-limiting, and is followed by rapid conversion of the intermediate into products.

---

*The symbol $\longrightarrow$ will be used subsequently to indicate an overall conversion that proceeds *via* more than one step.

3.3.2 Specific and general base catalysis

Exactly the same distinction can be made over catalysis by bases as was made above for acids. Thus in *specific base catalysis* the reaction rate is again found to be $\propto$ pH, this time rising as the pH rises, i.e. is $\propto [^{\ominus}OH]$. Thus in the reversal of an aldol condensation (*cf.* p. 220) it is found that,

$$\text{Rate} = k[^{\ominus}OH][Me_2C(OH)CH_2COMe]$$

and the reaction is believed to proceed:

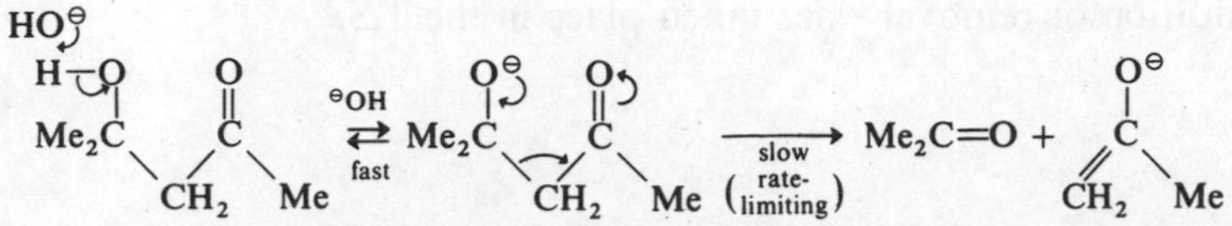

By analogy with acids above, specific basic catalysis is found to be characteristic of reactions in which there is rapid, reversible proton-removal from the substrate *before* the slow, rate-limiting step.

In *general base catalysis*, bases other than $^{\ominus}OH$ are involved. Thus in the base catalysed bromination of acetone (*cf.* p. 288) in an acetate buffer it is found that,

$$\text{Rate} = k_{\ominus OH}[^{\ominus}OH][MeCOMe] + k_{MeCO_2^{\ominus}}[MeCO_2^{\ominus}][MeCOMe]$$

and the reaction is believed to proceed:

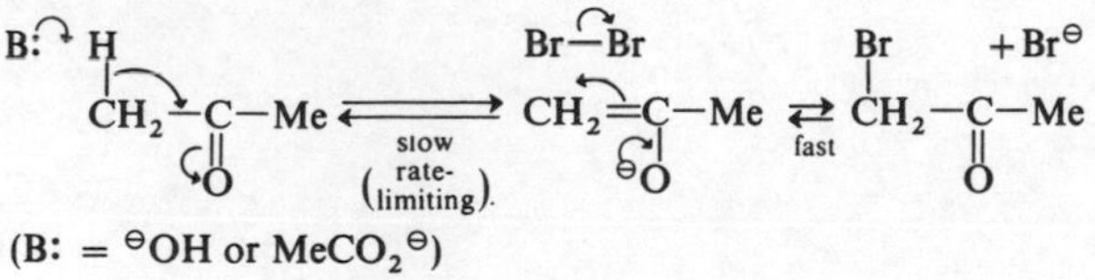

$(B: = {}^{\ominus}OH \text{ or } MeCO_2{}^{\ominus})$

Again by analogy with acids above, general base catalysis is found to be characteristic of reactions in which removal of proton from the substrate is slow, i.e. rate-limiting, and is followed by rapid conversion of the intermediate into products.

3.3.3 Brønsted catalysis law

In general acid catalysis it might well be expected that the effectiveness of a particular acid as a catalyst would be related to its acid strength—

and similarly for bases. Brønsted found, experimentally, that the relationship (Brønsted catalysis law) was,

General acid catalysis: $\log k_{HA} = \alpha p K_a + \text{const.}$

General base catalysis: $\log k_B = \beta p K_b + \text{const.}$

where k = rate of the catalysed reaction; a plot of k_{HA} *v.* pK_a gives a straight line, whose slope is α. α and β are thus *sensitivity parameters*, indicating the degree of sensitivity of the reaction being catalysed to acid (or base) catalysis. They are found to vary in magnitude from $\approx 0.1 \rightarrow 0.9$, and can give useful information about the transition state of the reaction being studied, i.e. the extent to which proton transfer—addition or removal—has taken place in the T.S.

4

Nucleophilic substitution at a saturated carbon atom

A type of reaction that has probably received more detailed study than any other—largely due to the monumental work of Ingold and his school—is nucleophilic substitution at a saturated carbon atom: the classical displacement reaction exemplified by the conversion of an alkyl halide into an alcohol by the action of aqueous base:

$$HO^\ominus + R-Hal \rightarrow HO-R + Hal^\ominus$$

Kinetic measurements on reactions in which alkyl halides are attacked by a wide variety of different nucleophiles, Nu:, have revealed two, essentially extreme, types: one in which,

$$\text{Rate} = k_2[\text{RHal}][\text{Nu:}] \qquad [1]$$

and another in which,

$$\text{Rate} = k_1[\text{RHal}] \qquad [2]$$

i.e. the rate is independent of [Nu:]. In some cases the rate equations are found to be 'mixed' or are otherwise complicated, but examples are known which exactly follow the simple relations above.

4.1 RELATION OF KINETICS TO MECHANISM

Hydrolysis of the primary halide bromomethane (methyl bromide) in aqueous base has been shown to proceed according to equation [1] above, and this has been interpreted as involving the participation of both alkyl halide and hydroxyl ion in the rate-limiting (i.e. slowest) step of the reaction. Ingold has suggested a transition state in which the attacking hydroxyl ion becomes partially bonded to the reacting carbon atom before the incipient bromide ion has become wholly detached from it; thus part of the energy necessary to effect the breaking of the C—Br bond is then supplied by that produced in forming the HO—C bond. Quantum mechanical calculation shows that an approach by the hydroxyl ion along the line of centres of the carbon and bromine atoms is that of lowest energy requirement. This can be represented:

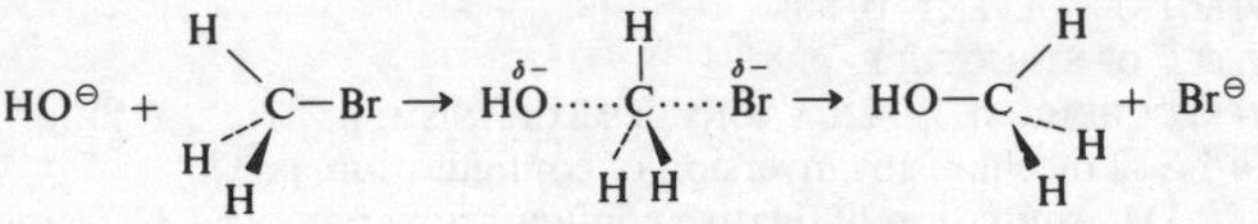

Transition state

The negative charge is spread in the transition state in the course of being transferred from hydroxyl to bromine, and the hydrogen atoms attached to the carbon atom attacked pass through a position in which they all lie in one plane (at right angles to the plane of the paper as drawn here). This type of mechanism has been named by Ingold S_N2, standing for <u>S</u>ubstitution <u>N</u>ucleophilic <u>bi</u>molecular.

By contrast, hydrolysis of the tertiary halide 2-chloro-2-methyl-propane (3,*t*-butyl chloride) in base is found kinetically to follow equation [2], i.e. as the rate is independent of [$^{\ominus}$OH], this can play no part in the rate-limiting step. This has been interpreted as indicating that the halide undergoes slow ionisation (in fact, completion of the R→Cl polarisation that has already been shown to be present in such a molecule) as the rate-limiting step to yield the ion pair $R^{\oplus}Cl^{\ominus}$ (4); followed by rapid, non rate-limiting attack by $^{\ominus}$OH or, if that is suitable, by solvent, the latter often predominating:

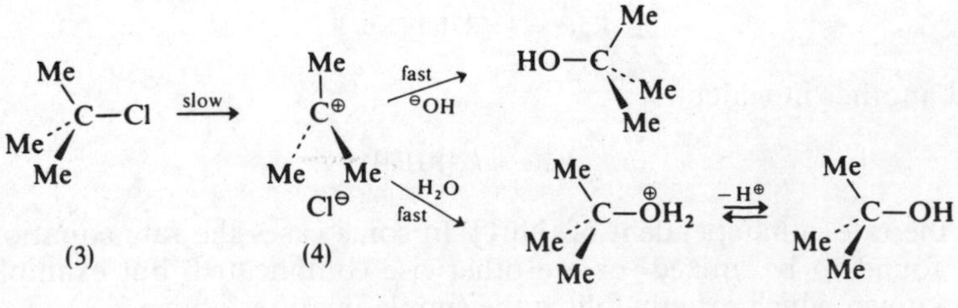

This type of mechanism has been named S_N1, i.e. Substitution Nucleophilic unimolecular. The energy necessary to effect the initial ionisation is largely recovered in the energy of solvation of the ions so formed. The cation, in the ion pair (4), in which the carbon atom carries a positive charge is a carbonium ion, and during its formation the initially tetrahedral carbon atom collapses to a more stable planar state in which the three methyl groups are as far apart as they can get; attack by $^\ominus OH$ or solvent (e.g. $H_2O\colon$) can then take place from either side. If this assumption of a planar state is inhibited by steric or other factors (*cf.* p. 86), the carbonium ion will be formed only with difficulty if at all, i.e. ionisation, and hence reaction by the S_N1 mechanism, may then not take place.

A certain element of confusion is to be met with both in textbooks, and in the literature, over the use and meaning of the terms *order* (*cf.* p. 39) and *molecularity* as applied to reactions. The order is an experimentally determined quantity, the overall order of a reaction being the sum of the powers of the concentration terms that appear in the rate equation:

$$\text{Rate} = k_3[A][B][C] \qquad \underline{\text{Third}} \text{ order overall}$$

$$\text{Rate} = k_3[A]^2[B] \qquad \underline{\text{Third}} \text{ order overall}$$

$$\text{Rate} = k_2[A]^2 \qquad \underline{\text{Second}} \text{ order overall}$$

Generally, however, it is the order with respect to a particular reactant (or reactants) that is of more interest and significance than the overall order, i.e. that the above reactions are first order, second order, and second order, respectively, with respect to A. Examples of both zero order, and non-integral orders, with respect to a particular reactant are also known.

The molecularity refers to the number of species (molecules, ions, etc.) that are undergoing bond-breaking and/or bond-making in one step of the reaction, usually in the rate-limiting step. It is important to realise that the molecularity is not an experimentally determined quantity, and has significance only in the light of the particular mechanism chosen for the reaction: it is an integral part of the mechanistic interpretation of the reaction and is susceptible to re-evaluation, in the light of additional experimental information about the reaction, in a way that the order cannot be. The molecularity of the reaction as a whole only has meaning if the reaction proceeds in a single step (an *elementary* reaction), as is believed to be the case with the hydrolysis of bromomethane above (p. 78); order and molecularity then coincide, the reaction being second order overall (first order in each reactant) and bimolecular. Order and molecularity do not always, or necessarily, have the same value, however.

Simple kinetic measurements can, however, be an inadequate guide to which of the above two mechanisms, S_N1 or S_N2, is actually operating in, for example, the hydrolysis of a halide. Thus, as we have seen (p. 45), where the solvent can act as a nucleophile (*solvolysis*), e.g. H_2O, we would expect for an S_N2 type reaction,

$$Rate = k_2[RHal][H_2O]$$

but as $[H_2O]$ remains effectively constant the rate equation actually observed will be,

$$Rate = k_{obs}[RHal]$$

and simple kinetic measurements in aqueous solution will thus suggest, erroneously, that the reaction is of the S_N1 type.

A kinetic distinction between the operation of the S_N1 and S_N2 modes can often be made by observing the effect on the overall reaction rate of adding a competing nucleophile, e.g. azide anion, $N_3^\ominus$. The total nucleophile concentration is thus increased, and for the S_N2 mode where [Nu:] appears in the rate equation, this will result in an increased reaction rate due to the increased [Nu:]. By contrast, for the S_N1 mode [Nu:] does <u>not</u> appear in the rate equation, i.e. is <u>not</u> involved in the rate-limiting step, and addition of $N_3^\ominus$ will thus <u>be</u> without significant effect on the observed reaction rate.

4.2 EFFECT OF SOLVENT

Changing the solvent in which a reaction is carried out often exerts a profound effect on its rate and may, indeed, even result in a change in its mechanistic pathway. Thus for a halide that undergoes hydrolysis by the S_N1 mode, increase in the polarity of the solvent (i.e. increase in ϵ, the dielectric constant) and/or its ion-solvating ability is found to result in a very marked <u>increase</u> in reaction rate. Thus the rate of solvolysis of the tertiary halide, Me_3CBr, is found to be 3×10^4 times faster in 50% aqueous ethanol than in ethanol alone. This occurs because, in the S_N1 mode, charge is developed and concentrated in the T.S. compared with the starting material:

$$R{-}Hal \rightarrow \overset{\delta+}{R} \cdots \overset{\delta-}{Hal} \rightarrow R^\oplus Hal^\ominus$$

The energy required to effect such a process decreases as ϵ rises; the process is also facilitated by increasing solvation, and consequent stabilisation, of the developing ion pair compared with the starting material. That such effects, particularly solvation, are of prime importance is borne out by the fact that S_N1 type reactions are extremely uncommon in the gas phase.

For the S_N2 mode, however, increasing solvent polarity is found to have a much less marked effect, resulting in a slight <u>decrease</u> in reaction rate. This occurs because in this particular example new charge is not developed, and existing charge is dispersed, in the T.S. compared with the starting materials;

$$Nu^{\ominus} + R-Hal \rightarrow \overset{\delta-}{Nu}\cdots R\cdots \overset{\delta-}{Hal} \rightarrow Nu-R + Hal^{\ominus}$$

thus, solvation of the T.S. is likely to be somewhat less effective than that of the initial nucleophile—hence the slight decrease. This differing behaviour of S_N1 and S_N2 modes to changes of solvent can be used to some extent diagnostically.

A very marked effect on the rate of S_N2 reactions is, however, effected on transferring them from polar hydroxylic solvents to polar non-hydroxylic solvents. Thus the reaction rate of the primary halide, MeI, with $N_3^{\ominus}$ at 0° increased 4.5×10^4 fold on transfer from methanol ($\epsilon = 33$) to N,N-dimethylmethanamide(dimethyl formamide, DMF), $HCONMe_2$, with very much the same polarity ($\epsilon = 37$). This very large rate difference stems from the fact that the attacking nucleophile, $N_3^{\ominus}$, is highly solvated though hydrogen-bonding in MeOH (*cf.* p. 56) whereas it is very much less strongly solvated—and not by hydrogen bonding—in $HCONMe_2$. The largely unsolvated $N_3^{\ominus}$ anion (in $HCONMe_2$) is a very much more powerful nucleophile than when surrounded (as in MeOH) by a very much less nucleophilic solvation envelope, hence the rise in reaction rate. Rate increases of as much as 10^9-fold have been observed on transferring S_N2 reactions from, e.g. MeOH, to another polar non-protic solvent, dimethyl sulphoxide (DMSO), Me_2SO ($\epsilon = 46$).

So far as actual changes of mechanistic pathway with change of solvent are concerned, increase in solvent polarity and ion-solvating ability may (but not necessarily will) change the reaction mode from $S_N2 \rightarrow S_N1$. Transfer from hydroxylic to polar, non-protic solvents (e.g. DMSO) can, and often do, change the reaction mode from $S_N1 \rightarrow S_N2$ by enormously increasing the effectiveness of the nucleophile in the system.

4.3 EFFECT OF STRUCTURE

An interesting sequence is provided by the reaction with base of the series of halides:

$$CH_3-Br \quad MeCH_2-Br \quad Me_2CH-Br \quad Me_3C-Br$$
$$(5) \qquad\quad (6) \qquad\qquad (7) \qquad\qquad (8)$$

The first and last members are described in the literature as undergoing ready hydrolysis, the two intermediate members being more resistant. Measurement of their rates of hydrolysis with dilute, aqueous ethanolic sodium hydroxide solution gives the plot* (Fig. 4.1),

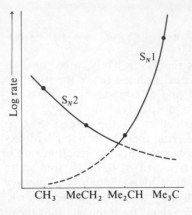

Fig. 4.1

and further kinetic investigation reveals a change in order of reaction, and hence presumably, of mechanism, as the series is traversed. Thus bromomethane (5) and bromoethane (6) are found to follow a second order rate equation, 2-bromopropane (7) a mixed second and first order equation, the relative proportion of the two depending on the initial [$^{\ominus}$OH] (the higher the initial concentration the greater the second order proportion) and the total rate here being a minimum for the series, while 2-bromo-2-methylpropane (8) is found to follow a first order rate equation.

In seeking an explanation for the implied changeover in mechanistic pathway we need to consider, in each case, the effect on the transition state of both electronic and steric factors. For S_N2 attack, the enhanced inductive effect of an increasing number of methyl groups, as we go across the series, might be expected to make the carbon atom that bears the bromine progressively less positively polarised, and hence less readily attacked by $^{\ominus}$OH. This effect is probably small, and steric factors are of more importance; thus $^{\ominus}$OH will find it progressively more difficult to attack the bromine-carrying carbon as the latter becomes more heavily substituted. More significantly, the resultant S_N2 transition state will have *five* groups around this carbon atom (compared with only *four* in the initial halide), there will thus be an increase in crowding on going from the initial halide to the transition state, and this relative crowding will increase as the size of the original substituents increases (H $\rightarrow$ Me). The more crowded the T.S. relative

---

* Based on Ingold, *Structure and Mechanism in Organic Chemistry,* by permission of Cornell University Press.

to the starting materials, the higher its energy will be, and the slower therefore will it be formed. The S_N2 reaction rate will thus <u>decrease</u> as the series is traversed.

For S_N1 attack, considerable charge separation has taken place in the T.S. (*cf.* p. 80), and the ion pair intermediate to which it gives rise is therefore often taken as a model for it. As the above halide series is traversed, there is increasing stabilisation of the carbonium ion moiety of the ion pair, i.e. increasing rate of formation of the T.S. This increasing stabilisation arises from the operation of both an inductive effect,

and hyperconjugation (p. 25), e.g.

via the hydrogen atoms attached to the α-carbons, the above series of carbonium ions having 0, 3, 6 and 9 such hydrogen atoms, respectively.

Support for such an interaction of the H—C bonds with the carbon atom carrying the positive charge is provided by substituting H by D in the original halide, the rate of formation of the ion pair is then found to be slowed down by $\approx 10\%$ per deuterium atom incorporated: a result compatible only with the H—C bonds being involved in the ionisation. This is known as a *secondary kinetic isotope effect*, secondary because it is a bond other than that carrying the isotopic label that is being broken (*cf.* p. 46). The relative contributions of hyperconjugation and inductive effects to the stabilisation of carbonium ions is open to debate, but it is significant that a number of carbonium ions will only form at all if they can take up a planar arrangement, the state in which hyperconjugation will operate most effectively (*cf.* p. 103).

In steric terms there is a relief of crowding on going from the initial halide, with a tetrahedral disposition of four substituents about the sp^3 hybridised carbon atom, to the carbonium ion, with a planar disposition of only *three* substituents (*cf. five* for the S_N2 T.S.) about the now sp^2 hybridised carbon atom. The three substituents are as far apart from each other as they can get in the planar carbonium ion, and the relative relief of crowding will increase as the substituents increase in size (H $\rightarrow$ Me). The S_N1 reaction rate will thus increase markedly (on both electronic and steric grounds) as the series is traversed.

As the S_N2 rate <u>decreases</u> and the S_N1 rate <u>increases</u> across the series of halides, the reasons for the mechanistic changeover become apparent.

A similar mechanistic changeover is observed, though considerably sooner, in traversing the series:

$$CH_3-Cl \qquad C_6H_5CH_2-Cl \qquad (C_6H_5)_2CH-Cl \qquad (C_6H_5)_3C-Cl$$
$$(9) \qquad\qquad (10) \qquad\qquad\qquad (11) \qquad\qquad\qquad (12)$$

Thus for hydrolysis in 50% aqueous acetone, a mixed second and first order rate equation is observed for phenylchloromethane (benzyl chloride, 10)—moving over almost completely to the S_N1 mode in water alone. Diphenylchloromethane (11) is found to follow a first order rate equation, with a very large increase in total rate, while with triphenylchloromethane (trityl chloride, 12) the ionisation is so pronounced that the compound exhibits electrical conductivity when dissolved in liquid SO_2. The main reason for the greater promotion of ionisation—with consequent earlier changeover to the S_N1 pathway in this series—is the considerable stabilisation of the carbonium ion, by delocalisation of its positive charge, that is now possible:

This is a classical example of an ion stabilised by charge delocalisation *via* the agency of the delocalised π orbitals of the benzene nucleus (*cf.* the negatively charged phenoxide ion, p. 23). In terms of overall reactivity, phenylchloromethane (benzyl chloride, 10) is rather similar to 2-chloro-2-methylpropane (*cf.* 8); the effect will become progressively more pronounced, and S_N1 attack further facilitated, with $(C_6H_5)_2CHCl$ (11) and $(C_6H_5)_3CCl$ (12) as the possibilities of delocalising the positive charge are increased in the carbonium ions obtainable from these halides.

S_N2 attack on the CH_2 in (10) is also found to be considerably faster, under comparable conditions, than on the CH_2 in, for example, $MeCH_2Cl$ (≈ 100 times as fast): the sp^2 hybridised carbon in the S_N2 transition state can use its unhybridised p orbital to interact not only with the entering $^\ominus OH$ and the leaving $Cl^\ominus$, but also with the π orbital system of the phenyl substituent thereby stabilising itself.

Similar carbonium ion stabilisation can also occur with allyl halides, e.g. 3-chloropropene:

$$CH_2=CH-CH_2Cl \rightarrow [CH_2=CH-\overset{\oplus}{C}H_2 \leftrightarrow \overset{\oplus}{C}H_2-CH=CH_2]\ Cl^{\ominus}$$

S_N1 attack is thus promoted and allyl, like benzyl, halides are normally extremely reactive, as compared with, for example, $CH_3CH_2CH_2Cl$ and $C_6H_5CH_2CH_2CH_2Cl$, respectively, in which such carbonium ion stabilisation cannot take place. S_N2 attack is also speeded up by interaction, in the T.S., of the unhybridised p orbital on the now sp^2 carbon atom attacked with the π orbital of the double bond, thus resembling the S_N2 T.S. with phenylchloromethane (10) above. The proportions of the total reaction proceeding by each of the two modes is found to depend on the conditions, more powerful nucleophiles promoting the S_N2 mode (*cf.* p. 95).

By contrast, vinyl halides such as chloroethene, $CH_2=CHCl$, and halogenobenzenes are very unreactive towards nucleophiles. This stems from the fact that the halogen atom is now bonded to an sp^2 hybridised carbon, with the result that the electron pair of the $C-Cl$ bond is drawn closer to carbon (p. 4) than in the bond to an sp^3 hybridised carbon. The $C-Cl$ is found to be stronger, and thus less easily broken, than in, for example, CH_3CH_2Cl, and the $C-Cl$ dipole is smaller; there is thus less tendency to ionisation (S_N1) and a less positive carbon for $^{\ominus}OH$ to attack (S_N2); the π electrons of the double bond also inhibit the close approach of an attacking nucleophile. The double bond would not help to stabilise either the S_N2 transition state or the carbonium ion involved in the S_N1 pathway. Very much the same considerations apply to halogenobenzenes, with their sp^2 hybridised carbons and the π orbital system of the benzene nucleus; their reactions, which though often bimolecular are not in fact simply S_N2 in nature, are discussed further below (p. 167).

The influence of steric factors on the reaction pathway is particularly observed when substitution takes place at the β-position. Thus for the series,

CH_3-CH_2-Br	$MeCH_2-CH_2-Br$	Me_2CH-CH_2-Br	Me_3C-CH_2-Br
(6) 1·0	(13) $2·8 \times 10^{-1}$	(14) $3·0 \times 10^{-2}$	(15) $4·2 \times 10^{-6}$

the figure quoted are relative rates of reaction (S_N2 throughout) with $EtO^{\ominus}$ in EtOH at 55°. Any differences in electronic effect of the Me groups through two saturated carbon atoms would be very small, and the reason for the rate differences is steric: increased difficulty of approach of $EtO^{\ominus}$ 'from the back' of the carbon atom carrying Br, and increased crowding in the resultant T.S. The reason for the particularly large drop in rate between 1-bromo-2-methylpropane (14) and 1-bromo-2,2-dimethylpropane(neopentyl bromide, 15) is that the T.S. for the former, though somewhat crowded, can, by rotation about the C_α—C_β bond, adopt one conformation (14a) in which the attacking $EtO^\ominus$ is interfered with only by H, while no such relief of crowding is open in the T.S. (15a) for the latter:

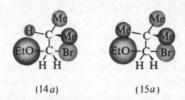

(14a)   (15a)

The T.S. (15a) will thus be at a much higher energy level, $\Delta G^\ddagger$ (p. 38) will be larger and the reaction rate correspondingly lower.

The effect of structure on relative reactivity may be seen particularly clearly when a halogen atom is located at the bridgehead of a bicyclic system. Thus the following rates were observed for solvolysis in 80% aqueous ethanol at 25°:

(8)	(16)	(17)
1	$\approx 10^{-6}$	$\approx 10^{-14}$

All are tertiary halides so that attack by the S_N2 mode would not be expected to occur on (16) or (17) any more than it did on (8, *cf.* p. 81). S_N2 attack 'from the back' on the carbon atom carrying Br would in any case be prevented in (16) and (17) both sterically by their cage-like structure, and also by the impossibility of forcing their fairly rigid framework through transition states with the required planar distribution of bonds to the bridgehead carbon atom (*cf.* p. 83). Solvolysis *via* rate-limiting formation of the ion pair (S_N1), as happens with (8), is also inhibited because the resultant carbonium ions from (16) and (17) would be unable, because of their rigid frameworks, to stabilise themselves by collapsing to the stable planar state. The carbonium ion intermediates are thus of very much higher energy level than usual, and therefore are formed only slowly and with reluctance. The very greatly reduced solvolysis rate of (17) compared with (16) reflects the greater rigidity about the bridgehead (carbonium ion) carbon with a one-carbon (17), than with a two-carbon (16), bridge.

This rigidity is carried even further in 1-bromotriptycene (19),

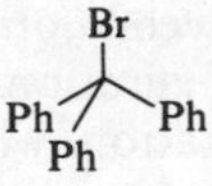

(18)   (19)

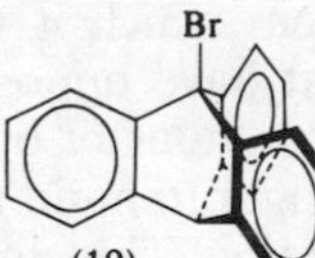

1   10^{-23}

in which the bromine atom is found to be virtually inert to nucleophiles. Despite the formal resemblance in the environment of the bromine atom in (19) to that in (18), they are found to differ in their rate of reaction under parallel conditions by a factor of $\approx 10^{-23}:1$! This is because stabilisation of the carbonium ion from (18) can occur by delocalisation of its charge through the π orbital systems of the three benzene rings, whereas in (19) attainment of the necessary coplanarity in the carbonium ion is totally prevented by its extremely rigid structure.

4.4 STEREOCHEMICAL IMPLICATIONS OF MECHANISM

Hydrolysis of an optically active form of a *chiral** halide presents some interesting stereochemical features. Thus considering each pathway in turn:

4.4.1 S_N2 mechanism: inversion of configuration

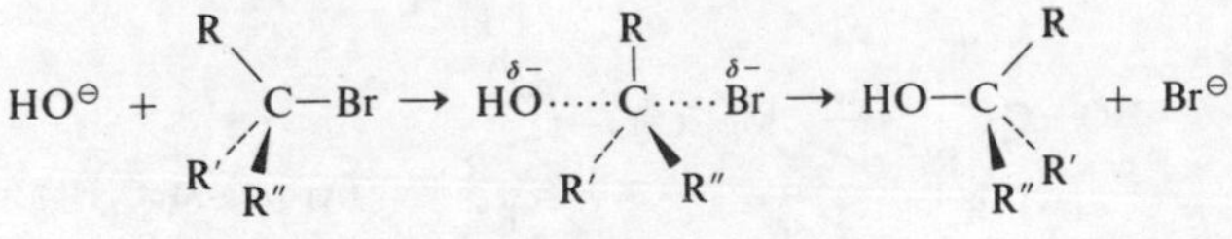

(+)   (?)

It will be seen that the spatial arrangement of the three residual groups attached to the carbon atom attacked has been effectively turned inside out. The carbon atom is said to have undergone *inversion* of its *configuration* (the arrangement in space of the groups attached to it). Indeed, if the product could be the bromide, instead of, as here, the corresponding alcohol, it would be found to rotate the plane of polarisation of plane polarised light in the opposite direction, i.e. $(-)$,

* A *chiral* compound is one that is not superimposable on its mirror image.

to the starting material, $(+)$, for it would, of course, be its mirror image (*cf.* p. 89). The actual product is the alcohol, however, and we are unfortunately not able to tell, merely by observing its direction of optical rotation, whether it has the same or the opposite configuration to the bromide from which it was derived: compounds, other than mirror images, that have opposite configurations do not necessarily exhibit opposite directions of optical rotation, while compounds that have the same configuration do not necessarily exhibit the same direction of optical rotation. Thus in order to confirm that the above S_N2 reaction is, in practice, attended by an inversion of configuration, as theory requires, it is necessary to have an independent method for relating the configuration of starting material and product, e.g., the bromide and corresponding alcohol above.

4.4.2   Determination of relative configuration

This turns essentially on the fact that if a chiral compound undergoes a reaction in which a bond joining one of the groups to the chiral centre is broken, then the centre may—though it need not of necessity—undergo inversion of configuration; while if the compound undergoes reaction in which no such bond is broken then the chiral centre will preserve its configuration intact.

Thus in the series of reactions on the optically active$(+)$ alcohol (20),

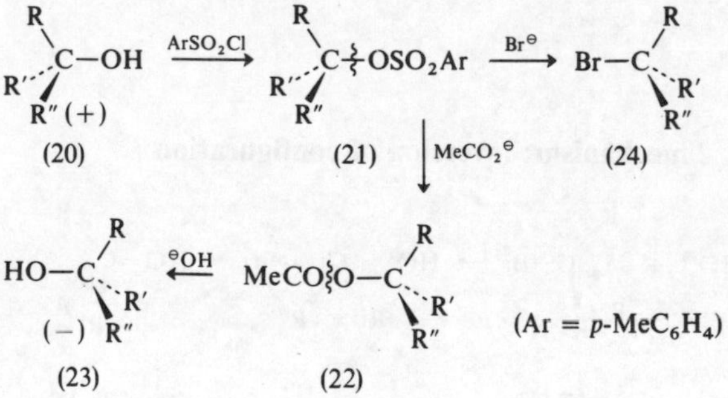

formation of an ester with 4-methylbenzenesulphonyl(tosyl) chloride is known not to break the C—O bond of the alcohol,* hence the tosylate (21) must have the same configuration as the original alcohol. Reaction of this ester (21) with $MeCO_2^\ominus$ is known to be a displacement in which $ArSO_3^\ominus$ $(Ar = p\text{-}MeC_6H_4)$ is expelled and $MeCO_2^\ominus$ introduced,* hence the C—O bond *is* broken in this reaction, and

---

* That such is the case may be shown by using an alcohol labelled with ^{18}O in its OH group, and demonstrating that this atom is not eliminated on forming the tosylate; it is, however, eliminated when the tosylate is reacted with $MeCO_2^\ominus$.

inversion of configuration can thus take place in forming the acetate (22). Alkaline hydrolysis of the acetate (22 → 23) can be shown not to involve fission of the alkyl-oxygen C—O linkage,* so the alcohol (23) must have the same configuration as the acetate (22). As (23) is found to be the mirror image of the starting material (20)—opposite direction of optical rotation—an inversion of configuration must have taken place during the series of reactions, and can have occurred <u>only</u> during reaction of $MeCO_2^{\ominus}$ with the tosylate (21). Reaction of this tosylate (21) with a number of other anions showed that inversion of configuration occurred in each case: it may thus be concluded with some confidence that it occurs on reaction with $Br^{\ominus}$ to yield the bromide (24), i.e. that the bromide (24), like the acetate (22), has the opposite configuration to the original alcohol (20).

The general principle—that bimolecular (S_N2) displacement reactions are attended by inversion of configuration—was established in an elegant and highly ingenious experiment, in which an optically active alkyl halide undergoes displacement by the same—though isotopically labelled—halide ion as nucleophile, e.g. radioactive $^{128}I^{\ominus}$ on (+)2-iodooctane (25):

$$^{128}I^{\ominus} + \quad \underset{\underset{H}{\overset{Me}{\diagdown}}{\overset{C_6H_{13}}{\diagup}}}{C-I} \longrightarrow \quad ^{128}I\overset{\delta-}{\cdots\cdots}\underset{\underset{Me \quad H}{}}{\overset{C_6H_{13}}{C}}\overset{\delta-}{\cdots\cdots}I \longrightarrow \quad ^{128}I-\underset{\underset{H}{\overset{Me}{(-)\diagup}}}{\overset{C_6H_{13}}{C}} + I^{\ominus}$$

(25)   (25a)

The displacement was monitored by observing the changing distribution of radioactive ^{128}I between the inorganic (sodium) iodide and 2-iodooctane, and it was found, under these conditions, to be second order overall (first order with respect to $^{128}I^{\ominus}$ and to 2-iodooctane) with $k_2 = 3.00 \pm 0.25 \times 10^{-5}$ (at 30°).

If *inversion* takes place, as S_N2 requires, the optical activity of the solution will decline to zero, i.e. *racemisation* will occur. This will happen because inversion of the configuration of a molecule of (+) (25) results in formation of a molecule of its mirror image (−) (25a), which 'pairs off' with a second molecule of (+) (25) to form a (±) racemate: the observed rate of *racemisation* will thus be <u>twice</u> the rate of inversion. The reaction was monitored polarimetrically, the rate of racemisation measured thereby, and the rate of inversion calculated from it: it was found to have $k = 2.88 \pm 0.03 \times 10^{-5}$ (at 30°).

The rate of displacement and of inversion are thus identical within the limits of experimental error, and it thus follows that <u>each</u> act of

---

*Hydrolysis of an acetate in which the alcohol-oxygen atom is ^{18}O labelled fails to result in the latter's replacement, thus showing that the alkyl-oxygen bond of the acetate is not broken during its hydrolysis (*cf.* p. 47).

bimolecular displacement must thus proceed with inversion of configuration. Having shown that S_N2 reactions are attended by inversion of configuration, independent demonstration that a particular reaction occurs *via* the S_N2 mode is often used to correlate the configuration of product and starting material in the reaction.

4.4.3 S_N1 mechanism: racemisation?

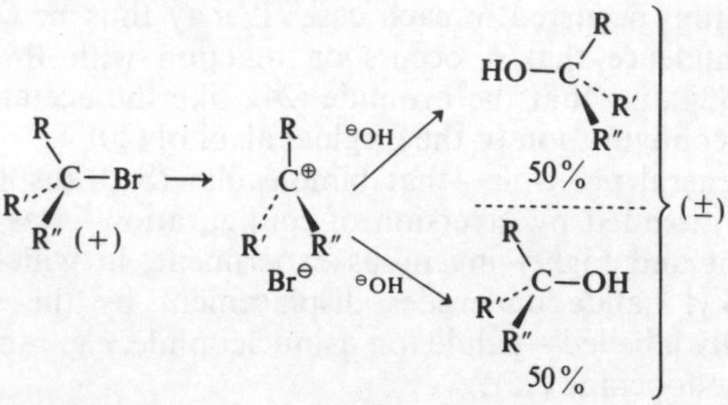

As the carbonium ion formed in the slow, rate-limiting step of the reaction is planar, it might be expected that subsequent attack by a nucleophile such as $^\ominus OH$, or the solvent ($H_2O:$), would take place with equal readiness from either side of the planar carbonium ion; leading to a 50/50 mixture of species having the same, and the opposite, configuration as the starting material, i.e. that *racemisation* would take place yielding an optically inactive ($\pm$) product.

In practice however the expected racemisation—and nothing but racemisation—is rarely observed, it is almost always accompanied by some degree of inversion. The relative proportions of the two are found to depend on: (*a*) the structure of the halide, in particular the relative stability of the carbonium ion to which it gives rise; and on (*b*) the solvent, in particular on its ability as a nucleophile. The more stable the carbonium ion, the greater is the proportion of racemisation; the more nucleophilic the solvent, the greater is the proportion of inversion. These observations become understandable if the rate-limiting S_N1 ionisation follows the sequence:

$$\overset{\delta+}{R}-\overset{\delta-}{Br} \rightleftarrows \boxed{R^\oplus Br^\ominus} \rightleftharpoons \boxed{R^\oplus | Br^\ominus} \rightleftharpoons \boxed{R^\oplus} + \boxed{Br^\ominus}$$

$$\qquad\qquad\quad (26)\qquad\qquad (27)\qquad\qquad (28)$$

Here (26) is an *intimate* ion pair in which the jointly solvated gegenions are in very close association with no solvent molecules between them, (27) is a *solvent-separated* ion pair, and (28) represents the now dissociated, and separately solvated, pair of ions.

In a solvolysis reaction, attack on R$^{\oplus}$ by a solvent molecule, e.g. H$_2$O:, in (26) is likely to lead to inversion, as attack can take place (by the solvent envelope) on the 'back' side of R$^{\oplus}$, but not on the 'front' side where there are no solvent molecules, and which is shielded by the Br$^{\ominus}$ gegen ion. Attack in (27) is more likely to lead to attack from either side, leading to racemisation, while attack on (28) can clearly happen with equal facility from either side. Thus the longer the life of R$^{\oplus}$, i.e. the longer it escapes nucleophilic attack, the greater the proportion of racemisation that we should expect to occur. The life of R$^{\oplus}$ is likely to be longer the more stable it is—(a) above—but the shorter the more powerfully nucleophilic the solvent—(b) above.

Thus solvolysis of (+)C$_6$H$_5$CHMeCl, which can form a stabilised benzyl type carbonium ion (*cf.* p. 84), leads to 98% racemisation while (+)C$_6$H$_{13}$CHMeCl, where no comparable stabilisation can occur, leads to only 34% racemisation. Solvolysis of (+)C$_6$H$_5$CHMeCl in 80% acetone/20% water leads to 98% racemisation (above), but in the more nucleophilic water alone to only 80% racemisation. The same general considerations apply to nucleophilic displacement reactions by Nu: as to solvolysis, except that R$^{\oplus}$ may persist a little further along the sequence because part at least of the solvent envelope has to be stripped away before Nu: can get at R$^{\oplus}$. It is important to notice that racemisation is clearly very much less of a stereochemical requirement for S$_N$1 reactions than inversion was for S$_N$2.

4.4.4 S$_N$i mechanism: retention of configuration

Despite what has been said above about displacement reactions leading to inversion of configuration, to racemisation, or to a mixture of both, a number of cases are known of reactions that proceed with actual retention of configuration, i.e. in which the starting material and product have the same configuration. One reaction in which this has been shown to occur is in the replacement of OH by Cl through the use of thionyl chloride, SOCl$_2$:

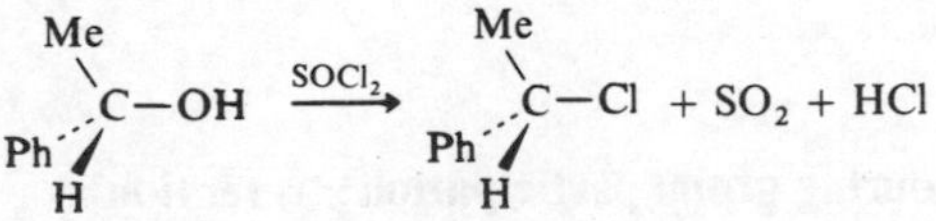

The reaction has been shown to follow a second order rate equation, rate = k_2[ROH][SOCl$_2$], but clearly cannot proceed by the simple S$_N$2 mode for this would lead to inversion of configuration (p. 87) in the product, which is not observed.

Carrying out the reaction under milder conditions allows of the isolation of an alkyl chlorosulphite, ROSOCl (31), and this can be shown to be a true intermediate. The chlorosulphite is formed with retention of configuration, the R—O bond not being broken during the reaction. The rate at which the alkyl chlorosulphite intermediate (31) breaks down to the product, RCl (30a), is found to increase with increasing polarity of the solvent, and also with increasing stability of the carbonium ion $R^{\oplus}$: an ion pair, $R^{\oplus\ominus}OSOCl$ (32), is almost certainly involved. Provided collapse of the ion pair to products then occurs rapidly, i.e. in the intimate ion pair (33) within a solvent cage (*cf.* p. 90), then attack by $Cl^{\ominus}$ is likely to occur on the same side of $R^{\oplus}$ from which $^{\ominus}OSOCl$ departed, i.e. with retention of configuration:

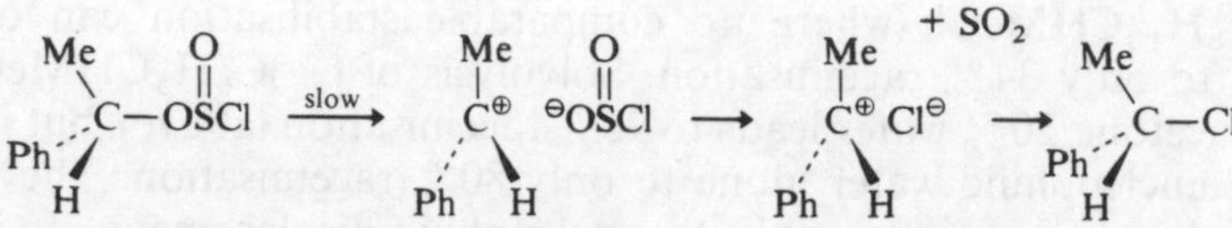

Whether the breaking of the C—O and the S—Cl bonds occurs simultaneously, or whether the former occurs first, is still a matter of debate.

It is interesting that if the $SOCl_2$ reaction on ROH (29) is carried out in the presence of pyridine, the product RCl is found now to have undergone inversion of configuration (30b). This occurs because the HCl produced during the formation of (31) from ROH and $SOCl_2$ is converted by pyridine into $C_5H_5NH^{\oplus}Cl^{\ominus}$ and $Cl^{\ominus}$, being an effective nucleophile, attacks (31) 'from the back' in a normal S_N2 reaction with inversion of configuration:

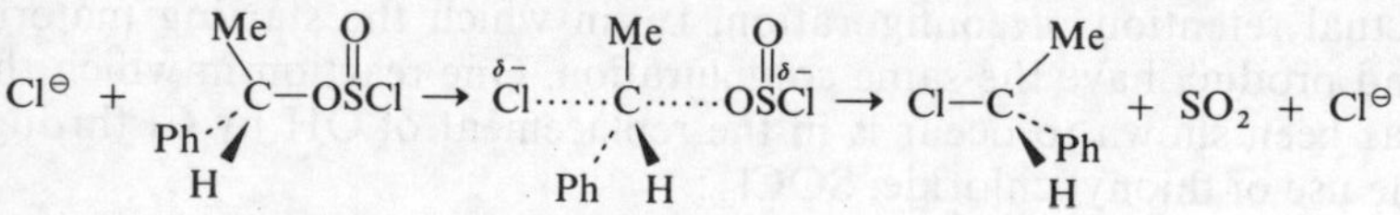

4.4.5   Neighbouring group participation: 'retention'

There are also some examples of retention of configuration in nucleophilic displacement reactions where the common feature is an atom or group—close to the carbon undergoing attack—which has an electron pair available. This *neighbouring group* participates by using its electron pair to shield the 'backside' of the carbon atom undergoing displacement from attack by the nucleophilic reagent; attack can thus take place only 'from the frontside', leading to retention of configuration. Thus base hydrolysis of the 1,2-chlorohydrin (34) is found to yield the 1,2-diol (35) with the same configuration (retention):

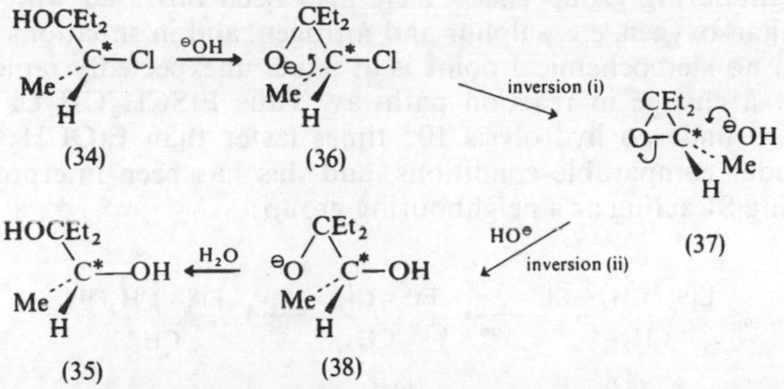

Initial attack by base on (34) yields the alkoxide anion (36), internal attack by this $RO^\ominus$ then yields the epoxide (37) with inversion of configuration at C* (these cyclic intermediates can actually be isolated in many cases); this carbon atom†, in turn, undergoes ordinary S_N2 attack by $^\ominus OH$, with a second inversion of configuration at C*. Finally, this second alkoxide anion (38) abstracts a proton from the solvent to yield the product 1,2-diol (35) with the same configuration as the starting material (34). This *apparent* retention of configuration has, however, been brought about by two successive inversions.

Another example of oxygen as a neighbouring group occurs in the hydrolysis of the 2-bromopropanoate anion (39) at low $[^\ominus OH]$, which is also found to proceed with retention of configuration (40). The rate is found to be independent of $[^\ominus OH]$, and the reaction is believed to proceed:

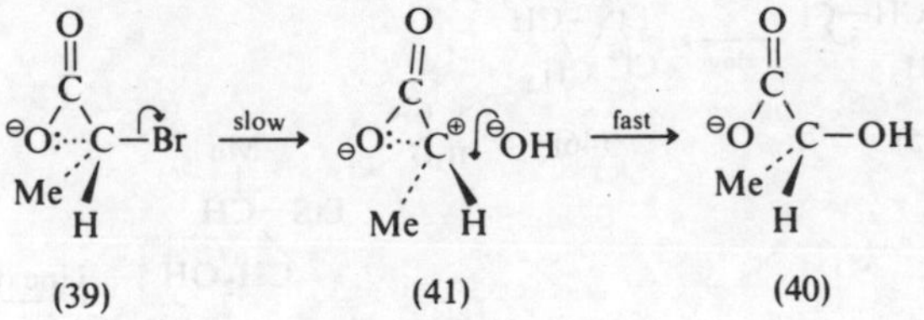

Whether the intermediate (41) is a zwitterion as shown or a highly labile α-lactone (41a)

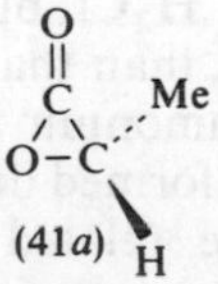

---

† Preferential attack takes place on this, rather than the other, carbon of the three-membered ring as it will be the more positive of the two, carrying as it does only one electron-donating alkyl group.

has not been clearly established. As the concentration of nucleophile, [$^{\ominus}$OH], is increased an increasing proportion of normal S_N2 'attack from the back', with inversion of configuration, is observed.

Neighbouring group effects have also been observed with atoms other than oxygen, e.g. sulphur and nitrogen, and in situations where, though no stereochemical point is at issue, unexpectedly rapid rates suggest a change in reaction pathway. Thus $EtSCH_2CH_2Cl$ (42) is found to undergo hydrolysis 10^4 times faster than $EtOCH_2CH_2Cl$ (43) under comparable conditions, and this has been interpreted as involving S: acting as a neighbouring group:

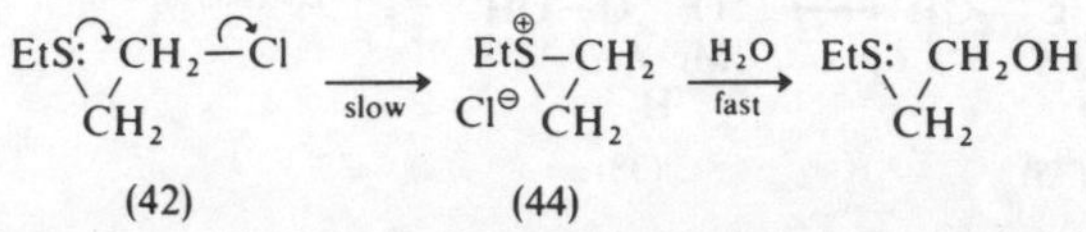

By contrast, O: in (43) is sufficiently electronegative not to donate an electron pair (unlike $O^{\ominus}$ in $RO^{\ominus}$ and $RCO_2{}^{\ominus}$ above), and hydrolysis of $EtOCH_2CH_2Cl$ thus proceeds *via* ordinary S_N2 attack by an external nucleophile—which is likely to be very much slower than the *internal* nucleophilic attack in (42) → (44). That a cyclic sulphonium salt such as (44) is involved is demonstrated by the hydrolysis of the analogue (45), which yields <u>two</u> alcohols (the unexpected one in greater yield) indicating the participation of the unsymmetrical intermediate (46):

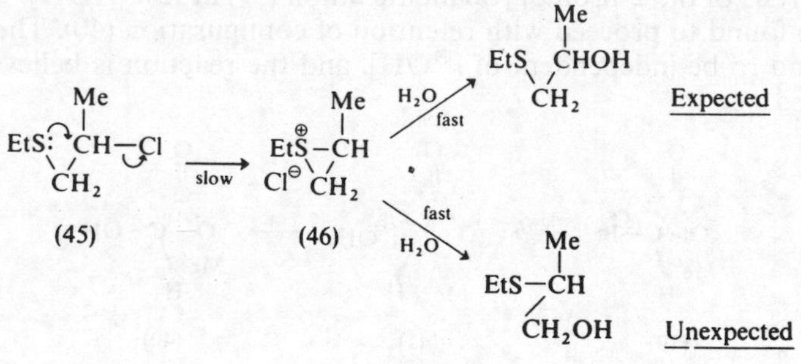

N: can act as a neighbouring group in similar circumstances, e.g. the hydrolysis of $Me_2NCH_2CH_2Cl$, but the rate is markedly slower, under comparable conditions, than that for (42) above, because of the greater stability of the cyclic immonium ion intermediate corresponding to (44). Such cyclic species are formed during the hydrolysis of mustard gas, $S(CH_2CH_2Cl)_2$ and the related nitrogen mustards, such as $MeN(CH_2CH_2Cl)_2$: the cyclic immonium salts derived from the latter are also powerful neurotoxins. The π orbital system of the benzene ring can also act as a neighbouring group (*cf.* p. 104).

4.5   EFFECT OF ENTERING AND LEAVING GROUPS

4.5.1   The entering group

Changing the nucleophilic reagent employed, i.e. the *entering group*, will not directly alter the rate of an S_N1 displacement reaction for this reagent does not take part in the rate-limiting step of the overall reaction. In an S_N2 displacement, however, the more strongly nucleophilic the reagent the more the reaction will be promoted. The *nucleophilicity* of a reagent might perhaps be expected to correlate with its basicity, as both involve the availability of electron pairs and the ease with which they are donated. The parallel is by no means exact, however, in that basicity involves electron pair donation to hydrogen, whereas nucleophilicity involves electron pair donation to another atom, very often carbon; basicity involves an equilibrium (thermodynamic), i.e. $\Delta G^{\ominus}$, situation, whereas nucleophilicity usually involves a kinetic, i.e. $\Delta G^{\ddagger}$, one; basicity is likely to be little affected by steric influences, whereas nucleophilicity may be markedly affected.

This distinction follows to some extent the recently introduced one between *hard* and *soft bases*: a hard base is one in which the donor atom is of high electronegativity, low polarisability, and is hard to oxidise, i.e. $^{\ominus}OH$, $^{\ominus}OR$, $R_3N:$; while a soft base is one in which the donor atom is of low electronegativity, high polarisability, and is easy to oxidise, e.g. $RS^{\ominus}$, $I^{\ominus}$, $SCN^{\ominus}$; for a given degree of basicity, softness promotes nucleophilicity. Basicity data is often the more readily available, however, and can be used as a guide to nucleophilicity provided like is being compared with like. Thus if the attacking atom is the same (*cf.* electronegativity above), then the two run reasonably in parallel, and we find the stronger the base the more powerful the nucleophile:

$$EtO^{\ominus} > PhO^{\ominus} > MeCO_2^{\ominus} > NO_3^{\ominus}$$

A shift in mechanistic type can also occur with change of nucleophile, thus a displacement that is S_N1 with, for example, $H_2O:$, $HCO_3^{\ominus}$, $MeCO_2^{\ominus}$ etc. may become S_N2 with $^{\ominus}OH$ or $EtO^{\ominus}$.

Nucleophilicity is found to be very much affected by the *size* of the attacking atom in the nucleophile, at least for comparisons within the same group or sub-group of the periodic table; thus we find:

$$I^{\ominus} > Br^{\ominus} > Cl^{\ominus} \qquad RS^{\ominus} > RO^{\ominus}$$

Size as well as electronegativity, governs polarisability (*cf.* soft bases, above): as the atom increases in size the hold the nucleus has on the peripheral electrons decreases, with the result that they become more readily polarisable, leading to the initiation of bonding at increasing nuclear separations. Also the larger the nucleophilic ion or group the less its solvation energy, i.e. the more readily is it converted into the effective, largely non-solvated, nucleophile. It is the combination of these factors that make the large, highly polarisable, poorly solvated iodide ion, $I^{\ominus}$, a very much better nucleophile than the small, difficultly polarisable, highly solvated (H-bonding with a hydroxylic solvent) fluoride ion, $F^{\ominus}$, despite the fact that the latter is much the stronger base of the two. We should, on this basis, expect the increase in reaction rate on transfer from a hydroxylic to a polar non-protic solvent (*cf.* p. 81) to be much less for $I^{\ominus}$ than, for example, for $Br^{\ominus}$ or $Cl^{\ominus}$: as is indeed found to be the case ($Br^{\ominus}$ is a better nucleophile than $I^{\ominus}$ in Me_2CO).

A further interesting point arises with nucleophiles which have more than one—generally two—suitable atoms through which they can attack the substrate, *ambident* nucleophiles:

$$[^{\ominus}X{=\!=}Y \leftrightarrow X{=\!=}Y^{\ominus}]$$

It is found in practice that in (highly polar) S_N1 reactions attack takes place on the carbonium ion intermediate, $R^{\oplus}$, through the atom in the nucleophile on which *electron density* is the higher. With, for example, halides that do not readily undergo S_N1 attack this can be promoted by use of the silver salt of the anion, e.g. AgCN, as $Ag^{\oplus}$ promotes $R^{\oplus}$ formation by precipitation of AgHal (*cf.* p. 101):

$$[^{\ominus}C{\equiv}\ddot{N} \leftrightarrow C{=}\ddot{N}^{\ominus}]$$

$$R{-}Br + Ag^{\oplus}[CN]^{\ominus} \underset{slow}{\rightarrow} AgBr\downarrow + R^{\oplus} + [CN]^{\ominus} \underset{fast}{\rightarrow} R{-}\overset{\oplus}{N}{\equiv}C^{\ominus}$$

In the absence of such promotion by $Ag^{\oplus}$, e.g. with $Na^{\oplus}[CN]^{\ominus}$, the resulting S_N2 reaction is found to proceed with preferential attack on the atom in the nucleophile which is the more *polarisable*:

$$NC^{\ominus} + R{-}Br \rightarrow \overset{\delta-}{N}C\cdots R\cdots\overset{\delta-}{Br} \rightarrow N{\equiv}C{-}R + Br^{\ominus}$$
$$\text{T.S.}$$

This is understandable as, unlike S_N1, bond formation is now taking place in the T.S. for the rate-limiting step, for which ready polarisability of the bonding atom of the nucleophile is clearly important—the beginning of bonding at as great an internuclear separation as possible (*cf.* above). This AgCN/NaCN dichotomy has long been exploited preparatively. Similarly, nitrite ion $[NO_2]^{\ominus}$ is found to result in the formation of alkyl nitrites, $R{-}O{-}N{=}O$, under S_N1 conditions (O is the atom of higher electron density) and nitroalkanes, $R{-}NO_2$, under S_N2 conditions (N is the more readily polarisable atom).

4.5.2 The leaving group

Changing the *leaving group* will clearly alter the rate of both S_N1 and S_N2 reactions, as breaking the bond to the leaving group is involved in the rate-limiting step of both (but not necessarily in S_N2 (*aromatic*), *cf.* p. 168). In either case the strength of the bond is clearly a matter of some significance, but in S_N2 reactions re-distribution of negative charges is taking place in the T.S. and, as with the entering group, polarisability in the leaving group's bonded atom will be at a premium in forming the T.S. This is the main reason for the halide S_N2 reactivity sequence:

$$R-I > R-Br > R-Cl > R-F$$

High polarisability makes $I^\ominus$ both a good entering <u>and</u> a good leaving group, it can thus often be used as a catalyst to promote an otherwise slow displacement reaction, e.g.:

$$H_2O: + R-Cl \xrightarrow[slow]{} HO-R + H^\oplus Cl^\ominus$$

$$I^\ominus + R\!-\!\!\!-Cl \atop \searrow \text{fast}$$
$$\uparrow \qquad\qquad I-R + Cl^\ominus \atop \swarrow \text{fast}$$
$$H^\oplus I^\ominus + R-OH$$

This is known as *nucleophilic catalysis*. The stronger, and harder, as a base a leaving group is, the less readily can it be displaced; thus groups such as $^\ominus OH$, $^\ominus OR$, $^\ominus NH_2$ bonded to carbon by small, highly electronegative atoms of low polarisability (*cf.* hard bases, above) cannot normally be displaced directly by other nucleophiles. The best leaving groups are found to be the anions of strong acids, e.g. the halide anions above and species such as $p\text{-MeC}_6H_4SO_3{}^\ominus$ (tosylate anion, *cf.* p. 88), $p\text{-BrMeC}_6H_4SO_3{}^\ominus$('bromosylate' anion), etc. S_N1 requirements are very similar, e.g. the same R—Hal reactivity sequence is observed, and the main requirements in the leaving group are weakness as a base (anion of a strong acid) and ready solvation as an anion.

Displacements that are otherwise difficult, or even impossible, to accomplish directly may sometimes be effected by modification of the potential leaving group—often through protonation—so as to make it weaker, and/or softer, as a base. Thus $^\ominus OH$ cannot be displaced directly by $Br^\ominus$, but is displaced readily if protonated first:

$$Br^\ominus + R\!-\!\overset{..}{\underset{..}{O}}H \nrightarrow Br-R + {}^\ominus OH$$

$$H^\oplus \updownarrow$$

$$Br^\ominus + R\!-\!\overset{\oplus}{\underset{\overset{..}{H}}{O}}H \rightarrow Br-R + H_2O$$

There are two main reasons for this : (a) $Br^{\ominus}$ is now attacking a positively charged, as opposed to a neutral, species, and (b) the very weakly basic H_2O is a very much better leaving group than the strongly basic $^{\ominus}OH$. The well known use of HI to cleave ethers results from $I^{\ominus}$ being about the most nucleophilic species that can be generated in the strongly acid solution required for the initial protonation :

$$R-\underset{..}{\overset{..}{O}}Ph \overset{H^{\oplus}}{\rightleftharpoons} R-\overset{\oplus}{\underset{H}{O}}Ph \overset{I^{\ominus}}{\rightarrow} RI + PhOH$$

4.6   OTHER NUCLEOPHILIC DISPLACEMENTS

In this discussion of nucleophilic displacement at a saturated carbon atom, interest has tended to centre on attack by nucleophilic anions Nu: $^{\ominus}$, especially $^{\ominus}OH$, on polarised neutral species, especially alkyl halides, $^{\delta+}R-Hal^{\delta-}$. In fact this general type of displacement is extremely common involving, in addition to the above, attack by non-charged nucleophiles Nu: on polarised neutral species,

$$Me_3N: + Et-Br \rightarrow Me_3\overset{\oplus}{N}Et + Br^{\ominus}$$

$$Et_2S: + Me-Br \rightarrow Et_2\overset{\oplus}{S}Me + Br^{\ominus}$$

nucleophilic anions on positively charged species,

$$I^{\ominus} + C_6H_{13}-\overset{\oplus}{\underset{H}{O}}H \rightarrow C_6H_{13}-I + H_2O:$$

$$Br^{\ominus} + Me-\overset{\oplus}{N}Me_3 \rightarrow Me-Br + :NMe_3$$

and non-charged nucleophiles on positively charged species (N_2 is an excellent leaving group):

$$H_2O: + Ph\overset{\oplus}{N_2} \rightarrow PhOH + N_2 + H^{\oplus}$$

We have also seen good leaving groups other than halide ion, e.g. tosylate anion (cf. p. 88),

$$MeCO_2^{\ominus} + ROSO_2C_6H_4Me\text{-}p \rightarrow ROCOMe + p\text{-}MeC_6H_4SO_3^{\ominus}$$

and 'internal' leaving groups (cf. p. 93):

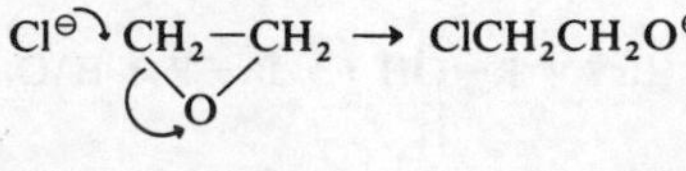

$$Cl^{\ominus} \overset{\frown}{} CH_2-CH_2 \rightarrow ClCH_2CH_2O^{\ominus}$$

There are also a group of nucleophilic displacement reactions, of considerable synthetic importance, in which the attacking nucleophile is a carbanion (p. 282) or a source of positively polarised carbon (*cf.* p. 219), resulting in the formation of carbon–carbon bonds:

$$HC{\equiv}CH \overset{^{\ominus}NH_2}{\rightleftharpoons} HC{\equiv}C^{\ominus} + Pr{-}Br \rightarrow HC{\equiv}C{-}Pr + Br^{\ominus}$$

$$CH_2(CO_2Et)_2 \overset{EtO^{\ominus}}{\rightleftharpoons} (EtO_2C)_2CH^{\ominus} + PhCH_2{-}Br \rightarrow (EtO_2C)_2CH{-}CH_2Ph + Br^{\ominus}$$

$$BrMgPh + \overset{\delta+}{C_6H_{13}}{-}\overset{\delta-}{Br} \rightarrow MgBr_2 + Ph{-}C_6H_{13}$$

It should be remembered that in the above examples what is nucleophilic attack from the viewpoint of one participant is electrophilic attack from the viewpoint of the other. Any designation of the process as a whole tends therefore to be somewhat arbitrary, reflecting as it does our preconceptions about what constitutes a reagent as opposed to a substrate (*cf.* p. 30).

Hardly surprisingly, not all nucleophilic displacement reactions proceed so as to give 100 % yields of the desired products! Here, as elsewhere, side-reactions occur yielding unexpected, and in preparative terms unwanted, products. A major side-reaction is *elimination* to yield unsaturated compounds: this is discussed in detail below (p. 240).

5

Carbonium ions, and electron-deficient N and O atoms

Reference has already been made in the last chapter to the generation of carbonium ions, in ion pairs, as intermediates in some displacement reactions at a saturated carbon atom, e.g. the solvolysis of an alkyl halide *via* the S_N1 mechanism. Carbonium ions are, however, fairly widespread in occurrence and, although their existence is often only transient, they are of considerable importance in a wide variety of chemical reactions.

5.1 METHODS OF FORMING CARBONIUM IONS

5.1.1 Heterolytic fission of neutral species

The most obvious example is simple *ionisation* to form ion pairs that has already been referred to in the last chapter:

100

$$Me_3C-Br \rightleftharpoons Me_3C^\oplus Br^\ominus$$

$$Ph_2CH-Cl \rightleftharpoons Ph_2CH^\oplus Cl^\ominus$$

$$MeOCH_2-Cl \rightleftharpoons MeOCH_2^\oplus Cl^\ominus$$

In each case a highly polar (high ϵ), powerful ion-solvating medium is generally necessary. In a similar context the effect of $Ag^\oplus$ in catalysing reactions, often by a shift from $S_N2 \rightarrow S_N1$ mode,

$$Ag^\oplus + R-Br \rightarrow AgBr\downarrow + R^\oplus$$

has already been referred to (p. 96). The catalytic effect of $Ag^\oplus$ can be complicated, however, by the fact that the precipitated silver halide may itself act as a heterogeneous catalyst.

Ionisation may also be induced by Lewis acids, e.g. BF_3,

$$MeCOF + BF_3 \rightleftharpoons MeCO^\oplus BF_4^\ominus$$

to yield in this case an *acyl cation*; the equilibrium here being considerably influenced by the very great stability of the anion, $BF_4^\ominus$. Also with $AlCl_3$,

$$Me_3CCOCl + AlCl_3 \rightleftharpoons Me_3CCO^\oplus AlCl_4^\ominus \rightarrow Me_3C^\oplus AlCl_4^\ominus + CO\uparrow$$

here the relatively unstable acyl cation decomposes to yield the very stable $Me_3C^\oplus$, the equilibrium being driven over to the right by the escape of CO.

Particularly striking examples are provided by the work of Olah with SbF_5 as a Lewis acid, with either liquid SO_2 or excess SbF_5 as solvent,

$$R-F + SbF_5 \rightleftharpoons R^\oplus SbF_6^\ominus$$

leading to the formation of simple alkyl cations in conditions that allow of their detailed study by n.m.r. spectroscopy and other means. The use of the same investigators 'super acids', such as SbF_5/FSO_3H, allows of the formation of alkyl cations even from alkanes:

$$Me_3C-H + SbF_5/FSO_3H \rightarrow H_2 + Me_3C^\oplus SbF_5FSO_3^\ominus$$

The relative stability of $Me_3C^\oplus$ is shown by the fact that under these conditions the isomeric carbonium ion, $Me\overset{\oplus}{C}HCH_2Me$, obtained from $MeCH_2CH_2Me$, was found to rearrange virtually instantaneously to $Me_3C^\oplus$. The relation between the relative stability of carbonium ions and their structure is discussed below (p. 103).

5.1.2 Addition of cations to neutral species

The most common cation is $H^\oplus$, adding to unsaturated linkages, i.e. *protonation*, in for example the acid-catalysed hydration of alkenes (p. 184):

$$-CH{=}CH- \underset{\rightleftarrows}{\overset{H^\oplus}{}} \underset{\underset{\oplus OH_2}{|}}{-CH{-}CH-} \underset{\rightleftarrows}{\overset{H_2O}{}} \underset{\underset{\oplus OH_2}{|}}{-CH{-}CH-} \underset{\rightleftarrows}{\overset{-H^\oplus}{}} \underset{\underset{OH}{|}}{-CH{-}CH-}$$

The reaction is reversible, the reverse being the perhaps better known acid-catalysed dehydration of alcohols (p. 241). Protonation can also occur on oxygen in a carbon-oxygen double bond,

$$\overset{\delta+ \quad \delta-}{\underset{/}{\overset{\backslash}{C}}}{=}O \overset{H^\oplus}{\rightleftarrows} \left[\underset{/}{\overset{\backslash}{C}}{=}\overset{\oplus}{O}H \leftrightarrow \underset{/}{\overset{\backslash}{\overset{\oplus}{C}}}{-}OH \right] \overset{H_2O}{\rightleftarrows} \underset{/}{\overset{\backslash}{C}}\!\!\overset{OH}{\underset{OH}{}} + H^\oplus$$

thus providing a more positive carbon atom for subsequent attack by a nucleophile, in this case $H_2O\!:$ in acid-catalysed hydration of carbonyl compounds (*cf.* p. 204). That such protonation does indeed occur may be demonstrated, in the absence of water, by the two-fold depression of freezing point observed with ketones in concentrated sulphuric acid:

$$\underset{/}{\overset{\backslash}{C}}{=}O + H_2SO_4 \rightleftarrows \underset{/}{\overset{\backslash}{\overset{\oplus}{C}}}{-}OH + HSO_4{}^\ominus$$

Carbonium ions may also be generated by protonation of lone pair electrons, if the protonated atom is converted into a better leaving group thereby and ionisation thus promoted:

$$Ph_3C{-}\underset{\cdot\cdot}{O}H \overset{H_2SO_4}{\rightleftharpoons} HSO_4{}^\ominus + Ph_3C{-}\overset{\oplus}{\underset{H}{O}}H \overset{H_2SO_4}{\rightleftharpoons} Ph_3C^\oplus + H_3O^\oplus + 2HSO_4{}^\ominus$$

cf. the reverse of the acid-catalysed hydration of alkenes above. Lewis acids may also be used,

$$\underset{/}{\overset{\backslash}{C}}{=}O\!: + AlCl_3 \rightleftarrows \underset{/}{\overset{\backslash}{\overset{\oplus}{C}}}{-}\overset{\cdot\cdot}{O}AlCl_3{}^\ominus$$

and other cations, e.g. $^\oplus NO_2$ in the nitration of benzene (p. 133),

(1)

where the intermediate (1) is a delocalised carbonium ion.

5.1.3 From other cations

Carbonium ions may be obtained from the decomposition of other cations, e.g. diazonium cations from the action of $NaNO_2/HCl$ on RNH_2 (*cf.* p. 118),

$$[R-N\overset{\oplus}{=}N \leftrightarrow R-N\overset{\oplus}{\equiv}N] \rightarrow R^{\oplus} + N\equiv N\uparrow$$

and also by the use of a readily available carbonium ion to generate another that is not so accessible (*cf.* p. 105):

$$Ph_3C^{\oplus} + \quad \rightleftarrows \quad Ph_3C-H + \quad$$

5.2 STABILITY AND STRUCTURE OF CARBONIUM IONS

The simple alkyl carbonium ions have already been seen (p. 83) to follow the stability sequence,

$$Me_3C^{\oplus} > Me_2CH^{\oplus} > MeCH_2^{\oplus} > CH_3^{\oplus}$$

due to increasing substitution of the carbonium ion carbon atom resulting in increasing delocalisation of the positive charge (with consequent progressive stabilisation) by both inductive and hyper-conjugative effects. The particular stability of $Me_3C^{\oplus}$ is borne out by the fact that it may often be formed, under vigorous conditions, by the isomerisation of other first-formed carbonium ions (*cf.* p. 101), and also by the observation that it remained unchanged after heating at 170° in SbF_5/FSO_3H for four weeks!

An essential requirement for such stabilisation is that the carbonium ion should be planar, for it is only in this configuration that effective delocalisation can occur. Quantum mechanical calculations for simple alkyl cations do indeed suggest that the planar (sp^2) configuration is more stable than the pyrimidal (sp^3) by ≈ 84 kJ (20 kcal) mol^{-1}. As planarity is departed from, or its attainment inhibited, instability of the cation and consequent difficulty in its formation increases very rapidly. This has already been seen in the extreme inertness of 1-bromo-triptycene (p. 87) to S_N1 attack, due to inability to assume the planar configuration preventing formation of the carbonium ion. The expected planar structure of even simple cations has been confirmed by analysis of the n.m.r. and i.r. spectra of species such as $Me_3C^{\oplus}SbF_6^{\ominus}$; they thus parallel the trialkyl borons, R_3B, with which they are isoelectronic.

A major factor influencing the stability of less simple cations is again the possibility of delocalising the charge, particularly where this can involve π orbitals:

$$CH_2=CH-\overset{\oplus}{C}H_2 \leftrightarrow \overset{\oplus}{C}H_2-CH=CH_2$$

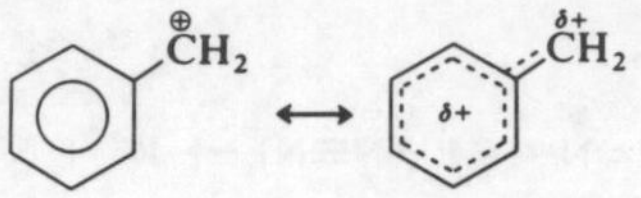

$$Me-O-\overset{\oplus}{C}H_2 \leftrightarrow Me-\overset{\oplus}{O}=CH_2$$

Thus the S_N1 reactivity of allyl and benzyl halides has already been referred to, and the particular effectiveness of the lone pair on the oxygen atom above is reflected in the fact that $MeOCH_2Cl$ is solvolysed at least 10^{14} faster than CH_3Cl.

Stabilisation can also occur, again by delocalisation, through the operation of a neighbouring group effect resulting in the formation of a 'bridged' carbonium ion. Thus the action of SbF_5 in liquid SO_2 on $p\text{-}MeOC_6H_4CH_2CH_2Cl$ (2) results in the formation of (3) rather than the expected cation (4), phenyl acting as a neighbouring group (*cf.* p. 92):

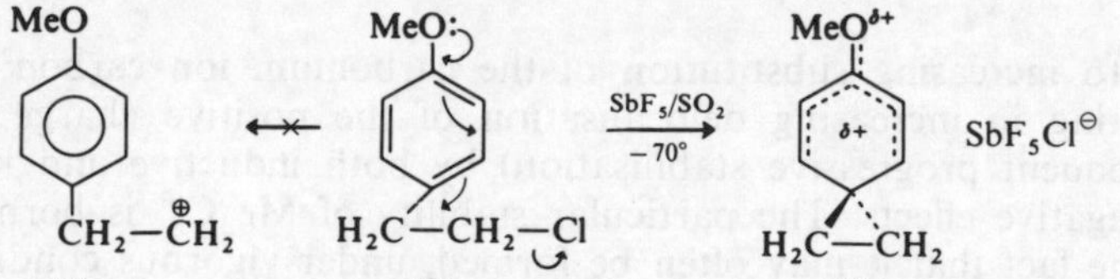

Such species with a bridging phenyl group are known as *phenonium* ions. The neighbouring group effect is even more pronounced with an OH rather than an OMe substituent in the *p*-position. Solvolysis is found to occur $\approx 10^6$ times more rapidly under comparable conditions, and matters can be so arranged as to make possible the isolation of a bridged intermediate (5), albeit not now a carbonium ion:

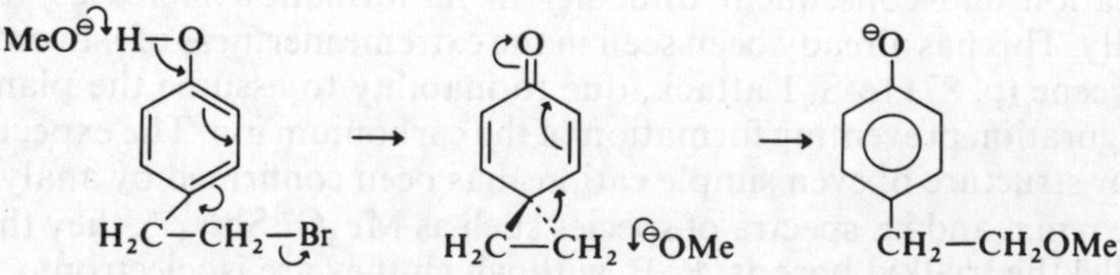

Stabilisation, through delocalisation, can also occur through *aromatisation*. Thus 1-bromocyclohepta-2,4,6-triene(tropylium bromide, 6),

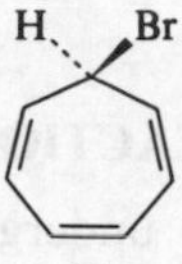

(6)

which is isomeric with $C_6H_5CH_2Br$, is found, unlike the latter compound, to be a crystalline solid (m.p. 208°) which is highly soluble in water yielding bromide ions in solution, i.e. it has not the above covalent structure but is an ion pair. The reason for this behaviour resides in the fact that the cyclic cation (7) has $6\pi e$ which can be accommodated in three delocalised molecular orbitals spread over the seven atoms. It is thus a Hückel $4n + 2$ system ($n = 1$) like benzene (*cf.* p. 17) and exhibit quasi-aromaticity stability:

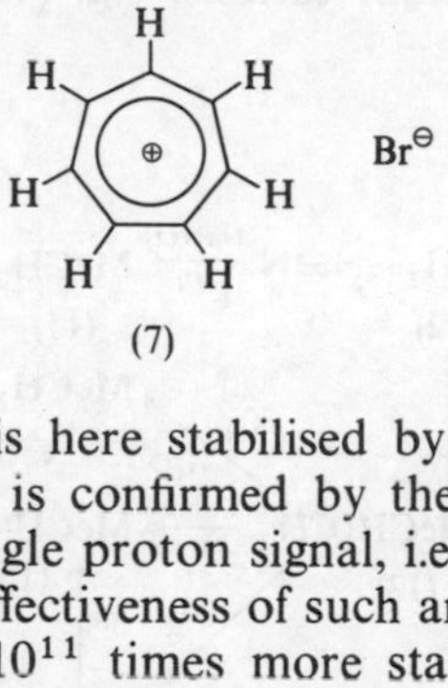

(7)

the planar carbonium ion is here stabilised by aromatisation. The above delocalised structure is confirmed by the fact that its n.m.r. spectrum exhibits only a single proton signal, i.e. all seven hydrogen atoms are equivalent. The effectiveness of such aromatic stabilisation is reflected in its being $\approx 10^{11}$ times more stable than the highly delocalised $Ph_3C^\oplus$. The generation of (7) by the action of $Ph_3C^\oplus$ on cycloheptatriene itself has already been referred to (p. 103).

A particularly interesting case of carbonium ion stabilisation occurs with Hückel $4n + 2$ systems when $n = 0$, i.e. cyclic systems with $2\pi e$ (p. 18). Thus derivatives of 1,2,3-tripropylcyclopropene (8) are found to yield ion pairs containing the corresponding cyclopropenyl cation (9) extremely readily,

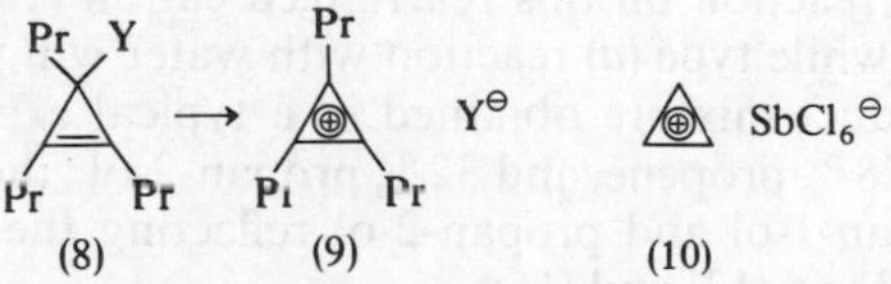

(8)   (9)   (10)

and the latter is found to be even stabler ($\approx 10^3$ times) than (7) above: it is still present as a carbonium ion to the extent of $\approx 50\%$ in water at pH 7! More recently it has also proved possible to isolate an ion pair containing the parent cyclopropenium cation itself (10) as a white crystalline solid.

5.3   CARBONIUM ION REACTIONS

Carbonium ions are found to undergo four basic types of reaction:
   (*a*) Combination with a nucleophile.
   (*b*) Elimination of a proton.
   (*c*) Addition to an unsaturated linkage.
   (*d*) Rearrangement of their structure.
The first two reaction types often lead to the formation of stable end products, but (*c*) and (*d*) lead to the formation of new carbonium ions to which the whole spectrum of reaction types is still open. Most of these possibilities are neatly illustrated in the reaction of 1-amino-propane (11) with sodium nitrite and dilute hydrochloric acid [the behaviour of diazonium cations, e.g. (12), will be discussed further below, p. 118]:

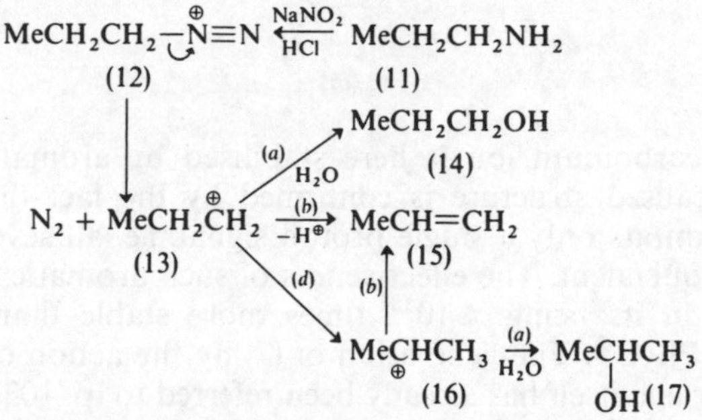

$$MeCH_2CH_2-\overset{\oplus}{\underset{\frown}{N}}{\equiv}N \xleftarrow[HCl]{NaNO_2} MeCH_2CH_2NH_2$$
$$(12) \qquad\qquad\qquad (11)$$

$$\downarrow \qquad\qquad\qquad\xrightarrow[H_2O]{(a)} MeCH_2CH_2OH \atop (14)$$

$$N_2 + MeCH_2\overset{\oplus}{CH_2} \xrightarrow[-H^{\oplus}]{(b)} MeCH{=}CH_2$$
$$(13) \qquad\qquad\qquad\uparrow (15)$$

$$MeCHCH_3 \atop \underset{\oplus}{(16)} \xrightarrow[H_2O]{(a)} MeCHCH_3 \atop \underset{OH}{(17)}$$

Thus reaction of the 1-propyl cation (13) with water (reaction type *a*) will yield propan-1-ol (14), elimination of a proton from (13) will yield propene (15, reaction type *b*), while rearrangement of (13, reaction type *d*)—in this case migration of $H^{\ominus}$—will yield the 2-propyl cation (16). Type (*b*) reaction on this rearranged cation (16) will yield more propene (15), while type (*a*) reaction with water will yield propan-2-ol (17). The product mixture obtained in a typical experiment was 7% propan-1-ol, 28% propene, and 32% propan-2-ol: the relative proportions of propan-1-ol and propan-2-ol reflecting the relative stability of the two cations, (13) and (16).

The sum of the above products still represents only 67% conversion of the original 1-aminopropane, however, and we have clearly not exhausted the reaction possibilities. There are indeed other nucleophiles present in the system, e.g. $Cl^\ominus$ and $NO_2{}^\ominus$, capable of reacting with either cation, (13) or (16), the latter nucleophile leading to the possible formation of both RNO_2 and $RONO$ (nitrite esters may also arise from direct esterification of first formed ROH). The cations (13) and (16) may also react with first formed ROH to yield ethers, ROR, or with as yet unchanged RNH_2 to yield $RNHR$ (which may itself undergo further alkylation, or nitrosation *cf.* p. 118). Finally, either cation may add to the double bond of first formed propene, $MeCH{=}CH_2$ (reaction type *c*, *cf.* p. 185), to yield further cations, $Me\overset{\oplus}{C}H{-}CH_2R$, which can themselves undergo the whole gamut of reactions. The mixture of products actually obtained is considerably influenced by the reaction conditions, but it will come as no surprise that this reaction is seldom a satisfactory preparative method for the conversion: $RNH_2 \rightarrow ROH$!

Reaction type (*d*) also complicates the Friedel–Crafts alkylation of benzene (a type *c/b* reaction, p. 140) by 1-bromopropane, $MeCH_2\text{-}CH_2Br$, in the presence of gallium bromide, $GaBr_3$, as Lewis acid catalyst. The attacking electrophile is here a highly polarised complex, $\overset{\delta++}{R}GaBr_4{}^{\delta--}$, and the greater stability of the complex in which $R^{\delta++}$ carries its positive charge mainly on the secondary, rather than on a primary, carbon atom, i.e. $Me_2\overset{\delta++}{C}HGaBr_4^{\delta--}$ rather than $Me\overset{\delta++}{C}H_2CH_2\text{-}GaBr_4^{\delta--}$, again results in a hydride shift (*cf.* above) so that the major product of the reaction is $Me_2CHC_6H_5$.

That such rearrangements are not necessarily quite as simple as they look, i.e. mere migration of $H^\ominus$, is illustrated by the behaviour of $^{13}CH_3CH_2CH_3$ with $AlBr_3$, when the label is found to become statistically scrambled: the product consists of 2 parts $^{13}CH_3CH_2CH_3$ to 1 part $CH_3{}^{13}CH_2CH_3$, as determined by analysis of the fragments produced in the mass spectrometer. It is possible that the scrambling proceeds through the agency of a protonated cyclopropane intermediate (18):

(18)

An explanation that would also account for the similar statistical scrambling of the ^{13}C label that is found to occur (over several hours) in the initial 2-propyl cation, $CH_3{}^{13}CH^{\oplus}CH_3$, generated from $CH_3{}^{13}CH(Cl)CH_3$ with SbF_5 at $-60°$.

The elimination reactions of carbonium ions (type *b*) will be discussed in more detail subsequently (p. 242), but the rearrangement reactions (type *d*) are of sufficient interest and importance to merit further study now.

5.4 CARBONIUM ION REARRANGEMENTS

Despite the apparent confusion above, rearrangements involving carbonium ions may be usefully divided into those in which an actual change in carbon skeleton itself does, or does not, take place; the former are much the more important but the latter will first be briefly referred to.

5.4.1 Without change in carbon skeleton

We have already seen one example of this type (p. 106), in which the 1-propyl cation rearranged to the 2-propyl cation by the migration of a hydrogen atom with its electron pair from C_2 to the carbonium ion carbon C_1, a 1,2-*hydride shift*:

$$\overset{\displaystyle H}{\underset{\displaystyle |}{CH_3\overset{\oplus}{CH}CH_2}} \rightarrow \overset{\displaystyle H}{\underset{\displaystyle |}{CH_3\overset{\oplus}{CH}CH_2}}$$

This reflects the greater stability of a secondary than a primary carbonium ion; shifts in the reverse direction can, however, take place where this makes available the greater delocalisation possibilities of the π orbital system of a benzene ring (i.e. tertiary → secondary):

$$\underset{\displaystyle C_6H_5CH_2\overset{|}{C}Me_2}{\overset{\displaystyle OH}{|}} \xrightarrow{\underset{SbF_5}{FSO_3H}} C_6H_5\overset{|}{\overset{\oplus}{C}H}CMe_2 \longrightarrow C_6H_5\overset{\oplus}{C}H\overset{|}{C}Me_2$$

There are more interesting rearrangement possibilities inherent in delocalised cations, e.g. *allylic* rearrangements.

5.4.1.1 Allylic rearrangements

Thus in the S_N1 solvolysis in EtOH of 3-chlorobut-1-ene (19), not one but a mixture of two isomeric ethers is obtained; and the same mixture (i.e. the same ethers in approximately the same proportion) is also obtained from the similar solvolysis of 1-chlorobut-2-ene (20):

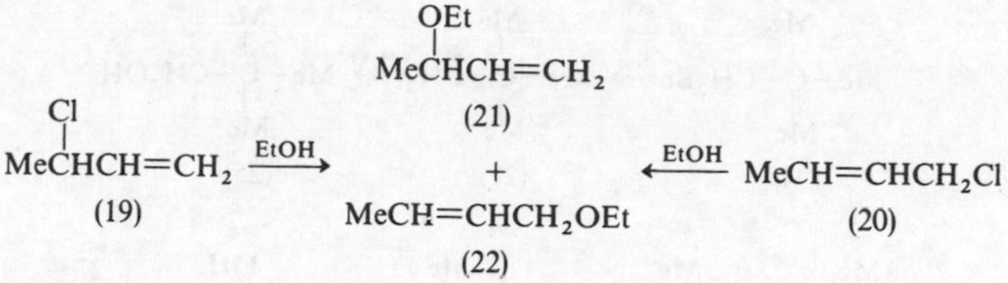

$$
\begin{array}{c}
\overset{\text{OEt}}{\underset{|}{\text{MeCHCH=CH}_2}}\\
(21)
\end{array}
$$

$$
\underset{(19)}{\overset{\text{Cl}}{\underset{|}{\text{MeCHCH=CH}_2}}} \xrightarrow{\text{EtOH}} \quad + \quad \xleftarrow{\text{EtOH}} \underset{(20)}{\text{MeCH=CHCH}_2\text{Cl}}
$$

$$
\underset{(22)}{\text{MeCH=CHCH}_2\text{OEt}}
$$

This clearly reflects formation of the same, delocalised allylic cation (23, *cf.* p. 104) as an ion pair intermediate from each halide, capable of undergoing subsequent rapid nucleophilic attack by EtOH at either C_1 or C_3:

$$
[\text{MeCH}\overset{\oplus}{-}\text{CH=CH}_2 \leftrightarrow \text{MeCH=CH}\overset{\oplus}{-}\text{CH}_2]\ \text{Cl}^{\ominus}
$$

$$
(23)
$$

It is interesting that when $EtO^{\ominus}$, in fairly high concentration, is used as the nucleophile in preference to EtOH, the reaction of (19) becomes S_N2 in type and yields only the one ether (21). Allylic rearrangements have been observed, however, in the course of displacement reactions that are proceeding by a bimolecular process. Such reactions are referred to as S_N2' and are believed to proceed:

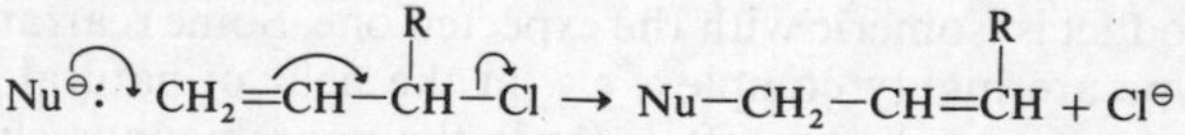

$$
\text{Nu}^{\ominus}:\ \overset{\frown}{\text{CH}_2}=\text{CH}\overset{R}{\underset{|}{-}}\text{CH}\overset{\frown}{-}\text{Cl} \rightarrow \text{Nu}-\text{CH}_2-\text{CH}=\overset{R}{\underset{|}{\text{CH}}} + \text{Cl}^{\ominus}
$$

This process tends to occur when substituents (R) on the α-carbon atom are bulky enough to markedly reduce the rate of direct S_N2 displacement at C_α. Allylic rearrangements are of quite common occurrence, but disentangling the detailed pathway by which they proceed is a matter of considerable difficulty.

5.4.2 With change in carbon skeleton

5.4.2.1 Neopentyl rearrangements

We have already noticed (p. 86) that the S_N2 hydrolysis of 1-bromo-2,2-dimethylpropane (neopentyl bromide, 24) is slow due to steric hindrance. Carrying out the reaction under conditions favouring the S_N1 mode can result in an increased 'reaction rate but the product alcohol is found to be 2-methylbutan-2-ol (26) and not the expected

2,2-dimethylpropanol (neopentyl alcohol, 25); a *neopentyl* rearrangement has taken place:

$$
\underset{(24)}{Me-\underset{\underset{Me}{|}}{\overset{\overset{Me}{|}}{C}}-CH_2Br} \xrightarrow{S_N1} \underset{(27)}{Me-\underset{\underset{Me}{|}}{\overset{\overset{Me}{|}}{C}}-\overset{\oplus}{CH_2}} \xrightarrow{H_2O} \underset{(25)}{Me-\underset{\underset{Me}{|}}{\overset{\overset{Me}{|}}{C}}-CH_2OH}
$$

$$
\underset{(29)}{\underset{Me}{\overset{Me}{>}}C=C\underset{H}{\overset{Me}{<}}} \xleftarrow{-H^\oplus} \underset{(28)}{Me-\underset{\underset{Me}{|}}{\overset{\overset{Me}{|}}{C}}-\overset{\oplus}{CH_2}} \xrightarrow{H_2O} \underset{(26)}{Me-\underset{\underset{Me}{|}}{\overset{\overset{OH}{|}}{C}}-CH_2Me}
$$

The greater stability of the tertiary carbonium ion (28), compared with the initially formed primary one (27), provides the driving force for the necessary C—C bond-breaking involved in the migration of the Me group. Such changes in carbon skeleton, involving carbonium ions, are known collectively as Wagner–Meerwein rearrangements. Further confirmation of the involvement of (28) is the simultaneous formation of the alkene, 2-methylbut-2-ene (29) by loss of proton: a product not obtainable from (27).

The possible occurrence of such major rearrangement of a compound's carbon skeleton, during the course of apparently unequivocal reactions, is clearly of the utmost significance in interpreting the results of experiments aimed at structure elucidation: particularly when the actual product is isomeric with the expected one. Some rearrangements of this type are highly complex, e.g. in the field of natural products such as terpenes, and have often made the unambiguous elucidation of reaction pathways extremely difficult. The structure of reaction products should <u>never</u> be assumed but <u>always</u> confirmed as a routine measure: n.m.r. spectroscopy has proved of enormous value in this respect.

It is interesting to note that while the neopentyl-type bromide (30) undergoes rearrangement during S_N1 hydrolysis, no such rearrangement takes place with its phenyl analogue (31):

$$
\underset{(30)}{Me-\underset{\underset{Me}{|}\ \underset{Br}{|}}{\overset{\overset{Me}{|}}{C}-CHMe}} \xrightarrow{S_N1} Me-\underset{\underset{Me}{|}}{\overset{\overset{Me}{|}}{C}}-\overset{\oplus}{CHMe} \rightarrow Me-\underset{\underset{Me}{|}}{\overset{\overset{Me}{|}}{C}}-CHMe \longrightarrow \text{Products}
$$

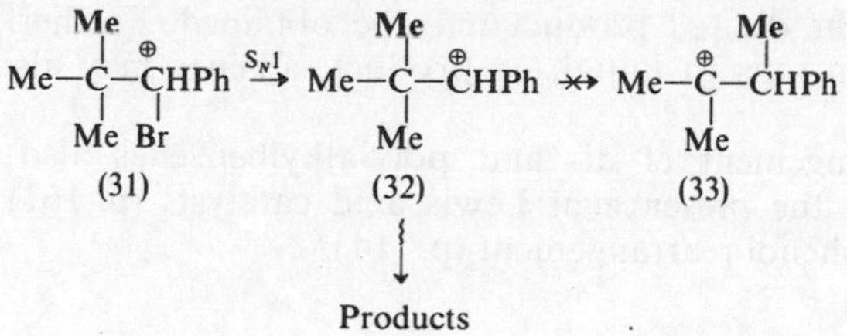

$$\begin{array}{ccc}
\underset{Me}{\overset{Me}{\underset{|}{\underset{Me\ Br}{Me-\overset{\oplus}{C}-CHPh}}}} \xrightarrow{S_N 1} & \underset{Me}{\overset{Me}{\underset{|}{Me-\overset{\oplus}{C}-CHPh}}} \nleftrightarrow & \underset{Me}{\overset{Me}{\underset{|}{Me-\overset{\oplus}{C}-CHPh}}} \\
(31) & (32) & (33)
\end{array}$$

$$\downarrow$$

Products

This reflects the greater stability of the benzylic cation (32), though only secondary, compared with the tertiary cation (33) that would be—but in fact is not—obtained by its rearrangement (*cf.* p. 104).

5.4.2.2 Rearrangement of hydrocarbons

Wagner–Meerwein type rearrangements are also encountered in the cracking of petroleum hydrocarbons when catalysts of a Lewis acid type are used. These generate carbonium ions from the straight chain alkanes (*cf.* the isomerisation of ^{13}C labelled propane, p. 107), which then tend to rearrange to yield branched-chain products. Fission also takes place, but this branching is important because the resultant alkanes cause less knocking in the cylinders of internal combustion engines than do their straight-chain isomers. It should be mentioned, however, that petroleum cracking can also be induced by catalysts that promote reaction *via* radical intermediates (p. 297).

Rearrangement of alkenes takes place readily in the presence of acids:

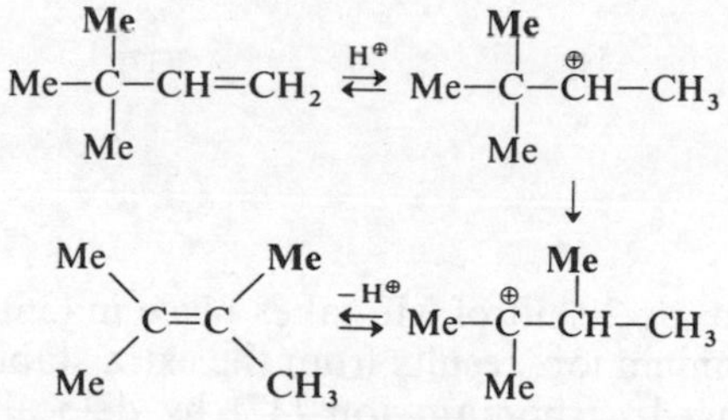

$$\underset{Me}{\overset{Me}{\underset{|}{Me-\overset{|}{C}-CH=CH_2}}} \underset{\rightleftarrows}{\overset{H^\oplus}{}} \underset{Me}{\overset{Me}{\underset{|}{Me-\overset{|}{C}-\overset{\oplus}{CH}-CH_3}}}$$

$$\downarrow$$

$$\underset{Me}{\overset{Me}{\diagdown}}C=C\underset{CH_3}{\overset{Me}{\diagup}} \overset{-H^\oplus}{\underset{\leftrightharpoons}{}} \underset{Me}{\overset{Me}{\underset{|}{Me-\overset{|}{C}-\overset{\oplus}{CH}-CH_3}}}$$

This relatively ready rearrangement can be a nuisance in the preparative addition of acids, e.g. hydrogen halides (p. 181) to alkenes, or in their acid-catalysed hydration (p. 184): mixed products that are difficult to separate may result or, in unfavourable cases, practically none of the desired product may be obtained. Further, addition of carbonium ions to initial, or product, alkenes may also take place (p. 185).

Rearrangement of di- and poly-alkylbenzenes also takes place readily in the presence of Lewis acid catalysts (p. 161), and in the dienone/phenol rearrangement (p. 114).

5.4.2.3   Pinacol/pinacolone rearrangements

Another example of migration of a group, in the original case Me, to a carbonium ion carbon atom occurs in the acid-catalysed rearrangement of 1,2-diols, e.g. pinacol (*cf.* p. 214) $Me_2C(OH)C(OH)Me_2$ (34) to ketones, e.g. pinacolone, $MeCOCMe_3$ (35):

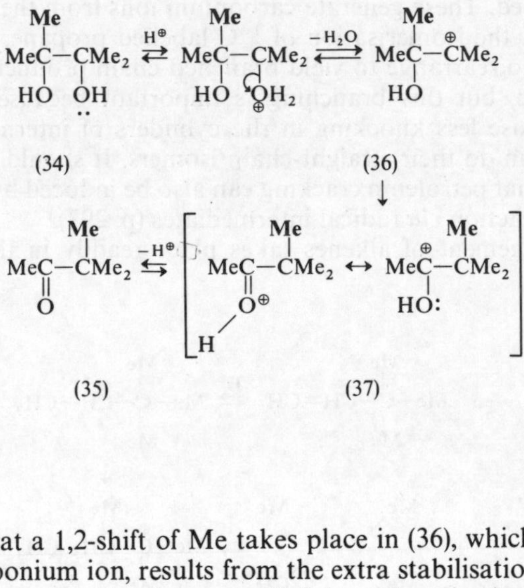

(34)   (36)

(35)   (37)

The fact that a 1,2-shift of Me takes place in (36), which is already a tertiary carbonium ion, results from the extra stabilisation conferred on the rearranged carbonium ion (37) by delocalisation of charge through an electron pair on the oxygen atom; (37) can also readily lose a proton to yield a stable end product (35). It might be expected that an analogous reaction would occur with other compounds capable of forming the crucial carbonium ion (36): this is, in fact, found to be the case. Thus the corresponding 1,2-bromohydrin (38) and 1,2-amino-alcohol (39) are found to yield pinacolone (35) when treated with $Ag^{\oplus}$ and $NaNO_2/HCl$, respectively:

$$
\underset{(38)}{\overset{\textbf{Me}}{\underset{\text{HO}\quad\text{Br}}{\text{MeC}-\text{CMe}_2}}}
\xrightarrow[-\,AgBr\downarrow]{Ag^{\oplus}}
\underset{(36)}{\overset{\textbf{Me}}{\underset{\text{HO}}{\text{MeC}-\overset{\oplus}{\text{CMe}_2}}}}
\rightsquigarrow
\underset{(35)}{\overset{\textbf{Me}}{\underset{\text{O}}{\text{MeC}-\text{CMe}_2}}}
$$

$\uparrow$

$$
\underset{(39)}{\overset{\textbf{Me}}{\underset{\text{HO}\quad\text{NH}_2}{\text{MeC}-\text{CMe}_2}}}
\xrightarrow[\text{HCl}]{NaNO_2}
\overset{\textbf{Me}}{\underset{\text{HO}\ \overset{\oplus}{\curvearrowleft\text{N}}\equiv\text{N}}{\text{MeC}-\text{CMe}_2}}
$$

A number of experiments have been carried out to determine the *relative migratory aptitude* of groups in pinacol/pinacolone type rearrangements, and in general the relative ease of migration is found to be:

$$Ph > Me_3C > MeCH_2 > Me$$

It should be realised that there are considerable difficulties involved in choosing suitable models for such experiments, and in interpreting the results when we have got them. Thus in the rearrangement of the 1,2-diol, $Ph_2C(OH)C(OH)Me_2$ (40), it is Me that is found to migrate and *not* C_6H_5 as might have been expected from the sequence above. However, the reaction is here controlled by preferential protonation on that ÖH group which will lead to the more stable initial carbonium ion (41 rather than 42), and the migration of Me rather than Ph is thereby predetermined:

$$
\underset{\text{(40)}}{\overset{}{\underset{\text{HO}\quad\text{OH}}{\text{Ph}_2\text{C}-\text{CMe}_2}}}
\begin{array}{c}\xrightarrow{\;H^{\oplus}\;}\\[2mm]\xcancel{\xrightarrow{\;H^{\oplus}\;}}\end{array}
$$

$$
\overset{\textbf{Me}}{\underset{\overset{\oplus}{\text{H}_2\text{O}}\quad\text{OH}}{\text{Ph}_2\text{C}-\text{CMe}}}
\xrightarrow{-H_2O}
\overset{\textbf{Me}}{\underset{\text{OH}}{\text{Ph}_2\overset{\oplus}{\text{C}}-\text{CMe}}}
\qquad (41)
$$

$$
\overset{\textbf{Ph}}{\underset{\text{HO}\ \overset{\oplus}{\curvearrowright\text{OH}_2}}{\text{PhC}-\text{CMe}_2}}
\xrightarrow{H_2O}
\overset{\textbf{Ph}}{\underset{\text{HO}}{\text{Ph}\overset{\oplus}{\text{C}}-\text{CMe}_2}}
\qquad (42)
$$

This particular problem can be avoided by chosing symmetrical 1,2-diols such as PhArC(OH)C(OH)PhAr (43) and it has been possible to establish by experiments on such compounds,

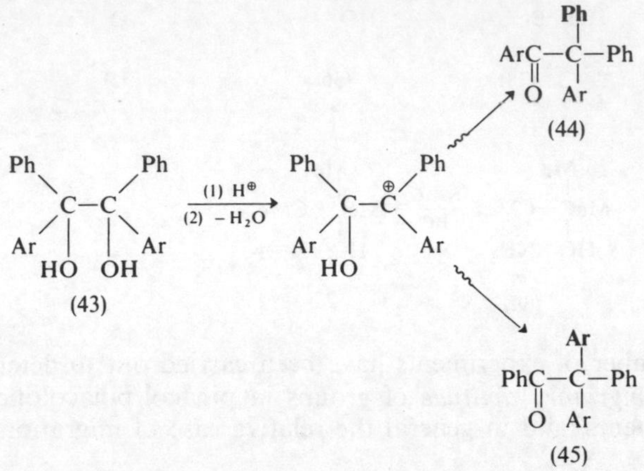

$$
\begin{array}{c}
\overset{\text{Ph}}{\underset{\text{O Ar}}{\text{ArC}-\text{C}-\text{Ph}}} \\
(44)
\end{array}
$$

i.e. by determining the relative proportions of the two ketones (44) and (45) that are produced, the relative migratory aptitude sequence:

$$p\text{-MeOC}_6\text{H}_4 > p\text{-MeC}_6\text{H}_4 > \text{C}_6\text{H}_5 > p\text{-ClC}_6\text{H}_4 > o\text{-MeOC}_6\text{H}_4$$

This sequence could be interpreted (except for o-MeOC$_6$H$_4$) in terms of decreasing potential electron-donation in the group that is migrating, with its electron pair, to a positive centre, the carbonium ion carbon. A similar simple theory of potential electron release could also account for the observed alkyl group sequence mentioned above. The o-MeOC$_6$H$_4$ group, despite being electron-donating, is found to migrate > 1000 times *slower* than C$_6$H$_5$, and there is evidence that both the relative crowding of possible alternative transition states, and the conformation adopted by the starting material at reaction (see below) are also of considerable importance. These may, as with o-MeOC$_6$H$_4$ above, outweigh electronic effects.

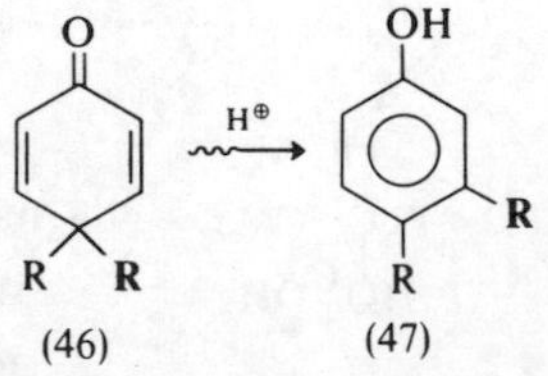

A rearrangement essentially akin to a reversal of the pinacol/pinacolone change, a *retro* pinacol reaction, is the dienone/phenol rearrangement, in which protonation of the initial dienone (46) allows reattainment of the wholly aromatic condition (47) through 1,2-migration of an alkyl group:

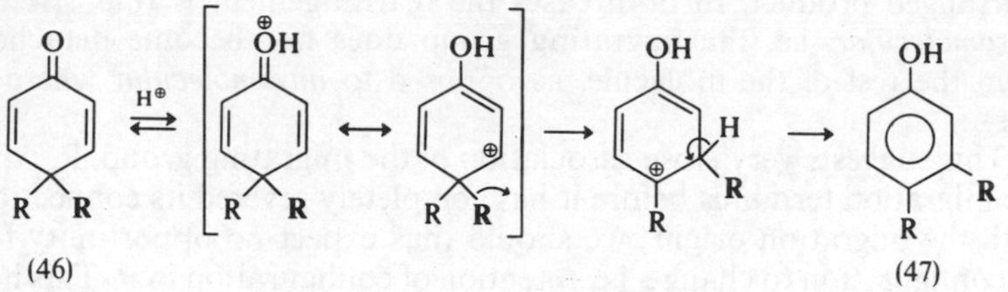

(46)   (47)

5.4.2.4   Stereochemistry of rearrangements

There are essentially three points of major stereochemical interest in carbonium ion rearrangements: what happens to the configuration at the carbon atom *from* which migration takes place (the *migration origin*), to the configuration at the carbon atom *to* which migration takes place (the carbonium ion carbon, the *migration terminus*), and to the configuration of the migrating group, if that is chiral, e.g. PhMe-CH. Interestingly enough, these three questions have never been answered for one and the same compound, despite the enormous body of work that has now been done on carbonium ions.

The last point has never, in fact, been directly established experimentally for a purely carbonium ion rearrangement; there is, however, a good deal of data that bears on it. Thus it has been established that the migrating group never becomes free during rearrangement. This is demonstrated by taking, for example, two pinacols (48 and 49) that are very similar in structure (and that rearrange at very much the same rate) but that have different migrating groups, and rearranging them simultaneously in the same solution (a *crossover experiment*): no cross migration is <u>ever</u> observed:

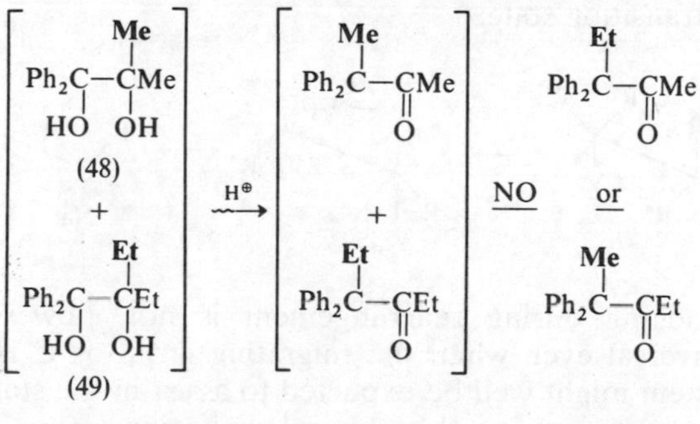

Similarly, if rearrangements in which there is a hydride shift (*cf.* p. 108) are carried out in a deuteriated solvent (e.g. D_2O, MeOD, etc.), no deuterium is incorporated into the new $C-H(D)$ bond in the final rearranged product. In both cases the rearrangement is thus strictly *intramolecular*, i.e. the migrating group does not become detached from the rest of the molecule, as opposed to *intermolecular* where it does.

This suggests very close association of the migrating group, R, with the migration terminus before it has completely severed its connection with the migration origin; we should thus expect no opportunity for its configuration to change, i.e. retention of configuration in R. This has been established experimentally in one reaction involving migration to an electron-deficient carbon, albeit not to a carbonium ion, in the Wolff rearrangement (p. 117), and also in the Hofmann and related reactions (p. 121) where R migrates to an electron-deficient nitrogen atom.

On the other two points the evidence supports predominant inversion of configuration at both the migration origin and terminus:

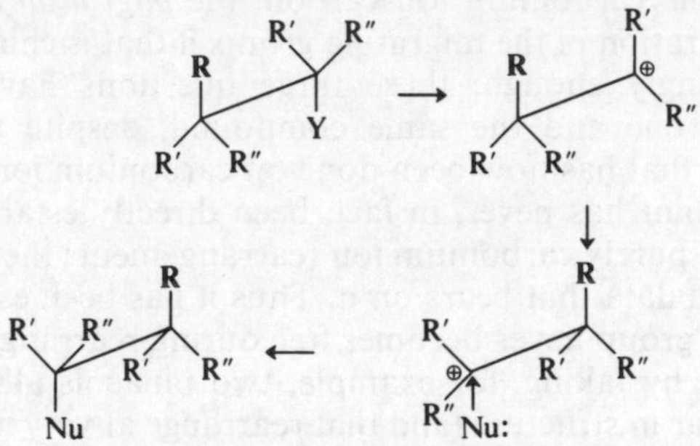

The inversion is often found to be almost complete in cyclic compounds where rotation about the C_1-C_2 bond is largely prevented, but also to a considerable extent in acyclic compounds as well. This could be explained on the basis of a '*bridged*' intermediate (*cf.* bromium ion, p, 177) or transition state:

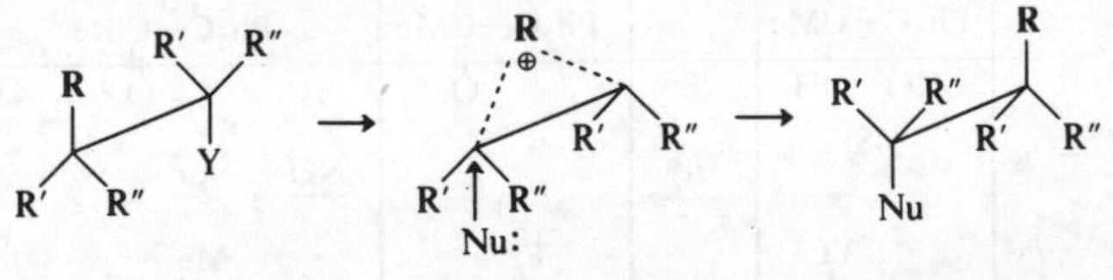

Actual bridging during rearrangement is not, however, by any means universal even when the migrating group is C_6H_5, whose π orbital system might well be expected to assist in the stabilisation of a bridged carbonium ion through delocalisation (*cf.* p. 104).

This is clearly demonstrated in the pinacolinic deamination (*cf.* p.

112) of an optically active form of the amino-alcohol (50). Such reactions proceed from a conformation (*antiperiplanar*; 50*a* or 50*b*) in which the migrating (Ph) and leaving (NH$_2$: as N$_2$; *cf.* p. 113) groups are TRANS to each other. Rearrangement *via* a bridged carbonium ion would necessarily lead to 100% *inversion* at the migration terminus in the product ketone (51*ab*), whichever initial conformation, (50*a*) or (50*b*), was involved:

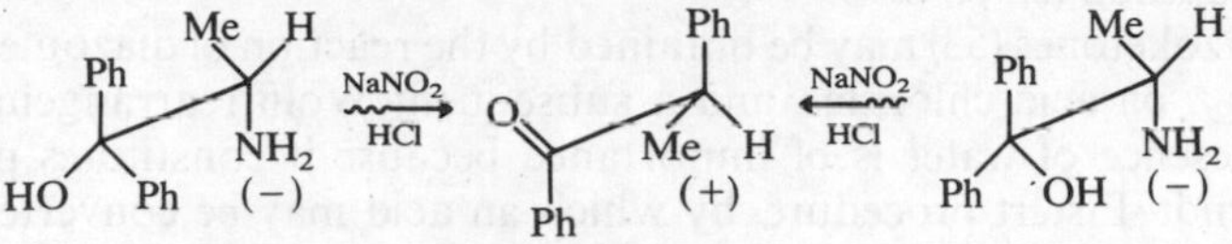

(50*a*)   (51*ab*)   (50*b*)

It was actually found, however, that though inversion was predominant (51*ab*: 88%), the product ketone contained a significant amount of the mirror image (51*d*: 12%): thus 12% of the total reaction can **not** have proceeded *via* a bridged carbonium ion. The simplest explanation is that part at least of the total rearrangement is proceeding *via* a *non*-bridged carbonium ion (52*c*), in which some rotation about the C$_1$—C$_2$ bond can take place (52*c* → 52*d*), thereby yielding ketone (51*d*) in which the original configuration has been retained:

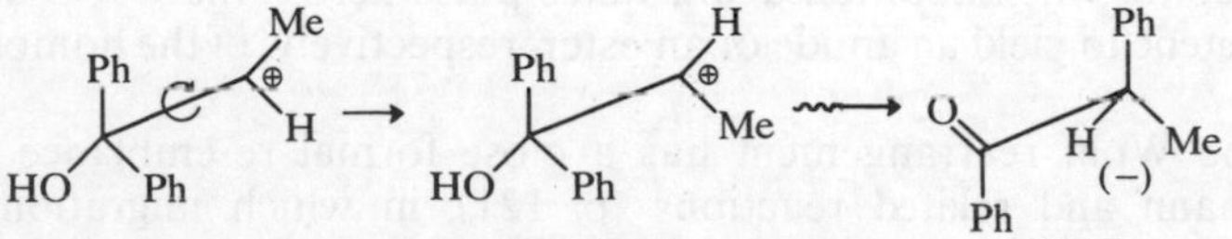

(52*c*)   (52*d*)   (51*d*)

The ratio of inversion (51*ab*) to retention (51*d*) in the product ketone would then be determined by the relative rate of rotation about C$_1$—C$_2$ in (52*c*) compared with the rate of migration of Ph.

5.4.2.5 Wolff rearrangements

This rearrangement has been separated from carbonium ion rearrangements proper as it involves migration to an uncharged, albeit electron-deficient, carbene-like carbon (*cf.* p. 260) atom rather than to a positively charged one. The reaction involves the loss of nitrogen from α-diazoketones (53), and rearrangement to highly reactive *ketenes* (54):

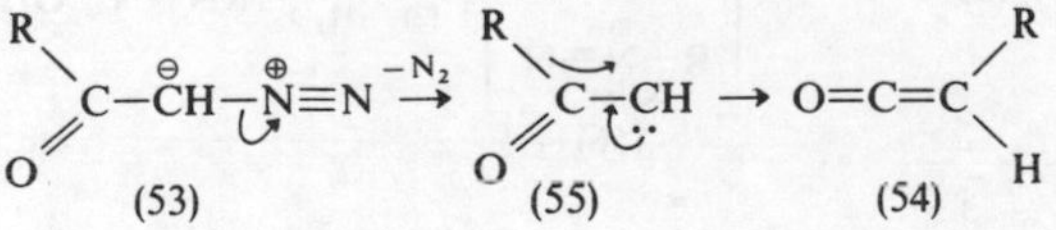

(53)   (55)   (54)

The ketenes will then react readily with any nucleophiles present in the system, e.g. H_2O below. The reaction can be brought about by photolysis, thermolysis, or by treatment with silver oxide. In the first two cases an actual carbene intermediate (55) is probably formed as shown above, in the silver catalysed reaction loss of nitrogen and migration of R may be more or less simultaneous. In the case where R is $C_4H_9C(Me)Ph$ it has been shown to migrate with retention of configuration (*cf.* p. 121).

Diazoketones (53) may be obtained by the reaction of diazomethane, CH_2N_2, on acid chlorides, and a subsequent Wolff rearrangement in the presence of water is of importance because it constitutes part of the Arndt–Eistert procedure, by which an acid may be converted into its homologue:

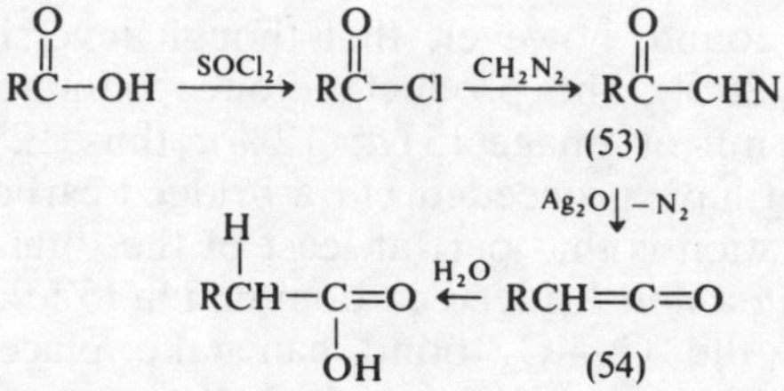

$$\underset{}{\overset{O}{\overset{\|}{R\overset{}{C}}}}-OH \xrightarrow{SOCl_2} \underset{}{\overset{O}{\overset{\|}{R\overset{}{C}}}}-Cl \xrightarrow{CH_2N_2} \underset{(53)}{\overset{O}{\overset{\|}{R\overset{}{C}}}}-CHN_2$$

$$\underset{\overset{|}{OH}}{\overset{H}{\overset{|}{RCH}}}-C{=}O \overset{H_2O}{\underset{}{\leftarrow}} RCH{=}C{=}O \qquad (54)$$

with $Ag_2O \downarrow -N_2$

As well as in water, the reaction can be carried out in ammonia or in an alcohol when addition again takes place across the C=C bond of the ketene to yield an amide or an ester, respectively, of the homologous acid.

The Wolff rearrangement has a close formal resemblance to the Hofmann and related reactions (p. 121), in which migration takes place to an electron-deficient nitrogen atom to form an isocyanate, RN=C=O, intermediate.

5.5 DIAZONIUM CATIONS

The nitrosation of primary amines, RNH_2, with, for example, sodium nitrite and dilute acid (*cf.* p. 106) leads to the formation of diazonium cations (56):

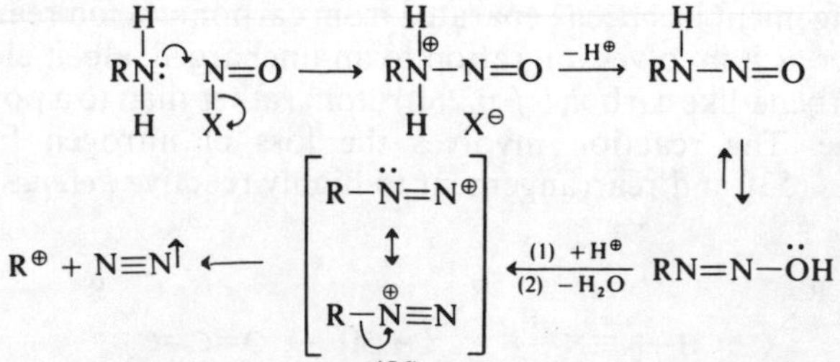

$$\underset{\overset{|}{H}\ \ X}{\overset{H}{\overset{|}{RN}}}{:}^{\curvearrowright}\,N{=}O \longrightarrow \underset{\overset{|}{H}\ \ X^\ominus}{\overset{H}{\overset{|\oplus}{RN}}}-N{=}O \xrightarrow{-H^\oplus} \underset{\overset{}{}}{\overset{H}{\overset{|}{RN}}}-N{=}O$$

$$R^\oplus + N{\equiv}N{\uparrow} \longleftarrow \left[\begin{array}{c} R-\overset{..}{N}{=}N^\oplus \\ \updownarrow \\ R{-}\!\!\underset{\curvearrowleft}{\overset{\oplus}{N}}{\equiv}N \end{array} \right] \overset{(1)\ +H^\oplus}{\underset{(2)\ -H_2O}{\longleftarrow}} RN{=}N{-}\overset{..}{O}H$$

$$(56)$$

The effective nitrosating agent is probably never HNO_2 itself; at relatively low acidity it is thought to be N_2O_3 (X = ONO) obtained by,

$$2HNO_2 \rightleftarrows ONO-NO + H_2O$$

while as the acidity is increased this is replaced by the more effective species, protonated nitrous acid $H_2\overset{\oplus}{O}-NO$ (X = H_2O), and finally by the nitrosonium ion, $^{\oplus}NO$ (*cf.* p. 136). A compromise has to be struck in nitrosation, however, between an increasingly effective nitrosating agent as the acidity of the solution is increased, and decreasing $[R\overset{..}{N}H_2]$, as the amine becomes increasingly protonated and so rendered unreactive.

With simple aliphatic amines, the initial diazonium cation (56) breaks down extremely readily to yield carbonium ions (*cf.* p. 106) which are, for reasons that are not wholly clear, markedly more reactive than those obtained from other fission processes, e.g. $RBr \rightarrow$ $R^{\oplus}Br^{\ominus}$. Where the prime purpose is the formation of carbonium ions, the nitrosation is better carried out on a derivative of the amine (to avoid formation of H_2O) under anhydrous conditions:

$$RNH_2 \xrightarrow{COCl_2} RNCO \xrightarrow{^{\oplus}NOSbF_6{^{\ominus}}} R^{\oplus}SbF_6{^{\ominus}} + N_2{^{\dagger}} + CO_2{^{\dagger}}$$

If R contains a powerful electron-withdrawing group, however, loss of $H^{\oplus}$—rather than loss of N_2—can take place to yield a substituted diazoalkane, e.g. ethyl aminoacetate → ethyl diazoacetate:

$$NH_2CH_2CO_2Et \xrightarrow{NaNO_2}{HCl} \left[N{\equiv}N{\overset{\oplus}{\underset{\underset{H}{|}}{-}}}CHCO_2Et \right] \rightarrow \overset{\ominus}{N}{=}\overset{\oplus}{N}{=}CHCO_2Et$$

The instability of aliphatic diazonium cations, in the absence of any stabilising structural feature, is due very largely to the effectiveness of N_2 as a leaving group; in aromatic diazonium cations, however, such a stabilising feature is provided by the π orbital system of the aromatic nucleus:

Because primary aromatic amines are weaker bases/nucleophiles than aliphatic (due to interaction of the electron pair on N with the π orbital system of the aromatic nucleus), a fairly powerful nitrosating agent is required, and the reaction is thus carried out at relatively high acidity. Sufficient equilibrium concentration of unprotonated $Ar\overset{..}{N}H_2$ remains (as it is a weak base), but the concentration is low enough to prevent as yet undiazotised amine undergoing a coupling reaction with the first formed $ArN_2^{\oplus}$ (*cf.* p. 146). Aromatic diazonium chlorides, sulphates, nitrates, etc., are reasonably stable in aqueous solution at room temperature or below, but cannot readily be isolated without decomposition. Fluoroborates, $ArN_2^{\oplus}BF_4^{\ominus}$, are more stable (*cf.* stabilising effect of $BF_4^{\ominus}$ on other ion pairs, p. 135) and can be isolated in the dry solid state: thermolysis of the dry solid is an important preparative method for fluoroarenes:

$$ArN_2^{\oplus}BF_4^{\ominus} \xrightarrow{\Delta} Ar{-}F + N_2{\uparrow} + BF_3{\uparrow}$$

As might be expected, substituents in the aromatic nucleus have a marked effect on the stability of $ArN_2^{\oplus}$, electron-donating groups having a marked stabilising effect:

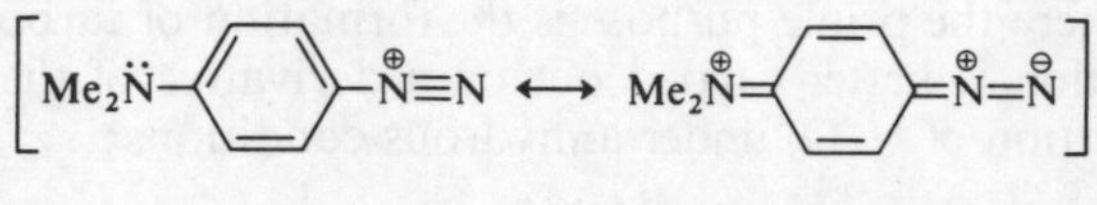

$$\left[Me_2\overset{..}{N}{-}\left\langle\overline{}\right\rangle{-}\overset{\oplus}{N}{\equiv}N \longleftrightarrow Me_2\overset{\oplus}{N}{=}\left\langle\overline{}\right\rangle{=}N{=}\overset{\ominus}{N} \right]$$

Nitrosation also occurs with secondary amines but stops at the stable N-nitroso stage, $R_2N{-}N{=}O$. Tertiary aliphatic amines are converted initially into the nitrosotrialkylammonium cation, $R_3\overset{\oplus}{N}{-}N{=}O$, but this then readily undergoes C${-}$N fission to yield relatively complex products. With aromatic tertiary amines, $ArNR_2$, nitrosation can take place not on N but at the activated *p*-position of the nucleus (*cf.* p. 136) to yield a C-nitroso compound:

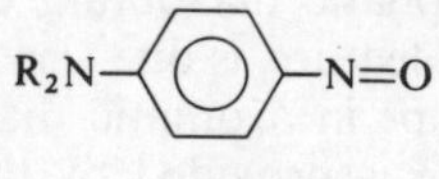

$$R_2N{-}\left\langle\bigcirc\right\rangle{-}N{=}O$$

5.6 MIGRATION TO ELECTRON-DEFICIENT N

The rearrangements that we have considered to date all have one feature in common: the migration of an alkyl or aryl group, with its electron pair, to a carbon atom which, whether a carbonium ion or not, is electron-deficient. Another atom that can similarly become electron-deficient is nitrogen in, for example, $R_2N^{\oplus}$ or $R\overset{..}{N}$ (a *nitrene*, *cf.* carbenes above), and it might be expected that alkyl or aryl migration to such centres would take place, just as it did to $R_3C^{\oplus}$ and $R_2\overset{..}{C}$:this is indeed found to be the case.

5.6.1 Hofmann, Curtius, Lossen and Schmidt reactions

A typical example is the conversion of an amide (57) to an amine (58), containing one carbon less, by the action of alkaline hypobromite, the Hofmann reaction:

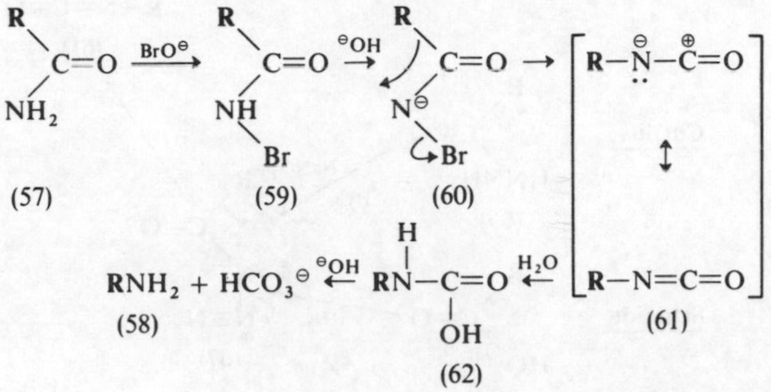

The formal end-product of the reaction is the isocyanate (61), corresponding exactly to the ketene in the Wolff reaction (p. 117), but this undergoes addition of water under the reaction conditions to yield the unstable carbamic acid (62) which, in turn, decarboxylates to yield the amine (58). By careful control of conditions it is possible actually to isolate N-bromoamide (59), its anion (60), and isocyanate (61) as intermediates: the suggested reaction pathway is thus unusually well documented. The rate-limiting step is probably loss of $Br^{\ominus}$ from (60), and the question arises whether this loss is concerted with the migration of R, or whether a carbonylnitrene intermediate, $RC\ddot{O}\ddot{N}$, is formed, which then rearranges. The fact that the rearrangement of $ArCONH_2$ is speeded up when Ar contains electron-donating substituents (*cf.* the pinacol/pinacolone rearrangement, p. 114), and that the formation of hydroxamic acids, RCONHOH (that would be expected from attack of solvent H_2O on $RC\ddot{O}\ddot{N}$), has never been detected, both support a concerted mechanism. Crossover experiments lead to no mixed products, i.e. the rearrangement is strictly intramolecular, and it is further found that when R is chiral, e.g. C_6H_5MeCH, it migrates with its configuration unchanged.

There are a group of reactions very closely related to that of Hofmann, all of which involve the formation of an isocyanate (61) by rearrangement of an intermediate analogous to (60):

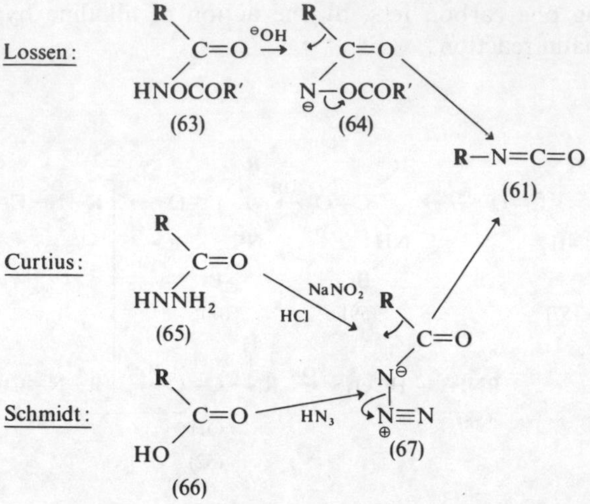

Lossen:

$$\underset{\substack{HNOCOR'\\(63)}}{\overset{R}{\underset{}{}}}C=O \overset{{}^{\ominus}OH}{\rightarrow} \underset{\substack{N-OCOR'\\\ominus\\(64)}}{\overset{R}{}}C=O$$

$$R-N=C=O$$
$$(61)$$

Curtius:

$$\underset{\substack{HNNH_2\\(65)}}{\overset{R}{}}C=O \xrightarrow[HCl]{NaNO_2} \underset{\substack{N^\ominus\\|\\N\equiv N\\\oplus\\(67)}}{\overset{R}{}}C=O$$

Schmidt:

$$\underset{\substack{HO\\(66)}}{\overset{R}{}}C=O \xrightarrow{HN_3}$$

The Lossen reaction involves the action of base on O-acyl derivatives (63) of hydroxamic acids, RCONHOH, and involves $R'CO_2{}^\ominus$ as the leaving group from the intermediate (64), as compared with $Br^\ominus$ from (60). The reaction also occurs with the hydroxamic acids themselves, but not as well as with their O-acyl derivatives as $R'CO_2{}^\ominus$ is a better leaving group than $^\ominus OH$. The concerted nature of the rearrangement is supported by the fact that not only is the reaction facilitated by electron-donating substituents in R (*cf*. Hofmann), but also by electron-withdrawing substituents in R', i.e. <u>both</u> are involved in the rate-limiting step of the reaction.

The Curtius and Schmidt reactions both involve N_2 as the leaving group from the azide intermediate (67), and here again the migration of R occurs in a concerted process. The azide may be obtained either by nitrosation of an acid hydrazide (65)—Curtius reaction, or by the reaction of hydrazoic acid, HN_3, on a carboxylic acid (66)—the Schmidt reaction.

5.6.2 Beckmann rearrangements

The most famous of the rearrangements in which R migrates from carbon to nitrogen is undoubtedly the conversion of ketoximes to N-substituted amides, the Beckmann rearrangement:

$$RR'C=NOH \rightarrow R'CONHR \quad or \quad RCONHR'$$

The reaction is catalysed by a wide variety of acidic reagents, e.g. H_2SO_4, SO_3, $SOCl_2$, P_2O_5, PCl_5, BF_3, etc., and takes place not only with ketoximes themselves but also with their O-esters. Only a very few aldoximes rearrange under these conditions, but more can be made to do so by use of polyphosphoric acid as catalyst. Perhaps the most interesting feature of the rearrangement is that, unlike those we have already considered, it is not the <u>nature</u> (e.g. relative electron-releasing ability) but the <u>stereochemical arrangement</u> of the R and R' groups that determines which of them migrates. Almost without exception it is found to be the R group *anti* to the OH group that migrates $C \rightarrow N$:

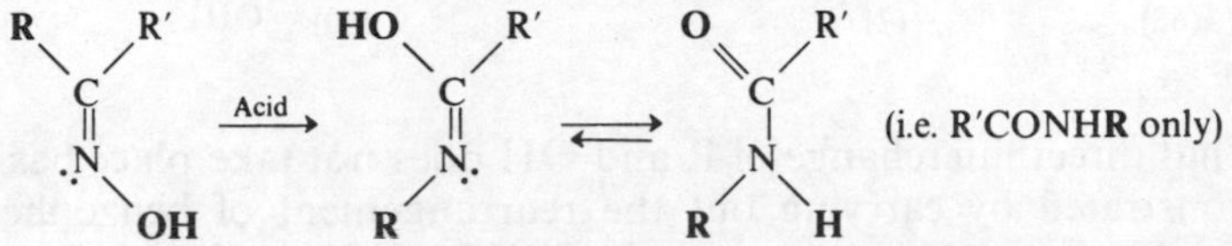

(i.e. R'CONHR only)

Confirmation that this is indeed the case requires an initial, unambiguous assignment of configuration to a pair of oximes. This was effected by working with the pair of oximes (68) and (69)—one of them cyclised to the benzisoxazole (70) with base even in the cold, while the other was little affected even under much more vigorous conditions. The one undergoing easy cyclisation was, on this basis, assigned the configuration (68) in which the nuclear Br atom and the H of the OH group, that are to be eliminated, are close together:

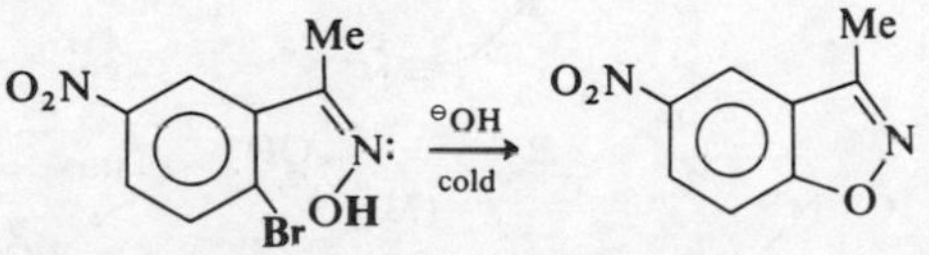

(68)   (70)

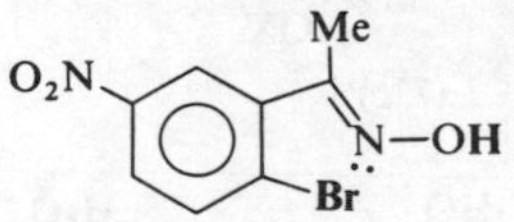

(69)

In (69) these atoms are far apart, and can be brought within reacting distance only by cleavage of the $C=N$ bond in the oxime.

Subsequently, configuration may be assigned to other pairs of ketoximes by correlation of their physical constants with those of oxime pairs whose configuration has already been established. Once it had been clearly demonstrated that the *anti*-R group migrates in Beckmann rearrangements, however, the structure of the amide produced is now normally used to establish the configuration of the oxime from which it was derived. Thus, as expected, (68) is found to yield only the N-methyl substituted benzamide (71), while (69) yields only the aryl substituted acetanilide (72):

$$\underset{\underset{HO}{\overset{Ar}{\diagdown}}\underset{(68)}{\underset{N}{\overset{\|}{\underset{.}{\overset{C}{\diagup}}}}}{} \rightarrow \underset{(71)}{Ar\overset{\overset{O}{\|}}{C}-NHMe} \qquad\qquad \underset{\underset{(69)}{\overset{Ar}{\diagdown}}\underset{OH}{\underset{N}{\overset{\|}{\underset{.}{\overset{C}{\diagup}}}}}{} \rightarrow \underset{(72)}{Me\overset{\overset{O}{\|}}{C}-NHAr}$$

That direct interchange of R and OH does not take place has been demonstrated by carrying out the rearrangement of benzophenone oxime, $Ph_2C{=}NOH$, to benzanilide, $PhCONHPh$, in $H_2{}^{18}O$. Provided that neither the initial oxime nor the product anilide exchanges its oxygen for ^{18}O when dissolved in $H_2{}^{18}O$—as has been confirmed—direct, intramolecular interchange of Ph and OH would result in the incorporation of *no* ^{18}O in the rearranged product. In fact, the product benzanilide is found to contain the same proportion of ^{18}O as did the original water, so the rearrangement must involve loss of the oxime OH group and subsequent re-introduction of oxygen from a water molecule.

The rearrangement is believed to proceed as follows:

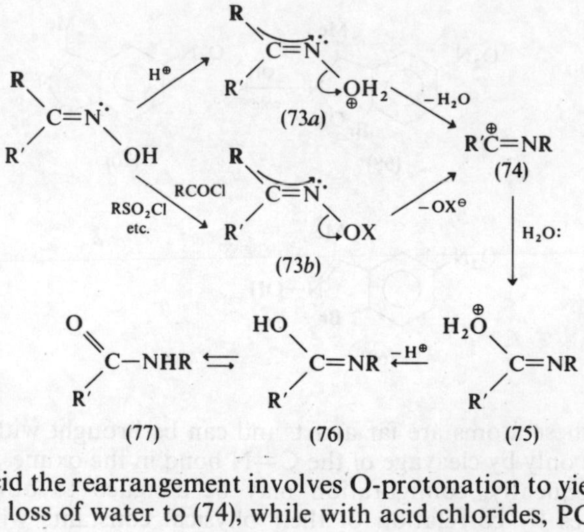

In strong acid the rearrangement involves O-protonation to yield (73a) followed by loss of water to (74), while with acid chlorides, PCl_5, etc., the intermediate ester (73*b*) is formed; the anion $XO^\ominus$ constitutes a good leaving group so that, again, (74) is obtained. A number of intermediate esters (73*b*) have indeed been prepared independently and shown to undergo rearrangement to the expected amides, in the absence of added catalysts and in neutral solvents. The stronger the acid XOH is, i.e. the more capable the anion is of independent existence, the better leaving group $XO^\ominus$ should be and hence the faster the rearrangement should occur. This is observed in the series where $XO^\ominus$ is $CH_3CO_2{}^\ominus < ClCH_2CO_2{}^\ominus < PhSO_3{}^\ominus$. That such ionisation is the rate-limiting step in the rearrangement is also suggested by the observation that the rate of reaction increases as the solvent polarity increases.

As with the rearrangements we have discussed previously, loss of leaving group and migration of R are believed to proceed essentially simultaneously in the conversion of (73) into (74). This is borne out by the strict intramolecularity of the reaction (no crossover products, *cf.* p. 115), the high stereoselectivity already referred to (i.e. only R, not R' migrates), and the fact that R, if chiral, e.g. PhCHMe, retains its configuration on migration. This probably also reflects the greater stability of the carbenium ion intermediate, $R'\overset{\oplus}{C}=NR$ (74), than the nitrenium ion, $RR'C=\overset{\oplus}{\underset{\cdot\cdot}{N}}$, that would be obtained if loss of leaving group preceded the migration of R. The rearrangement is completed by attack of water (it is, of course, at this stage that ^{18}O is introduced in the rearrangement of benzophenone oxime in $H_2{}^{18}O$ referred to above) on the carbenium ion carbon of (74) to yield (75), followed by loss of proton to form the enol (76) of the product amide (77).

The stereochemical use of the Beckmann rearrangement in assigning configuration to ketoximes has already been referred to, and it also has a large-scale application in the synthesis of the textile polymer Nylon-6 from cyclohexanone oxime (78) *via* the cyclic amide (*lactam*, 79):

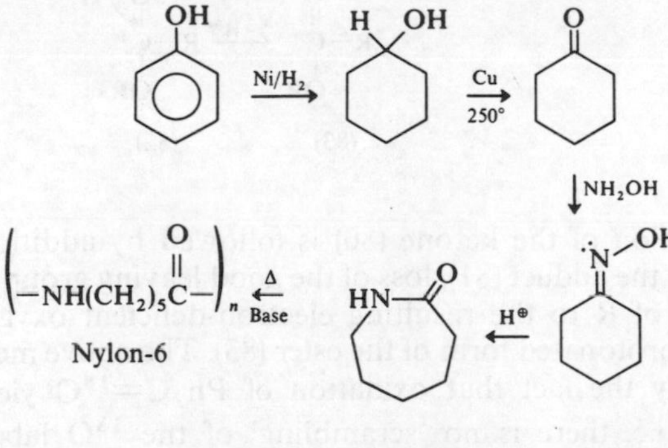

5.7 MIGRATION TO ELECTRON-DEFICIENT O

It might reasonably be expected that similar rearrangements would also occur in which the migration terminus was an electron-deficient oxygen atom: such rearrangements are indeed known.

5.7.1 Baeyer–Villiger oxidation of ketones

Oxidation of ketones with hydrogen peroxide or with a peroxyacid, RCO_2OH (*cf.* p. 320) results in their conversion into esters:

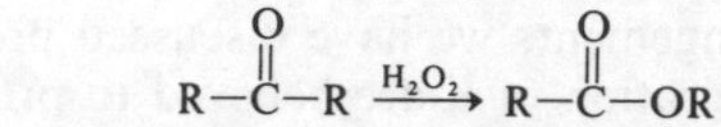

$$R\overset{\overset{\displaystyle O}{\|}}{-C}-R \xrightarrow{H_2O_2} R\overset{\overset{\displaystyle O}{\|}}{-C}-OR$$

Cyclic ketones are converted into *lactones* (cyclic esters):

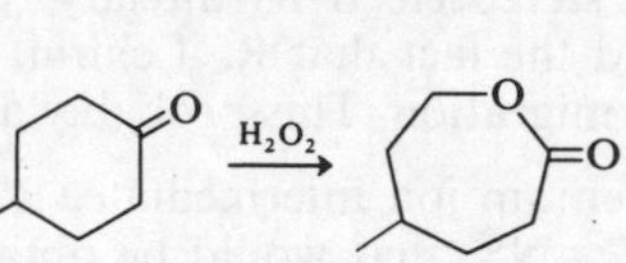

The reaction is believed to proceed as follows:

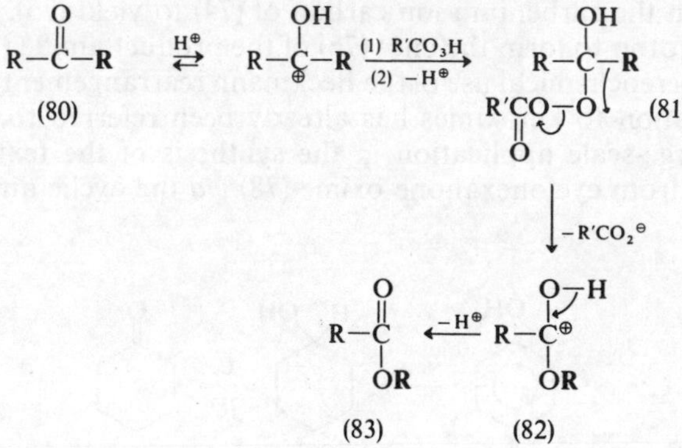

Initial protonation of the ketone (80) is followed by addition of the peracid to yield the adduct (81), loss of the good leaving group $R'CO_2^{\ominus}$ and migration of R to the resulting electron-deficient oxygen atom yields (82), the protonated form of the ester (83). The above mechanism is supported by the fact that oxidation of $Ph_2C{=}^{18}O$ yields only $PhC^{18}O{\cdot}OPh$, i.e. there is no 'scrambling' of the ^{18}O label in the product ester. That loss of $R'CO_2^{\ominus}$ and migration of R are concerted is supported by the fact that the reaction is speeded up by electron-withdrawing substituents in R′ of the leaving group, and by electron-donating substituents in the migrating group R: the concerted conversion of (81) into (82) thus appears to be the rate-limiting step of the reaction. Further, a chiral R is found to migrate with its configuration unchanged. When an unsymmetrical ketone, RCOR′, is oxidised either group could migrate, but it is found in practice that it is normally the more nucleophilic group, i.e. the group better able to stabilise a positive charge, that actually migrates, *cf.* the pinacol/pinacolone rearrangement (p. 114). As with the latter reaction, however, steric effects can also play a part, and may occasionally change markedly the expected order of migratory aptitude based on electron-releasing ability alone.

5.7.2 Hydroperoxide rearrangements

A very similar rearrangement takes place during the acid-catalysed decomposition of hydroperoxides, RO—OH, where R is a secondary or tertiary carbon atom carrying alkyl or aryl groups. A good example is the decomposition of the hydroperoxide (84) obtained by the air-oxidation of cumene [(1-methylethyl)benzene]; this is used on the large-scale for the preparation of phenol and acetone:

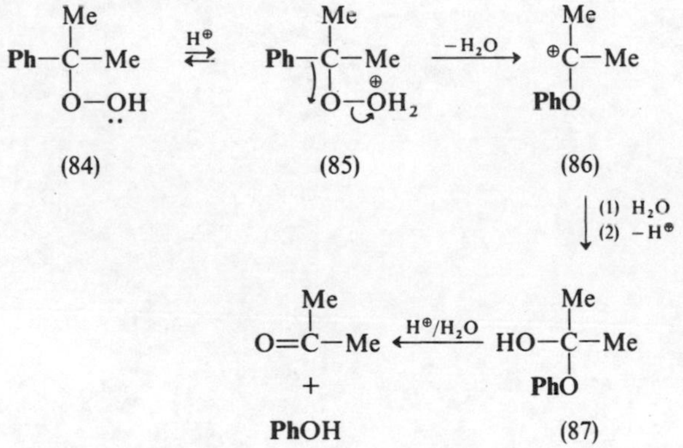

Here again loss of the leaving group (H_2O), and migration of Ph to the resulting electron-deficient oxygen atom in (85), are almost certainly concerted. Addition of water to the carbonium ion (86) yields the hemi-ketal (87), which undergoes ready hydrolysis under the reaction conditions to yield phenol and acetone. It will be observed that Ph migrates in preference to Me in (85) as, from previous experience, we would have expected. Electron-donating substituents in a migrating group are found to increase the rate of reaction, and to promote the migratory ability of a particular group with respect to its unsubstituted analogue. It may be that the superior migratory aptitude of Ph above results from its migrating *via* a bridged transition state:

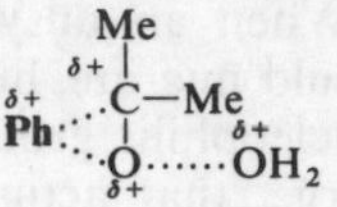

$$
\begin{array}{c}
\text{Me} \\
{}^{\delta+}| \\
\text{Ph}{}^{\delta+}\cdots\overset{\cdot}{\underset{\cdot}{\text{C}}}\text{—Me} \\
\phantom{\text{Ph}}\cdots\underset{\delta+}{\text{O}}\cdots\cdots\overset{\delta+}{\text{OH}_2}
\end{array}
$$

In these examples we have been considering the essentially hetero-lytic fission of peroxide linkages, $-\text{O}:\text{O}- \rightarrow -\text{O}^{\oplus}:\text{O}^{\ominus}-$, in polar solvents; homolytic fission can also occur, under suitable conditions, to yield radicals, $-\text{O}:\text{O}- \rightarrow -\text{O}\cdot\ \cdot\text{O}-$, as we shall see below (p. 296).

6

Electrophilic and nucleophilic substitution in aromatic systems

Reference has already been made to the structure of benzene and, in particular, to its delocalised π orbitals (p. 14); the concentration of negative charge above and below the plane of the ring-carbon atoms is thus benzene's most accessible feature:

This concentration of charge might be expected to shield the ring carbon atoms from the attack of nucleophilic reagents and, by contrast, to promote attack by cations, $X^\oplus$, or electron-deficient species, i.e. by electrophilic reagents; this is indeed found to be the case.

6.1 ELECTROPHILIC ATTACK ON BENZENE

6.1.1 π and σ* complexes

It might be expected that the first phase of reaction would be inter-action between the approaching electrophile and the delocalised π orbitals and, in fact, so-called *π complexes* such as (1) are formed:

(1)

Thus methylbenzene (toluene) forms a 1:1 complex with hydrogen chloride at $-78°$, the reaction being readily reversible. That no actual bond is formed between a ring-carbon atom and the proton from HCl is confirmed by repeating the reaction with DCl; this also yields a π complex, but its formation and decomposition does not lead to the exchange of deuterium with any of the hydrogen atoms of the nucleus, showing that no C—D bond has been formed in the complex. Aromatic hydrocarbons have also been shown to form π complexes with species such as the halogens, $Ag^\oplus$, and, better known, with picric acid, 2,4,6-$(O_2N)_3C_6H_2OH$, to form stable coloured crystalline adducts whose melting points may be used to characterise the hydrocarbons. These adducts are also known as *charge transfer complexes*. In the complex that benzene forms with bromine, it has been shown that the halogen molecule is located centrally, and at right angles to the plane of the benzene ring.

In the presence of a compound having an electron-deficient orbital e.g. a Lewis acid such as $AlCl_3$, a different complex is formed, however. If DCl is now employed in place of HCl, rapid exchange of deuterium with the hydrogen atoms of the nucleus is found to take place indicating the formation of a *σ complex* (2), also called a Wheland intermediate (*cf.* p. 41), in which $H^\oplus$ or $D^\oplus$, as the case may be, has actually become covalently bonded to a ring-carbon atom. The positive charge is shared over the remaining five carbon atoms of the nucleus *via* the π orbitals and the deuterium and hydrogen atoms are in a plane at

---

*The term *σ complex*, though still current, dates from a time when little was known about their structure; it is being replaced by the term *arenium ion*.

right angles to that of the ring:

(2a)   (2b)   (2c)   (2)

That the π and σ complexes with, e.g. methylbenzene and HCl, really are different from each other is confirmed by their differing behaviour. Thus formation of the former leads to a solution that is a non-conductor of electricity, to no colour change, and to but little difference in u.v. spectrum, indicating that there has been little disturbance of electron distribution in the original methylbenzene; while if $AlCl_3$ is present the solution becomes green, will conduct electricity and the u.v. spectrum of the original methylbenzene is modified, indicating the formation of a complex such as (2) as there is no evidence that aluminium chloride forms complexes of the type, $H^{\oplus}AlCl_4^{\ominus}$.

The reaction may be completed by $AlCl_4^{\ominus}$ removing a proton from the σ complex (2) $\rightarrow$ (4). This can lead only to exchange of hydrogen atoms when HCl is employed but to some substitution of hydrogen by deuterium with DCl, i.e. the overall process is electrophilic *substitution*. In theory, (2) could, as an alternative, react by removing $Cl^{\ominus}$ from $AlCl_4^{\ominus}$ resulting in an overall electrophilic *addition* reaction (2) $\rightarrow$ (3) as happens with a simple carbon–carbon double bond (p. 181); but this would result in permanent loss of the stabilisation conferred on the molecule by the presence of delocalised π orbitals involving all six carbon atoms of the nucleus, so that the product, an addition compound, would no longer be aromatic with all that implies. By expelling $H^{\oplus}$, i.e. by undergoing substitution rather than addition, the completely filled, delocalised π orbitals are reattained in the product (4) and characteristic aromatic stability recovered:

(3)   (2)   (4)

Addition   Substitution

The gain in stabilisation in going from (2) $\rightarrow$ (4) helps to provide the energy required to break the strong C—H bond that expulsion of $H^{\oplus}$ necessitates; in the reaction of, for example, HCl with alkenes (p. 181) there is no such factor promoting substitution and addition reactions are therefore the rule.

It might perhaps be expected that conversion of benzene into the σ complex (2), which has forfeited its aromatic stabilisation, would involve the expenditure of a considerable amount of energy, i.e. that the activation energy for the process would be high and the reaction rate correspondingly low: in fact, many aromatic electrophilic substitutions are found to proceed quite rapidly at room temperature. This is because there are two factors operating in (2) that serve to reduce the energy barrier that has to be surmounted in order to effect its formation: first, the energy liberated by the complete formation of the new bond to the attacking electrophile, and, second, the fact that the positively charged σ complex can stabilise itself, i.e. lower its energy level, by delocalisation

(2)

as has indeed been implied by writing its structure as (2). The use of (2) should not, however, be taken to imply a uniform distribution of electron density in the ion—that this could not be so is plain when the separate canonical states (*cf.* 2*a* → 2*c*, p. 131) contributing to (2) are written out.

If we are correct in our assumption that the electrophilic substitution of aromatic species involves such σ complexes as intermediates—and it has proved possible actually to isolate them in the course of some such substitutions (p. 140)—then what we commonly refer to as aromatic 'substitution' really involves initial *addition* followed by subsequent *elimination*. How this basic theory is borne out in the common electrophilic substitution reactions of benzene will now be considered.

6.2 NITRATION

The aromatic substitution reaction that has received by far the closest study is nitration and, as a result, it is the one that probably provides the most detailed mechanistic picture. Preparative nitration is most frequently carried out with a mixture of concentrated nitric and sulphuric acids, the so-called nitrating mixture. The 'classical' explanation for the presence of the sulphuric acid is that it absorbs the water formed in the nitration proper

$$C_6H_6 + HNO_3 \rightarrow C_6H_5NO_2 + H_2O$$

and so prevents the reverse reaction from proceeding. This explanation is unsatisfactory in a number of respects, not least in that nitrobenzene, once formed, appears not to be attacked by water under the conditions of the reaction! What is certain is that nitration is slow in the absence of sulphuric acid, yet sulphuric acid by itself has virtually no effect on benzene under the conditions normally employed. It would thus appear that the sulphuric acid is acting on the nitric acid rather than the benzene in the system. This is borne out by the fact that solutions of nitric acid in pure sulphuric acid show an almost four-fold molecular freezing-point depression (actually $i \approx 3.82$), which has been interpreted as being due to formation of the four ions:

$$\overset{..}{HO}-NO_2 \underset{\longleftarrow}{\overset{H_2SO_4}{\rightleftharpoons}} H_2\overset{\frown}{\underset{\oplus}{O}}-NO_2 \overset{H_2SO_4}{\rightleftharpoons} H_3O^\oplus + HSO_4{}^\ominus + {}^\oplus NO_2$$
$$+$$
$$HSO_4{}^\ominus$$

$$\textit{i.e. } HNO_3 + 2H_2SO_4 \rightleftharpoons \underbrace{{}^\oplus NO_2 + H_3O^\oplus + 2HSO_4{}^\ominus}$$

The slight shortfall of i below 4 is probably due to incomplete protonation of H_2O under these conditions.

The presence of ${}^\oplus NO_2$, the *nitronium ion*, both in this solution and in a number of salts (some of which, e.g. ${}^\oplus NO_2 \, ClO_4{}^\ominus$, have actually been isolated) has been confirmed spectroscopically: there is a line in the Raman spectrum of each of them at $1400 \, cm^{-1}$ which can only originate from a species that is both linear and triatomic. Nitric acid itself is converted in concentrated sulphuric acid virtually entirely into ${}^\oplus NO_2$, and there can be little doubt left that this is the effective electrophile in nitration under these conditions. If the purpose of the sulphuric acid is merely to function as a highly acid medium in which ${}^\oplus NO_2$ can be released from $HO-NO_2$, it would be expected that other strong acids, e.g. $HClO_4$, would also promote nitration. This is indeed found to be the case, and HF plus BF_3 are also effective. The poor performance of nitric acid by itself in the nitration of benzene is thus explained for it contains but little ${}^\oplus NO_2$; the small amount that is present is obtained by the two-stage process

$$\overset{..}{HO}-NO_2 + HNO_3 \overset{fast}{\rightleftharpoons} H_2\underset{\oplus}{O}-NO_2 + NO_3{}^\ominus$$

$$H_2\overset{\frown}{\underset{\oplus}{O}}-NO_2 + HNO_3 \overset{slow}{\rightleftharpoons} H_3O^\oplus + NO_3{}^\ominus + {}^\oplus NO_2$$

in which nitric acid is first converted rapidly into its conjugate acid, and that then more slowly into nitronium ion.

Many nitrations are found to conform to an idealised rate law of the form,

$$\text{Rate} = k[Ar-H][{}^\oplus NO_2]$$

but, in practice, the actual kinetics are not always easy to follow or to interpret for a variety of reasons. Thus the solubility of, for example, benzene itself in nitrating mixture is sufficiently low for the rate of nitration to be governed by the rate at which the immiscible hydrocarbon dissolves in the acid layer. With nitrating mixture, $[^{\oplus}NO_2]$ is related directly to $[HNO_3]$ added, as HNO_3 is converted rapidly and completely into $^{\oplus}NO_2$, but with nitrations in other solvents complex equilibria may be set up. The relation of the concentration of the effective nucleophile—probably always $^{\oplus}NO_2$—to the concentration of HNO_3, or other potential nitrating agent, actually added may then be far from simple.

The above general, idealised rate law is compatible with at least three different potential pathways for nitration: one-step, concerted pathway [1] that involves a single transition state (5),

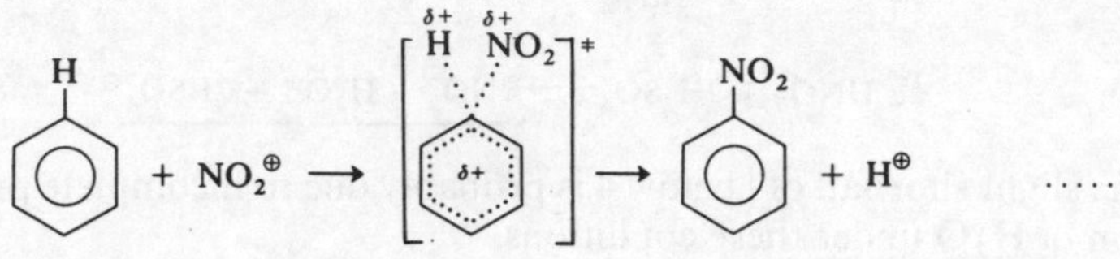

(5)

in which $C-NO_2$ bond-formation and $C-H$ bond-breaking are occurring simultaneously; or two-step pathways [2] involving a Wheland intermediate or σ complex (6),

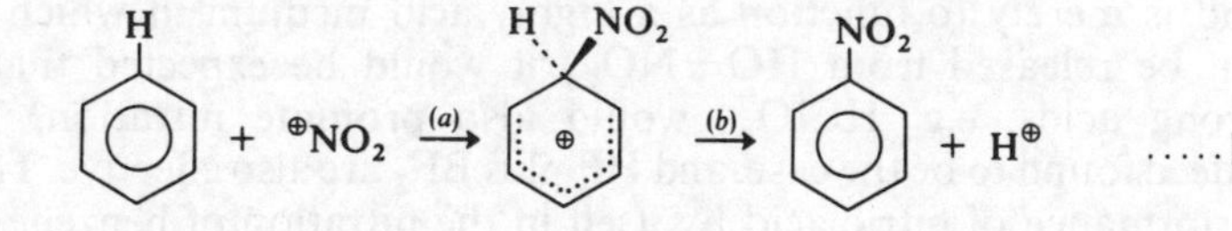

(6)

in which *either* step (*a*)— $C-NO_2$ bond-formation—*or* step (*b*)— $C-H$ bond-breaking—could be slow and rate-limiting. The $C-H$ bond must, of course, be broken at some stage in all three of the above pathways, but a partial distinction between them is that it must be broken in the slow, rate-limiting step in [1] (only one step, anyway) and in [2*b*], but *not* in [2*a*]. If the $C-H$ bond is, in fact, broken in a rate-limiting step, then the reaction will exhibit a primary kinetic isotope effect (*cf.* p. 46) if C_6H_6 is replaced by C_6D_6. The comparison was in fact (for experimental reasons) made on $C_6H_5NO_2$ and $C_6D_5NO_2$—this makes no difference to the argument—and it was found that k_H/k_D at 25° $\approx$ 1·00, i.e. that there is *no* primary kinetic isotope effect. The C—H bond is thus *not* being broken in the rate-limiting step of the reaction: pathways [1] and [2b] are therefore ruled out. This does not of course *prove* that nitration proceeds by pathway [2a]—slow, rate-limiting formation of the C—NO$_2$ bond—

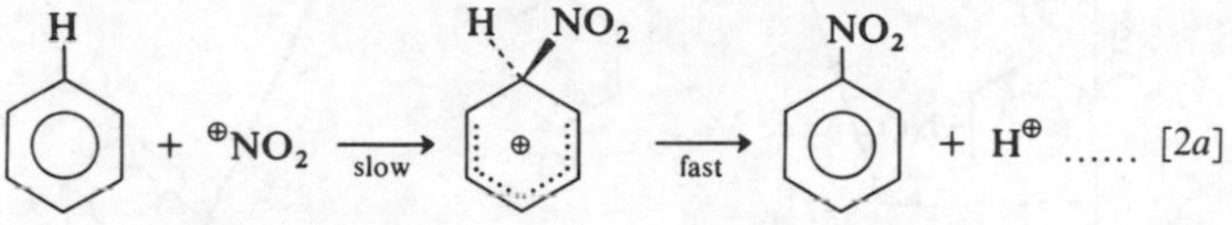

(6)

but this is the only one of those considered that is compatible with the experimental data.

That the cleavage of the C—H bond—a strong one—should be fast seems less surprising when we realise that by loss of H$^{\oplus}$ the intermediate (6) is able to reattain the highly stabilised aromatic condition in the product nitrobenzene. The incipient proton is removed from (6) by the attack of bases, e.g. probably $HSO_4^{\ominus}$ in nitrating mixture, but sometimes by solvent molecules. The credibility of species such as (6) as intermediates is enhanced by the actual detection (n.m.r.) of analogous examples, e.g. (8), in which this loss of proton is prevented, and the intermediate thus 'stranded', i.e. with hexamethylbenzene (7):

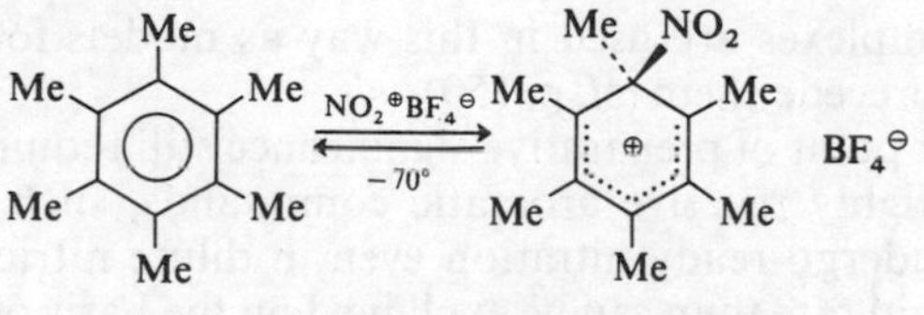

The particular stability of (8) probably owes something to the presence of $BF_4^{\ominus}$ in the ion pair (*cf.* p.101). The characterisation of (8) is not in itself conclusive evidence that analogous intermediates are formed during the course of nitration with, for example, nitrating mixture; but coupled with the kinetic and other evidence it does make the general involvement of such species seem very much more plausible.

In discussing rates of aromatic substitution reactions it is, of course, the formation of the transition state (T.S.$_1$) immediately preceding (6) that exerts the controlling influence (Fig. 6.1):

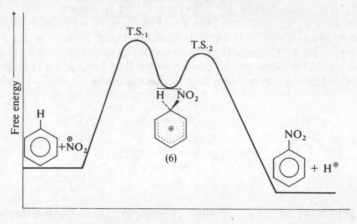

Fig. 6.1

It is often very difficult to obtain detailed information about such species, and the intermediates, of which the transition states are the immediate predecessors, are thus often taken as models for them, because detailed information about such intermediates is much more readily come by. This may be justified on the basis of *Hammond's principle* that in a sequence, immediately succeeding species that closely resemble each other in energy level are likely to resemble each other in structure also; certainly the intermediate (6) in the sequence above is likely to be a better model for T.S.₁, than is the starting material. We shall see a number of examples subsequently where σ complexes are used in this way as models for the transition states that precede them (*cf.* p. 150).

A further point of preparative significance still requires explanation, however. Highly reactive aromatic compounds, such as phenol, are found to undergo ready nitration even in dilute nitric acid, and at a far more rapid rate than can be explained on the basis of the concentration of $^{\oplus}NO_2$ that is present in the mixture. This has been shown to be due to the presence of nitrous acid in the system which nitrosates the reactive nucleus *via* the *nitrosonium ion*, $^{\oplus}NO$ (or other species capable of effecting nitrosation, *cf.* p. 119):

$$HNO_2 + 2HNO_3 \rightleftharpoons H_3O^{\oplus} + 2NO_3^{\ominus} + {}^{\oplus}NO$$

The nitroso-phenol (10) so obtained is known to be oxidised very rapidly by nitric acid to yield the nitrophenol (11) and nitrous acid; more nitrous acid is produced thereby and the process is progressively speeded up. No nitrous acid need be present initially in the nitric acid for a little of the latter attacks phenol oxidatively to yield HNO_2. The rate-determining step is again believed to be the formation of the intermediate (9). Some direct nitration of such reactive aromatic compounds by $^\oplus NO_2$ also takes place simultaneously, the relative amount by the two routes depending on the conditions.

Many other aromatic electrophilic substitution reactions are found to follow the general pathway [2] discussed above, usually corresponding to [2a] though a number of [2b] examples are known. The major point still requiring elucidation is very often the exact nature of the electrophilic species that is involved in attack on the aromatic nucleus.

6.3 HALOGENATION

In contrast to nitration, halogenation can involve a variety of different electrophiles in attack on the aromatic system. The free halogens, e.g. Cl_2 and Br_2, will readily attack an activated nucleus (*cf.* p. 149) such as phenol, but are unable to substitute benzene itself (photochemical activation can lead to *addition*, however, through the agency of free halogen atoms, p. 307): a Lewis acid catalyst such as $AlCl_3$ is required to assist in polarising the attacking halogen molecule, so as to provide it with an 'electrophilic' end. The rate law for halogenation is often of the form:

$$\text{Rate} = k[\text{Ar}-\text{H}][\text{Hal}_2][\text{Lewis acid}]$$

It seems likely that benzene forms a π complex (12) with, for example, Br_2 (*cf.* p. 130), and that the Lewis acid then interacts with this. The catalyst probably polarises $Br-Br$, assists in the formation of a σ bond between the bromine molecule's now electrophilic end and a ring carbon atom, and finally helps to remove the incipient bromide ion so as to form a σ complex (13):

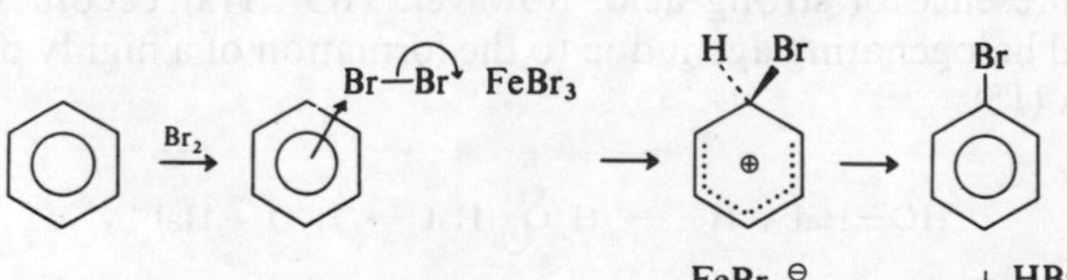

The anion $FeBr_4^\ominus$ assists in the removal of a proton from the σ complex (13). The classical halogen carrier iron filings does of course act by being converted into the Lewis acid FeX_3.

Kinetic isotope effects have not been observed for chlorination, and only rarely for bromination, i.e. the reactions normally follow pathway [2a] like nitration. In iodination, which only takes place with iodine itself on activated species, kinetic isotope effects are the rule. This presumably arises because the reaction is readily reversible (unlike other halogenations), loss of I occurring more often from the σ complex (14) than loss of H, i.e. $k_{-1} \gtrsim k_2$:

$$\text{OH} + \text{I}_2 \underset{k_{-1}}{\overset{k_1}{\rightleftarrows}} \overset{\oplus}{\text{OH}} \text{I}^{\ominus} \overset{k_2}{\longrightarrow} \text{OH} + \text{HI}$$

(14)

Thus k_H/k_D for the iodination of phenol and 2,4,6-trideuteriophenol is found to be ≈ 4, i.e. pathway [2b]. Iodination is often assisted by the presence of bases or of oxidising agents, which remove HI and thus displace the above equilibrium to the right. Oxidising agents also tend to produce $I^\oplus$, or a complex containing positively polarised iodine, from I_2, thus providing a more effective electrophile. Halogenation may also be carried out by use of interhalogen compounds $\overset{\delta+}{\text{Br}}{-}\overset{\delta-}{\text{Cl}}, \overset{\delta+}{\text{I}}{-}\overset{\delta-}{\text{Cl}}$, etc., attack occurring through the less electronegative halogen as this will constitute the 'electrophilic' end of the molecule. The two species above are thus found to effect bromination and iodination, respectively.

Halogenation may be effected by hypohalous acids, $\overset{\delta-}{\text{HO}}{-}\overset{\delta+}{\text{Hal}}$, also. This is markedly slower than with molecular halogens as $HO^\ominus$ is a poorer leaving group from $\overset{\delta-}{\text{HO}}{-}\overset{\delta+}{\text{Hal}}$ than $Hal^\ominus$ is from $\overset{\delta+}{\text{Hal}}{-}\overset{\delta-}{\text{Hal}}$. The reaction is speeded in the presence of $Hal^\ominus$, however, as $HO{-}Hal$ is then converted into the more reactive Hal_2, e.g.:

$$^\ominus\text{OCl} + \text{Cl}^\ominus + 2\text{H}^\oplus \rightarrow \text{Cl}_2 + \text{H}_2\text{O}$$

In the presence of strong acid, however, $HO{-}Hal$ becomes a very powerful halogenating agent due to the formation of a highly polarised complex (15):

$$\text{H}\overset{..}{\text{O}}{-}\text{Hal} + \text{H}^\oplus \rightarrow \text{H}_2\overset{\oplus}{\text{O}}{-}\text{Hal} \leftrightarrow \text{H}_2\text{O} + \text{Hal}^\oplus$$

(15)

The evidence is that this species is the effective electrophile under these conditions, and does not support the further conversion of (15) into $Hal^\oplus$, i.e. unlike the case with $H_2O^\oplus{-}NO_2$ (p. 133); HOCl + acid can still be a more effective chlorinating agent than $Cl_2 + AlCl_3$, however.

6.4 SULPHONATION

The mechanistic details of sulphonation have been less closely explored than those of nitration or halogenation. Benzene itself is sulphonated fairly slowly by hot concentrated sulphuric acid, but rapidly by oleum, the rate being related to the latter's SO_3 content. There is still some doubt about the attacking electrophile, e.g. whether it is SO_3 or the latter's conjugate acid, $^\oplus SO_3H$; this may vary with the conditions but is probably normally SO_3, of which small equilibrium concentrations exist in H_2SO_4:

$$2H_2SO_4 \rightleftarrows SO_3 + H_3O^\oplus + HSO_4^\ominus$$

Attack takes place through S as this is highly positively polarised, i.e. electron deficient:

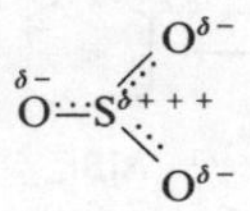

Sulphonation, like iodination, is reversible and is believed to take place in concentrated sulphuric acid *via* the pathway:

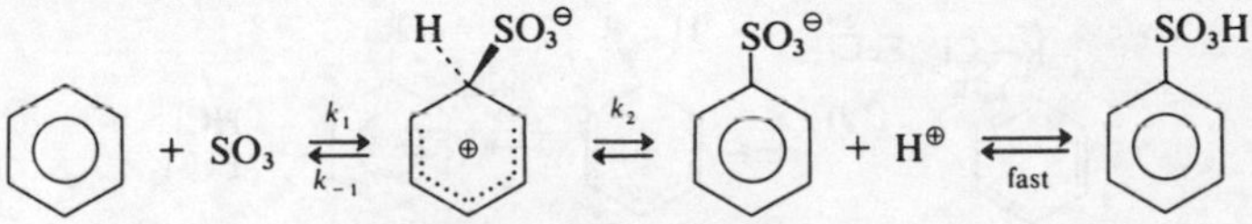

In oleum, the σ complex (16) is believed to undergo protonation of the $SO_3^\ominus$ *before* undergoing C—H fission to yield the SO_3H analogue of (17). Like iodination, sulphonation exhibits a kinetic isotope effect, indicating that C—H bond-breaking is involved in the rate-limiting step of the reaction, i.e. that $k_{-1} \gtrsim k_2$.

Practical use is made of the reversibility of the reaction in order to replace SO_3H by H on treating sulphonic acids with steam. It may thus be possible to introduce an SO_3H group for its directive influence (*cf.* p. 149), and then eliminate it subsequently. The sulphonation of naphthalene presents some interesting features (p. 162).

6.5 FRIEDEL CRAFTS REACTIONS

This can be conveniently divided into alkylation and acylation.

6.5.1 Alkylation

The carbon atom of alkyl halides, $\overset{\delta+}{R}-\overset{\delta-}{Hal}$, is electrophilic, but rarely is it sufficiently so to effect the substitution of aromatic species: the presence of a Lewis acid catalyst, e.g. $AlHal_3$ is also required. That alkyl halides do react with Lewis acids has been demonstrated by the exchange of radioactive bromine into EtBr from $AlBr_3^*$ on mixing and re-isolation; also the actual isolation of solid 1:1 complexes, e.g. $CH_3Br \cdot AlBr_3$, at low temperatures ($-78°$). These complexes, though polar, are only faintly conducting. Where R is capable of forming a particularly stable carbonium ion, e.g. Me_3C-Br, it is probable that the attacking electrophile in alkylation is the actual carbonium ion, $Me_3C^\oplus$:

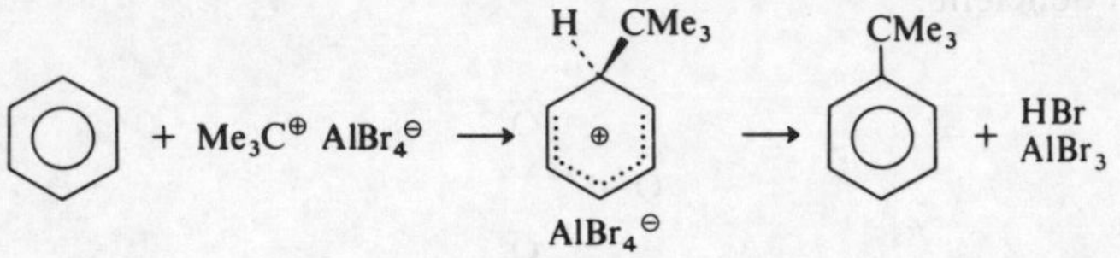

(18)

In other cases it seems more likely that the attacking electrophile is a polarised complex (19), the degree of polarisation in a particular case depending on R in R—Hal and the Lewis acid employed:

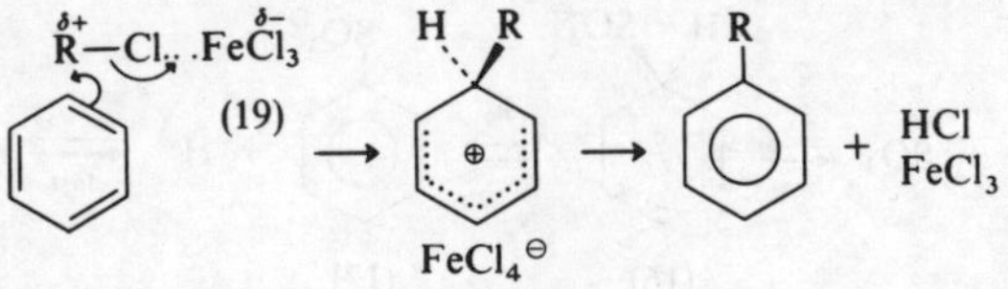

(19)   (20)

Either pathway is, of course, compatible with the commonly observed rate law:

$$Rate = k[ArH][RX][MX_3]$$

The order of effectiveness of Lewis acid catalysts has been shown to be:

$$AlCl_3 > FeCl_3 > BF_3 > TiCl_3 > ZnCl_2 > SnCl_4$$

The validity of Wheland intermediates such as (18) and (20) in Friedel–Crafts alkylation has been established by the actual isolation of some of them, c.g. (21), at low temperatures (the stabilising effect of $BF_4^\ominus$ on ion pairs has already been referred to p. 135):

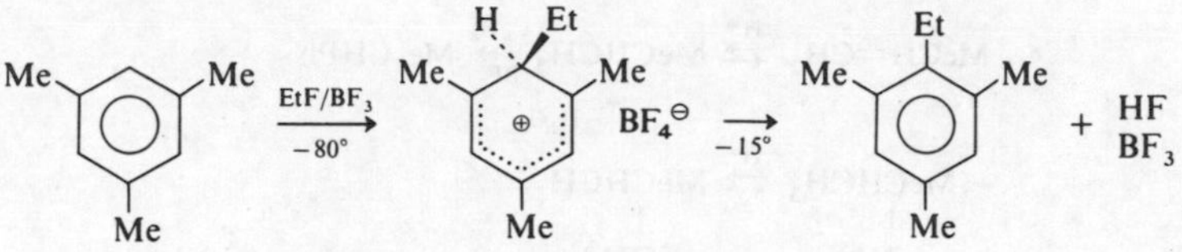

(21)

Thus (21) is an orange, crystalline solid that melts with decomposition at $-15°$ to yield the expected alkylated product in essentially quantitative yield.

In a number of cases of Friedel–Crafts alkylation the final product is found to contain a rearranged alkyl group. Thus the action of $Me_3CCH_2Cl/AlCl_3$ on benzene is found to yield almost wholly the rearranged product, $PhCMe_2CH_2Me$, which would be explainable on the basis of the initial electrophilic complex being polarised enough to allow the rearrangement of $[Me_3CCH_2]^{\delta+} \cdots Cl \cdots AlCl_3^{\delta-}$ to the more stable $[Me_2CCH_2Me]^{\delta+} \cdots Cl \cdots AlCl_3^{\delta-}$ (*cf.* relative stability of the corresponding carbonium ions, p. 103). By contrast $Me_3CCH_2Cl/FeCl_3$ on benzene is found to yield almost wholly the unrearranged product, Me_3CCH_2Ph; the presumption being that the complex with the weaker Lewis acid, $FeCl_3$, is not now polarised enough to allow of isomerisation taking place. Temperature is also found to have an effect, the amount of rearranged product from a given halide and Lewis acid being less at lower temperatures.

The actual proportions of products obtained in many cases are not necessarily found to reflect the relative stabilities of the incipient carbonium ions, unrearranged and rearranged, however. This follows from the fact that their relative rates of reaction with the aromatic species almost certainly does not follow the order of their relative stabilities, and may well be diametrically opposed to it. Attack on the aromatic species by the first formed polarised complex may be faster than its rearrangement. The study of these rearrangements is also complicated by the fact that Lewis acids are found to be capable of rearranging both the original halides, and the final, alkylated end products, e.g.:

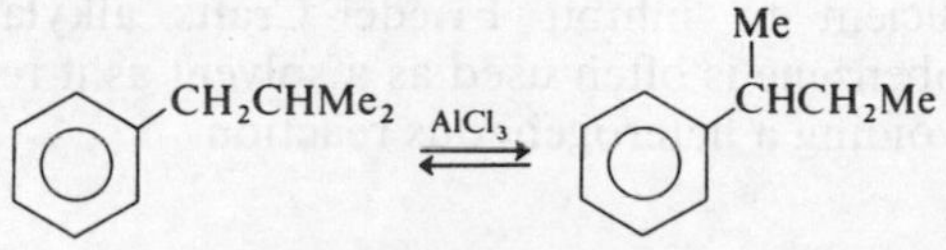

Alkenes and alcohols can also be used in place of alkyl halides for alkylating aromatic species. The presence of a proton acid is required to protonate the alkene or alcohol; BF_3 is then often used as the Lewis acid catalyst:

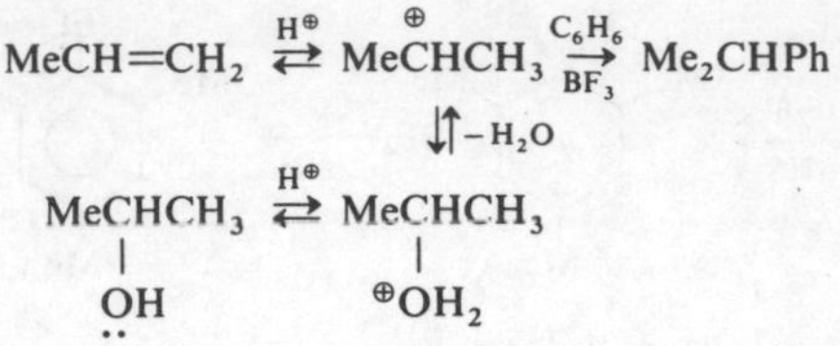

$$MeCH=CH_2 \overset{H^\oplus}{\rightleftarrows} \overset{\oplus}{MeCHCH_3} \overset{C_6H_6}{\underset{BF_3}{\rightarrow}} Me_2CHPh$$

$$\updownarrow -H_2O$$

$$\underset{\underset{\overset{|}{OH}}{\overset{|}{\cdots}}}{MeCHCH_3} \overset{H^\oplus}{\rightleftarrows} \underset{\overset{|}{\oplus OH_2}}{MeCHCH_3}$$

Lewis acid catalysts can also effect dealkylation, i.e. the reaction is reversible. Thus ethylbenzene (22) with BF_3 and HF, is found to disproportionate:

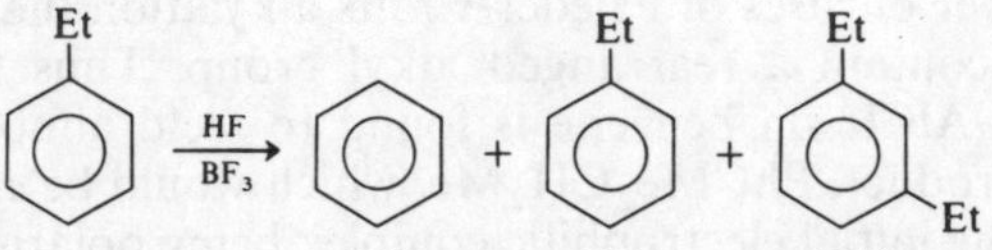

(22)   45%   10%   45%

This reaction must of course be intermolecular, but rearrangements involving change in the relative positions of substituents in the benzene ring are also known, and these are found to be intramolecular. Thus heating *p*-dimethylbenzene (*p*-xylene, 23) with $AlCl_3$ and HCl results in the conversion of the majority of it into the more stable (*cf.* p. 161) *m*-dimethylbenzene (*m*-xylene, 24). The presence of HCl is essential, and the change is believed to involve migration of an Me group in the initially protonated species (25):

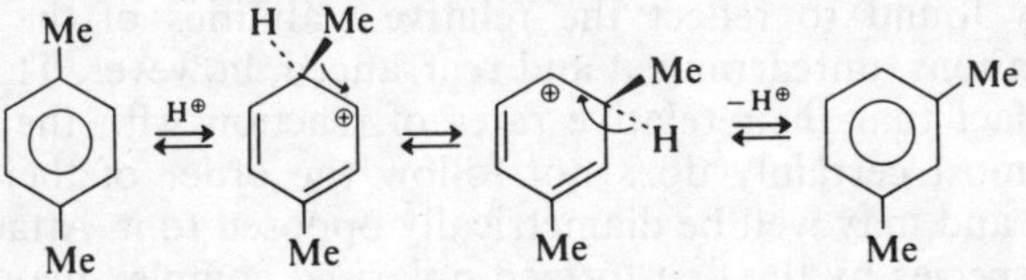

(23)   (25)   (24)

Apart from the possibility of rearrangement, the main drawback in the preparative use of this Friedel–Crafts reaction is polyalkylation (*cf.* p. 152). The presence of an electron-withdrawing substituent is generally sufficient to inhibit Friedel–Crafts alkylation and, for example, nitrobenzene is often used as a solvent as it readily dissolves $AlCl_3$, thus avoiding a heterogeneous reaction.

6.5.2 Acylation

Friedel–Crafts acylation, in cases where the kinetics can readily be monitored, is often found to follow the same general rate law as alkylation:

$$\text{Rate} = k[\text{ArH}][\text{RCOCl}][\text{AlCl}_3]$$

There is also the similar general dilemma of whether the effective electrophile is the acylium ion (26) as a constituent of an ion pair or a polarised complex (27):

$$\overset{\oplus}{RC}{=}O \quad AlCl_4{}^{\ominus}$$
$$(26)$$

$$\overset{\delta+}{RC}{\cdots}O{\cdots}AlCl_3{}^{\delta-}$$
$$\underset{Cl}{|} \quad (27)$$

Acylium ions have been detected in a number of solid complexes, in the liquid complex between MeCOCl and $AlCl_3$ (by i.r. spectroscopy), in solution in polar solvents, and in a number of cases where R is very bulky. In less polar solvents, and under a number of other circumstances, acylium ions are not detectable, however, and it must be the polarised complex that acts as the electrophile.

The direct chemical evidence clearly indicates that either (26) or (27) can be involved depending on the circumstances. Thus in the benzoylation of toluene, the same mixture of products (1 % *m*-, 9 % *o*- and 90 % *p*-) is obtained no matter what the Lewis acid catalyst is, and with either benzoyl chloride or benzoyl bromide, though the reaction rates do of course differ: this suggests a common attacking species in all cases, i.e. $Ph{-}\overset{\oplus}{C}{=}O$. On the other hand, in many cases the proportion of *o*-product is very small compared with other electrophilic substitutions, e.g. nitration, suggesting a very bulky electrophile: a role better filled by the complex (27) than by the linear $R{-}\overset{\oplus}{C}{=}O$ (26). The nature of the electrophile in any given case clearly depends very much on the conditions.

The reaction may thus be represented:

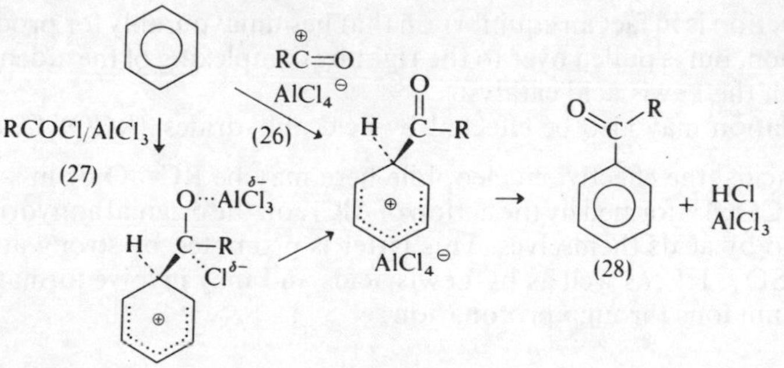

One significant difference of acylation from alkylation is that in the former rather more than one mole of Lewis acid is required, compared with the catalytic quantity only that is required in the latter. This is because the Lewis acid complexes (29) with the product ketone (28) as it is formed,

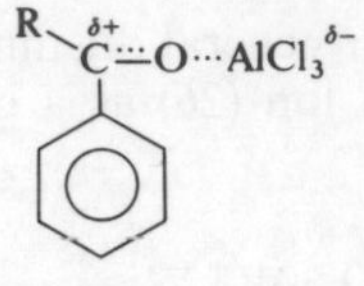

$$R-\overset{\delta+}{C}\!\!\!=\!\!O\cdots AlCl_3^{\ \delta-}$$

(29)

and is thereby removed from further participation in the reaction. Polyacylation does not take place (*cf.* alkylation, p. 142) as the product ketone is much less reactive than the original hydrocarbon. Rearrangement of R does not take place, as in alkylation, but decarbonylation can take place, especially where R would form a stable carbonium ion, so that the end result is then alkylation rather than the expected acylation:

$$Me_3C-\overset{\oplus}{C}\!\!=\!\!O \longrightarrow CO + Me_3C^{\oplus} \xrightarrow{C_6H_6} PhCMe_3$$

Formylation may be carried out by use of CO, HCl, and $AlCl_3$ (the Gattermann–Koch reaction); it is doubtful whether HCOCl is ever formed, the most likely electrophile being the acylium ion, $\overset{\oplus}{HC}\!\!=\!\!O$, i.e. protonated CO:

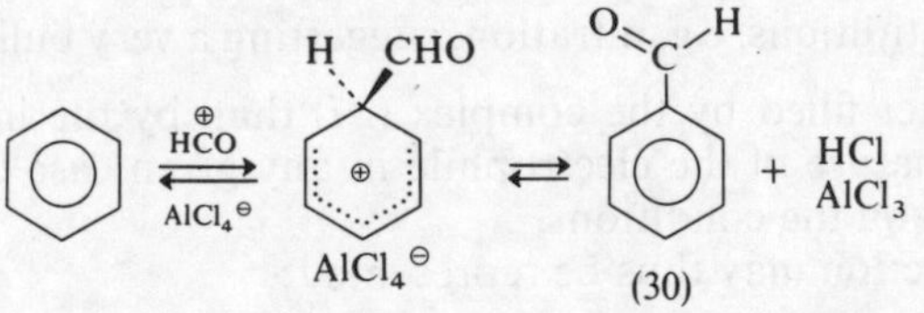

(30)

The reaction is in fact an equilibrium that lies unfavourably for product formation, but is pulled over to the right by complexing of the aldehyde (30) with the Lewis acid catalyst.

Acylation may also be effected by acid anhydrides, $(RCO)_2O$, and Lewis acids (the effective nucleophile here may be $\overset{\oplus}{RC}\!\!=\!\!O$ or in some cases RCOCl is formed by the action of $AlCl_3$ on the original anhydride), and also by acids themselves. This latter is promoted by strong acids, e.g. H_2SO_4, HF, as well as by Lewis acids and may involve formation of acylium ions through protonation:

$$RC\overset{\displaystyle O}{\underset{\displaystyle OH}{<}} + H_2SO_4 \rightleftarrows RC\overset{\displaystyle O}{\underset{\displaystyle \overset{\oplus}{O}H_2}{<}} \rightleftarrows \overset{\oplus}{RC}\!\!=\!\!O + H_2O$$

This latter acylation is used particularly in ring-closures:

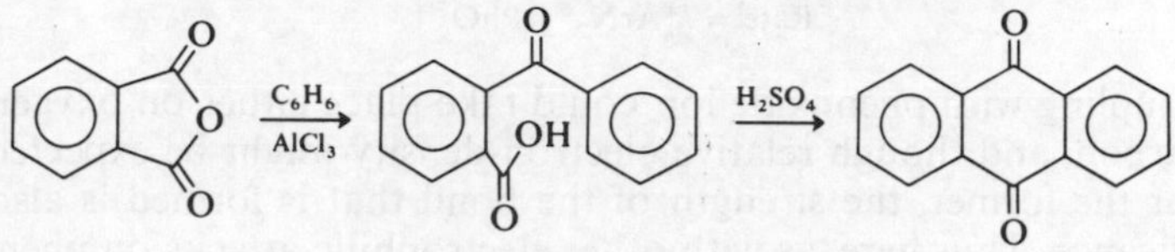

Because polyacylation does not occur (*cf.* p. 144), it is often preferable to prepare alkyl-benzenes by acylation, followed by Clemmensen or other reduction, rather than by direct alkylation:

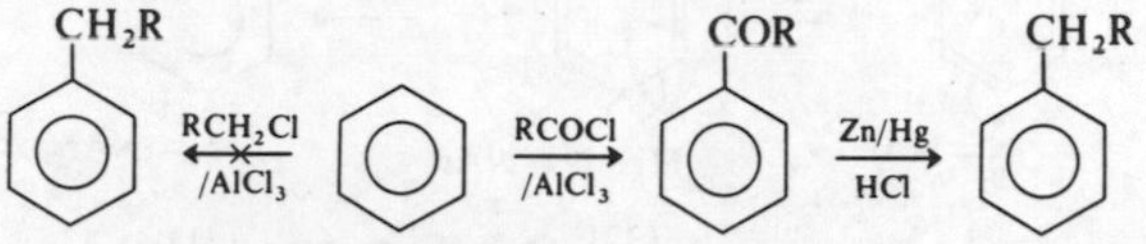

6.6   DIAZO COUPLING

Another classical electrophilic aromatic substitution reaction is diazo coupling, in which the effective electrophile has been shown to be the diazonium cation (*cf.* p. 119):

$$PhN\overset{\oplus}{\equiv}\underset{\cdot\cdot}{N} \leftrightarrow PhN\underset{\cdot\cdot}{=}\overset{\oplus}{N}$$

This is, however, a weak electrophile compared with species such as $^{\oplus}NO_2$ and will normally only attack highly reactive aromatic compounds such as phenols and amines; it is thus without effect on the otherwise highly reactive PhOMe. Introduction of electron-withdrawing groups into the *o*- or *p*-positions of the diazonium cation enhance its electrophilic character, however, by increasing the positive charge on the diazo group:

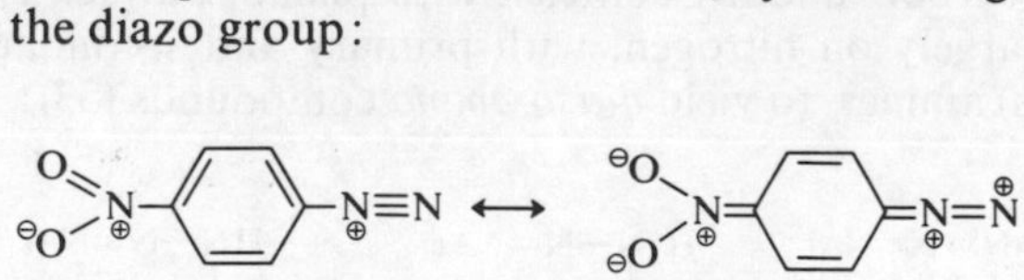

Thus the 2,4-dinitrophenyldiazonium cation will couple with PhOMe and the 2,4,6-compound with even the hydrocarbon 2,4,6-trimethylbenzene (mesitylene). Diazonium cations exist in acid and slightly alkaline solution (in more strongly alkaline solution they are converted into diazohydroxides, PhN=N—OH and further into diazotate anions, PhN=N—O$^{\ominus}$) and coupling reactions are therefore carried out under these conditions, the optimum pH depending on the species being attacked. With phenols this is at a slightly alkaline pH as it is PhO$^\ominus$, and not PhOH, that undergoes attack by ArN$_2$$^\oplus$:

$$\text{Rate} = k[\text{ArN}_2{}^\oplus][\text{PhO}^\ominus]$$

Coupling with phenoxide ion could take place either on oxygen or on carbon, and though relative electron-density might be expected to favour the former, the strength of the bond that is formed is also of significance. Thus here, as with other electrophilic attacks on phenols, it is found to be the C-substituted product (31) that is formed:

Removal of the proton (usually non rate-limiting) from (32) is assisted by one or other of the basic species present in solution. Coupling normally takes place largely in the *p*-, rather than the *o*-, position (*cf.* p. 152)—provided this is available—because of the considerable bulk of the attacking electrophile, ArN$_2$$^\oplus$ (*cf.* p. 158).

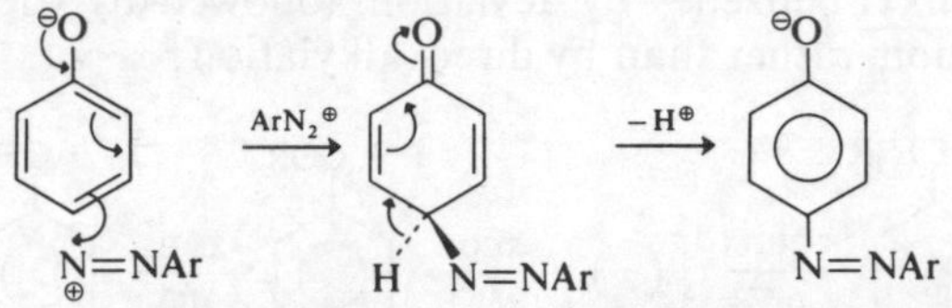

Aromatic amines are in general somewhat less readily attacked than phenols and coupling is often carried out in slightly acid solution, thus ensuring a high [PhN$_2$$^\oplus$] without markedly converting the amine, Ar$\overset{..}{\text{N}}$H$_2$, into the unreactive, protonated cation, Ar$\overset{\oplus}{\text{N}}H_3$—such aromatic amines are very weak bases (*cf.* p. 68). The initial diazotisation of aromatic primary amines is carried out in strongly acid media to ensure that as yet unreacted amine *is* converted to the cation and so prevented from coupling with the diazonium salt as it is formed.

With aromatic amines there is the possibility of attack on either nitrogen or carbon, and, by contrast with phenols, attack is found to take place largely on nitrogen, with primary and secondary (i.e. N-alkylanilines) amines, to yield *diazo-amino* compounds (33):

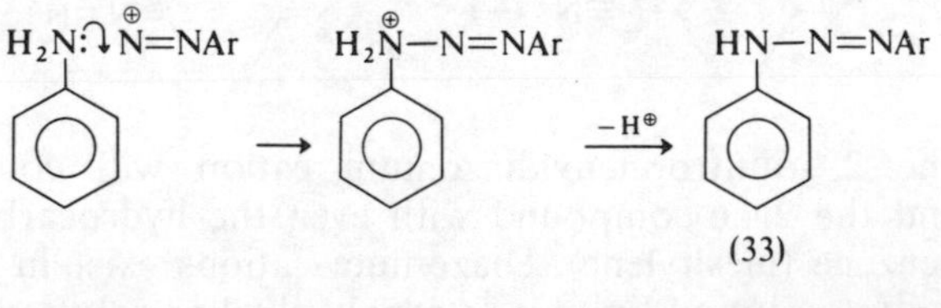

With most primary amines this is virtually the sole product, but with secondary amines (i.e. N-alkylanilines) some coupling may also take place on a carbon atom of the nucleus, while with tertiary amines (i.e. N,N-dialkylanilines) only the product coupled on carbon (34) is obtained:

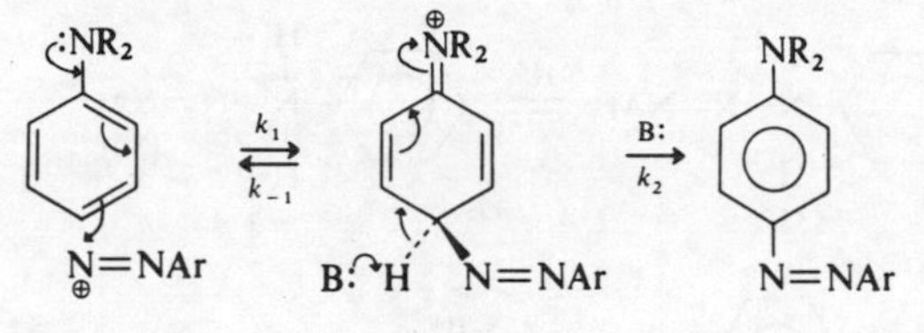

(35)   (34)

The reaction is usually found to follow the general rate law:

$$Rate = k[ArN_2^{\oplus}][PhNR_2]$$

In some cases the coupling reaction is found to be base-catalysed, and this is found to be accompanied by a kinetic isotope effect, i.e. $k_{-1} \gtrsim k_2$, and the breaking of the C—H bond in (35) is now involved in the rate-limiting step of the reaction.

An interesting example of an internal coupling reaction is provided by the diazotisation of *o*-diaminobenzene (36):

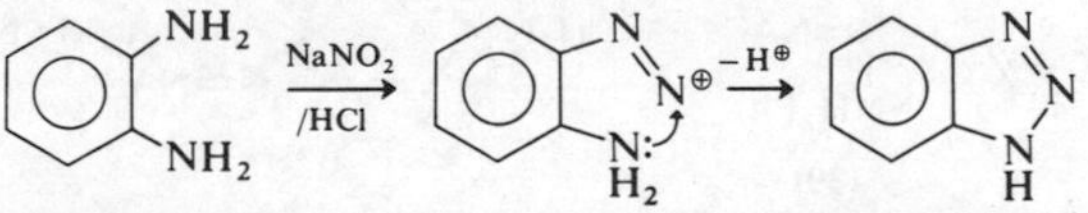

(36)   (37)

Benzotriazole (37) may be obtained preparatively (75% yield) in this way.

The difference in position of attack on primary and secondary aromatic amines, compared with phenols, probably reflects the relative electron-density of the various positions in the former compounds exerting the controlling influence for, in contrast to a number of other aromatic electrophilic substitution reactions, diazo coupling is sensitive to relatively small differences in electron density (reflecting the rather low ability as an electrophile of $PhN_2^{\oplus}$). Similar differences in electron-density do of course occur in phenols but here control over the position of attack is exerted more by the relative strengths of the bonds formed in the two products: in the two alternative coupled products derivable from amines, this latter difference is much less marked.

The formation of diazoamino compounds, on coupling $ArN_2^{\oplus}$ with primary amines, does not constitute a total preparative bar to obtaining products coupled on the benzene nucleus for diazoamino compounds (33) may be rearranged to the corresponding *amino-azo* compounds (38) by warming in acid:

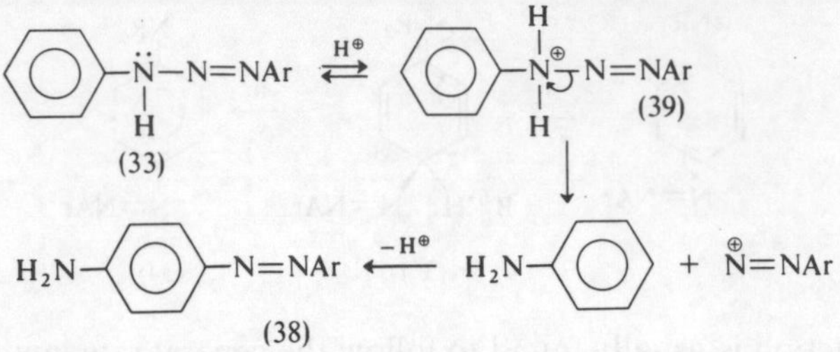

The rearrangement has been shown under these conditions to be an *inter*molecular process, i.e. the diazonium cation becomes free, for the latter may be transferred to phenols, aromatic amines or other suitable species added to the solution. It is indeed found that the rearrangement proceeds most readily with an acid catalyst *plus* an excess of the amine that initially underwent coupling to yield the diazoamino compound (33). It may then be that this amine attacks the protonated diazoamino compound (39) directly with expulsion of PhNH$_2$ and loss of a proton:

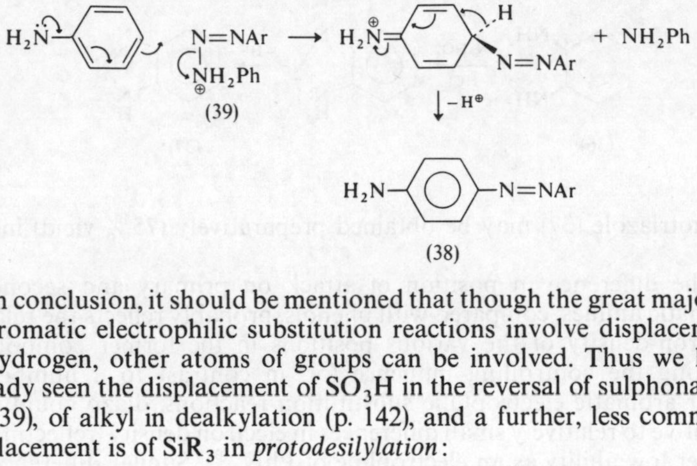

In conclusion, it should be mentioned that though the great majority of aromatic electrophilic substitution reactions involve displacement of hydrogen, other atoms of groups can be involved. Thus we have already seen the displacement of SO$_3$H in the reversal of sulphonation (p. 139), of alkyl in dealkylation (p. 142), and a further, less common, displacement is of SiR$_3$ in *protodesilylation*:

$$ArSiR_3 + H^\oplus \rightarrow Ar-H + {}^\oplus SiR_3$$

Displacements such as this show all the usual characteristics of electrophilic aromatic substitution (substituent effects, etc., see below), but they are normally of much less preparative significance than the examples we have already considered. In face of all the foregoing discussion of polar intermediates it is pertinent to point out that homolytic aromatic substitution reactions, i.e. by radicals, are also known (p. 321); as too is attack by nucleophiles (p. 164).

6.7 ELECTROPHILIC ATTACK ON C_6H_5Y

When a mono-substituted benzene derivative, C_6H_5Y, undergoes further electrophilic substitution, e.g. nitration, the incoming substituent may be incorporated at the *o*-, *m*- or the *p*-position, and the overall rate at which substitution takes place may be faster or slower than with benzene itself. What is found in practice is that substitution occurs so as to yield <u>either</u> predominantly the *m*-isomer, <u>or</u> predominantly a mixture of *o*- and *p*-isomers; in the former case the overall rate of attack is <u>always</u> slower than on benzene itself, in the

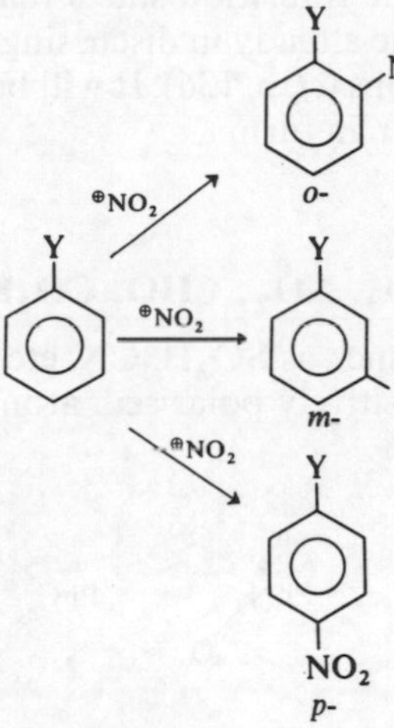

latter case the overall rate of attack is <u>usually</u> faster than on benzene itself. The major controlling influence is found to be exerted by Y, the substituent already present, and this can be explained in detail on the basis of the electronic effects that Y can exert. It can, of course, also exert a steric effect, but the operation of this factor is confined essentially to attack at the *o*-position; this influence will be discussed separately below (p. 158).

Substituents, Y, are thus classed as being *m*-, <u>or</u> *o-/p-directing*; if they induce faster overall attack than on benzene itself they are said to be *activating*, if slower, then *deactivating*. It should be emphasised that these directing effects are relative rather than absolute: some of all three isomers are nearly always formed in a substitution reaction, though the proportion of *m*-product with an *o-/p*-directing Y or of *o-/p*-products with a *m*-directing Y may well be very small. Thus nitration of nitrobenzene (Y = NO_2) is found to result in a mixture of 93% *m*-, 6% *o*- and 1% *p*-isomers, i.e. NO_2 is classed as a *m*-directing (deactivating) substituent. By contrast nitration of methoxybenzene (anisole, Y = OMe) yields 56% *p*-, 43% *o*- and 1% *m*-isomers, i.e. OMe is an *o-/p*-directing (activating) substituent.

6.7.1 Electronic effects of Y

What we shall be doing in the discussion that follows is comparing the effect that a particular Y would be expected to have on the rate of attack on positions *o-*/*p-* and *m-*, respectively, to the substituent Y. This assumes that the proportions of isomers formed is determined entirely by their relative rates of formation, i.e. that the control is wholly *kinetic* (*cf.* p. 160). Strictly we should seek to compare the effect of Y on the different transition states for *o-*, *m-* and *p*-attack, but this is not usually possible. Instead we shall use Wheland intermediates as models for the transition states that immediately precede them, just as we have done already in discussing the individual electrophilic substitution reactions (*cf.* p. 136). It will be convenient to discuss several different types of Y in turn.

6.7.1.1 Y = $^{\oplus}NR_3$, CCl_3, NO_2, CHO, CO_2H, etc.

These groups, and other such as SO_3H, CN, etc., all have in common a positively charged, or positively polarised, atom adjacent to a carbon atom of the benzene ring:

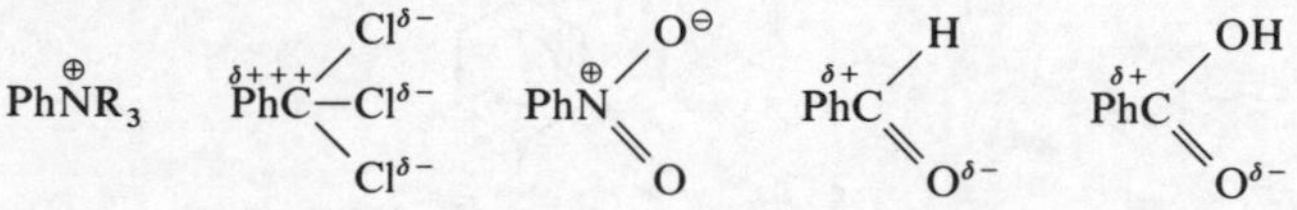

They are thus all electron-withdrawing with respect to the benzene ring, i.e. aromatic species containing them all have a dipole with the positive end located on the benzene nucleus. Taking Y = $^{\oplus}NR_3$ as exemplar of the rest, we can write the σ complexes for attack by an electrophile, $E^{\oplus}$ (e.g. $^{\oplus}NO_2$), *o-*, *m-* and *p-* to the original $^{\oplus}NR_3$ substituent:

o-attack:

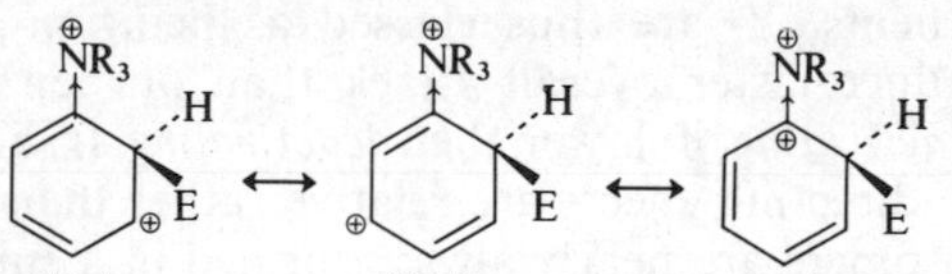

(40*a*) (40*b*) (40*c*)

m-attack:

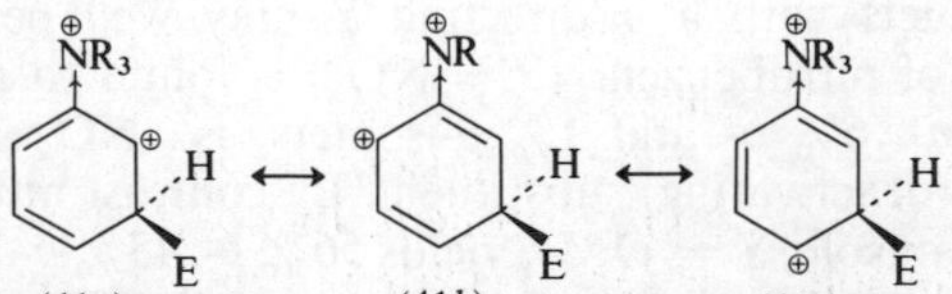

(41*a*) (41*b*) (41*c*)

p-attack:

$$\overset{\oplus}{NR_3} \quad\longleftrightarrow\quad \overset{\oplus}{NR_3} \quad\longleftrightarrow\quad \overset{\oplus}{NR_3}$$

| H   E | H   E | H   E |
| (42a) | (42b) | (42c) |

The $\overset{\oplus}{N}R_3$ substituent will exert an electron-withdrawing, i.e. destabilising, inductive (polar) effect on all three positively charged σ complexes (40, 41 and 42). There can be no comparable effect on the σ complex for attack on benzene itself, so that attack on *any* position (o-, m- or p-) in $C_6H_5\overset{\oplus}{N}R_3$ will be slower than attack on benzene ($k_{C_6H_5Y}/k_{C_6H_6} = 1.6 \times 10^{-5}$ for bromination when R = Me). The $\overset{\oplus}{N}R_3$ group will, however, exert a specific destabilising effect on one of the canonical states (40c) of the σ complex for o-attack, and on one (42b) of the σ complex for p-attack, for in each of these $\oplus$ charges are located on adjacent atoms. The $\oplus$ charge will thus be delocalised less well in (40) and (42) than in (41), in which there is no such disability. The transition state for which (41) is taken as a model will thus be at a lower energy level than those corresponding to (40) and (42); its free energy of activation (ΔG^+) will be lower and it will therefore be formed more rapidly: the m-isomer will thus predominate in the reaction product.

Where the positive charge on the atom adjacent to the nucleus is real rather than formal, i.e. $\overset{\oplus}{N}R_3$ rather than NO_2, there is evidence that its effect on σ complex stability is exerted through a *field effect* operating through space, in addition to any polar (inductive) effect operating through the bonds. The deactivating effect of Y on the nucleus declines, i.e. the overall rate of substitution increases, in the approximate order:

$$\overset{\oplus}{N}R_3 < NO_2 < CN < SO_3H < C{=}O < CO_2H$$

The order is approximate only as it is found to vary slightly from one substitution process to another, depending to some extent on the nature of the attacking electrophile. Thus, hardly surprisingly, substituents such as $\overset{\oplus}{N}R_3$ will be particularly deactivating in substitution reactions where the attacking electrophile is itself positively charged, e.g. $\overset{\oplus}{N}O_2$ ($k_{C_6H_5Y}/k_{C_6H_6} = 1.5 \times 10^{-8}$ for nitration when R = Me).

6.7.1.2   Y = Alkyl, phenyl

Alkyl groups are electron-donating compared with hydrogen, and those canonical states for o- and p-attack, respectively, in which a positive charge is located on the adjacent nuclear carbon atom—(43c) and (44b)—will thus be selectively stabilised;

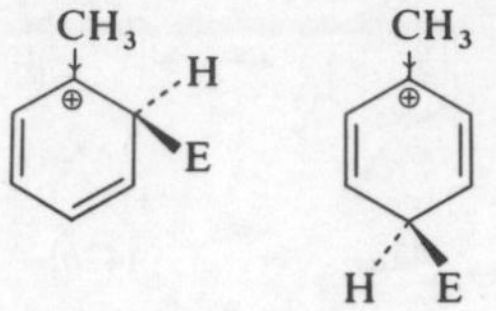

(43c)   (44b)

in contrast to (40c) and (42b) above which were selectively destabilised. No such factor operates in the σ-complex for *m*-attack, *cf.* ($\overline{41a} \rightarrow 41c$), and *o-/p-* substitution is thus promoted at the expense of *m*-. Because of the overall electron-donating inductive (polar) effect, attack on any position will be faster than in benzene itself ($k_{C_6H_5Me}/k_{C_6H_6} = 3.4 \times 10^2$ for chlorination).

A further point of interest is that though CMe_3 is known to exert a more powerful electron-donating inductive effect than CH_3, nevertheless toluene is found to undergo more rapid *o-/p-* attack than does 2,2-dimethylethylbenzene. This is because electron-donation from CH_3 can occur through hyperconjugation (*cf.* p. 25), e.g. ($45b \leftrightarrow 45d$) as well as through an inductive effect, e.g. (44b), whereas with CMe_3 only an inductive effect can operate (46b):

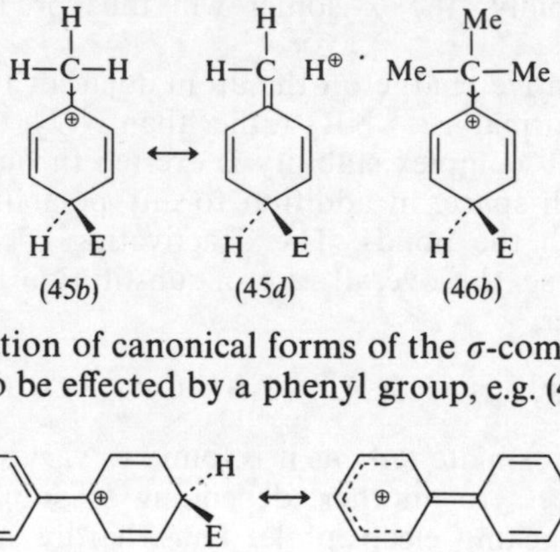

(45b)   (45d)   (46b)

Specific stabilisation of canonical forms of the σ-complexes for *o-* and *p*-attack can also be effected by a phenyl group, e.g. ($47b \leftrightarrow 47d$),

(47b)   (47d)

and the overall rate of attack on biphenyl is found to be faster than on benzene itself ($k_{C_6H_5Y}/k_{C_6H_6} = 4.2 \times 10^2$ for chlorination).

6.7.1.3   Y = OCOR, NHCOR, OR, OH, NH_2, NR_2

These groups all have in common an atom adjacent to the nucleus that can exert an electron-withdrawing inductive (polar) effect (*cf.* N in, e.g., NO₂), but they also possess an electron pair (e.g. ÖMe) that can effect the specific stabilisation of the σ complexes for *o*- and *p*-attack, (48*c* ↔ 48*d*) and (49*b* ↔ 49*d*), respectively:

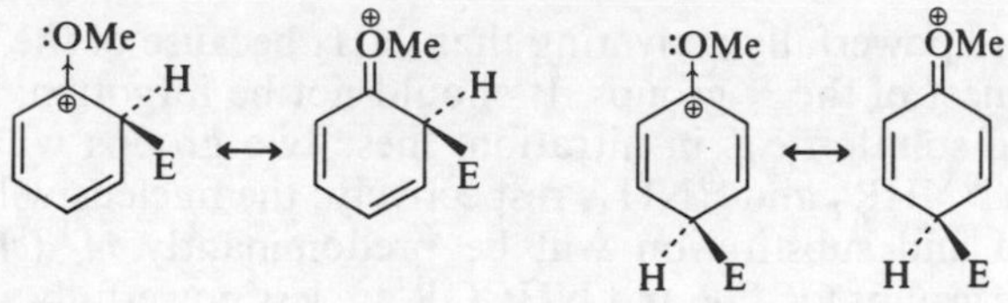

The stabilisation is particularly marked in that not only is an extra (fourth) canonical state involved in the stabilisation of the *o*- and *p*-σ complexes, but these forms (48*d* and 49*d*, respectively), in which the positive charge is located on oxygen, are inherently more stable than their other three complementary forms, in which the positive charge is located on carbon (*cf.* 40*a* → 40*c*, and 42*a* → 42*c*). This effect is sufficiently pronounced to outweigh by far the electron-withdrawing inductive (polar) effect also operating in these two σ complexes, substitution is thus almost completely *o*-/*p*- (≪1% of the *m*-isomer is obtained in the nitration of PhOMe), and much more rapid than on benzene itself ($k_{C_6H_5OMe}/k_{C_6H_6} = 9.7 \times 10^6$ for chlorination).

The operation of the electron-withdrawing inductive effect can, however, be seen in the fact that the very small amount of *m*-attack (for which there is no specific stabilisation of the σ complex by delocalisation) occurs more slowly than attack on benzene itself (*cf.* p. 157). In the case of the phenoxide ion (50),

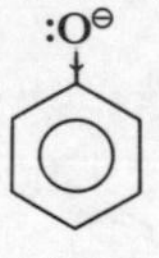

(50)

the inductive effect will be reversed in direction because of the negative charge now carried by the oxygen atom, thereby making it even more rapidly attacked than phenol. Even the *m*-position will now be attacked more readily than benzene itself (little or no *m*-product is formed, however). Many electrophilic substitution reactions take place under acid conditions so that (50) cannot be involved, but an exception is diazo-coupling (p. 145) on phenols which is carried out in slightly basic solution.

The activating effect of Y on the nucleus is found to increase, i.e. the overall rate of substitution increases, in the approximate order:

$$OCOR < NHCOR < OR < OH < NH_2 < NR_2 < O^\ominus$$

NR_2 is more powerfully activating than NH_2 because of the electron-donating effect of the R groups. It should not be forgotten, however, that in acid solution e.g. in nitration, these two groups will be converted into $^\oplus NHR_2$ and $^\oplus NH_3$, respectively; the nucleus will then be deactivated and substitution will be predominantly *m*- (*cf.* $^\oplus NR_3$, p. 150). The groups OCOR and NHCOR are less powerfully activating than OH and NH_2, respectively, because of the reduction in electron-availability on O and N by delocalisation over the adjacent, electron-withdrawing carbonyl group:

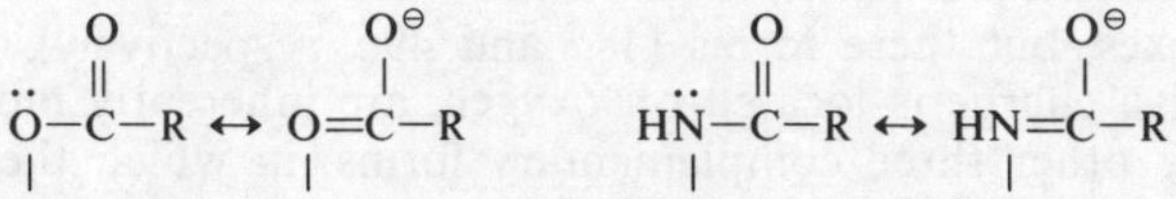

$$
\begin{array}{cccc}
\overset{..}{\underset{|}{O}}-\overset{O}{\overset{\|}{C}}-R & \leftrightarrow & \overset{O^\ominus}{\overset{|}{O}}=\overset{}{\underset{|}{C}}-R & \qquad
HN-\overset{O}{\overset{\|}{C}}-R \leftrightarrow HN=\overset{O^\ominus}{\overset{|}{C}}-R
\end{array}
$$

The NHCOR group is not protonated in acid solution, and nitration of aniline to yield *o*-/*p*-products can thus be carried out by using, for example, COMe as a protecting group.

6.7.1.4   Y = Cl, Br, I

The halobenzenes also have an atom adjacent to the nucleus that carries an electron-pair; thus specific stabilisation of the σ complexes for *o*- and *p*-attack, $(51c \leftrightarrow 51d)$ and $(52b \leftrightarrow 52d)$ respectively, can again take place,

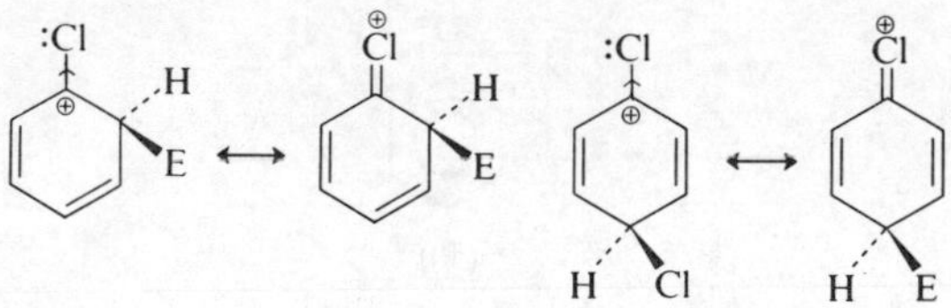

(51c)   (51d)   (52b)   (52d)

i.e. the halogens are *o*-/*p*-directing. The electron-withdrawing inductive effects of the halogens are such that attack is *slower* than on benzene itself, i.e. they are *deactivating* substituents ($k_{C_6H_5Cl}/k_{C_6H_6} = 3 \times 10^{-2}$ for nitration). This nett electron-withdrawal by the halogens is reflected in the *ground state* by a dipole in chlorobenzene (53) with its +ve end on the nucleus, compared with anisole (54) in which the dipole is in the opposite direction:

$$\mu = 1\cdot6\ D \qquad (53) \qquad (54) \qquad \mu = 1\cdot2\ D$$

That the nett effect in the positively charged σ complexes is not quite so straightforward is reflected in the fact that the approximate rate order for electrophilic attack on the halogenobenzenes is:

$$C_6H_5F > C_6H_5Cl \approx C_6H_5Br > C_6H_5I$$

Fluorobenzene is in fact attacked at very nearly the same rate as benzene itself.

One might have expected a rate order essentially the reverse of this on the basis of (*a*) the electron-withdrawing inductive effect order (F > Cl > Br > I), and (*b*) an expected electron-donating mesomeric effect order (F < Cl < Br < I) reflecting the readiness with which the atom concerned will support a positive charge. Clearly the mesomeric effect does not follow the expected order based on relative electronegativity, and this is believed to be due to *p* orbital overlap (with a carbon atom of the nucleus in each case) being greatest when the overlapping orbitals are of comparable size, i.e. F > Cl > Br > I. The *nett* electron-donating mesomeric effect is thus made up of electronegativity and *p*-orbital overlap effects, which vary along the halogen series in opposite directions: the observed rate order (above) thus becomes more readily understandable.

A very similar situation is encountered in the electrophilic *addition* of unsymmetrical adducts (e.g. HBr) to vinyl halides (e.g. $CH_2{=}CHBr$), where the inductive effect of halogen controls the *rate*, but relative mesomeric stabilisation of the carbonium ion intermediate controls the *orientation*, of addition (p. 182).

6.7.2 Partial rate factors and selectivity

More refined kinetic methods, and the ability to determine very precisely the relative proportions of *o*-, *m*- and *p*-isomers formed—by, for example, spectroscopic methods rather than by isolation as in the past—now allow of a much more quantitative approach to aromatic substitution. One very useful concept here is that of *partial rate factors*: the rate at which <u>one</u> position, e.g. the *p*-, in C_6H_5Y is attacked compared with the rate of attack on <u>one</u> position in benzene; it is written as f_{p^-}.

Partial rate factors may be obtained by separate kinetic measurements of the overall rate constants $k_{C_6H_5Y}$ and $k_{C_6H_6}$ under analogous conditions (or by a *competition experiment* in which equimolar quantities of C_6H_5Y and C_6H_6 compete for an inadequate supply of an electrophile, thus giving the ratio $k_{C_6H_5Y}/k_{C_6H_6}$), and analysis of the relative amounts of *o*-, *m*-, and *p*-products obtained from C_6H_5Y— the *isomer distribution* (generally quoted as percentages of the total substitution product obtained). Then, remembering that there are **6** positions available for attack in C_6H_6 compared with **2***o*- **2***m*- and **1***p*-positions in C_6H_5Y, we have:

$$f_{o-} = \frac{k_{o-}}{k_H} = \frac{k_{C_6H_5Y/2}}{k_{C_6H_6/6}} \times \frac{\% \, o\text{-isomer}}{100} \qquad \text{(2 } o\text{-positions } v. \text{ 6 H positions)}$$

$$f_{m-} = \frac{k_{m-}}{k_H} = \frac{k_{C_6H_5Y/2}}{k_{C_6H_6/6}} \times \frac{\% \, m\text{-isomer}}{100} \qquad \text{(2 } m\text{-positions } v. \text{ 6 H positions)}$$

$$f_{p-} = \frac{k_{p-}}{k_H} = \frac{k_{C_6H_5Y/1}}{k_{C_6H_6/6}} \times \frac{\% \, p\text{-isomer}}{100} \qquad \text{(1 } p\text{-position } v. \text{ 6 H positions)}$$

Thus for the nitration of toluene by nitric acid in acetic anhydride at 0° $k_{C_6H_5Me}/k_{C_6H_6}$ was found to be 27, and the isomer distribution (%): *o*-, 61·5; *m*-, 1·5; *p*-, 37·0; the partial rate factors for nitration, under these conditions, are thus:

| Nitration | Chlorination | Bromination |

Comparison of the partial rate factors for nitration of toluene with those for chlorination and bromination (above) show that these differ, both absolutely and relatively, with the attacking electrophile: in other words relative directive effects in C_6H_5Y *do* depend on $E^\oplus$ as well as on Y. We notice above that the absolute values of the partial rate factors, i.e. k_Y/k_H, increase in the order,

Nitration < Chlorination < Bromination

i.e. as the reactivity of the attacking electrophile <u>decreases</u>. This apparent paradox is seen on reflection to be reasonable enough: if $E^\oplus$ was reactive enough every collision would lead to substitution, the attacking reagent would thus be quite undiscriminating, and each partial rate factor would be unity. As the reactivity of $E^\oplus$ decreases, however, every collision will no longer lead to reaction, which will increasingly depend on the relative ability of *o*-, *m*- and *p*-positions in C_6H_5Y, and positions in C_6H_6, to supply an electron pair to bond with $E^\oplus$. The reagent will thus become increasingly more discriminating—

its *selectivity* will rise—the absolute values of the partial rate factors will increase, as will the relative difference between these values: exactly what is seen in the figures quoted above. This relative selectivity is best considered by comparing f_{p-} and f_{m-}, only, as f_{o-}, will be influenced by steric effects (size of Y, and relative size of attacking reagent, *cf.* p. 158) in addition to the electronic effects that influence all three.

The use of partial rate factors allows of a more precise investigation of directive effects than has been possible to date. Thus all the partial rate factors for toluene above are > 1, indicating that the CH_3 group (p. 152) activates all positions in the nucleus compared with benzene. The same is true for Y = CMe_3 but here f_{m-} for nitration is 3·0, compared with 1·3 for toluene, indicating that CMe_3 exerts a larger electron-donating inductive (polar) effect than does CH_3. By contrast, when Y = C_6H_5 in biphenyl (p. 152), f_{m-} for chlorination is found to be 0·7, i.e. attack on this position is slower than on benzene (although $k_{C_6H_5Y}/k_{C_6H_6} = 4·2 \times 10^2$), because the sp^2 carbon atom by which the C_6H_5 substituent is attached to the benzene ring exerts an electron-withdrawing inductive (polar) effect (55):

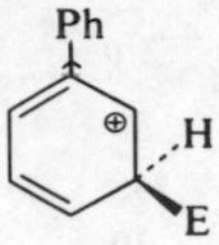

(55)

A similar effect is also seen with *o-/p*-directing, activating substituents when a reaction can be investigated that produces enough *m*-product to measure, e.g. deuteration (deuterium exchange) with the strong acid, CF_3CO_2D, on C_6H_5OPh(56):

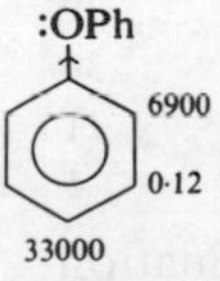

(56)

The enormous f_{o-} and f_{p-} values reflect the ability of the electron pair on O to stabilise, selectively, the transition states for *o*- and *p*-attack (*cf.* p. 153), while the f_{m-} value of <1 reflects the destabilisation (compared with attack on benzene) of the transition state for *m*-attack by the electron-withdrawing inductive (polar) effect of the oxygen atom.

Partial rate factors, and hence the isomer distribution in a particular substitution reaction, are also affected by temperature. Increasing temperature has the greatest relative effect on the substitution reaction of highest $\Delta G^{\ddagger}$ (out of the three possible, alternative attacks on C_6H_5Y), i.e. on the slowest. The effect of a rise in temperature is thus, like the effect of an increase in the reactivity of $E^\oplus$, to 'iron out' differences between partial rate factors, and to make the isomer distribution in the product move a little more towards the statistical.

6.7.3   Steric effects and *o*-/*p*-ratios

After what has gone before, it will come as no surprise to find that the ratio of *o*-:*p*-product is seldom, if ever, the statistical one of 2:1. Apart from any difference in the electronic effects that may operate in the transition states (or in their Wheland intermediate models) for *o*- and for *p*-attack, the former will in addition be susceptible to steric effects, that can be of little or no significance in the latter. It is found in general that the proportion of *o*-isomer decreases as the bulk of Y, or the bulk of the attacking electrophile, increases.

Thus in the nitration of alkylbenzenes ($Y = CH_3 \rightarrow CMe_3$) under comparable conditions we find:

Y	%*o*-	%*p*-	*o*-/*p*-ratio
CH_3	58	37	1·57
CH_2Me	45	49	0·92
$CHMe_2$	30	62	0·48
CMe_3	16	73	0·22

Similarly for several different electrophilic attacks on chlorobenzene we find:

Reaction	%*o*-	%*p*-	*o*-/*p*-ratio
Chlorination	39	55	0·71
Nitration	30	70	0·43
Bromination	11	87	0·14
Sulphonation	1	99	0·01

The latter sequence, chlorination $\rightarrow$ sulphonation, corresponds to increasing bulk in the actual attacking electrophilic species. These results are explainable on the basis of increasing crowding in the transition state for *o*-attack, this representing a higher energy level (larger ΔG^+) and resulting in a correspondingly slower rate of formation of the *o*-isomer.

That a steric factor may not be the only one in operation can be seen in the nitration of the halogenobenzenes:

Y	%*o*-	%*p*-	*o*-/*p*-ratio
F	12	88	0·14
Cl	30	69	0·44
Br	37	62	0·60
I	38	60	0·63

Despite the increase in the size of the substituent Y on going from F → I, the proportion of *o*-isomer—and the *o*-/*p*-ratio—is actually found to increase. There is no reason to suppose that an increasing steric effect, like that with the alkylbenzenes above, is not operating and its effect must in this case be outweighed by the electron-withdrawing inductive (polar) effect of Y. This will fall off with distance, being exerted less strongly on the distant *p*-position than on the adjacent *o*-position; this effect will be particularly strong *o*- to the very highly electronegative F, and relatively little *o*-attack will thus take place despite its small size. The electron-withdrawing inductive effect decreases considerably on going from F → I (the biggest change being on going from F → Cl), resulting in increasing attack at the *o*-position despite the increasing bulk of Y.

Even where the reaction is of such low steric demand (e.g. deuterium exchange) as to virtually exclude steric effects in attack at the position *o*- to Y, there may still be an apparent preference for attack *p*- to Y. This becomes easier to understand on the basis of the relative distribution of +ve charge (from calculation and n.m.r. spectra) in the cyclohexadienyl cation (57, the Wheland intermediate for proton exchange on benzene):

$$0.25 \qquad 0.16 \qquad \oplus \qquad 0.22$$

$$H \quad H$$

(57)

Thus for proton exchange in C_6H_5Y, attack *p*- to Y should be favoured where Y can exert an electron-donating inductive effect, e.g. Y = alkyl. This may also apply when $E^\oplus$ is an electrophile other than $H^\oplus$, and the increasing proportion of *p*-attack that we have seen above in the series Y = CH_3 → CMe_3 may involve some increasing stabilisation of the Wheland intermediate for *p*-attack (increasing inductive effect on going CH_3 → CMe_3), as well as the increasing *de*stabilisation of that for *o*-attack (increasing crowding on going CH_3 → CMe_3), that we have already discussed.

There are some cases where *o*-substitution occurs to the almost total exclusion of any *p*-attack. These commonly arise from complexing of the substituent already present with the attacking electrophile so that the latter is 'steered' into the adjacent *o*-position. Thus when the ether 1-methoxy-2-phenylethane (58) is nitrated with nitrating mixture, 32 % *o*- and 59 % *p*-isomers are obtained (quite a normal distribution); but nitration with N_2O_5 in MeCN results in the formation of 69 % *o*- and 28 % *p*-isomers. This preferential *o*-attack in the second case is believed to proceed:

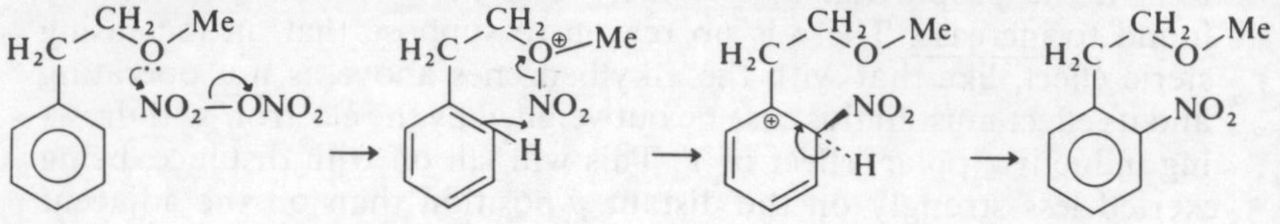

(58)

Finally it should be said that *o-/p*-ratios can be considerably influenced by the solvent in which the reaction is carried out. This can arise from changes in the relative stabilisation by solvent molecules of the transition states for *o-* and *p*-attack, but it may also involve the actual attacking electrophile being different in two different solvents: the species actually added complexing with solvent molecules to form the electrophile proper—a different one in each case. This almost certainly occurs in halogenation without Lewis acid catalysts, e.g. in the chlorination of toluene at 25° where *o-/p*-ratios of 1·5 → 0·67 have been observed depending on the solvent employed.

6.8   KINETIC versus THERMODYNAMIC CONTROL

In all that has gone before a tacit assumption has been made: that the proportions of alternative products formed in a reaction, e.g. *o-*, *m-* and *p*-isomers, are determined by their relative rates of formation, i.e. that the control is kinetic (p. 42). This is not, however, always what is observed in practice; thus in the Friedel–Crafts alkylation of methylbenzene with benzyl bromide and $GaBr_3$ (as Lewis acid catalyst) at 25°, the isomer distribution is found to be:

Time (sec)	% *o-*	% *m-*	% *p-*
0·01	40	21	39
10	23	46	31

Even after a very short reaction time (0·01 sec) it is doubtful whether the isomer distribution (in the small amount of product that has as yet been formed) is *purely* kinetically controlled—the proportion of *m*-isomer is already relatively large—and after 10 sec it clearly is not: *m*-benzyltoluene, the thermodynamically most stable isomer, predominating and the control now clearly being equilibrium or thermodynamic (p. 43).

This is a situation that must rise where the alternative products are mutually interconvertible under the conditions of the reaction, either by direct isomerisation or by reversal of the reaction to form the starting material which then undergoes new attack to yield a more thermodynamically stable isomer. It is important to emphasise that the relative proportions of alternative products formed will be defined by their relative thermodynamic stabilities *under the conditions of the reaction*, which may possibly differ from those of the isolated molecules. Thus if *m*-dimethylbenzene is heated at 82° with HF and a catalytic amount of BF_3 the proportions of the three isomeric dimethylbenzenes in the product resemble very closely those calculated thermodynamically:

	Experimental	Calculated
%*o*-	19	18
%*m*-	60	58
%*p*-	21	24

If, however, an excess of BF_3 is used the reaction product is found to contain >97% of *m*-dimethylbenzene; this is because the dimethylbenzenes can now be converted to the corresponding salts, e.g.

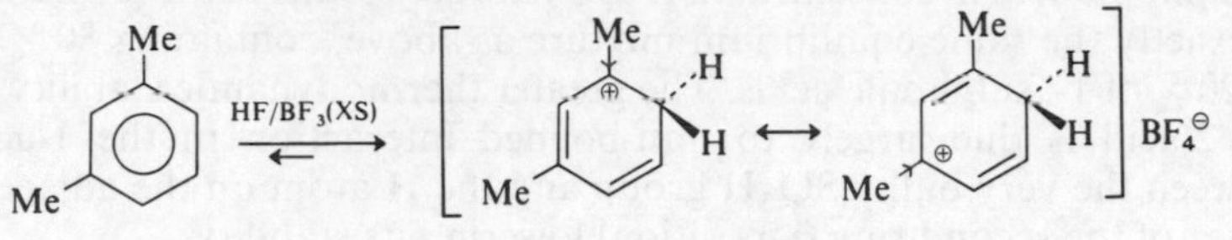

(59)

The equilibrium will therefore be shifted towards the most *basic* isomer, i.e. the one (*m*-) that forms the most stabilised cation (59) in the ion pair. Cases are also known in which the type of control that is operative is dependent on temperature (p. 162).

6.9 ELECTROPHILIC SUBSTITUTION OF OTHER AROMATIC SPECIES

With naphthalene, electrophilic substitution (e.g. nitration) is found to take place preferentially at the 1- (α-), rather than the alternative 2- (β-), position. This can be accounted for by the more effective delocalisation, and hence stabilisation, that can take place in the Wheland intermediate for 1- attack (60*a* ↔ 60*b*) compared with that for 2-attack (61):

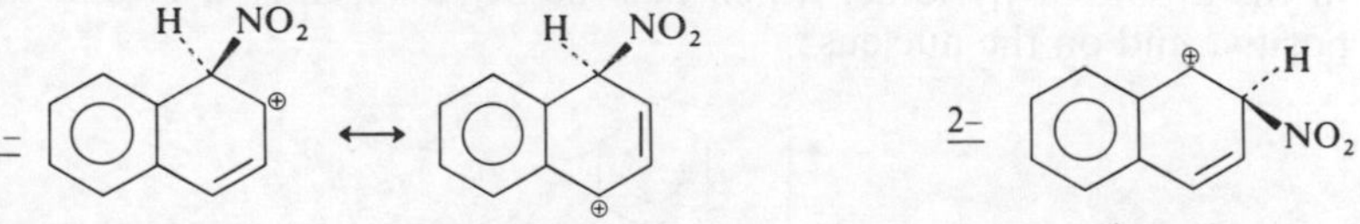

(60*a*)   (60*b*)   (61)

More forms can also be written in each case in which the positive charge is now delocalised over the second ring, leading to a total of seven forms for the 1-intermediate as against six for the 2-, but the above, in which the second ring retains intact, fully delocalised π orbitals, are probably the most important and the contrast, between two contributing forms in the one case and one in the other, correspondingly more marked. The possibility of the charge becoming more widely delocalised in the naphthalene intermediate, as compared with benzene, would lead us to expect more ready electrophilic attack on naphthalene which is indeed observed.

The sulphonation of naphthalene with concentrated H_2SO_4 at 80° is found to lead to almost complete 1-substitution, the rate of formation of the alternative 2-sulphonic acid being very slow at this temperature, i.e. kinetic control. Sulphonation at 160°, however, leads to the formation of no less than 80% of the 2-sulphonic acid, the remainder being the 1-isomer. That we are now seeing thermodynamic control is confirmed by the observation that heating pure naphthalene 1- or 2-sulphonic acid in concentrated H_2SO_4 at 160° results in the formation of exactly the same equilibrium mixture as above, containing 80% of 2-, 20% of 1-, sulphonic acids. The greater thermodynamic stability of the 2-acid is due largely to non-bonded interaction in the 1-acid between the very bulky SO_3H group and the H atom on the adjacent vertex of the second ring (8-position) lowering its stability.

The interconversion of 1- and 2-acids in H_2SO_4 at 160° could result either from a direct intramolecular isomerisation, or by reversal of sulphonation to yield naphthalene which undergoes new attack at the other position. It should be possible to distinguish between these alternatives by carrying out the reaction in $H_2{}^{35}SO_4$, for the former should lead to no incorporation of ^{35}S in the product sulphonic acids, whereas the latter should lead to such incorporation. Experimentally it is found that incorporation of ^{35}S does take place but at a rate slower than that at which the conversion occurs. This could imply either that both routes are operative simultaneously, or that, after reversal of sulphonation, new attack takes place on the resultant naphthalene by the departing H_2SO_4 molecule faster than by surrounding $H_2{}^{35}SO_4$ molecules—the question is still open.

Pyridine (62), like benzene, has six π electrons (one being supplied by nitrogen) in delocalised π orbitals but, unlike benzene, the orbitals will be deformed by being attracted towards the nitrogen atom because of the latter's being more electronegative than carbon. This is reflected in the dipole of pyridine, which has the negative end on N and the positive end on the nucleus:

$$\mu = 2 \cdot 3 \, D \quad (62)$$

Pyridine is thus referred to as a *π-deficient* heterocycle and, by analogy with a benzene ring that carries an electron-withdrawing substituent, e.g. NO_2 (p. 150), one would expect it to be deactivated towards electrophilic attack. Substitution takes place, with difficulty, at the 3-position because this leads to the most stable Wheland intermediate (63); the intermediates for 2- and 4-attack (64 and 65, respectively) each has a canonical state in which the $\oplus$ charge is located on divalent N—a highly unstable, i.e. high energy, state:

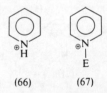

$\underline{3}-$ (63) $\underline{2}-$ (64) $\underline{4}-$ (65)

There are certain formal analogies here to *m-* attack on nitrobenzene (*cf.* p. 151), but pyridine is *very* much more difficult to substitute than the former. Thus nitration, chlorination, bromination and Friedel–Crafts reactions cannot really be made to take place usefully, and sulphonation only occurs on heating with oleum for 24 hours at 230°, with an $Hg^{2\oplus}$ catalyst. This difficulty of attack is due partly to the fact that pyridine has an available electron pair on nitrogen, and can thus protonate (66), or interact with an electrophile (67):

(66)   (67)

The positive charge will clearly further destabilise any of the σ complexes for electrophilic substitution, as did a substituent such as $\oplus NR_3$ on the benzene nucleus (p. 151), but the destabilisation will be much more marked than with $\oplus NR_3$ as the $\oplus$ charge is now on an atom of the ring itself and not merely on a substituent.

Pyrrole (68) also has 6π electrons in delocalised π orbitals, but here the nitrogen atom has to contribute two electrons to make up the six (thus becoming essentially non-basic in the process, *cf.* p. 73), and the dipole of pyrrole is found to be in the opposite direction to that of pyridine, i.e. with the positive end on nitrogen and the negative end on the nucleus:

$\mu = 1 \cdot 8\ D$ (68)

Pyrrole is thus referred to as a *π-excessive* heterocycle and behaves rather like a reactive benzene derivative, e.g. aniline (p. 152), undergoing very ready electrophilic attack. This may be complicated by the fact that in strongly acid solution protonation (69) is forced even on the weakly basic pyrrole (it takes place on the 2-carbon atom rather than on N, *cf.* p. 72):

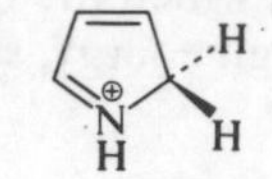

(69)

Aromatic character is thus lost, and the cation behaves like a conjugated diene in undergoing very ready polymerisation.

Electrophilic substitution of pyrrole can, however, be carried out under specialised conditions (e.g. acylation with $(MeCO)_2O/BF_3$, sulphonation with a pyridine/SO_3 complex, $C_5H_5N \cdot SO_3$, *cf.* (67) leading to preferential attack at the 2-, rather than the 3-, position. This reflects the slightly greater stabilisation of the Wheland intermediate for the former (70) compared with that for the latter (71):

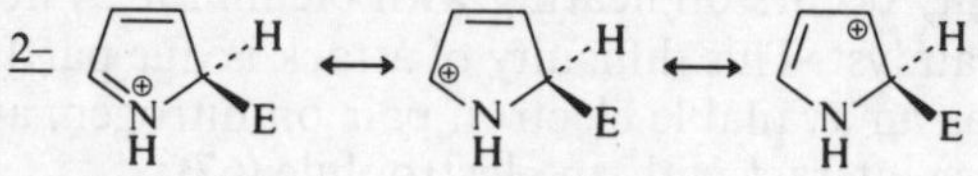

(70*a*)   (70*b*)   (70*c*)

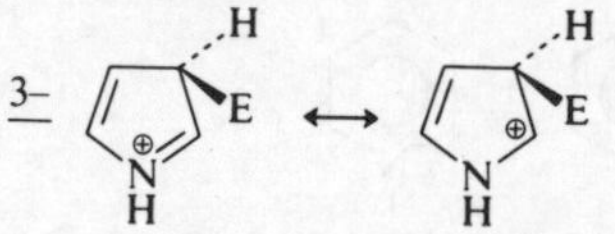

(71*a*)   (71*b*)

The difference in stability between the two is not very marked, however, reflecting the highly activated state of the nucleus, and ready substitution will take place at the 3-position if the 2- is blocked. It is, indeed, not uncommon to get substitution on all four carbon atoms, e.g. on bromination with bromine in alcohol.

6.10 NUCLEOPHILIC ATTACK ON AROMATIC SPECIES

6.10.1 Substitution of hydrogen

It is to be expected that attack by nucleophiles on an unsubstituted benzene nucleus will be much more difficult than attack by electrophiles. This is so (*a*) because the π electron cloud of the nucleus (p. 129) is likely to repel an approaching nucleophile, and (*b*) because its π orbital system is much less capable of delocalising (and so stabilising) the two extra electrons in the negatively charged (72), than the positively charged Wheland intermediate (73):

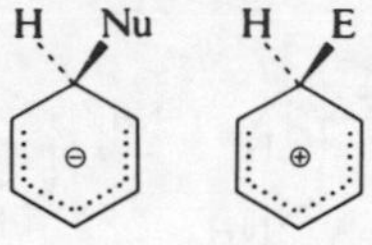

(72)   (73)

Both (*a*) and (*b*) would be overcome to some extent if a sufficiently powerful electron-withdrawing substituent was present, and nucleophilic attack might then become possible (*cf.* the *addition* of nucleophiles to alkenes carrying electron-withdrawing substituents, p. 195). It is found in practice that nitrobenzene can be fused with KOH, in the presence of air, to yield *o*- (plus a little *p*-) nitrophenol (74):

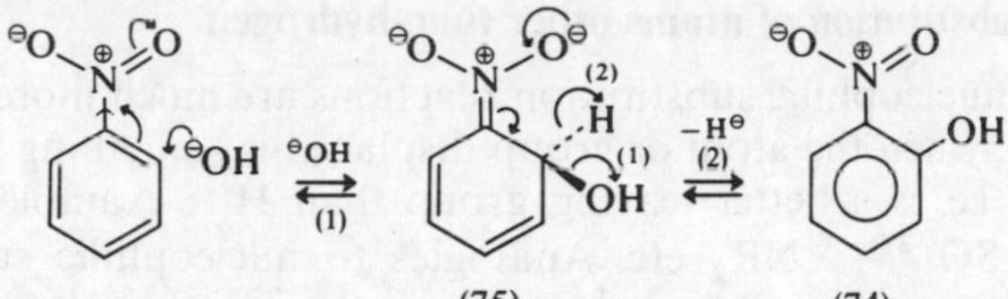

(75)   (74)

Other canonical states can be written for the anionic species (75, *cf.* Wheland intermediates), but by far the most significant one is that shown above in which the $\ominus$ charge is accommodated (and stabilised) by an oxygen atom of the nitro group. This can occur only if the attacking $^{\ominus}$OH enters the positions *o*- and *p*- to the NO_2 group (*cf.* specific stabilisation of σ-complexes for electrophilic attack *o*- and *p*- to OMe, p. 153). The species (75) can regain the aromatic condition by *either* $^{\ominus}$OH (1) *or* $^{\ominus}$H (2) acting as a leaving group: the former resulting in recovery of the starting material (nitrobenzene), the latter resulting in the formation of product (74). $H^{\ominus}$ is a poor leaving group (contrast the very much better leaving group $H^{\oplus}$ in electrophilic attack) so the equilibrium tends to lie over to the left—$^{\ominus}$OH, being a better leaving group, is lost a lot more often than $H^{\ominus}$—unless an oxidising agent, e.g. air, KNO_3, or $K_3Fe(CN)_6$, is present to encourage the elimination of hydride ion, and to destroy it as formed. Some conversion does occur in the absence of any added oxidising agent because nitrobenzene can act as its own oxidising agent (being reduced to azoxybenzene in the process), but the yield of nitrophenol is then very poor.

As we might have expected, the electron-withdrawing substituent, NO_2, that we have already seen to direct *electrophilic* attack *m*- to itself (p. 150), directs *nucleophilic* attack into the *o*- and *p*-positions.

Pyridine (76) requires no more than its own in-built capacity for electron withdrawal and is itself attacked by powerful nucleophiles, e.g. by $^{\ominus}NH_2$ (sodamide) in liquid ammonia—the Tschitschibabin reaction:

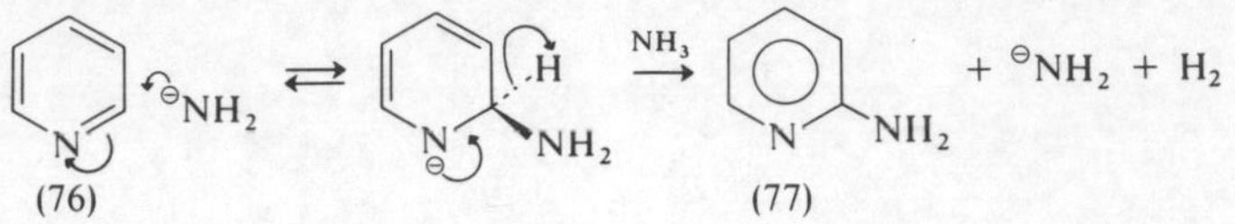

The $H^{\ominus}$ leaving group is 'helped off' by NH_3, H_2 being evolved and $^{\ominus}NH_2$ regenerated. The product, 2-aminopyridine (77), was at one time much used in the synthesis of the sulphonamide drug, sulphapyridine.

6.10.2   Substitution of atoms other than hydrogen

Aromatic nucleophilic substitution reactions are much more common, and useful, when the atom or group displaced is something other than hydrogen, i.e. is a better leaving group than $H^{\ominus}$; examples are $Cl^{\ominus}$, $Br^{\ominus}$, N_2, $SO_3{}^{2\ominus}$, $^{\ominus}NR_2$ etc. Analogies to nucleophilic substitution reactions at a saturated carbon atom (p. 77) then become more apparent.

One very common example is the displacement of N_2 in the reactions of diazonium salts, $ArN_2{}^{\oplus}$, a very useful preparative series:

$$ArN_2{}^{\oplus} + Y^{\ominus} \rightarrow ArY + N_2$$

This is found to follow the rate law,

$$\text{Rate} = k[ArN_2{}^{\oplus}]$$

i.e. the rate is independent of $[Y^{\ominus}]$, and analogies to S_N1 (p. 78) immediately spring to mind. The observed rate law has been interpreted in terms of the slow, rate-limiting, formation of an aryl cation, e.g. (78), followed by its rapid reaction with any nucleophile present:

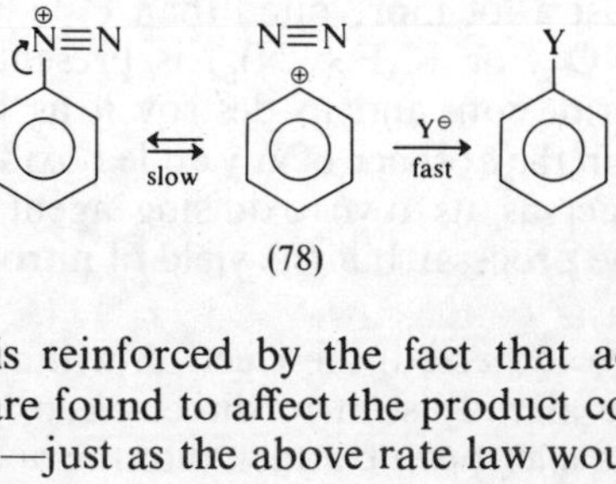

The S_N1 analogy is reinforced by the fact that added nucleophiles, $Cl^{\ominus}$, MeOH, etc., are found to affect the product composition but <u>not</u> the rate of reaction—just as the above rate law would require.

The formation of the highly unstable phenyl cation (78, the $\oplus$ charge cannot be delocalised by the π orbital system) is at first sight somewhat surprising, but the driving force is provided by the extreme effectiveness of N_2 as a leaving group [$N{\equiv}N$ bond energy = 946 kJ (226 kcal) mol^{-1}]. It is significant that this appears to be the only reaction by which simple aryl cations can be generated in solution. The aryl cations are highly reactive, and thus unselective, towards nucleophiles: thus the selectivity between $Cl^{\ominus}$ and H_2O ($k_{Cl^{\ominus}}/k_{H_2O}$) is only 3 for $C_6H_5{}^{\oplus}$ compared with 180 for $Me_3C^{\oplus}$. The very high reactivity of $C_6H_5{}^{\oplus}$ is reflected in its ability to recombine with N_2, i.e. the decomposition of the diazonium cation is reversible; this was demonstrated by observing the partial scrambling of the ^{15}N label in (79):

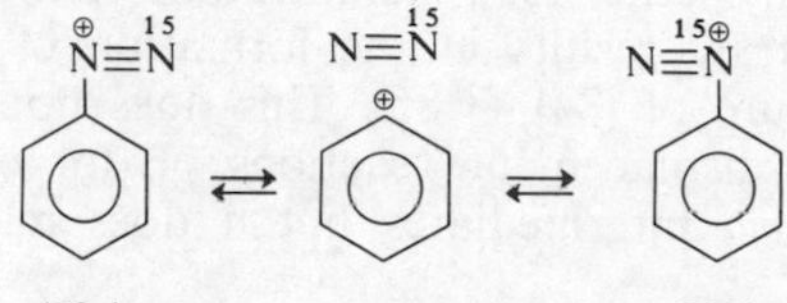

(79a)   (79b)

A number of the reactions of diazonium salts, particularly in less polar solvents, may proceed *via* the initial generation of an aryl radical, however (*cf.* p. 323).

Probably the most common aromatic nucleophilic displacement reactions involve the displacement of $Hal^{\ominus}$ from a halide activated by electron-withdrawing groups, e.g. (80):

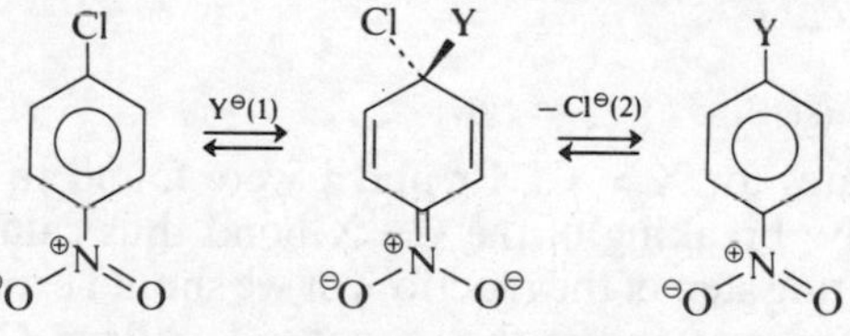

(80)   (81)

These reactions are generally found to follow the rate law.

$$\text{Rate} = k[\text{ArX}][\text{Y}^{\ominus}]$$

so that there is some formal resemblance to S_N2. The above pathway must, however, differ in that attack by $Y^{\ominus}$ cannot take place from the back of the carbon atom carrying the leaving group (*cf.* S_N2, p. 78), but must occur from the side; it is thus often referred to as S_N2(*aromatic*). Further, on the basis of the above rate law the reaction could be concerted (like S_N2)—in which case (81) is a transition state—or it could proceed by a stepwise pathway with either step (1) or step (2) as the slow, rate-limiting one—in which case (81) is an intermediate.

In support of the latter interpretation it has proved possible to isolate, and to characterise by n.m.r. spectroscopy and by X-ray diffraction, a number of species closely analogous to (81), e.g. (82);

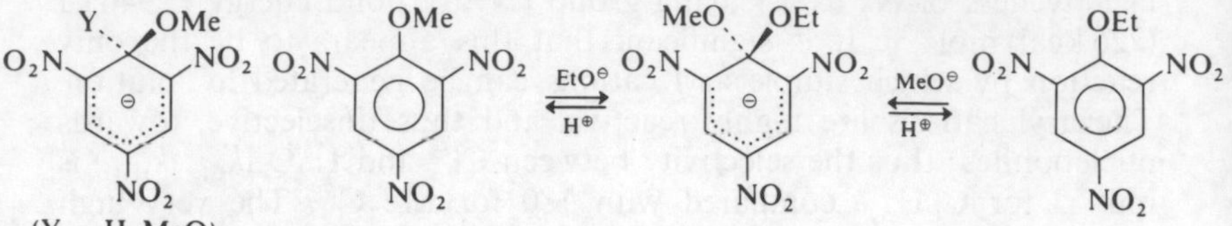

(Y = H, MeO)
(82)   (84)   (83)   (85)

and including the so-called Meisenheimer complex (83), a red crystalline solid obtainable by the action of EtO$^{\ominus}$ on the methyl ether (84) or of MeO$^{\ominus}$ on the ethyl ether (85). Acidification of the reaction mixture from *either* substrate results in the formation of exactly the same equilibrium mixture of (84) + (85). This does not, of course, prove that the normal displacement reactions of, for example, aromatic halides proceed *via* intermediates but it does make is seem more likely.

Direct support for a stepwise pathway is, however, provided by comparison of the rates of reaction of a series of substrates, having different leaving groups, with the same nucleophile, e.g. 2,4-dinitro-halogenobenzenes (86) with piperidine (87):

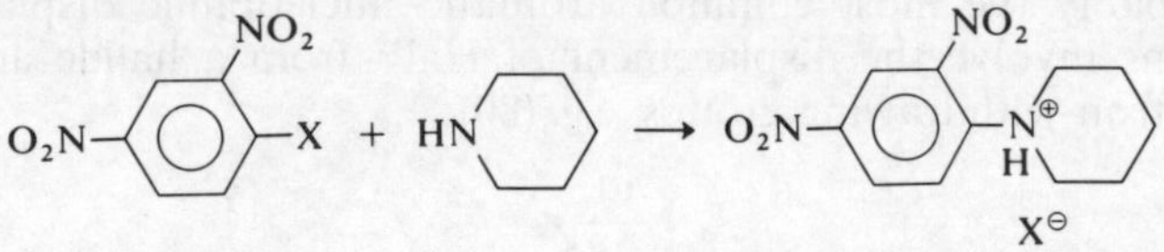

(86)   (87)

The relative rates for X = Cl, Br and I were found to be 4·3, 4·3 and 1·0, respectively; breaking of the C—X bond thus cannot be involved in the rate-limiting step of the reaction, or we should expect significantly bigger rate differences and in the sequence I > Br > Cl. The reaction, in this case, cannot therefore be one-step, i.e. concerted (*cf.* S$_N$2), and in the two-step pathway suggested above, step (1)—attack by the nucleophile—would have to be rate-limiting. It is interesting too, to observe that the rate of the above reaction when X = F is 3300. This results from the very powerfully electron-withdrawing F speeding up step (1): (*a*) by making the nuclear carbon to which it is attached more positive and hence more readily attacked by a nucleophile, and (*b*) by helping to stabilise the anionic intermediate (88):

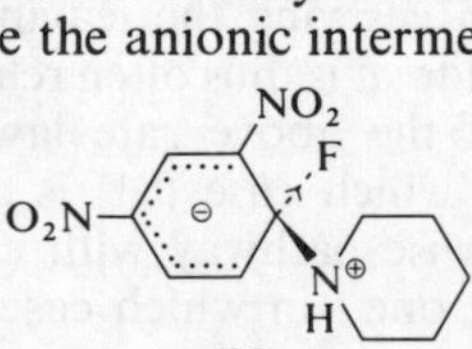

(88)

2,4-Dinitrofluorobenzene (86, X = F) is, because of its reactivity, much used for 'tagging' the NH_2 group of amino-acids in protein end group analysis. Once it has reacted with the NH_2 it is very difficult to remove again and will thus withstand the subsequent hydrolysis of the protein.

What difference there is in rate here is dependent on the effect of the halogen, through electron-withdrawal, in influencing the relative ease of attack on the substrate by the nucleophile: it is in the reverse order of the relative ability of the halide ions as leaving groups. When the same series of halides is reacted with C_6H_5NHMe (in nitrobenzene at 120°), however, the relative rates for X = F, Cl and Br were found to be 1, 15 and 46, e.g. in the order of their relative ability as leaving groups, so that in this latter reaction it would appear that step (2) is now involved, to some extent at least, in the rate-limiting state.

The first pathway above is much the more common, however, and we can add it [$S_N2(aromatic)$]—bond-breaking by the leaving group *after* bond-formation to the nucleophile—to the S_N2—bond-breaking by the leaving group and bond-formation to the nucleophile *simultaneous*—and the S_N1—bond-breaking by the leaving group *before* bond-formation to the nucleophile—pathways that we have already encountered. Thus nucleophilic aromatic 'substitution' is in fact an addition/elimination process very similar to electrophilic aromatic 'substitution', except for the different attacking species. Other important examples of nucleophilic aromatic substitutions of preparative significance are the displacement of $SO_3{}^{2\ominus}$ from the alkali-metal salts of sulphonic acids, e.g. $ArSO_3{}^\ominus Na^\oplus$, by $^\ominus OH$ and $^\ominus CN$ and, less importantly, the displacement of $^\ominus NR_2$ from *p*-nitroso-N,N-dialkylanilines by $^\ominus OH$.

Significant electron-withdrawal by a substituent to stabilise the anionic intermediate, e.g. (81), only occurs through a mesomeric effect, i.e. when the nitro group, for example, is *o*- and/or *p*- to the leaving group. Thus we observe the reactivity sequence:

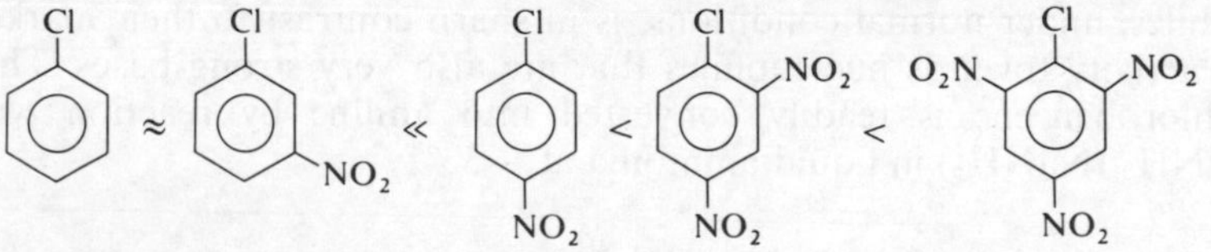

2- and 4-, but not 3-, halogenopyridines undergo ready nucleophilic displacement reactions for exactly the same reason. Mesomeric interaction with an electron-withdrawing substituent will be reduced or inhibited if the *p* orbital on the atom adjacent to the nucleus, e.g. N in NO_2, is prevented from becoming parallel to the *p* orbital on the nuclear carbon to which it is attached (steric inhibition of delocalisation, *cf*. p. 70). Thus the following relative rates of nucleophilic attack are observed:

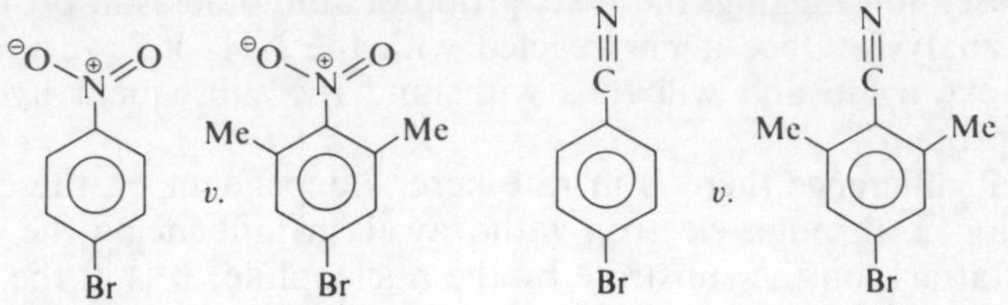

Br	Br	Br	Br
(89)	(90)	(91)	(92)
40	1	4	1

The rate difference between (91) and (92) is very small as the Me groups do not prevent mesomeric electron-withdrawal by the linear CN group. The rate difference is much more pronounced between (89) and (90), however, as the Me groups prevent the oxygen atoms of the nitro group lying in the same plane as the nucleus, p overlap between N and the adjacent C is thus markedly reduced.

Finally it should be mentioned that a number of nucleophilic substitution reactions of unactivated halides can be made to proceed in bipolar non-protic solvents such as dimethyl sulphoxide (DMSO), $Me_2S^{\oplus}-O^{\ominus}$. No hydrogen-bonded solvent envelope, as in for example MeOH, then needs to be stripped from $Y^{\ominus}$ before it can function as a nucleophile; $\Delta G^{\ddagger}$ is thus much lower and the reaction correspondingly faster. Rate differences of as much as 10^9 have been observed on changing the solvent from MeOH to Me_2SO. Chlorobenzene will thus react readily under these conditions with $Me_3CO^{\ominus}$:

$$Me_3CO^{\ominus} + Ph-Cl \xrightarrow{\text{DMSO}} Ph-OCMe_3 + Cl^{\ominus}$$

6.10.3   'Substitution' *via* aryne intermediates

The relative inertness of unactivated aromatic halides towards nucleophiles, under normal conditions, is in sharp contrast to their marked reactivity towards nucleophiles that are also very strong bases. Thus chlorobenzene is readily converted into aniline by reaction with $^{\ominus}NH_2$ (NaNH$_2$) in liquid ammonia at $-33°$:

$$PhCl + {}^{\ominus}NH_2 \xrightarrow[-33°]{\text{liq. NH}_3} PhNH_2 + Cl^{\ominus}$$

This surprising difference in reactivity suggests the possibility of a reaction pathway other than S$_N$2(*aromatic*), and some clue to what it might be is provided by the observation that *p*-chloromethylbenzene (93) undergoes the same reaction (equally readily) to give not only the expected *p*-aminomethylbenzene (94), but also the unexpected *m*-

aminomethylbenzene (95), and that in the larger relative yield:

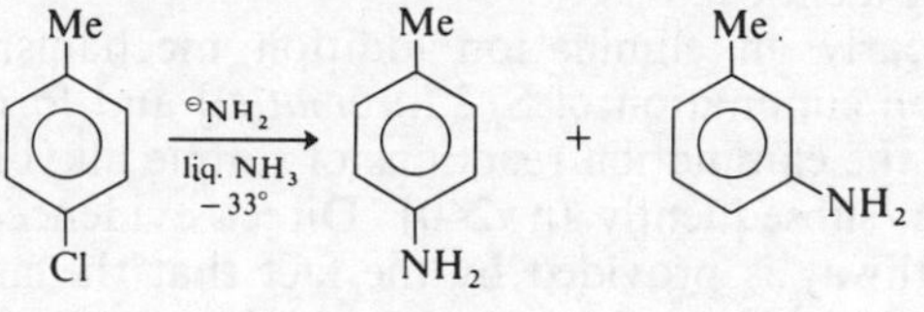

(93)   (94)   (95)

Expected: 38%   Unexpected: 62%

No *o*-isomer is ever obtained, and (94) and (95) are found not to be interconvertible under the conditions of the reaction. This, coupled with the fact that $^{\ominus}NH_2$ is known to be able to remove protons (deuterons) from a benzene ring [it removes proton (deuteron) 10^6 times faster from fluorobenzene with an *o*-deuterium substituent than from deuterobenzene itself],

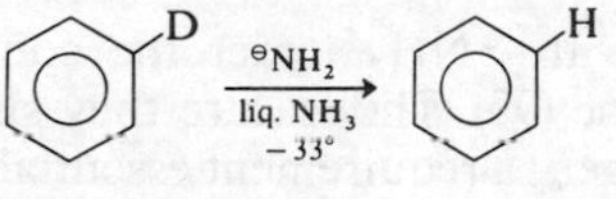

suggests that the initial attack of $^{\ominus}NH_2$ may be as a base rather than as a nucleophile:

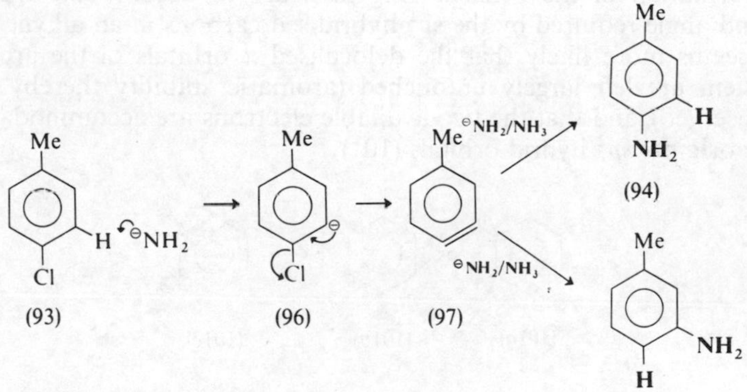

(93)   (96)   (97)

(94)

(95)

The loss of proton from (93) could be concerted with, or followed by, loss of $Cl^{\ominus}$ to yield the *aryne* intermediate (97). The latter has two alternative positions that could undergo attack by $^{\ominus}NH_2$, product formation then being completed by abstraction of a proton from the solvent NH_3; the nett effect is of the formal addition of NH_3 in two alternative ways round. We should not expect the relative proportions of the two alternative products to be the same because (97) is not a symmetrical intermediate, i.e. the two possible positions of attack by $^{\ominus}NH_2$ are not identical.

This is clearly an elimination/addition mechanism [in contrast to the addition/elimination of S_N2 (*aromatic*)] and formally parallels, in its genesis, the elimination reactions of simple alkyl halides that we shall consider subsequently (p. 240). Direct evidence in support of the aryne pathway is provided by the fact that the halides (98), (99) and (100),

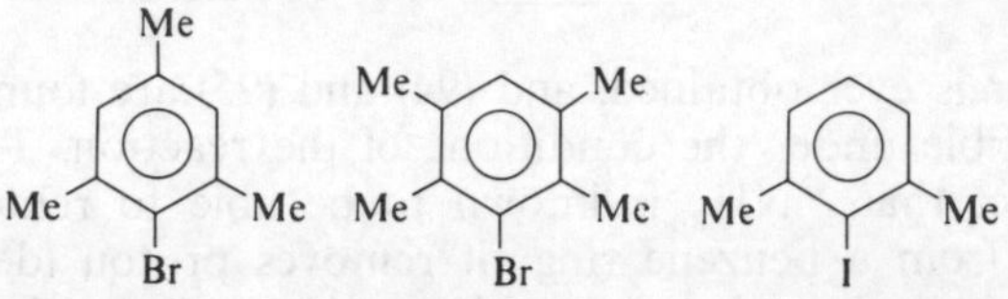

Me — (ring) — Me, Me, Br   (98)

Me, Me — (ring) — Me, Mc, Br   (99)

Me — (ring) — Me, I   (100)

do not react at all with $^{\ominus}NH_2$ under the conditions employed for *p*-chloromethylbenzene (93). The feature they share in common is no H atom *o*- to the halogen: a requirement essential for initiating reaction *via* an aryne intermediate, as we saw above.

Arynes present structural features of some interest. They clearly cannot be acetylenic in the usual sense as this would require enormous deformation of the benzene ring in order to accomodate the 180° bond angle required by the sp^1 hybridised carbons in an alkyne (p. 9). It seems more likely that the delocalised π orbitals of the aromatic system are left largely untouched (aromatic stability thereby being conserved), and that the two available electrons are accommodated in the original sp^2 hybrid orbitals (101):

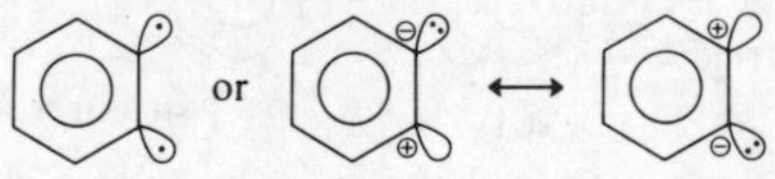

(101a)   or   (101b) ↔ (101c)

Overlap between these orbitals will, on spatial grounds, be very poor, and the resultant bonding correspondingly weak: arynes are thus likely to be highly reactive towards nucleophiles (and electrophiles), though they are found not to be entirely unselective in this.

No arynes have actually been isolated, but there is considerable evidence for their existence from 'trapping' experiments (*cf.* p. 49) and from spectroscopy. Thus generation of benzyne (101) in the presence of furan (102) leads to the formation of the Diels–Alder (p. 193) adduct (103), which undergoes ready acid-catalysed ring fission to yield the more familiar 1-naphthol (104):

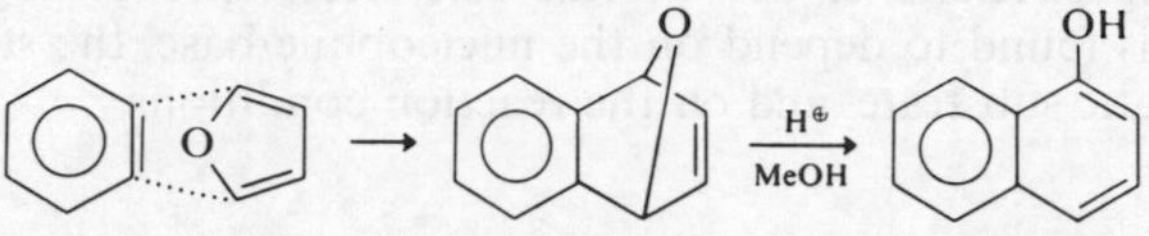

(101)   (102)   (103)   (104)

If benzyne is produced under conditions where there is no suitable species for it to react with, then it dimerises ('self-trapping') very rapidly to the stable biphenylene (105):

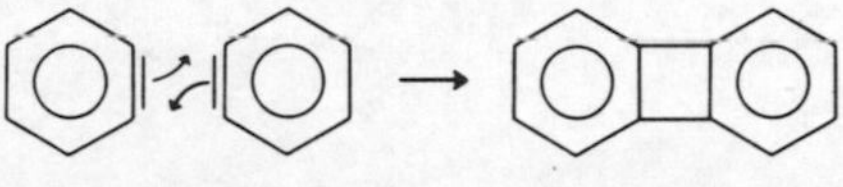

(101)   (105)

A very convincing demonstration of the existence of benzyne by physical methods involves the introduction into the heated inlet of a mass spectrometer of the zwitterion ion (106), a salt of diazotised *o*-aminobenzoic(anthranilic) acid. The mass spectrum is found to be a very simple one exhibiting *m/e* peaks at 28, 44, 76 and 152:

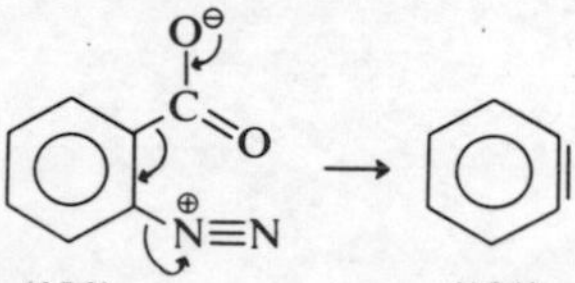

(106)   (101)
                          m/e 76

CO_2 *m/e* 44

N_2 *m/e* 28

The *m/e* 76 peak declined and the *m/e* 152 peak increased rapidly with time, indicating the progressive dimerisation of benzyne to the more stable biphenylene (105, above).

Methods such as the pyrolysis of (106), that do not require strongly basic conditions, have been used to generate arynes in bulk for preparative purposes, and another, even better, method is oxidation of 1-aminobenzotriazole (107) with lead tetraacetate:

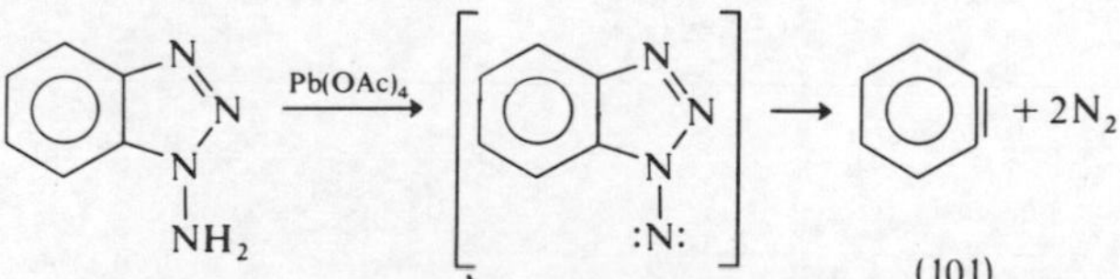

(107)   (101)

$+ 2N_2$

Reactions of unactivated halides with the weaker base $^{\ominus}OH$, that only proceed under considerably more vigorous conditions, may well involve both aryne intermediates and $S_N2(aromatic)$ pathways; the relative proportions of the overall conversion proceeding by each pathway is found to depend on the nucleophile/base, the structure of the aromatic substrate, and on the reaction conditions.

7
Electrophilic and nucleophilic addition to C=C

As we have already seen (p. 8), a carbon–carbon double bond consists of a strong σ bond plus a weaker π bond differently situated (1):

(1)

The pair of electrons in the π orbital are more diffuse and less firmly held by the carbon nuclei, and so more readily polarisable, than those of the σ bond, leading to the characteristic reactivity of such unsaturated compounds. As the π electrons are the most readily accessible feature of the carbon–carbon double bond, we should expect them to shield the molecule from attack by nucleophilic reagents and this is indeed found to be the case (cf. p. 195, however). The most characteristic reactions of the system are, hardly surprisingly, found to be initiated by electron-deficient species such as $X^{\oplus}$ and $X\cdot$ (radicals can be considered electron-deficient species as they are seeking a further electron with which to form a bond), cations inducing heterolytic, and radicals homolytic, fission of the π bond. The former is usually found to predominate in polar solvents, the latter in non-polar solvents especially in the presence of light. Radical induced additions are discussed subsequently (p. 304).

7.1 ADDITION OF HALOGENS

The decolorisation of bromine, usually in CCl_4 solution, is one of the classical tests for unsaturation, and probably constitutes the most familiar of the addition reactions of alkenes. It normally proceeds readily in the absence of added catalysts, and one is tempted to assume that it proceeds by a simple, one-step pathway;

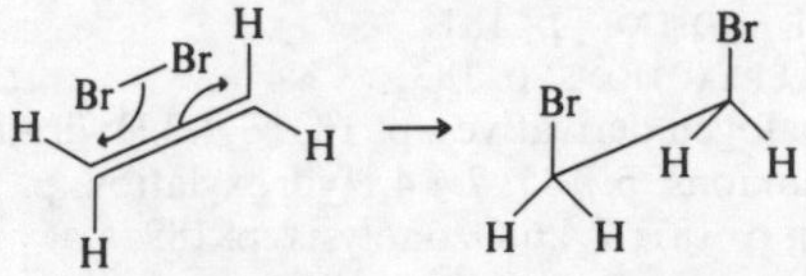

(1)   (2)

there are, however, two highly significant pieces of experimental evidence that serve to refute this.

Firstly, if bromine addition is carried out in the presence of added nucleophiles Y: or $Y^\ominus$ (e.g. $Cl^\ominus$, $NO_3{}^\ominus$, H_2O:) then, in addition to the expected 1,2-dibromide (3), products are also obtained in which one bromine atom and one Y atom, or group, have been added to the double bond (4):

$$CH_2{=}CH_2 \xrightarrow{\ Br_2\,+\,Y^\ominus\ } \underset{\substack{| \quad\;\; | \\ Br \quad Br}}{CH_2{-}CH_2} + \underset{\substack{| \quad\;\; | \\ Br \quad\; Y}}{CH_2{-}CH_2}$$

(3)   (4)

This is clearly incompatible with a one-step pathway like the above, in which there would be no opportunity for attack by $Y^\ominus$. It is, of course, important to establish that (4) does not arise merely by subsequent attack of $Y^\ominus$ on first formed (3), but it is found in practice that the formation of (4) is much more rapid than nucleophilic substitution reactions would be under these conditions. A possible explanation is competition by $Y^\ominus$ and $Br^\ominus$ (derived from Br_2) for a common intermediate (see below).

Secondly it is found—with those simple alkenes in which it can be detected, e.g. *trans* 2-butene (5)—that the two bromine atoms add on from opposite sides of the planar alkene, i.e. ANTI addition:

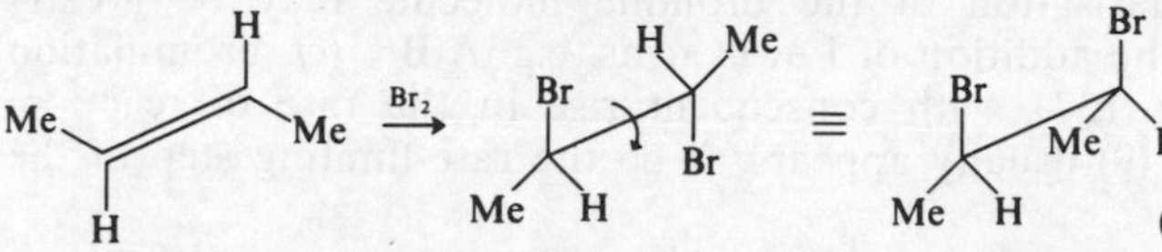

(5)   (6)   (meso)   (ANTI addition)

The product is the symmetrical *meso* dibromide (6), whereas if addition
had been SYN (both bromine atoms adding from the same side) it
would have been the unsymmetrical ($\pm$) dibromide (7):

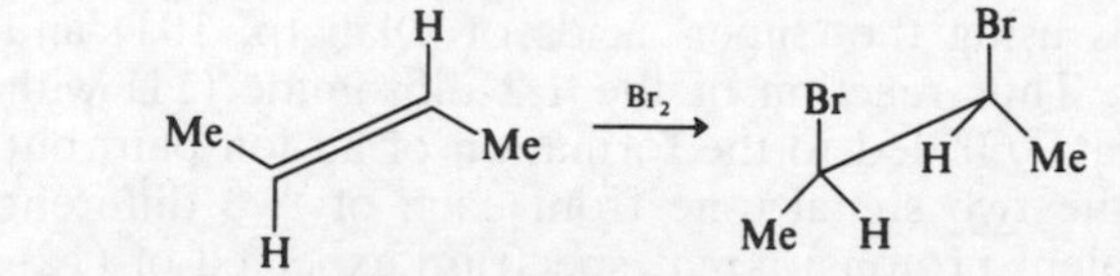

(5)   (7) ($\pm$)   (SYN addition)

It is found in practice with (5), and with other simple acyclic alkenes,
that the addition is almost completely *stereoselective*, i.e. $\approx 100\%$
ANTI addition. This result also is incompatible with a one-step
pathway, as the atoms in a bromine molecule are too close to each
other to be able to add, simultaneously, ANTI.

These observations are explainable by a pathway in which one end
of a bromine molecule becomes positively polarised through electron
repulsion by the π electrons of the alkene, thereby forming a π complex
with it (8; *cf.* Br_2 + benzene, p. 130). This then breaks down to form
a *cyclic bromonium ion* (9)—an alternative canonical form of the
carbonium ion (10). Addition is completed through nucleophilic
attack by the residual $Br^{\ominus}$ (or added $Y^{\ominus}$) on either of the original
double bond carbon atoms, from the side opposite to the large
bromonium ion $Br^{\oplus}$, to yield the *meso* dibromide (6):

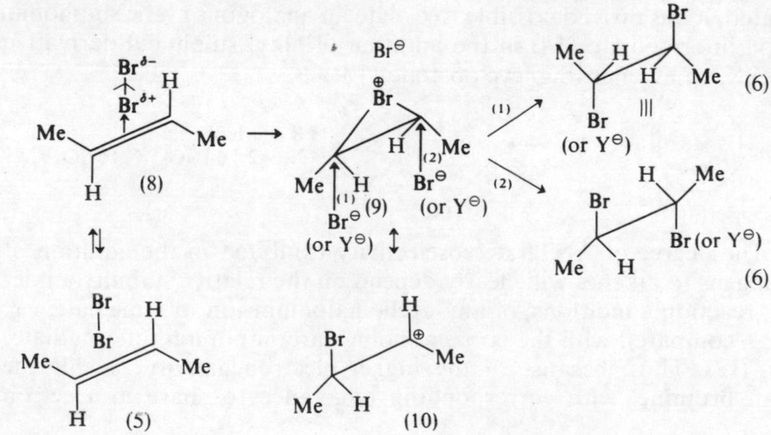

Enough mutual polarisation can apparently result, in (8), for (9) to form, but polarisation of the bromine molecule may be greatly increased by the addition of Lewis acids, e.g. $AlBr_3$ (*cf.* bromination of benzene, p. 137), with consequent rise in the rate of reaction. Formation of (9) usually appears to be the rate-limiting step of the reaction.

Bromonium ion intermediates such as (9) were suggested (as long ago as 1938) to account for the highly stereoselective (ANTI) addition often observed with simple, acyclic alkenes. One very unusual example has actually been isolated, and there has been supporting evidence from other fields: thus it has proved possible to detect one by physical methods using the 'super' acids of Olah (p. 101) and n.m.r. spectroscopy. Thus reaction of the 1,2-dibromide (11) with SbF_5 in liquid SO_2 at $-60°$ led to the formation of an ion pair, but this exhibited <u>not</u> the <u>two</u> signal (one from each of two different groups of six equivalent protons) n.m.r. spectrum expected of (12). Instead <u>one</u> signal only ($\tau\ 7 \cdot 1$) was observed, indicating that all twelve protons were equivalent, i.e. what is being observed is almost certainly the bromonium ion (9a):

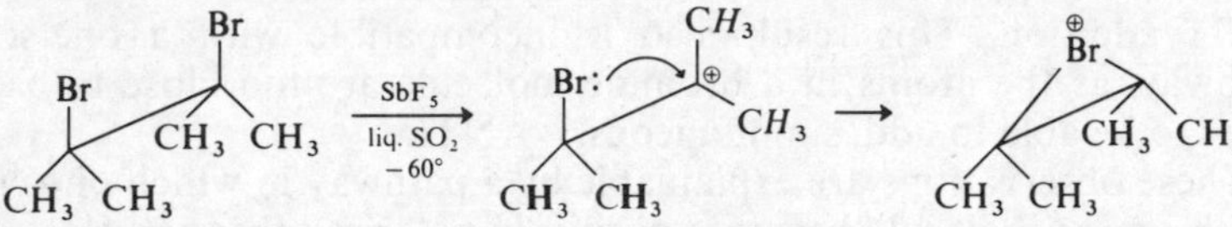

This neighbouring group participation by bromine (*cf.* p. 92) does not of course prove that addition to alkenes proceeds *via* cyclic bromonium ions, but it does mean that such species are no longer merely *ad hoc* assumptions, and to that extent are correspondingly more plausible as intermediates.

Although no simple bromonium ion intermediates have been isolated, it has proved possible to isolate an analogous cyclic sulphonium ion intermediate (14) in the addition of alkyl sulphenyl derivatives, RSX, to alkenes, e.g. cyclooctene (13):

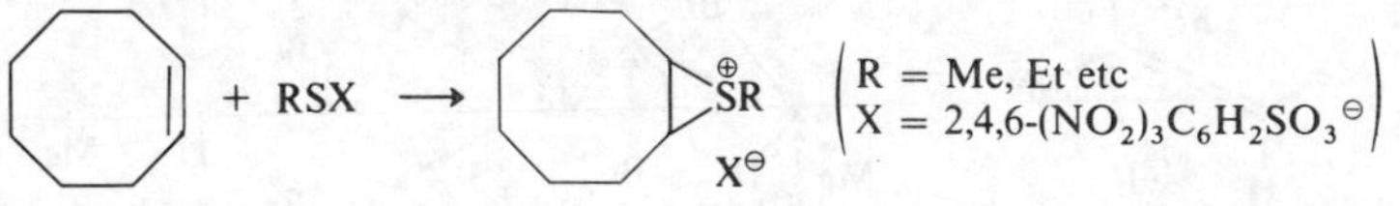

The degree of ANTI stereoselectivity exhibited in the addition of halogens to alkenes will clearly depend on the relative stability, under the reaction conditions, of any cyclic halonium ion intermediate, e.g. (9a), compared with the corresponding carbonium ion intermediate, e.g. (12). Thus, because of the higher electronegativity of chlorine than bromine, with corresponding reluctance to share its electron pairs, it might be expected that with some alkenes the addition of chlorine would be less stereoselective than that of bromine: this is found to be the case. It might also be expected that structural features leading to specific stabilisation of carbonium ions, might also lead to less ANTI stereoselectivity; this is observed with, for example, *trans* 1-phenylpropene (15):

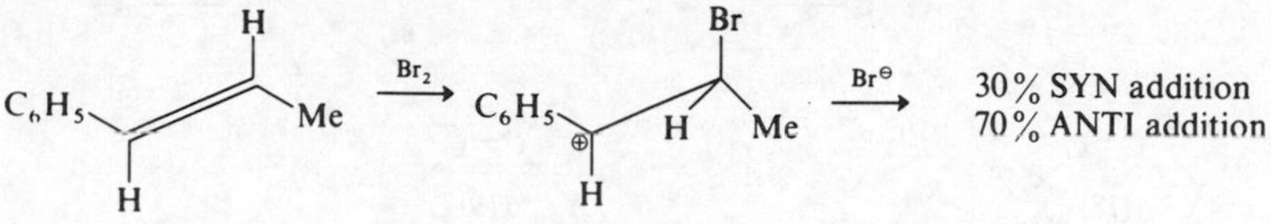

(15)   (16)

The possible formation of a delocalised benzyl type carbonium ion (16) results in much lower (70%) ANTI stereoselectivity than with *trans* 2-butene (5; ≈100% ANTI stereoselectivity, p. 177), where no such delocalisation is possible. It is also found that increasing the polarity, and ion-solvating ability, of the solvent also stabilises the carbonium ion, relative to the bromium ion, intermediate with consequent decrease in ANTI stereoselectivity. Thus addition of bromine to 1,2-diphenylethene (stilbene) was found to proceed 90–100% ANTI in solvents of low dielectric constant, but ≈50% ANTI only in a solvent with $\epsilon = 35$.

It is not normally possible to add fluorine directly to alkenes as the reaction is so exothermic that bond fission occurs. Many alkenes will not add iodine directly either, and when the reaction does occur it is usually readily reversible. Alkynes are also found to undergo preferential, though not exclusive, ANTI addition of halogens, e.g. with butyne-1,2-dioic acid (17):

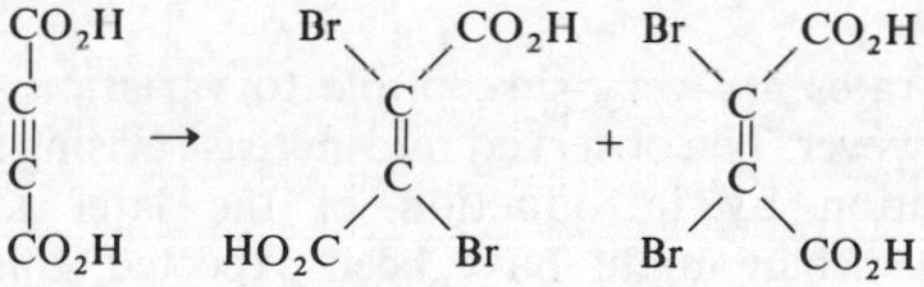

(17)   70%   30%

7.2 EFFECT OF SUBSTITUENTS ON RATE OF ADDITION

Whether the intermediate in bromination is a bromonium or carbonium ion, it is certainly positively charged. In so far as its formation is rate-limiting we should expect, by analogy with electrophilic aromatic substitution (p. 152), that it—and the transition state that precedes it—would be stabilised by electron-donating substituents; i.e. that such substituents (18) would speed up the rate of electrophilic addition, and vice versa with electron-withdrawing substituents (19):

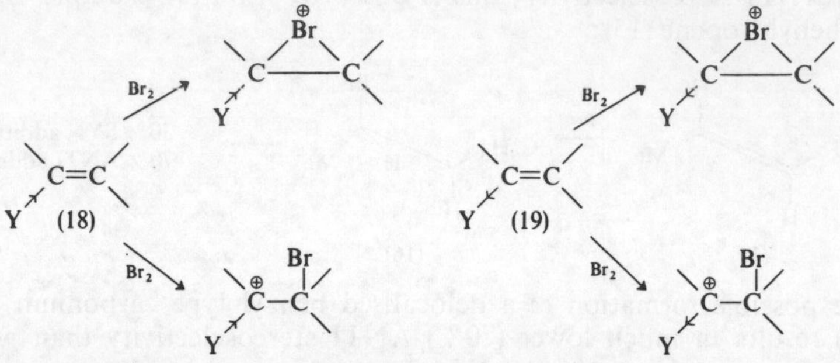

The following relative rates are actually observed in practice under analogous conditions;

$$CH_2=CH\!-\!Br \approx CH_2=CH\!-\!CO_2H < CH_2=CH_2 < Et\!-\!CH=CH_2$$

$$3\times10^{-2} \qquad\qquad\qquad 1 \qquad\qquad 9{\cdot}6\times10$$

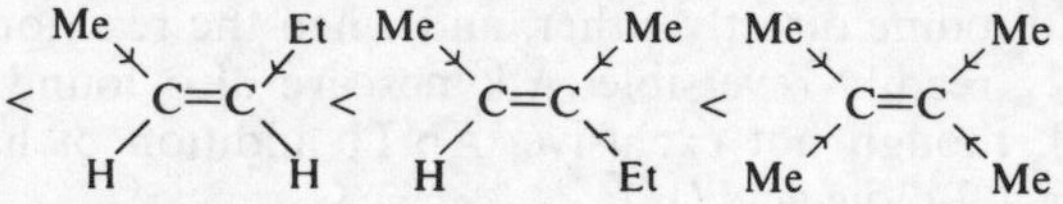

$$4{\cdot}16\times10^3 \qquad\qquad 1{\cdot}19\times10^5 \qquad\qquad 9{\cdot}25\times10^5$$

these relative rates are very susceptible to variation in the reaction conditions, however. The observed rate increase arising from increasing electron donation by introduction of the later alkyl groups is perhaps smaller than might have been expected; this is due to the increasing crowding in the transition state introduced by these later alkyl groups. A phenyl group also increases the rate of electrophilic addition considerably (4×10^3), due to the stabilisation that it can induce in the intermediate (20), and in the transition state that precedes it:

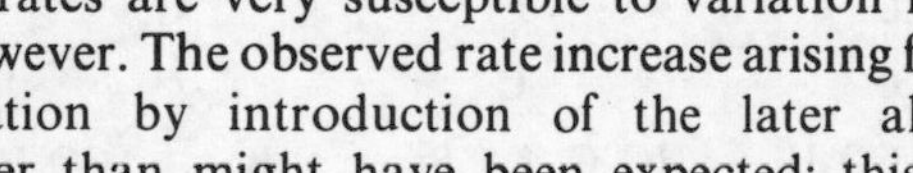

(20)

7.3 ORIENTATION OF ADDITION

When the electrophile being added is, unlike the halogens, non-symmetrical then with a non-symmetrical alkene, e.g. propene, the problem of *orientation* of addition arises: this will be the case with the hydrogen halides. These are found to add to a given alkene in the rate order: HF > HCl > HBr > HI, i.e. in order of their acid strengths. This suggests rate-limiting addition of proton to the alkene, followed by rapid nucleophilic attack by $Hal^\ominus$ to complete the addition. In non-polar solvents the proton is no doubt provided by HHal, but in polar, and especially in hydroxylic solvents, more likely by its conjugate acid, e.g. by $H_3O^\oplus$ in H_2O.

A bridged intermediate exactly analogous to a bromonium ion cannot be formed as H has no electron pair available, but it may be that in some cases a π complex (21) is the intermediate. We shall, however, normally write the intermediate as a carbonium ion, and it is the relative stability of possible, alternative, carbonium ions (e.g. 23 and 24) that determines the overall orientation of addition, e.g. in the addition of HBr to propene (22) under polar conditions:

$$
\begin{array}{ccc}
\overset{\displaystyle H}{\underset{\displaystyle |}{Me-CH}} \overset{\oplus}{\rightarrow} CH_2 & \xrightarrow{\ Br^\ominus\ } & \overset{\displaystyle H}{\underset{\displaystyle |}{Me-CH}} - CH_2 \\
& & | \\
& & Br \\
(23) & & (25)
\end{array}
$$

$$H^\oplus \nearrow \kern-1.5em \diagup$$

$$
\overset{H^\oplus\ \uparrow}{Me-CH=CH_2} \qquad Me-CH=CH_2
$$

$$(21) \qquad\qquad (22)$$

$$H^\oplus \searrow$$

$$
\begin{array}{ccc}
\overset{\displaystyle H}{\underset{\displaystyle |}{Me}} \overset{\oplus}{\rightarrow} CH \overset{\displaystyle}{\leftarrow} CH_2 & \xrightarrow{\ Br^\ominus\ } & \overset{\displaystyle H}{\underset{\displaystyle |}{Me-CH}} - CH_2 \\
& & | \\
& & Br \\
(24) & & (26)
\end{array}
$$

As we have seen already (p. 103) secondary carbonium ions are more stable than primary, and in so far as this also applies to the transition states that precede them, (24) will be formed in preference to (23). In fact it appears to be formed exclusively, as the only addition product obtained is 2-bromopropane (26). Addition, as here, in which halogen (or the more negative moiety of any other unsymmetrical adduct) becomes attached to the more highly substituted of the two alkene carbon atoms is known as *Markownikov* addition.

The study of the addition of HBr to alkenes presents a number of experimental difficulties. Thus the possible rearrangement of structure of the carbonium ion intermediates has already been referred to (p. 111). Further, in solution in water or in other hydroxylic solvents, acid catalysed hydration (p. 184) or solvation may constitute a competing reaction; while in less polar solvents radical formation may be encouraged, resulting in *anti*-Markownikov addition to give 1-bromopropane (MeCH$_2$CH$_2$Br, 25), *via* the preferentially formed radical intermediate, MeĊHCH$_2$Br. This is discussed in detail below (p. 307).

Electrophilic addition to 1-haloalkenes (e.g. 27), presents a number of parallels to the electrophilic substitution of halobenzenes (p. 154). Thus it is the involvement of the electron pairs on Br that controls the orientation of addition (*cf. o-/p*-direction in C$_6$H$_5$Br) ;

$$\text{CH}_2=\text{CH}-\text{Br} \xrightarrow{\text{H}^\oplus} \overset{\text{H}}{\underset{(28)}{\text{CH}_2-\overset{\oplus}{\text{CH}}-\text{Br}}} \xrightarrow{\text{Br}^\ominus} \overset{\text{H}}{\underset{\text{Br}}{\text{CH}_2-\text{CH}-\text{Br}}}$$

(27)   (28)

$$\text{H}^\oplus \downarrow$$

$$\left[\underset{\text{H}}{\text{CH}_2-\overset{\oplus}{\text{CH}}-\overset{\cdot\cdot}{\text{Br}}} \leftrightarrow \underset{\text{H}}{\text{CH}_2-\text{CH}=\overset{\oplus}{\text{Br}}} \right] \xrightarrow{\text{Br}^\ominus} \overset{\text{Br}}{\underset{\text{H}}{\text{CH}_2-\text{CH}-\text{Br}}}$$

(29a)   (29b)   (30)

(29) is stabilised compared with (28), is therefore formed preferentially, and 1,1-dibromoethane (30) is in fact the only product obtained. The rate of addition is, however, controlled by the electron-withdrawing inductive effect of the halogen atom, and (27) is found to add HBr about 30 times more slowly than does ethene (*cf.* bromobenzene is attacked more slowly by electrophiles than is benzene), i.e. (29) is less stable, and is formed more slowly, than (31):

$$\underset{(29)}{\overset{\text{H}}{\text{CH}_2 \rightarrow \overset{\oplus}{\text{CH}} \rightarrow \text{Br}}} \qquad \underset{(31)}{\overset{\text{H}}{\text{CH}_2 \rightarrow \overset{\oplus}{\text{CH}_2}}}$$

(29)   (31)

The addition of halogen hydracids to simple alkenes is found to be a good deal less stereoselective than was the addition of halogens, being much more dependent on the alkene and on the reaction conditions.

7.4 OTHER ADDITION REACTIONS

7.4.1 Further halogen derivatives

Interhalogen compounds, hardly surprisingly, add to alkenes very much as do the halogens themselves, and the following order of reactivity has been observed:

$$BrCl > Br_2 > ICl > IBr > I_2$$

Addition is initiated by the positively polarised end (the less electronegative halogen atom) of the unsymmetrical molecule, and a cyclic halonium ion intermediate probably results. With an unsymmetrical alkene, e.g. 2-methylpropene (32), the charge distribution in the intermediate will be unsymmetrical; the more heavily alkylated carbon will be the more positive (33), will be attacked by the residual halide nucleophile, and so will determine the orientation (Markownikov) of overall addition (34):

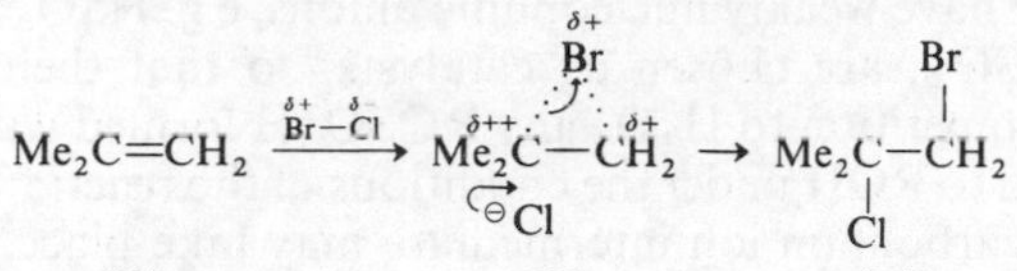

$$Me_2C{=}CH_2 \xrightarrow{\overset{\delta+}{Br}{-}\overset{\delta}{Cl}} \underset{\underset{\ominus}{\overset{\nearrow}{Cl}}}{\overset{\delta++}{Me_2}\overset{}{C}{-}\overset{\delta+}{CH_2}} \rightarrow \underset{\underset{Cl}{|}}{\overset{\overset{Br}{|}}{Me_2C{-}CH_2}}$$

$$(32) \qquad\qquad (33) \qquad\qquad (34)$$

Hypohalous acids, e.g. $HO^{\delta-}{-}Br^{\delta+}$ (bromine water), were thought to add on in very much the same way, but there is now evidence that the actual electrophile is probably the halogen itself, e.g. Br_2, and that both 1,2-dibromide (35a) and 1,2-bromhydrin (35b) are then obtained by competition of $Br^{\ominus}$ and $H_2O\colon$ for the initial bromonium ion intermediate (36):

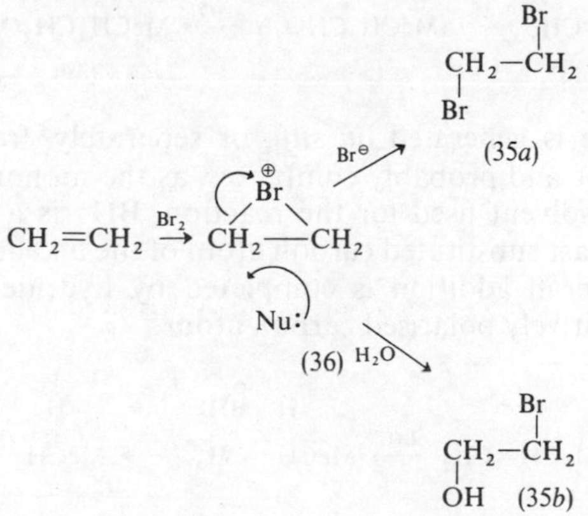

$$CH_2{=}CH_2 \xrightarrow{Br_2} \underset{(36)\ H_2O}{\overset{\overset{\oplus}{Br}}{\underset{Nu\colon}{CH_2{-}CH_2}}}$$

$$\underset{\overset{|}{Br}}{\overset{\overset{Br}{|}}{CH_2{-}CH_2}} \quad (35a)$$

$$\underset{OH}{\overset{\overset{Br}{|}}{CH_2{-}CH_2}} \quad (35b)$$

7.4.2 Hydration

Acid catalysed hydration of an alkene is the reversal of the similarly acid-catalysed dehydration (by the E1 pathway, *cf.* p. 242) of alcohols to alkenes:

$$\text{MeCH=CH}_2 \overset{\text{H}^\oplus}{\rightleftharpoons} \overset{\oplus}{\text{MeCH—CH}_2} \overset{\text{H}_2\text{O}}{\underset{\text{H}}{\rightleftharpoons}} \overset{\overset{\oplus}{\text{OH}_2}}{\underset{\text{H}}{\text{MeCH—CH}_2}} \overset{-\text{H}^\oplus}{\rightleftharpoons} \overset{\text{OH}}{\underset{\text{H}}{\text{MeCH—CH}_2}}$$

(37)

The formation of the carbonium ion intermediate (37), either directly or *via* an initial π complex, appears to be rate-limiting, and the overall orientation of addition is Markownikov. There is evidence of some ANTI stereoselectivity, but this is not very marked and is dependent on the alkene and on the reaction conditions.

Acids that have weakly nucleophilic anions, e.g. $HSO_4^\ominus$ from dilute aqueous H_2SO_4, are chosen as catalysts, so that their anions will offer little competition to H_2O; any $ROSO_3H$ formed will in any case be hydrolysed to ROH under the conditions of the reaction. Rearrangement of the carbonium ion intermediate may take place, and electrophilic addition of it to as yet unprotonated alkene is also known (p. 185). The reaction is used on the large scale to convert 'cracked' petroleum alkene fractions to alcohols by vapour phase hydration with steam over heterogeneous acid catalysts. Also under acid catalysis, ROH may be added to alkenes to yield ethers, and RCO_2H to yield esters.

Anti-Markownikov hydration of alkenes may be effected indirectly by addition of B_2H_6 (*hydroboration*), followed by oxidation of the resultant trialkylboron (38) with alkaline H_2O_2:

$$\text{MeCH=CH}_2 \xrightarrow{\text{B}_2\text{H}_6} (\text{MeCH}_2\text{CH}_2)_3\text{B} \xrightarrow{\text{H}_2\text{O}_2} \text{MeCH}_2\text{CH}_2\text{OH} + \text{B(OH)}_3$$

(38)   (39)

The diborane is generated (*in situ*, or separately, from $NaBH_4$ and $Et_2O^\oplus BF_3^\ominus$), and probably complexes, as the monomeric BH_3, with the ethereal solvent used for the reaction. BH_3 is a Lewis acid and adds to the least substituted carbon atom of the alkene (Markownikov addition), overall addition is completed by hydride transfer to the adjacent, positively polarised carbon atom:

$$\text{MeCH=CH}_2 \xrightarrow{\text{BH}_3} \overset{\overset{\ominus}{\text{H—BH}_2}}{\underset{\oplus}{\text{MeCH—CH}_2}} \longrightarrow \overset{\text{H}\quad\text{BH}_2}{\text{MeCH—CH}_2}$$

(40)

It may be that (40) has some cyclic character as the overall addition of BH_3 is found, in suitable cases, to be stereoselectively SYN. The first-formed RBH_2 then reacts further with the alkene to yield the trialkylboron, R_3B (38). H_2O_2 oxidation results in fission of the C—B bond to yield the alcohol (39), the nett result being overall anti-Markownikov hydration that is often stereoselectively SYN; yields are usually very good.

7.4.3 Carbonium ions

Protonation of alkenes yields carbonium ions, as we have seen, and in the absence of other effective nucleophiles (e.g. H_2O, p. 184) these ions can act as electrophiles towards as yet unprotonated alkene, (*cf.* p. 107) e.g. with 2-methylpropene (41):

$$Me_2C{=}CH_2$$

$$\text{(41)} \quad \updownarrow H^{\oplus}$$

$$Me_3C^{\oplus} / \ CH_2{-}CMe_2 \ \rightarrow \ Me_3C{-}CH_2{-}\overset{\oplus}{C}Me_2$$

$$\text{(42)} \qquad \text{(41)} \qquad \qquad \text{(43)} \ CH_2{=}CMe_2$$

$$Me_3C{-}CH{=}CMe_2$$

$$\text{(44)}$$

$$\text{(41)} \qquad Me_3C{-}CH_2{-}CMe_2{-}CH_2{-}\overset{\oplus}{C}Me_2 \ \longrightarrow$$

$$\text{(45)}$$

The first-formed cation (42) can add to a second molecule of 2-methyl-propene (41) to yield the new (dimeric) cation (43); this in turn can lose a proton to yield the C_8 alkene (44) or, alternatively, add to a third molecule of alkene to yield the (trimeric) cation (45), and so on. It will be noticed that protonation, and subsequent addition, occurs to give the most stable cation in each case.

2-Methylpropene can be made to continue the process to yield high polymers—*cationic polymerisation*—but most simple alkenes will go no further than di- or tri-meric structures. The main alkene monomers used on the large scale are 2-methylpropene ($\rightarrow$ 'butyl rubber'), and vinyl ethers, $ROCH{=}CH_2$ ($\rightarrow$ adhesives). Cationic polymerisation is often initiated by Lewis acid catalysts, e.g. BF_3, plus a source of initial protons, the co-catalyst, e.g. traces of H_2O etc.; polymerisation occurs readily at low temperatures and is usually very rapid. Many more alkenes are polymerised by a radical induced pathway, however (p. 311).

7.4.4 Hydroxylation

There are a number of reagents that, overall, add two OH groups to alkenes. Thus osmium tetroxide, OsO_4, adds to yield cyclic osmic

esters (46), which can be made to undergo ready hydrolytic cleavage
of their Os—O bonds to yield the 1,2-diol (47):

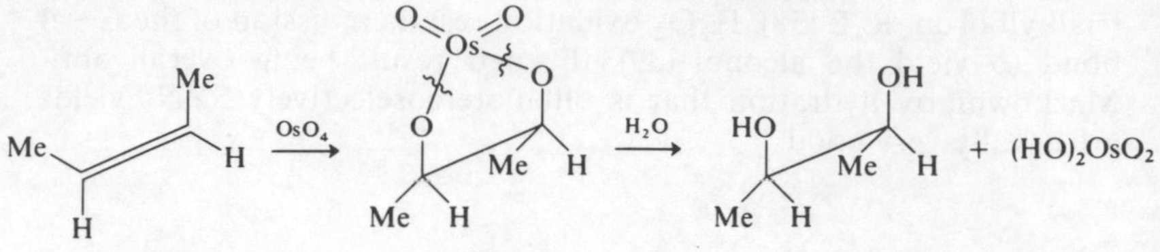

Cis 2-butene (48*a*) thus yields the *meso* 1,2-diol (47), i.e. the overall
hydroxylation is stereoselectively SYN, as would be expected from
Os—O cleavage in a necessarily *cis* cyclic ester (46). The disadvantage
of this reaction as a preparative method is the expense and toxicity
of OsO_4. This may, however, be overcome by using it in catalytic
quantities only, but in association with H_2O_2 which re-oxidises the
osmic acid, $(HO)_2OsO_2$, formed to OsO_4.

Alkaline permanganate $MnO_4^\ominus$, the classical reagent for the
hydroxylation of alkenes, also effects stereoselective SYN addition,
and this by analogy with the above, is thought to proceed *via* cyclic
(*cis*) permanganic esters. It has not proved possible actually to isolate
such species (some of them are detectable spectroscopically), but use
of $Mn^{18}O_4^\ominus$ was found to lead to a 1,2-diol (e.g. 47) in which *both*
oxygen atoms were ^{18}O labelled. Thus both were derived from $MnO_4^\ominus$,
and neither from the H_2O solvent, which provides support for a
permanganic analogue of (46) as an intermediate, provided that
$Mn^{18}O_4^\ominus$ undergoes no ^{18}O exchange with the solvent H_2O under
these conditions—as was shown to be the case. The disadvantage
of $MnO_4^\ominus$ for hydroxylation is that the resultant 1,2-diol (47) is
very susceptible to further oxidation by it.

Peroxyacids, RCO·OOH, will also oxidise alkenes, e.g. *trans*
2-butene (48*b*), by adding an oxygen atom across the double bond to
form an *epoxide* (49):

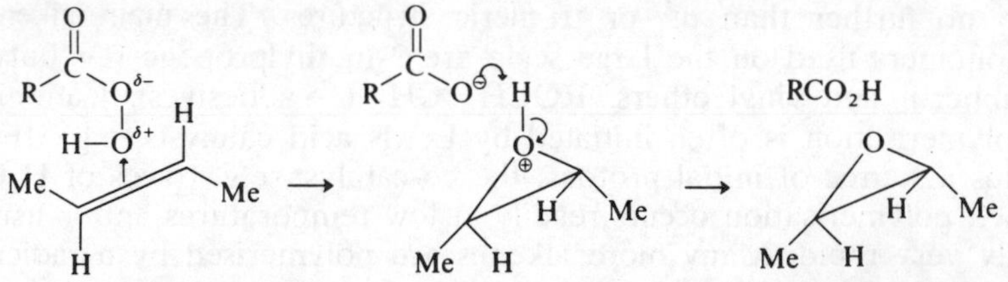

Epoxides, though uncharged, have a formal resemblance to cyclic
bromonium ion intermediates (*cf.* p. 177), but unlike them are stable
and may readily be isolated. They do, however, undergo nucleophilic

attack under either acid- or base-catalysed conditions to yield the 1,2-diol. In either case attack by the nucleophile on a carbon atom will be on the side opposite to the oxygen bridge in (49); such attack on the epoxide will involve inversion of configuration (*cf.* p. 93):

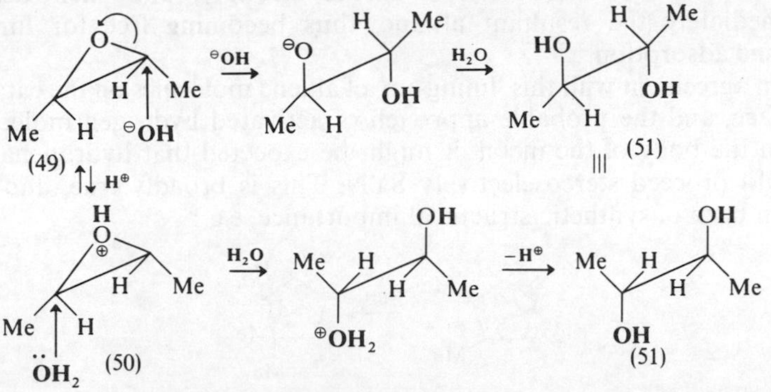

Attack has been shown on only one of the two possible carbon atoms in (49) and (50), though on different ones in the two cases. Attack on the other carbon, in each case, will lead to the same product, the *meso* 1,2-diol (51). By comparing the configuration of (51) with that of the original alkene (48*b*), it will be seen that stereoselective ANTI hydroxylation has occurred.

Thus by suitable choice of reagent, the hydroxylation of alkenes can be made stereoselectively SYN or ANTI at will.

7.4.5 Hydrogenation

The addition of hydrogen to unsaturated compounds is among the commonest, and almost certainly the most useful, of all their addition reactions; because of this it is considered here—though it is not polar in nature—rather than under the reactions of radicals. Direct addition of hydrogen normally involves heterogeneous catalysis by metals such as Ni, Pt, Pd, Ru, Rh. The atoms in the surface of a crystal of metal catalyst will clearly differ from their fellows in the body of the crystal in having 'residual combining power' directed away from the surface. It is significant in this context that both alkenes, e.g. ethene, and hydrogen, react exothermically, and reversibly, with the catalytic metals, e.g. nickel. With the alkene this presumably involves its π electrons as alkanes are not similarly adsorbed. No π electrons are available in the hydrogen molecule either, and its adsorption must involve considerable weakening of its σ bond, though not necessarily complete fission to yield H· atoms.

The actual spacings of the metal atoms in the surface will clearly be of importance in making one face of a metal crystal catalytically effective, and another not, depending on how closely the actual atom spacings approximate to the bond distances in alkene and hydrogen molecules. In practice only a relatively small proportion of the total metal surface is found to be catalytically effective—the so-called 'active points'. These adsorb alkene strongly, and then desorb immediately the resultant alkane, thus becoming free for further alkene adsorption.

In agreement with this 'lining-up' of alkene molecules on the catalyst surface, and the probable approach of activated hydrogen molecules from the body of the metal, it might be expected that hydrogenation would proceed stereoselectively SYN. This is broadly true, and has often been of synthetic/structural importance, e.g.:

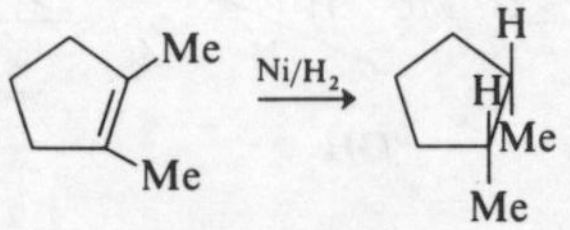

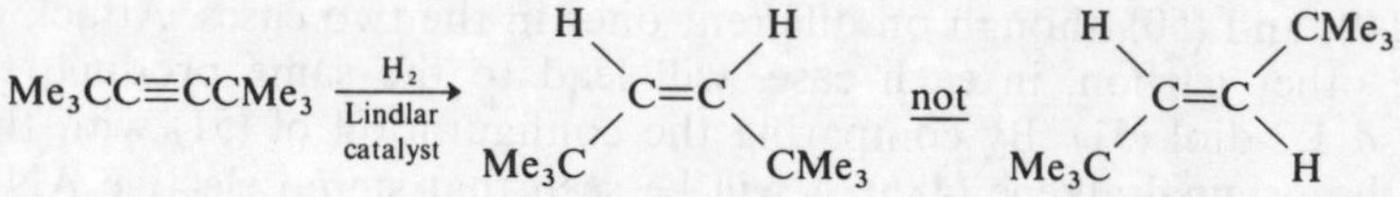

(52)   (53)

Alkynes can often be reduced selectively to the alkene by use of the *Lindlar* catalyst [Pd on $CaCO_3$, partly 'poisoned' with $Pb(OAc)_2$]. Here again SYN stereoselectivity is observed despite the fact that this will lead to the more crowded, thermodynamically less stable, *cis*-alkene, i.e. (52) rather than (53).

Stereoselectivity is often short of being 100% SYN, and can be influenced by reaction conditions, sometimes being very far short of 100% SYN. The actual mechanism of hydrogenation has received a good deal of detailed study, by use of D_2 etc., and is in fact highly complex; among other things, hydrogen exchange takes place with the alkene. It has been established that the two hydrogens are not added to the alkene simultaneously, however, and the reason for <100% SYN stereoselectivity thus becomes apparent. It has also been shown that *cis* alkenes, e.g. *cis* 2-butene, are usually hydrogenated much more rapidly than *trans*, e.g. *trans* 2-butene; in either case, the rate of hydrogenation falls with increasing substitution in the alkene.

Overall addition of hydrogen—stereoselectively SYN—can also be effected by reaction of trialkyl borons (*cf.* p. 184) with carboxylic acids:

$$Me_2C=CH_2 \xrightarrow{B_2H_6} (Me_2CHCH_2)_3B \xrightarrow{RCO_2H} Me_2CHCH_3 + (RCO_2)_3B$$

7.4.6 Ozonolysis

The addition of ozone to alkenes to form ozonides, and the subsequent decomposition of the latter to yield carbonyl compounds, has long been known;

$$R_2C{=}CR_2' \xrightarrow{O_3} \begin{bmatrix} R_2C{=}CR_2' \\ + \\ O_3 \end{bmatrix} \xrightarrow{H_2/Pt} R_2C{=}O + O{=}CR_2' + H_2O$$

ozonide

but the structure of the ozonide has been a matter of some debate. It is easy to envisage 1,3-dipolar addition of ozone initiated by its electrophilic end, and the crystalline adduct (54) has actually been isolated from the reaction of ozone at −70° with the alkene (55):

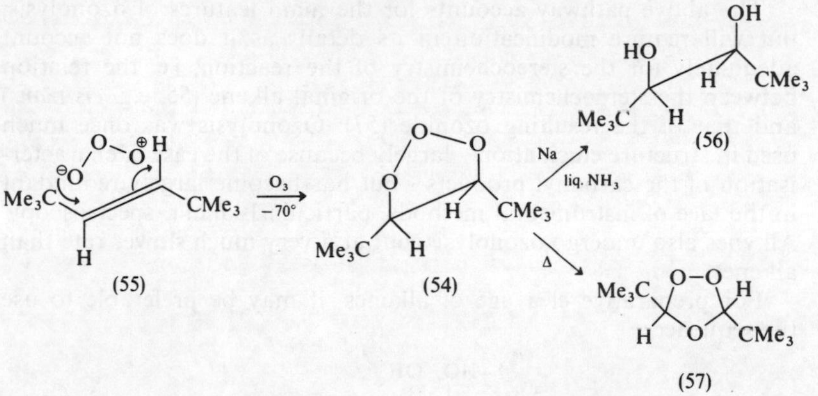

Its structure has been confirmed by n.m.r. spectroscopy, and by its reduction with sodium and liquid ammonia to the 1,2-diol (56). It is difficult to see how catalytic reduction of (54) could lead directly to the normal carbonyl end-products, however, and on raising the temperature it is found that (54) is converted into (57): which can (and *does*) yield the normal carbonyl products on catalytic reduction.

(54) is referred to as a *molozonide*, (57) as the normal ozonide, and the conversion of the former into the latter is believed to proceed by the pathway:

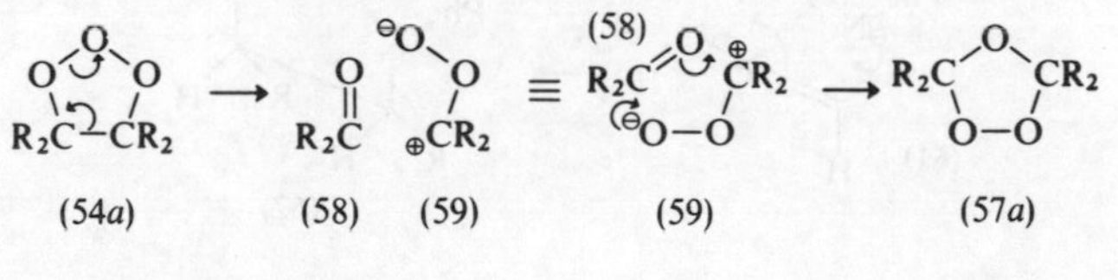

The suggested fragments from (54a) are a carbonyl compound (58) and a peroxy zwitterion (59), the latter then effecting a 1,3-dipolar addition on the former to yield the ozonide (57a). No such species as (58) and (59) have been isolated, but evidence in their support includes: (a) the isolation of cyclic bis peroxides (60)—obtainable by the 1,3-dipolar addition of (59) to itself, and (b) isolation of the ozonide (57b), in addition to the expected (57c), from the ozonolysis of $Me_2C=CMe_2$ in the presence of CH_2O—this resulting from reaction of (59) with the added aldehyde (a 'crossover' experiment):

$$R_2C \underset{O-O}{\overset{O-O}{<}} CR_2 \qquad Me_2C \underset{O-O}{\overset{O}{<}} CH_2 \qquad Me_2C \underset{O-O}{\overset{O}{<}} CMe_2$$

$$(60) \qquad\qquad (57b) \qquad\qquad (57c)$$

Less is known about the details of the conversion of ozonides (57) into the final carbonyl products.

The above pathway accounts for the main features of ozonolysis, but will require modification in its details as it does not account adequately for the stereochemistry of the reaction, i.e. the relation between the stereochemistry of the original alkene (55, e.g. *cis/trans*) and that of the resulting ozonide (57). Ozonolysis was once much used in structure elucidation—largely because of the ease of characterisation of the carbonyl products—but has become largely redundant in the face of instrumental methods, particularly n.m.r. spectroscopy. Alkynes also undergo ozonolysis, but at a very much slower rate than alkenes.

For preparative cleavage of alkenes, it may be preferable to use the sequence:

$$R_2C=CR'_2 \xrightarrow[\substack{or \\ OsO_4}]{MnO_4^{\ominus}} R_2C\overset{\overset{HO}{|}}{-}\overset{\overset{OH}{|}}{C}R'_2 \xrightarrow{NaIO_4} R_2C=O + O=CR'_2$$

The reaction may be carried out in one stage, the sodium metaperiodate used to cleave the 1,2-diol being present in sufficient excess to reoxidise the catalytic quantity only of $MnO_4^{\ominus}$ or OsO_4 needed for the fast stage.

1,3-Dipolar addition to alkenes also occurs with species other than ozone, often to give products much more stable than the labile molozonides (54), e.g. addition of azides (61) to give dihydrotriazoles (62):

(61) → (62)

1,3-Dipolar addition to alkenes is considered further subsequently (p. 339).

7.5 ADDITION TO CONJUGATED DIENES

Conjugated dienes, e.g. butadiene (63) are somewhat more stable than otherwise similar dienes in which the double bonds are not conjugated (*cf.* p. 11). This is reflected in their respective heats of hydrogenation (p. 16), though the delocalisation energy consequent on the extended π orbital system is only of the order of 17 kJ (4 kcal) mol^{-1}; conjugated dienes are found nevertheless to undergo addition reactions somewhat more rapidly than non-conjugated dienes. This occurs because the intermediates (and, more importantly, the transition states that precede them) arising from initial attack by *either* electrophiles (64) *or* radicals (65) are of the allylic type (*cf.* pp. 104, 302), and are stabilised by delocalisation to a considerably greater extent than was the initial diene. They are also stabilised with respect to the corresponding intermediates (66 and 67) obtained on similar addition to a simple alkene:

$$
\begin{array}{ccc}
\underset{\text{Br}}{\overset{|}{\text{CH}_2}}-\overset{\cdot}{\text{CH}}-\text{CH}=\text{CH}_2 & & \underset{\text{Br}}{\overset{|}{\text{CH}_2}}-\overset{\oplus}{\text{CH}}-\text{CH}=\text{CH}_2 \\
\updownarrow & \xleftarrow[\text{radical}]{\text{Br}_2} \quad \text{CH}_2=\text{CH}-\text{CH}=\text{CH}_2 \quad \xrightarrow[\text{electrophilic}]{\text{Br}_2} & \updownarrow \\
& (63) & \\
\underset{\text{Br}}{\overset{|}{\text{CH}_2}}-\text{CH}=\text{CH}-\overset{\cdot}{\text{CH}}_2 & & \underset{\text{Br}}{\overset{|}{\text{CH}_2}}-\text{CH}=\text{CH}-\overset{\oplus}{\text{CH}}_2 \\
(65) & & (64)
\end{array}
$$

$$
\underset{\text{Br}}{\overset{|}{\text{CH}_2}}-\overset{\cdot}{\text{CH}}_2 \quad \xleftarrow[\text{radical}]{\text{Br}_2} \quad \text{CH}_2=\text{CH}_2 \quad \xrightarrow[\text{electrophilic}]{\text{Br}_2} \quad \underset{\text{Br}}{\overset{|}{\text{CH}_2}}-\overset{\oplus}{\text{CH}}_2
$$
$$
(67) \hspace{7cm} (66)
$$

7.5.1 Electrophilic addition

Initial attack will always take place on a terminal carbon atom of the conjugated system, otherwise the carbonium ion intermediate (64), that is stabilised by delocalisation, would not be obtained. Completion of overall addition by nucleophilic attack on (64) can however take place at C_2 [*1,2-addition*, (*a*) → (68)] or C_4 [*1,4-addition*, (*b*) → (69)]:

$$
\underset{\overset{|}{\text{Br}}}{\overset{\overset{\text{Br}}{|}}{\text{CH}_2-\text{CH}-\text{CH}=\text{CH}_2}} \xleftarrow[\text{1,2 addition}]{(a)} \underset{(a)\ ^{\ominus}\text{Br}}{\overset{\overset{\text{Br}}{|}}{\text{CH}_2-\overset{\delta+}{\text{CH}}\text{---}\text{CH}\text{---}\overset{\delta+}{\text{CH}}_2}}_{(b)^{\ominus}\text{Br}} \xrightarrow[\text{1.4 addition}]{(b)} \underset{\overset{|}{\text{Br}}}{\overset{\overset{\text{Br}}{|}}{\text{CH}_2-\text{CH}=\text{CH}-\text{CH}_2}}
$$

(68)   (64)   (69)

Both products are commonly obtained, but their relative proportions depends very much on the reaction conditions, e.g. temperature. Thus HCl with butadiene (63) at −60° yields only 20–25 % of the 1,4-adduct (the rest being the 1,2-adduct), while at higher temperatures ≈ 75 % of the 1,4-adduct was obtained. There is reason to believe that at the lower temperature the control is *kinetic* (*cf.* p. 42), the 1,2-adduct being formed more rapidly from (64) than is the 1,4-adduct; while at higher temperatures, and/or with longer reaction times, *equilibrium* or *thermodynamic* control operates, and the thermodynamically more stable 1,4-adduct is then the major product. This is borne out by the fact that at higher temperatures *pure* 1,2- or 1,4-adduct can each be converted into the same equilibrium mixture of 1,2- + 1,4- under the conditions of the reaction. 1,4-addition is also favoured by increasing solvent polarity.

The question arises whether, in the addition of, for example, bromine, the 1,4-adduct might be obtained *via* an unstrained, five-membered cyclic bromonium ion (70). This would lead, on nucleophilic cleavage by $Br^{\ominus}$, to the *cis* 1,4-dibromide (71):

$$
\underset{(70)}{\overset{\overset{\text{HC}=\text{CH}}{\diagup\ \ \diagdown}}{\underset{\underset{\overset{\oplus}{\text{Br}}}{}}{\text{H}_2\text{C}\diagdown\ \ \diagup\text{CH}_2}}} \xrightarrow{Br^{\ominus}} \underset{(71)}{\overset{\overset{\text{HC}=\text{CH}}{\diagup\ \ \diagdown}}{\text{BrCH}_2\ \ \text{CH}_2\text{Br}}} \qquad \underset{(72)}{\overset{\overset{\text{CH}_2\text{Br}}{\diagup}}{\underset{\text{BrCH}_2}{\overset{\text{HC}=\text{CH}}{\diagup}}}}
$$

in fact, only the *trans* 1,4-dibromide (72) is obtained. Species such as (70) thus cannot be involved, and it seems likely that the common intermediate is the ion pair (73) involving a delocalised carbonium ion; interconversion of 1,2- and 1,4-adducts, (68) and (69) respectively, could also proceed *via* such an intermediate:

$$
\underset{\overset{|}{\text{Br}}}{\overset{\overset{\text{Br}}{|}}{\text{CH}_2-\text{CH}-\text{CH}=\text{CH}_2}} \rightleftarrows \underset{\overset{|}{\text{Br}}}{\overset{\overset{\text{Br}^{\ominus}}{}}{\text{CH}_2-\overset{\delta+}{\text{CH}}\text{---}\text{CH}\text{---}\overset{\delta+}{\text{CH}}_2}} \rightleftarrows \underset{\overset{|}{\text{Br}}}{\overset{\overset{\text{Br}}{|}}{\text{CH}_2-\text{CH}=\text{CH}-\text{CH}_2}}
$$

(68)   (73)   (69)

With unsymmetrical dienes (74*a* and 74*b*) and unsymmetrical adducts, the problem of orientation of addition (*cf.* p. 181) arises. Initial attack will still be on a terminal carbon atom of the conjugated system so that a delocalised allylic intermediate is obtained, but preferential attack will be on the terminal carbon that will yield the more stable of the two possible cations; i.e. (75) rather than (76), and (77) rather than (78):

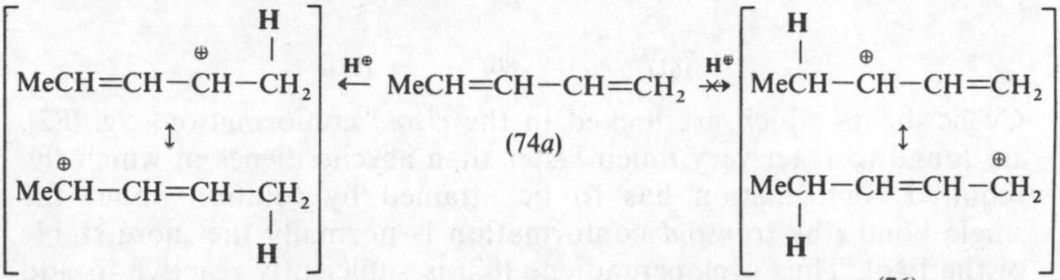

$$
\left[
\begin{array}{c}
\overset{\displaystyle H}{\underset{\displaystyle |}{}} \\
\overset{\oplus}{MeCH=CH-CH-CH_2} \\
\updownarrow \\
\overset{\oplus}{MeCH-CH=CH-CH_2} \\
\underset{\displaystyle |}{\underset{\displaystyle H}{}}
\end{array}
\right]
\overset{H^{\oplus}}{\leftarrow}
MeCH=CH-CH=CH_2
\overset{H^{\oplus}}{\nleftrightarrow}
\left[
\begin{array}{c}
\overset{\displaystyle H}{\underset{\displaystyle |}{}} \\
MeCH-\overset{\oplus}{CH}-CH=CH_2 \\
\updownarrow \\
MeCH-CH=CH-\overset{\oplus}{CH_2} \\
\underset{\displaystyle |}{\underset{\displaystyle H}{}}
\end{array}
\right]
$$

(74*a*)

(75)   (76)

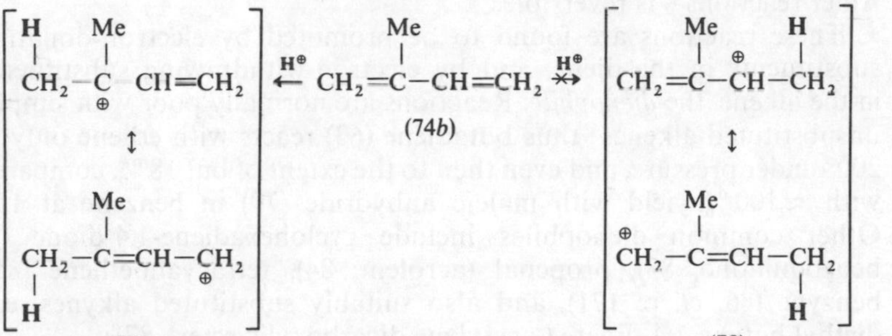

$$
\left[
\begin{array}{c}
\overset{H}{|} \quad \overset{Me}{|} \\
CH_2-\underset{\oplus}{C}\;\;CH=CH_2 \\
\updownarrow \\
\overset{Me}{|} \\
CH_2-C=CH-\underset{\oplus}{CH_2} \\
\underset{|}{H}
\end{array}
\right]
\overset{H^{\oplus}}{\leftarrow}
\overset{Me}{\underset{|}{CH_2=C-CH=CH_2}}
\overset{H^{\oplus}}{\nleftrightarrow}
\left[
\begin{array}{c}
\overset{Me}{|} \quad \overset{H}{|} \\
CH_2=C-\overset{\oplus}{CH}-CH_2 \\
\updownarrow \\
\overset{Me}{|} \\
\underset{\oplus}{CH_2}-C=CH-CH_2 \\
\underset{|}{H}
\end{array}
\right]
$$

(74*b*)

(77)   (78)

Among other addition reactions dienes undergo catalytic hydrogeneration (1,2- and 1,4-), epoxidation (1,2- only, and more slowly than the corresponding simple alkenes), but they seldom undergo hydration.

7.5.2   Diels–Alder reaction

This reaction, of which the classical example is between butadiene (63) and maleic anhydride (79),

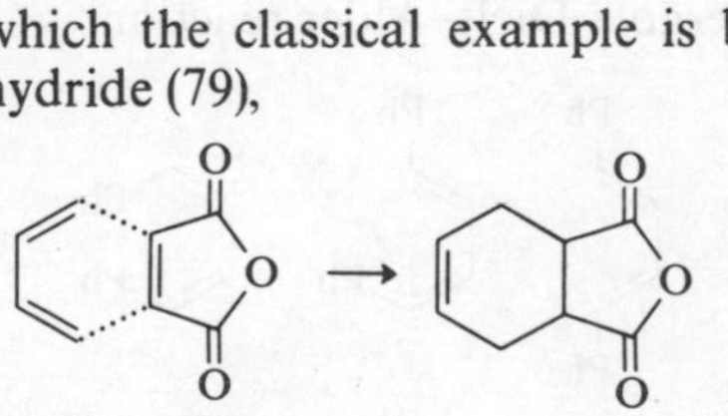

(63)   (79)

involves the 1,4-addition of an alkene to a conjugated diene. The reaction is usually easy and rapid, of very broad scope, and involves the formation of carbon–carbon bonds, hence its synthetic utility. The diene must react in the *cisoid* (80), rather than the *transoid* (81), conformation:

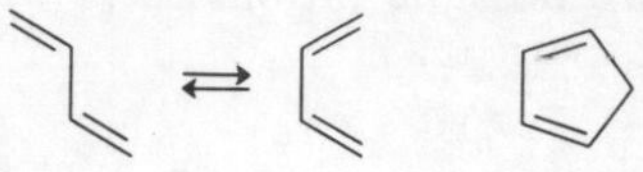

(81)   (80)   (82)

Cyclic dienes which are locked in the *cisoid* conformation, e.g. (82), are found to react very much faster than acyclic dienes in which the required conformation has to be attained by rotation about the single bond (the *transoid* conformation is normally the more stable of the two). Thus cyclopentadiene (82) is sufficiently reactive to add to itself to form a tricyclic dimer, whose formation—like most Diels–Alder reactions—is reversible.

These reactions are found to be promoted by electron-donating substituents in the diene, and by electron-withdrawing substituents in the alkene, the *dienophile*. Reactions are normally poor with simple, unsubstituted alkenes; thus butadiene (63) reacts with ethene only at 200° under pressure, and even then to the extent of but 18%, compared with ≈100% yield with maleic anhydride (79) in benzene at 15°. Other common dienophiles include cyclohexadiene-1,4-dione (*p*-benzoquinone, 83), propenal (acrolein, 84), tetracyanoethene (85), benzyne (86, *cf.* p. 171), and also suitably substituted alkynes, e.g. diethyl butyne-1,4-dioate ('acetylene dicarboxylic ester', 87):

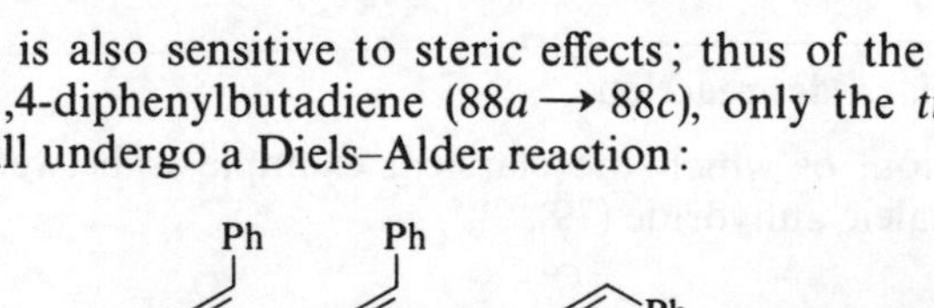

(83)   (84)   (85)   (86)   (87)

The reaction is also sensitive to steric effects; thus of the three iso-merides of 1,4-diphenylbutadiene (88a → 88c), only the *trans–trans* form (88a) will undergo a Diels–Alder reaction:

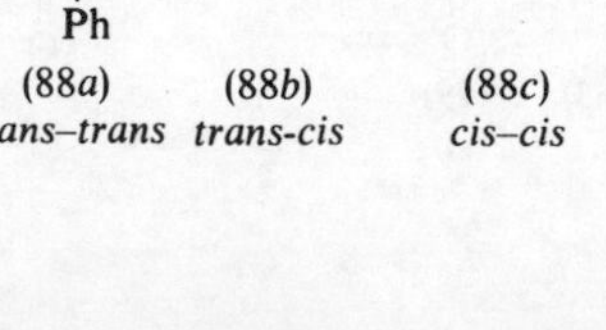

(88a)   (88b)   (88c)
trans–trans   *trans–cis*   *cis–cis*

Diels–Alder reactions are found to be little influenced by the introduction of radicals (*cf.* p. 292), or by changes in the polarity of the solvent: they are thus unlikely to involve either radical or ion pair intermediates. They are found to proceed stereoselectively SYN with respect both to the diene and to the dienophile, and are believed to take place *via* a concerted pathway in which bond-formation and bond-breaking occur more or less simultaneously, though not necessarily to the same extent, in the transition state. This cyclic transition state is a planar, aromatic type, with consequent stabilisation because of the cyclic overlap that can occur between the six *p* orbitals of the constituent diene and dienophile. Such *pericyclic* reactions are considered further below (p. 329).

7.6 NUCLEOPHILIC ADDITION

As we saw above, the introduction of electron-withdrawing groups into an aromatic nucleus tended to inhibit electrophilic substitution (p. 150), and to make nucleophilic substitution possible (p. 165): exactly the same is true of addition to alkenes. Thus we have already seen that the introduction of electron-withdrawing groups tends to inhibit addition initiated by electrophiles (p. 180); the same groups are also found to promote addition initiated by nucleophiles. A partial order of effectiveness is found to be,

$$CHO > COR > CO_2R > CN > NO_2$$

but SOR, SO$_2$R and F also act in the same way. Such substituents operate by reducing π electron density at the alkene carbon atoms, thereby facilitating the approach of a nucleophile, $Y^\ominus$, but more particularly by delocalising the -ve charge in the resultant carbanion intermediate, e.g. (89) and (90). This delocalisation is generally more effective when it involves mesomeric delocalisation (89), rather than only inductive electron-withdrawal (90):

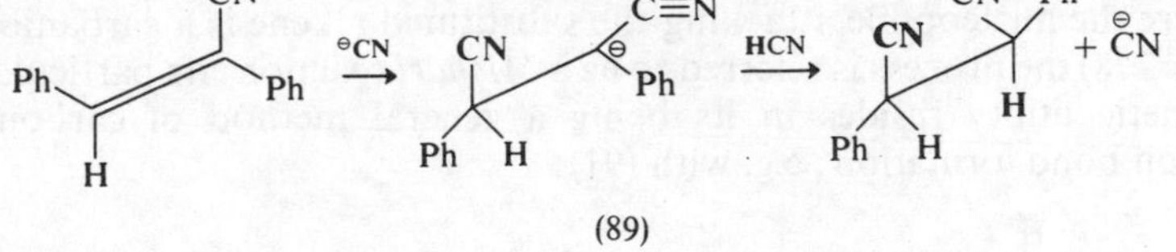

(89)

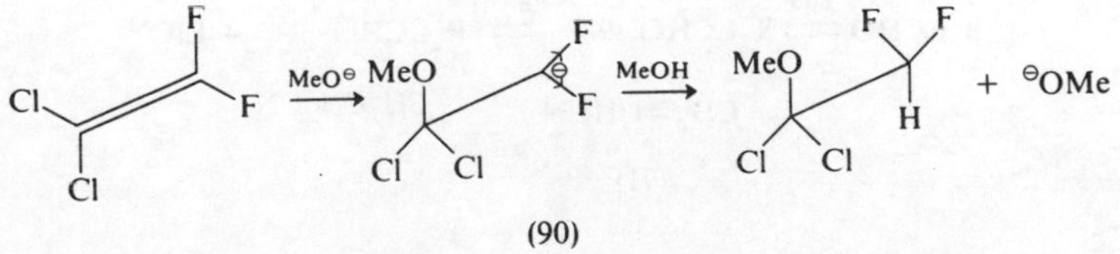

(90)

The orientation of addition of an unsymmetrical adduct, HY or XY, to an unsymmetrically substituted alkene will be defined by the preferential formation of the more stabilised carbanion, as seen above (*cf.* preferential formation of the more stabilised carbonium ion in electrophilic addition, p. 181). There is little evidence available about stereoselectivity in such nucleophilic additions to acyclic alkenes. Nucleophilic addition also occurs with suitable alkynes, generally more readily than with the corresponding alkenes.

A number of these nucleophilic addition reactions are of considerable synthetic importance:

7.6.1   Cyanoethylation

Among the more important of these reactions of general synthetic significance is one in which ethene carries a cyano-substituent (acrylonitrile, 91). Attack of $Y^{\ominus}$ or Y: on the unsubstituted carbon, followed by abstraction of a proton from the solvent, leads overall to the attachment of a 2-cyanoethyl group to the original nucleophile;

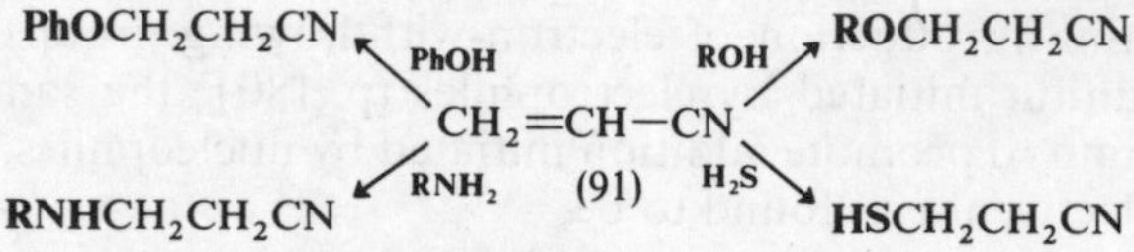

$$PhOCH_2CH_2CN \xleftarrow{\ PhOH\ } CH_2{=}CH{-}CN \xrightarrow{\ ROH\ } ROCH_2CH_2CN$$

$$RNHCH_2CH_2CN \xleftarrow{\ RNH_2\ } (91) \xrightarrow{\ H_2S\ } HSCH_2CH_2CN$$

the procedure is thus referred to as *cyanoethylation*. It is often carried out in the presence of base in order to convert HY into the more powerfully nucleophilic $Y^{\ominus}$. The synthetic utility of cyanoethylation resides in the incorporation of a three carbon unit, in which the terminal cyano group may be modified by reduction, hydrolysis, etc., preparatory to further synthetic operations.

7.6.2   Michael reaction

Where the nucleophile attacking the substituted alkene is a carbanion (*cf.* p. 278) the process is referred to as a *Michael* reaction; its particular synthetic utility resides in its being a general method of carbon–carbon bond formation; e.g. with (91):

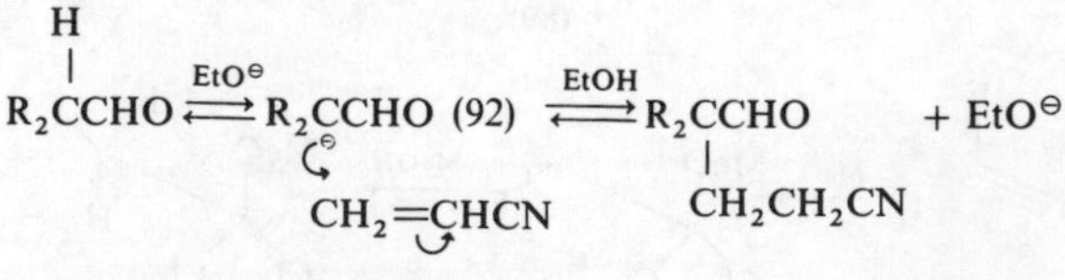

$$\underset{\substack{| \\ R_2CCHO}}{\overset{H}{}} \underset{EtO^{\ominus}}{\rightleftharpoons} R_2\overset{\ominus}{\underset{C}{C}}CHO\ (92) \underset{EtOH}{\overset{}{\rightleftharpoons}} \underset{\substack{| \\ CH_2CH_2CN}}{R_2CCHO} + EtO^{\ominus}$$

$$CH_2{=}CHCN$$

$$(91)$$

The reaction is promoted by a variety of bases, usually in catalytic quantities only, which generate an equilibrium concentration of carbanion (92); it is reversible, and the rate-limiting step is believed to be carbon–carbon bond formation, i.e. the reaction of the carbanion (92) with the substituted alkene (91). Its general synthetic utility stems from the wide variety both of substituted alkenes and of carbanions that may be employed; the most common carbanions are probably those from $CH_2(CO_2Et)_2$—see below, $MeCOCH_2CO_2Et$, $NCCH_2$-CO_2Et, RCH_2NO_2 etc. Many Michael reactions involve C=C—C=O as the substituted alkene.

7.6.3 Addition to C=C—C=O

Among the commonest substituents 'activating' an alkene to nucleophilic attack is the C=O group, in such $\alpha\beta$-unsaturated compounds as RCH=CHCHO, RCH=CHCOR′, RCH=CHCO$_2$Et etc. As the carbonyl group in such compounds can itself undergo nucleophilic attack (*cf.* p. 201), the question arises as to whether addition is predominantly to C=C, to C=O, or *conjugate* (1,4-) to the overall C=C—C=O system. In fact, the last type of addition (93) normally yields the same product (94) as would be obtained from addition to C=C, owing to tautomerisation of the first formed enol (95), e.g. with the Grignard reagent PhMgBr followed by acidification:

$$R_2C=CH-\underset{\underset{R}{|}}{C}=O \xrightarrow{\overset{\delta-}{Ph}\overset{\delta+}{MgBr}} R_2\overset{|}{\underset{Ph}{C}}-CH=\underset{\underset{R}{|}}{C}-O^{\ominus}\overset{\oplus}{MgBr}$$

(93)

$$\downarrow H^{\ominus}/H_2O$$

$$R_2\overset{H}{\underset{Ph}{\overset{|}{C}}}-\overset{|}{CH}-\underset{\underset{R}{|}}{C}=O \ \rightleftharpoons\ R_2\underset{Ph}{\overset{|}{C}}-CH=\underset{\underset{R}{|}}{C}-OH$$

(94)   (95)

Incidentally, 1,4-*electrophilic* addition (e.g. HBr) also yields the C=C adduct (96) for the same reason, and can be looked upon formally as acid-catalysed (97) addition of the nucleophile $Br^{\ominus}$:

$$R_2C=CH-\underset{\underset{R}{|}}{C}=O \ \overset{H^{\oplus}}{\rightleftharpoons}\ R_2\overset{\oplus}{C}-CH=\underset{\underset{R}{|}}{C}-OH$$

(97)

$$\downarrow Br^{\ominus}$$

$$R_2\overset{H}{\underset{Br}{\overset{|}{C}}}-\overset{|}{CH}-\underset{\underset{R}{|}}{C}=O \ \leftrightarrows\ R_2\underset{Br}{\overset{|}{C}}-CH=\underset{\underset{R}{|}}{C}-OH$$

(96)

Less powerful nucleophiles such as ROH can also be made to add (1,4-) under acid catalysis.

Whether nucleophilic addition is predominantly conjugate (1,4-) or to C═O may depend on whether the reaction is reversible or not; if it is reversible, then the control of product can be thermodynamic (equilibrium *cf.* p. 43), and this will favour 1,4-addition. This is so because the C═C adduct (98) obtained from 1,4-addition will tend to be thermodynamically more stable than the C═O adduct (99), because the former contains a residual C═O π bond, and this is stronger than the residual C═C π bond in the latter:

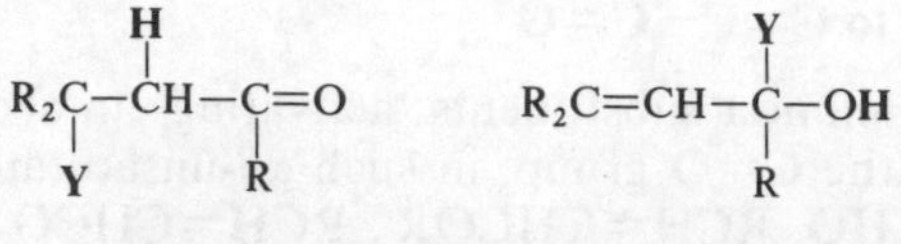

$$R_2C-CH-C=O \qquad R_2C=CH-C-OH$$

(98)   (99)

Steric hindrance at one site can, however, be very potent at promoting addition at the other; thus PhCH═CHCHO was found to undergo ≈ 100 % C═O addition with PhMgBr, whereas PhCH═CHCOCMe₃ underwent ≈ 100 % C═C addition with the same reagent. This also reflects decreasing 'carbonyl' reactivity of the C═O group in the sequence aldehyde > ketone > ester (*cf.* p. 202), with consequent increase in the proportion of C═C addition.

Amines, thiols, $^{\ominus}$OH (p. 222) etc. will also add to the β-carbon atom of αβ unsaturated carbonyl compounds and esters, but the most important reactions of C═C─C═O systems are in Michael reactions with carbanions: reactions in which carbon–carbon bonds are formed. A good example is the synthesis of 1,1-dimethylcyclohexan-3,5-dione (dimedone, 100) starting from 2-methylpent-2-ene-4-one (mesityl oxide, 101) and the carbanion $^{\ominus}$CH(CO₂Et)₂:

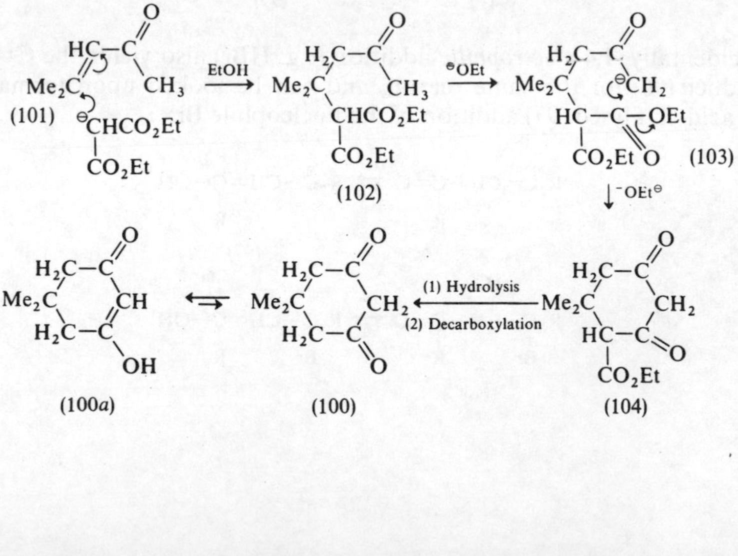

The Michael reaction as such is complete on formation of the adduct (102), but treatment of this with base ($^{\ominus}$OEt) yields the carbanion (103), which can, in turn, attack the carbonyl carbon atom of one of the CO_2Et groups; $^{\ominus}$OEt, a good leaving group, is expelled resulting in cyclisation to (104)—reminiscent of a *Dieckmann* reaction (*cf.* p. 226). Hydrolysis and decarboxylation of the residual CO_2Et group then yields the desired end-product dimedone (100), which exists to the extent of $\approx 100\%$ in the enol form (100*a*).

Dimedone is of value as a reagent for the differential characterisation, and separation, of aldehydes and ketones as it readily yields derivatives (105) with the former, but not with the latter, from a mixture of the two:

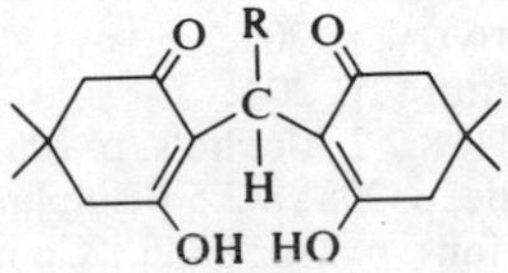

(105)

This selectivity is no doubt due largely to steric reasons.

8
Nucleophilic addition to C=O

Carbonyl compounds exhibit dipole moments (μ) because the oxygen
atom of the C=O group is more electronegative than the carbon:

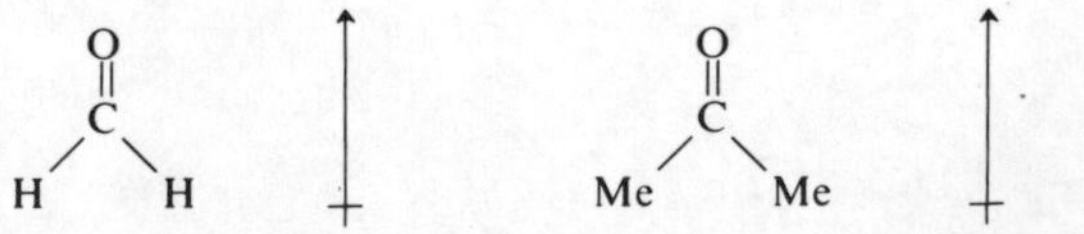

$$\mu = 2\cdot3\,\text{D} \qquad\qquad \mu = 2\cdot8\,\text{D}$$

As well as the C→O inductive effect in the σ bond joining the two
atoms, the more readily polarisable π electrons are also affected (*cf.*
p. 22) so that the carbonyl group is best represented by a hybrid

structure (1):

$$R_2C=O \leftrightarrow R_2\overset{\oplus}{C}-\overset{\ominus}{O} \quad \text{i.e. } R_2\overset{\delta+}{C}\overset{\delta-}{-}O \equiv R_2C \rightarrow O$$

$$(1a) \qquad\qquad (1b) \qquad\qquad\qquad (1ab)$$

We would expect the C=O linkage, by analogy with C=C (p. 175), to undergo addition reactions; but whereas polar attack on the latter is normally initiated only by electrophiles, attack on the former—because of its bipolar nature—could be initiated either by electrophilic attack of $X^\oplus$ or X on oxygen or by nucleophilic attack of $Y^\ominus$ or Y: on carbon (radical-induced addition reactions of carbonyl compounds are rare). In practice, initial electrophilic attack on oxygen is of little significance except where the electrophile is an acid (or a Lewis acid), when rapid, reversible protonation may be a prelude to slow, rate-limiting attack by a nucleophile on carbon, to complete the addition, i.e. the addition is then acid-catalysed.

Protonation will clearly increase the positive character of the carbonyl carbon atom (2),

$$R_2C=O: \overset{H^\oplus}{\rightleftarrows} R_2\overset{\oplus}{C}=OH \leftrightarrow R_2\overset{\oplus}{C}-OH$$

$$(2)$$

and thereby facilitate nucleophilic attack upon it. Similar activation, though to a lesser extent, can also arise through hydrogen bonding of an acid (3), or even of a hydroxylic solvent (4), to the carbonyl oxygen atom:

$$R_2\overset{\delta+}{C}=O: \qquad\qquad R_2\overset{\delta+}{C}=O:$$
$$H-A^{\delta-} \qquad\qquad H-\overset{\delta-}{O} \diagup R'$$

$$(3) \qquad\qquad\qquad (4)$$

In the absence of such activation, weak nucleophiles, e.g. H_2O:, may react only very slowly, but strong ones, e.g. $^\ominus CN$, do not require such aid. Additions may also be base-catalysed, the base acting by converting the weak nucleophile HY into the stronger one, $Y^\ominus$, e.g. HCN + base $\rightarrow {}^\ominus CN$. Further, while acids may activate the carbonyl carbon atom to nucleophilic attack, they may simultaneously reduce the effective concentration of the nucleophile, e.g. $^\ominus CN + HA \rightarrow HCN + A^\ominus$, $RNH_2 + HA \rightarrow RNH_3^\oplus + A^\ominus$. Many simple addition reactions of carbonyl compounds are thus found to have an optimum pH; this can be of great importance for preparative purposes.

8.1 STRUCTURE AND REACTIVITY

In simple nucleophilic additions where the rate-limiting step is attack by $Y^{\ominus}$, the positive character of the carbonyl carbon atom is reduced on going from the starting material (5) to the transition state (6):

$$\overset{\delta+}{R_2}\overset{\delta-}{C}=\overset{Y^{\ominus}}{O} \rightleftarrows \left[\overset{\delta-}{R_2}\overset{\delta-}{\underset{\underset{Y^{\delta-}}{\vdots}}{C}}\overset{\delta-}{O} \right]^{*} \rightleftarrows R_2\underset{Y}{\overset{|}{C}}-O^{\ominus} \overset{HY}{\rightleftarrows} R_2\underset{Y}{\overset{|}{C}}-OH + Y^{\ominus}$$

$$(5) \qquad\qquad (6)$$

We should thus expect the rate of addition to be reduced by electron-donating R groups and enhanced by electron-withdrawing ones; this is borne out by the observed sequence:

$$\overset{H}{\underset{H}{>}}C=O > \overset{R}{\underset{H}{>}}C=O > \overset{R}{\underset{R}{>}}C=O$$

R groups in which the C=O group is conjugated with C=C (1,4-addition can also compete here, *cf.* p. 197), or with a benzene ring, also exhibit slower addition reactions than their saturated analogues. This is because the stabilisation, through delocalisation, in the initial carbonyl compounds (7 and 8) is lost on proceeding to the adducts (9 and 10), and to the transition states that precede them:

$$\left[R_2C=CH-\overset{R}{\overset{|}{C}}=O \leftrightarrow R_2\overset{\oplus}{C}-CH=\overset{R}{\overset{|}{C}}-O^{\ominus} \right] \rightarrow R_2C=CH-\overset{R}{\underset{Y}{\overset{|}{\underset{|}{C}}}}-O^{\ominus}$$

$$(7) \qquad\qquad\qquad\qquad (9)$$

$$\left[\langle\!\!\!\bigcirc\!\!\!\rangle-\overset{R}{\overset{|}{C}}=O \leftrightarrow \langle\!\!\!\overset{\oplus}{\bigcirc}\!\!\!\rangle=\overset{R}{\overset{|}{C}}-O^{\ominus} \right] \rightarrow \langle\!\!\!\bigcirc\!\!\!\rangle-\overset{R}{\underset{Y}{\overset{|}{\underset{|}{C}}}}-O^{\ominus}$$

$$(8) \qquad\qquad\qquad\qquad (10)$$

In the above examples steric, as well as electronic, effects could be influencing relative rates of reaction, but the influence of electronic effects alone may be seen in the series of compounds (11):

$$X-\langle\!\!\!\bigcirc\!\!\!\rangle-\overset{H}{\overset{|}{C}}=O \qquad \text{Relative rates: } X=NO_2 > H > OMe$$

$$(11)$$

So far as steric effects are concerned, the least energy-demanding direction of approach by the nucleophile to the carbonyl carbon atom will be from above, or below, the substantially planar carbonyl compound. It is also likely to be from slightly to the rear of the carbon atom (*cf.* 12), because of potential coulombic repulsion between the approaching nucleophile and the high electron density at the carbonyl oxygen atom:

$$
\begin{array}{c}
R \\
\diagdown \\
R \diagup \\
\end{array}
\overset{Y^{\ominus}}{\underset{Y^{\ominus}}{C - O}}
$$

(12)

Increasing bulk in the R groups will slow the reaction as the sp^2 hybridised carbon atom in the original carbonyl compound (R—C—R bond angle $\approx 120°$) is converted to an sp^3 hybridised carbon atom in the adduct—and in the preceding T.S.—(R—C—R bond angle $\approx 109°$). The R groups thus move closer together as the reaction proceeds, i.e. the T.S. becomes more crowded, its energy level therefore increases and the reaction rate drops, as R increases in size. The observed drop in reaction rate, $H_2C{=}O > RHC{=}O > R_2C{=}O$, is thus determined by both electronic and steric effects. Increase in size of the nucleophile, with a given carbonyl compound, may also lead to a drop in reaction rate for the same reason.

Apart from reaction with the strongest nucleophiles, e.g. $AlH_4^{\ominus}$ (p. 210), $\overset{\delta-}{R}\overset{\delta+}{Mg}Br$ (p. 217), many additions to C=O are reversible. In general, the factors that we have seen to affect the rate of reaction (k) influence the position of equilibrium (K) in much the same way; this is because the T.S. for simple addition reactions probably resembles the adduct a good deal more closely than it does the original carbonyl compound. Thus the Ks for cyanohydrin formation (*cf.* p. 208) are found to reflect this operation of both steric and electronic factors:

	K
CH_3CHO	very large
$p\text{-}NO_2C_6H_4CHO$	1420
C_6H_5CHO	210
$p\text{-}MeOC_6H_4CHO$	32
$CH_3COCH_2CH_3$	38
$C_6H_5COCH_3$	0·8
$C_6H_5COC_6H_5$	very small indeed

Highly hindered ketones, such as $Me_3CCOCMe_3$, may not react at all except possibly with very small, highly reactive nucleophiles.

For a given carbonyl compound, K will be influenced by the size of the nucleophile; thus the K value for bisulphite addition ($S_2O_3^{2\ominus}$, *cf*. p. 209) to $(MeCH_2)_2C=O$ is only 4×10^{-4} (*cf*. $K = 38$ for HCN addition to $MeCOCH_2Me$ above). K is also influenced by the nature of the bond that is formed by the nucleophile to the carbonyl carbon atom, as observed in the following sequence of equilibrium concentrations of product:

$$\ominus CN > RNH_2 > ROH$$

A number of the more characteristic addition reactions will now be studied in greater detail; they will be grouped under the heads: (*a*) simple addition, (*b*) addition/elimination, and (*c*) addition of carbon nucleophiles.

8.2   SIMPLE ADDITION REACTIONS

8.2.1   Hydration

Many carbonyl compounds undergo reversible hydration in aqueous solution.

$$R_2C=O + H_2O \rightleftarrows R_2C(OH)_2$$

thus the K values at 20° for $H_2C=O$, $MeHC=O$ and $Me_2C=O$ are 2×10^3, 1·4, and 2×10^{-3}, respectively; this sequence reflects the progressive effect of increasing electron-donation. The ready reversibility of hydration is reflected in the fact that $H_2C=O$ can be distilled, as such, out of its aqueous solution. That a dynamic equilibrium actually is set up with $Me_2C=O$, though the ambient concentration of the hydrate is so low (its presence has been demonstrated in frozen Me_2CO/H_2O mixtures, however), may be demonstrated by working in $H_2^{18}O$:

$$Me_2C=O + H_2^{18}O \rightleftarrows Me_2C\genfrac{}{}{0pt}{}{\;^{18}OH}{OH} \rightleftarrows Me_2C=^{18}O + H_2O$$

(13)

Incorporation of ^{18}O into the ketone occurs hardly at all under these conditions, i.e. at pH 7, but in the presence of a trace of acid or base it occurs [*via* the hydrate (13)] very rapidly indeed. The fact that a carbonyl compound is hydrated will not influence nucleophilic additions that are irreversible; it may, however, influence the position of equilibrium in reversible addition reactions, and also the reaction rate, as the effective concentration of free carbonyl compound, $[R_2C{=}O]$, is naturally reduced.

Hydration is found to be susceptible to both general acid and general base (p. 73) catalysis, i.e. the rate-limiting step of the reaction involves *either* protonation of the carbonyl compound (general acid, 14), *or* conversion of H_2O into the more nucleophilic $^{\ominus}OH$ (general base, 15):

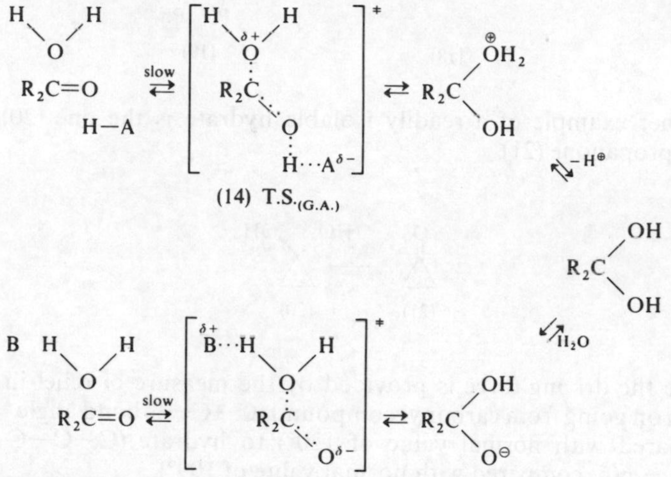

(14) T.S.$_{(G.A.)}$

(15) T.S.$_{(G.B.)}$

In contrast to Me_2CO, H_2CO hydrates quite readily at pH 7, reflecting the fact that its more positive carbonyl carbon atom undergoes attack by H_2O: without first requiring protonation of its carbonyl oxygen atom: it nevertheless hydrates very much faster at pH 4 or 11!

Just as electron-donating substituents inhibit hydrate formation, electron-withdrawing ones promote it. Thus K for the hydration of Cl_3CCHO (16) is 2.7×10^4, and this aldehyde (tri-chloroethanal, chloral) does indeed form an isolable, crystalline hydrate (17). The powerfully electron-withdrawing chlorine atoms destabilise the original carbonyl compound, but not the hydrate whose formation is thus promoted:

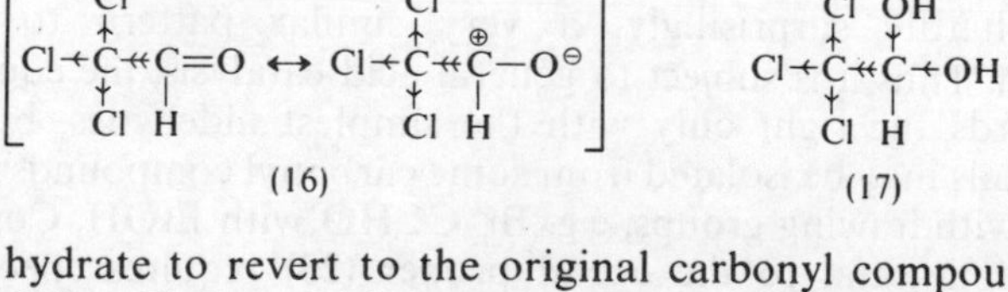

For the hydrate to revert to the original carbonyl compound it has to lose $^{\ominus}OH$ or H_2O:, which is rendered more difficult by the electron-withdrawing groups. Carbonyl groups are also effective at stabilising hydrates, possibily through hydrogen bonding as well as through electron-withdrawal, e.g. with diphenylpropantrione (18) which crystallise from water as the hydrate (19):

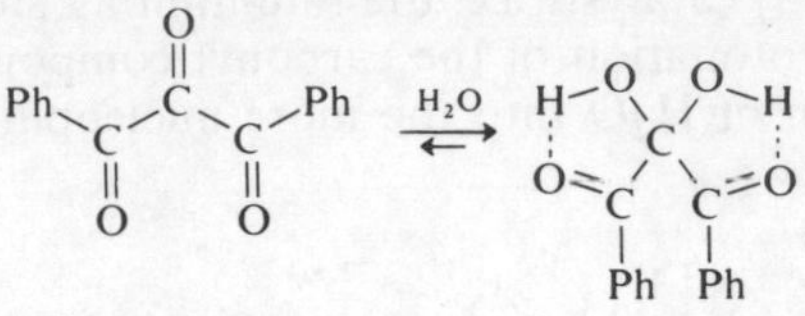

(18)   (19)

Another example of a readily isolable hydrate is the one (20) from cyclopropanone (21),

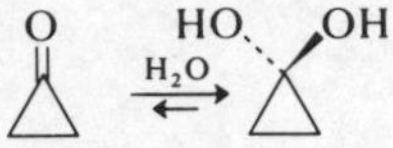

(21)   (20)

where the driving force is provided by the measure of relief in bond strain on going from carbonyl compound (C—C—C bond angle = 60°, compared with normal value of 120°) to hydrate (C—C—C bond angle = 60°, compared with normal value of 109°).

8.2.2 Alcohols

The reactions of carbonyl compounds with alcohols, R'OH, to yield hemi-acetals (22),

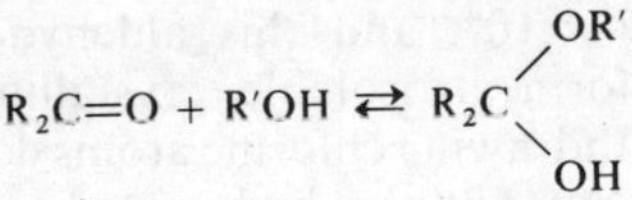

$$R_2C=O + R'OH \rightleftarrows R_2C \underset{OH}{\overset{OR'}{\diagup}}$$

(22)

follows—hardly surprisingly—a very similar pattern to hydrate formation. Thus it is subject to general acid catalysis, the equilibrium lies towards the right only with the simplest aldehydes, but stable hemi-acetals may be isolated from some carbonyl compounds carrying electron-withdrawing groups, e.g. Br_3CCHO with EtOH. Conversion of the hemi-acetal to the acetal proper (23) requires specific acid catalysis, however (*cf.* p. 73), i.e. it is loss of H_2O (S_N1, *cf.* p. 79) from (24) that is slow and rate limiting, followed by fast nucleophilic attack by R'OH:

$$
\underset{(22)}{\overset{\displaystyle RC\overset{OR'}{\underset{OH}{\Big\langle}}H}{}} \quad \underset{fast}{\overset{H^{\oplus}}{\rightleftharpoons}} \quad \underset{(24)}{\overset{\displaystyle RC\overset{OR'}{\underset{\overset{\oplus}{OH}_H}{\Big\langle}}H}{}} \quad \underset{slow}{\overset{-H_2O}{\rightleftharpoons}} \quad RCH\overset{\oplus}{-}OR'
$$

$$\uparrow\downarrow \begin{array}{c} R'OH \\ fast \end{array}$$

$$
\underset{(23)}{\overset{\displaystyle RC\overset{OR'}{\underset{OR'}{\Big\langle}}H}{}} \quad \underset{fast}{\overset{-H^{\oplus}}{\rightleftharpoons}} \quad RC\overset{OR'}{\underset{\overset{\oplus}{OR'}_H}{\Big\langle}}H
$$

The reaction does not normally take place with ketones under these conditions (i.e. with simple alcohols), but they can often be made to react with 1,2-diols, e.g. (25), to form cyclic acetals (26):

$$
R_2C{=}O + \overset{HO-CH_2}{\underset{HO-CH_2}{\big|}} \overset{H^{\oplus}}{\rightleftharpoons} R_2C\overset{O-CH_2}{\underset{O-CH_2}{\big\langle\big|}} + H_2O
$$

$$(25) \qquad\qquad\qquad (26)$$

The fact that reaction can be made to go with (25), but not with the simple R'OH, is due to the $\Delta S^{\ominus}$ (*cf.* p. 36) value for the former being more favourable than that for the latter, which involves a decrease in the number of molecules on going from starting material to product. Both aldehydes and ketones that are otherwise difficult to convert into acetals may often be transformed by use of orthoesters, e.g. $HC(OEt)_3$, triethoxymethane ('ethyl orthoformate'), with NH_4Cl as catalyst.

Acid-catalysed acetalisation is reversible and the position of equilibrium for a given aldehyde is determined by the relative proportions of R'OH and H_2O present. Preparative acetal formation is thus normally carried out in excess R'OH with an anhydrous acid catalyst, e.g. HCl gas; this also displaces the equilibrium by converting the liberated $H_2O: \rightarrow H_3O^{\oplus}$. Hydrolysis of the acetal back to the parent carbonyl compound may be effected readily with dilute acid. The hydrolysis of acetals is not susceptible to base-catalysis—there is no proton that can be removed from an oxygen atom, *cf.* the base-induced hydrolysis of hydrates: this results in acetals being very useful *protecting groups* for the $C{=}O$ function, which is itself very susceptible to the attack of bases (*cf.* p. 220). Such protection thus allows base-catalysed elimination of HBr from the acetal (27), followed by ready hydrolysis of the resultant unsaturated acetal (28) to the unsaturated carbonyl compound (29); a reaction that could not have been carried out directly on the bromoaldehyde (30), because of its polymerisation by base:

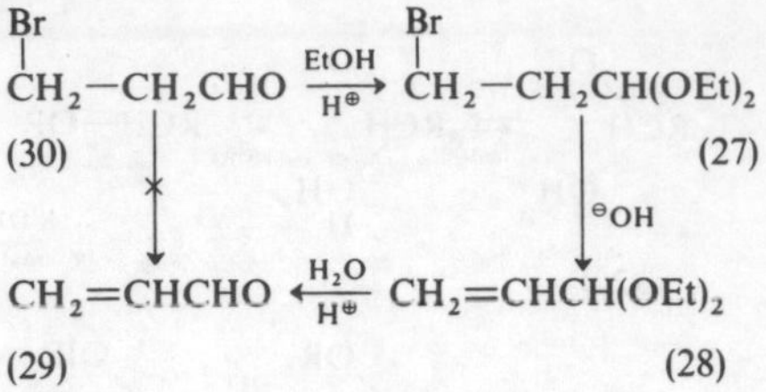

$$\underset{(30)}{\overset{\overset{\displaystyle Br}{|}}{CH_2-CH_2CHO}} \xrightarrow[H^\oplus]{EtOH} \underset{(27)}{\overset{\overset{\displaystyle Br}{|}}{CH_2-CH_2CH(OEt)_2}}$$

$$\underset{(29)}{CH_2=CHCHO} \xleftarrow[H^\oplus]{H_2O} \underset{(28)}{CH_2=CHCH(OEt)_2}$$

Acetals exhibit the three major requirements of an effective protecting group: (*a*) is easy to put on, (*b*) stays on firmly when required, and (*c*) is easy to take off finally.

8.2.3 Thiols

Carbonyl compounds react with thiols, RSH, to form hemi-thioacetals and thioacetals, rather more readily than with ROH; this reflects the greater nucleophilicity of sulphur compared with similarly situated oxygen. Thioacetals offer, with acetals, differential protection for the C=O group as they are relatively stable to dilute acid; they may, however, be decomposed readily by $HgCl_2/CdCO_3$. They also undergo desulphurisation with Raney nickel catalyst, thus providing a useful synthetic procedure for the indirect conversion of C=O → CH_2:

$$R_2C=O \xrightarrow{R'SH} R_2C(SR')_2 \xrightarrow{H_2/Ni} R_2CH_2$$

This is a conversion that is usually difficult to effect directly for preparative purposes.

8.2.4 Hydrogen cyanide

Although addition of HCN could be looked upon as a carbanion reaction, it is commonly regarded as involving a simple anion. It is of unusual interest in that it was almost certainly the first organic reaction to have its mechanistic pathway established (Lapworth 1903). HCN is not itself a powerful enough nucleophile to attack C=O, and the reaction requires base-catalysis in order to convert HCN into the more nucleophilic $^\ominus CN$; the reaction then obeys the rate-law:

$$Rate = k[R_2C=O][^\ominus CN]$$

The addition of $^\ominus CN$ is reversible, and tends to lie over in favour of starting materials unless a proton donor is present; this pulls the reaction over to the right, as the equilibrium involving the cyanohydrin

(31) is generally more favourable than that involving the anion (32):

$$R_2C{=}O \underset{\text{slow}}{\rightleftarrows} R_2C\overset{\overset{\ominus \text{CN}}{\diagup}O^{\ominus}}{\underset{\diagdown \text{CN}}{}} \underset{\text{fast}}{\overset{\text{HY}}{\rightleftarrows}} R_2C\overset{\diagup \text{OH}}{\underset{\diagdown \text{CN}}{}} + Y^{\ominus}$$

$$\qquad\qquad\qquad (32) \qquad\qquad\qquad (31)$$

Attack by $^{\ominus}$CN is slow (rate-limiting), while proton transfer from HCN or a protic solvent, e.g. H_2O, is rapid. The effect of the structure of the carbonyl compound on the position of equilibrium in cyanohydrin formation has already been referred to (p. 203): it is a preparative proposition with aldehydes, and with simple aliphatic and cyclic ketones, but is poor for ArCOR, and does not take place at all with ArCOAr. With ArCHO the benzoin reaction (p. 227) may compete with cyanohydrin formation; with C=C—C=O, 1,4-addition may compete (*cf*. p. 197).

8.2.5 Bisulphite and other anions

Another classic anion reaction is that with bisulphite ion to yield crystalline adducts. The structure of these was long a matter of dispute before it was established that they were indeed salts of sulphonic acids (33), reflecting the greater nucleophilicity of sulphur rather than oxygen in the attacking anion. The effective nucleophile is almost certainly $SO_3^{2\ominus}$ (34) rather than $HSO_3^{\ominus}$ ($HO^{\ominus} + HSO_3^{\ominus} \rightleftarrows H_2O + SO_3^{2\ominus}$), as though the latter will be present in higher relative concentration the former is a much more effective nucleophile:

$$R_2C{=}\overset{\frown}{O} \rightleftarrows R_2C\overset{\diagup O^{\ominus}}{\underset{\diagdown SO_3^{\ominus}}{}} \overset{H_2O}{\rightleftarrows} R_2C\overset{\diagup OH}{\underset{\diagdown SO_3^{\ominus}}{}}$$

$$\overset{\ominus}{O}{-}\overset{\displaystyle|}{\underset{\underset{\ominus}{C}O}{S}}{=}O$$

$$\qquad (34) \qquad\qquad\qquad (33)$$

The attacking anion is already present in solution as such so no base catalysis is required, and $SO_3^{2\ominus}$ is a sufficiently powerful nucleophile not to require activation (by protonation) of the carbonyl group, so no acid catalysis is required either. This nucleophile is a large one, however, and the K values for product formation are normally considerably smaller than those for cyanohydrin formation with the same carbonyl compound (*cf*. p. 203). Preparative bisulphite compound formation is indeed confined to aldehydes, methyl ketones and some cyclic ketones. Such carbonyl compounds can be separated from mixtures and/or purified by isolation, purification, and subsequent decomposition of their bisulphite adducts.

Halide ions will also act as nucleophiles towards aldehydes under acid catalysis, but the resultant, for example, 1,1-hydroxychloro compound (35) is highly unstable, the equilibrium lying over in favour of starting material. With HCl in solution in an alcohol, ROH, the equilibrium is more favourable, and 1,1-alkoxychloro compounds may be prepared, e.g. 1-chloro-1-methoxymethane (36, 'α-chloromethyl ether') from CH_2O and MeOH (*cf.* acetal formation, p. 206), provided the reaction mixture is neutralised before isolation is attempted:

$$H_2C=O \underset{}{\overset{H^\oplus}{\rightleftarrows}} H_2\overset{\oplus}{C}-OH \underset{}{\overset{Cl^\ominus}{\rightleftarrows}} H_2C\overset{\displaystyle Cl}{\underset{\displaystyle \overset{..}{O}H}{\big\backslash}}$$

(35)

$$\Updownarrow H^\oplus$$

$$H_2C\overset{\displaystyle Cl}{\underset{\displaystyle OMe}{\big\backslash}} \overset{(1)\ MeOH}{\underset{(2)\ -H^\oplus}{\rightleftarrows}} H_2\overset{\oplus}{C}-Cl \overset{-H_2O}{\rightleftarrows} H_2C\overset{\displaystyle Cl}{\underset{\displaystyle \overset{\oplus}{O}H \atop H}{\big\backslash}}$$

(36)

8.2.6 Hydride ions

Carbonyl groups may be reduced catalytically just as carbon–carbon unsaturated linkages were (p. 187). It is, however, normally more difficult to effect the catalytic reduction of C=O than of C=C, C≡C, C=N, or C≡N, so that <u>selective</u> reduction of the former—in the presence of any of the latter—cannot normally be achieved catalytically. This can, however, be done with various, usually complex, metal hydrides.

8.2.6.1 Complex metal hydride ions

Among the most powerful of these is lithium aluminium hydride, $Li^\oplus AlH_4{}^\ominus$, which will reduce the C=O group in aldehydes, ketones, acids, esters, and amides to CH_2, while leaving untouched any C=C or C≡C linkages also present in the compound (C=C conjugated with C=O is sometimes affected). The effective reducing agent is $AlH_4{}^\ominus$ which acts as a powerful hydride ion, $H^\ominus$, donor; such being the case, reduction cannot be carried out in protic solvents, e.g. H_2O, ROH, as preferential proton abstraction would then take place.

Ethers, in a number of which $Li^{\oplus}AlH_4{}^{\ominus}$ is significantly soluble, are thus commonly employed as solvents.

The nucleophilic $AlH_4{}^{\ominus}$ donates $H^{\ominus}$, irreversibly, to the carbonyl carbon atom, and the residual AlH_3 then complexes with its oxygen atom to form (37);

$$R_2C{=}O \xrightarrow{\;AlH_4{}^{\ominus}\;} R_2\underset{H}{\overset{\ominus}{C}}{-}OAlH_3 \longrightarrow \left[R_2\underset{H}{C}{-}O \right]_4 Al^{\ominus} \xrightarrow{\;R'OH\;} R_2\underset{H}{C}{-}OH$$
$$H{-}AlH_3$$
$$\hspace{3cm} (37) \hspace{3cm} (38) \hspace{2cm} (39)$$
$$+ AlH_4{}^{\ominus}$$

this then disproportionates to form (38) and $AlH_4{}^{\ominus}$, which is believed to be the only species actually involved in hydride donation. The complex (38) is converted into the product alcohol (39) on treatment with a protic solvent: thus one of the two H atoms in (39) is provided by $AlH_4{}^{\ominus}$ and the other by $R'OH$.

In the reduction of acids there is a tendency for the lithium salt, $RCO_2{}^{\ominus}Li^{\oplus}$ to separate from the ethereal solution, and thus bring reduction to a halt; this can be avoided by first converting the acid to a simple, e.g. Me or Et, ester. In the reduction of the latter, the initial nucleophilic attack by $AlH_4{}^{\ominus}$ results in an addition/elimination reaction—OR′ is a good leaving group in (40)—followed by normal attack, as above, on the resultant carbonyl compound (41) to yield the primary alcohol (42):

$$\underset{H{-}AlH_3}{\overset{OR'}{RC{=}O}} \xrightarrow{\;AlH_4{}^{\ominus}\;} \underset{H}{\overset{OR'}{RC{-}O^{\ominus}}} \xrightarrow{\;-OR'\;} \underset{H}{\overset{}{RC{=}O}} \xrightarrow[(2)\,R'OH]{(1)\,AlH_4{}^{\ominus}} \underset{H}{\overset{H}{RC{-}OH}}$$
$$\hspace{4cm} (40) \hspace{2cm} (41) \hspace{2cm} (42)$$

A less powerful complex metal hydride is $Na^{\oplus}BH_4{}^{\ominus}$ which will reduce aldehydes and ketones only, and does not attack carboxylic acid derivatives; nor does it—as $Li^{\oplus}AlH_4{}^{\ominus}$ does—attack NO_2 or $C{\equiv}N$ present in the same compound. It has the great advantage of being usable in hydroxylic solvents. A wide variety of other reagents of the $MH_4{}^{\ominus}$, $MH_3OR^{\ominus}$, $MH_2(OR)_2{}^{\ominus}$ type have been developed: their relative effectiveness is related to both the nucleophilicity and size of $MH_4{}^{\ominus}$, etc.

8.2.6.2 Meerwein–Ponndorf reaction

Hydride transfer from carbon to a carbonyl carbon atom occurs, reversibly, in the above reaction of which the classical example is the reduction of ketones, e.g. (43), with $Al(OCHMe_2)_3$ (44) in propan-2-ol, an equilibrium being set up:

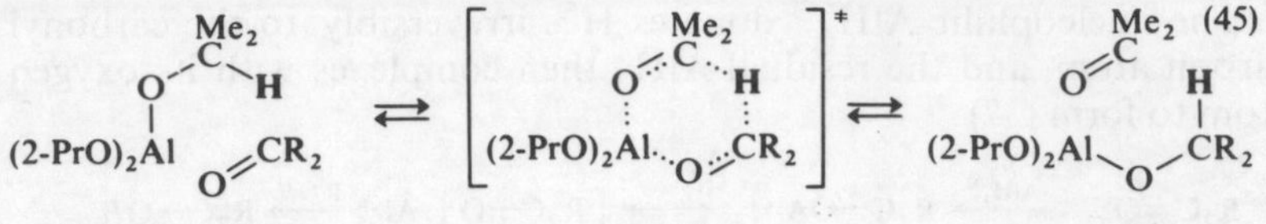

(44)   (43)   (47)   (46)

Propanone (45) is the lowest boiling constituent of the system, and the equilibrium can be displaced, essentially completely, to the right by distilling this continuously out of the system. An excess of propan-2-ol is employed, and this exchanges with the mixed Al alkoxide product (46) to liberate the desired reduction product R_2CHOH: again one hydrogen atom has been supplied by hydride transfer and one by a hydroxylic solvent. Because of the specific nature of the equilibrium, and of the way in which it is set up, no other groups present in the original ketone are reduced.

That specific hydride transfer from carbon to carbon does occur, was established by showing that use of labelled $(Me_2CDO)_3Al$ led to the formation of R_2CDOH. The reaction probably proceeds *via* a cyclic T.S. such as (47), though some cases have been observed in which two moles of alkoxide are involved—one to transfer hydride ion, while the other complexes with the carbonyl oxygen atom. The reaction has now been essentially superseded by $MH_4^{\ominus}$ reductions, but can sometimes be made to operate in the reverse direction (oxidation) by use of $Al(OCMe_3)_3$ as catalyst, and with a large excess of propanone to drive the equilibrium over to the left.

8.2.6.3   Cannizzaro reaction

This involves hydride transfer from an aldehyde molecule lacking an α-H atom, e.g. HCHO, R_3CCHO, ArCHO, to a second molecule of either the same aldehyde (disproportionation) or sometimes to a molecule of a different aldehyde ('crossed' Cannizzaro). The reaction requires the presence of strong bases, and with, for example, PhCHO the rate law is found to be,

$$Rate = k[PhCHO]^2[^{\ominus}OH]$$

and the reaction is believed to follow the pathway:

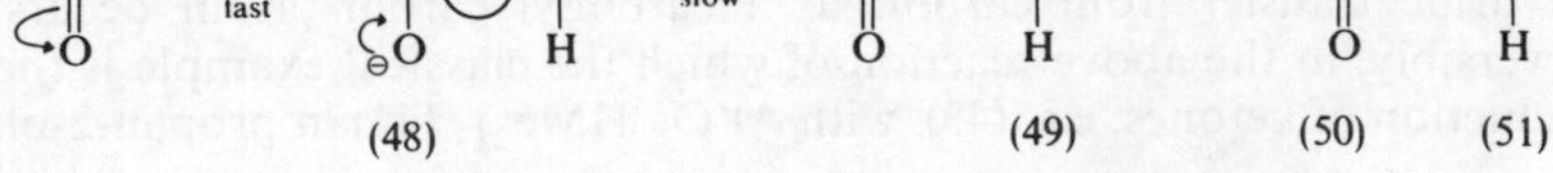

(48)   (49)   (50)   (51)

Rapid, reversible addition of $^{\ominus}$OH to PhCHO yields the potential hydride donor (48), this is followed by slow, rate-limiting hydride transfer to the carbonyl carbon atom (49) of a second molecule of PhCHO, and the reaction is completed by rapid proton exchange to yield the stabler pair (50) and (51). Mutual oxidation/reduction of two molecules of aldehyde has thus taken place to yield one molecule each of the corresponding carboxylate anion (50) and of the primary alcohol (51).

That <u>direct</u> hydride transfer has taken place, from carbon to carbon, is confirmed by carrying out the reaction in D_2O: <u>no</u> D is incorporated into the CH_2 group of (51), as it would have been if $H^{\ominus}$ had become free, and so able to equilibrate with the D_2O solvent. In very concentrated base, the rate law may, e.g. with HCHO (52), approach the form:

$$Rate = k[HCHO]^2[^{\ominus}OH]^2$$

This corresponds to the removal of a second proton from the species (53), corresponding to (48), to yield the dianion (54), which will clearly be a much more powerful hydride donor than (53)—or (48):

$$\underset{(52)}{\overset{\overset{\displaystyle ^{\ominus}OH}{\diagdown}}{\underset{\overset{\displaystyle \|}{O}}{\overset{\displaystyle |}{HC}}-H}} \quad \overset{^{\ominus}OH}{\rightleftarrows} \quad \underset{(53)}{\overset{^{\ominus}OH}{\underset{_{\ominus}O}{\overset{|}{HC}-H}}} \quad \overset{^{\ominus}OH}{\rightleftarrows} \quad \underset{(54)}{\overset{O^{\ominus}}{\underset{_{\ominus}O}{\overset{|}{HC}-H}}}$$

Suitable dialdehydes can also undergo intramolecular hydride transfer, as in the Cannizzaro reaction of ethan-1,2-dial (55, 'glyoxal') → hydroxyethanoate ('glycollate,' 56) anion,

$$\underset{(55)}{\overset{\displaystyle H}{\underset{\displaystyle H}{O=C-C=O}}} \quad \overset{^{\ominus}OH}{\longrightarrow} \quad \underset{(56)}{\overset{\displaystyle H}{\underset{\overset{\displaystyle |}{\underset{\ominus}{O}}\ \ H}{O=C-C-OH}}}$$

for which the observed rate law is found, as expected, to be:

$$Rate = k[OHCCHO][^{\ominus}OH]$$

Aldehydes that possess H atoms on the carbon atom adjacent to the CHO group (the α-carbon atom) do not undergo the Cannizzaro reaction with base, as they undergo the aldol reaction (p. 220) very much faster.

8.2.7   Electrons

Atoms of a number of the more strongly electropositive metals, e.g. Na, K, etc., can under suitable conditions yield solvated electrons in solution:

$$Na\cdot \underset{NH_3}{\overset{liq}{\rightleftarrows}} Na^{\oplus} + e^{\ominus}(NH_3)_n$$

Such electrons may act as nucleophiles, and add to the carbonyl carbon atom of a C=O group to yield a *radical anion* (57), often as an ion pair with the metal cation, $M^{\oplus}$:

$$R_2C=O + M^{\oplus} + e^{\ominus} \rightleftarrows R_2\overset{\cdot}{C}-O^{\ominus}M^{\oplus}$$

$$(57)$$

Thus when Na is dissolved, in the absence of air, in ethereal solutions of aromatic ketones, a blue colour is seen, due to the presence of the delocalised (over Ar as well as over the C=O system) species (58), a sodium *ketyl*;

$$Na^{\oplus}[Ar_2\overset{\cdot}{C}-O^{\ominus} \leftrightarrow Ar_2\underset{\ominus}{C}-O\cdot] \rightleftarrows \begin{bmatrix} Ar_2C-O^{\ominus} \\ | \\ Ar_2C-O^{\ominus} \end{bmatrix}2Na^{\oplus}$$

$$(58) \hspace{3cm} (59)$$

this latter is also in equilibrium with its dimer (59), the dianion of a 1,2-diol or *pinacol*. Under the right conditions, addition of a proton donor, e.g. ROH, can lead to the preparative formation of the pinacol itself. This tends to work better with aromatic rather than with aliphatic ketones, but propanone (60) may be converted readily by magnesium into 2,3-dimethylbutan-2,3-diol (61), so-called pinacol itself:

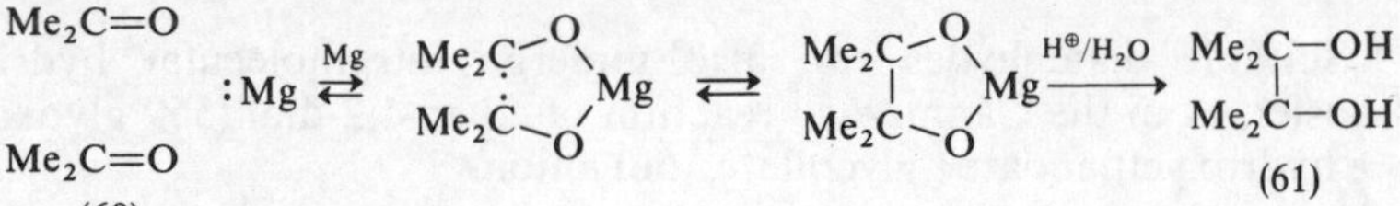

Similar nucleophilic addition of electrons can also occur to the carbonyl carbon atom of diesters such as (62), e.g. from sodium in solvents such as xylene, but the resultant dianion (63) differs from (59) in possessing excellent leaving groups, e.g. $^{\ominus}OEt$, and the overall result is the *acyloin condensation*:

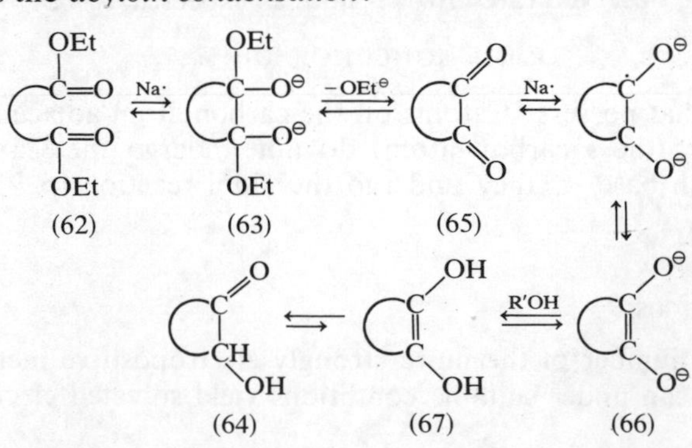

The end product is the 1-hydroxyketone, or acyloin (64). The reaction possibily proceeds as above *via* a 1,2-diketone (65) which can itself accept further electrons from sodium. The end product of the reaction in xylene is the Na salt of (66), but subsequent addition of R'OH effects protonation to yield the 1,2-enediol (67); the final acyloin (64) is merely the more stable tautomeric form of this. The reaction is of considerable preparative value for the cyclisation of long chain diesters, $EtO_2C(CH_2)_nCO_2Et$, in the synthesis of large-ring hydroxy-ketones. The yields are very good: 60–95% over the range $n = 8$–18, i.c. for 10–20 membered rings, respectively.

8.3 ADDITION/ELIMINATION REACTIONS

There are a number of nucleophilic additions to C=O known in which the added nucleophile still carries an acidic proton (68); a subsequent elimination then becomes possible, leading overall to nett substitution (69):

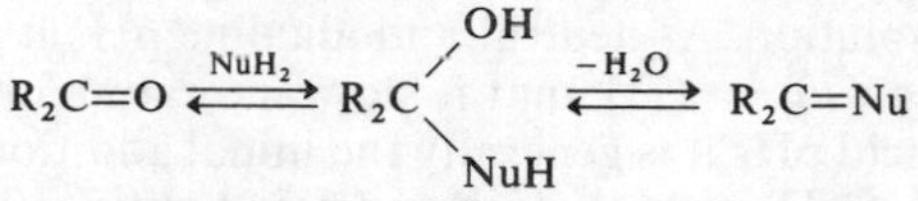

$$R_2C=O \underset{}{\overset{NuH_2}{\rightleftharpoons}} R_2C\begin{smallmatrix} OH \\ \\ NuH \end{smallmatrix} \underset{}{\overset{-H_2O}{\rightleftharpoons}} R_2C=Nu$$

$$(68) \qquad\qquad (69)$$

By far the most common examples of this are with derivatives of NH_3, particularly those like $HONH_2$, $NH_2CONHNH_2$, $PhNHNH_2$ etc., which have long been used to convert liquid carbonyl compounds into solid derivatives, for their better characterisation.

8.3.1 Derivatives of NH₃

If, for example, the reaction at pH 7 between pyruvate anion (70) and hydroxylamine, NH_2OH, is followed by monitoring the infra-red spectrum of the reaction mixture, then the absorption characteristic of C=O (v_{max} 1710 cm^{-1})—in the starting material (70)—is found to disappear completely before any absorption characteristic of C=N (v_{max} 1400 cm^{-1})—in the product oxime (71)—appears at all. Clearly an intermediate must thus be formed, and it seems likely to be (72), or something of that general nature [*cf.* (68) above]:

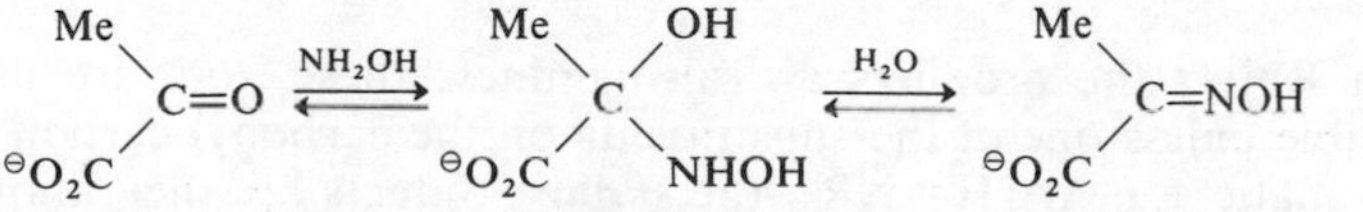

$$(70) \qquad\qquad (72) \qquad\qquad (71)$$

$$v_{max} \text{ 1710 cm}^{-1} \qquad\qquad\qquad v_{max} \text{ 1400 cm}^{-1}$$

Increasing the acidity of the reaction mixture is found to decrease the rate at which C=O absorption disappears—: NH_2OH is being progressively converted into $H\overset{\oplus}{N}H_2OH$, which is not nucleophilic—and to increase very markedly the rate at which the C=N absorption appears—increasing acid-catalysis of the dehydration of (72) → (71). This is compatible with a reaction pathway of the general form:

$$R_2\overset{\frown}{C=O} \quad \rightleftharpoons \quad R_2C\overset{O^\ominus}{\underset{H_2\overset{\oplus}{N}Y}{\diagdown}} \quad \rightleftharpoons \quad R_2C\overset{\overset{\cdot\cdot}{O}H}{\underset{HNY}{\diagdown}} \quad \overset{H^\oplus}{\rightleftharpoons} \quad R_2C\overset{\overset{H}{\overset{|}{\overset{\oplus}{C}OH}}}{\underset{\overset{NY}{\underset{H}{|}}}{\diagdown}} \quad \overset{H_2O}{\rightleftharpoons} \quad R_2C=NY$$

Strong nucleophiles such as NH_2OH (Y = OH) do not require catalysis of their initial addition to C=O, but weaker ones such as $PhNHNH_2$ (Y = PhNH) and $NH_2CONHNH_2$ (Y = $NHCONH_2$) often require acid catalysis to activate the C=O group (*cf.* p. 201, it is in fact *general* acid catalysis). Often, either the initial addition step or the dehydration step can be made rate-limiting at will, depending on the pH of the solution. At neutral and alkaline pHs it is generally the dehydration, e.g. (72) → (71), that is slow and rate-limiting (*cf.* above), while at more acid pHs it is generally the initial addition of the nucleophile, e.g. (70) → (72), that is slow and rate-limiting. This clearly has significance in preparative terms, and formation of such derivatives of carbonyl compounds tend to exhibit pH optima—the value depending on the nature of the particular carbonyl compound and of the ammonia derivative employed: thus for the formation of an oxime from propanone, Me_2CO, the optimum pH is found to be ≈4·5.

Ammonia itself yields imines, $R_2C=NH$, with carbonyl compounds but these derivatives are unstable and react with each other to form polymers of varying size. The classical 'aldehyde ammonias' are found to be hydrated cyclic trimers, but from aldehydes carrying powerfully electron-withdrawing substituents it is possible to isolate the simple ammonia adduct [73, *cf.* (72), and hydrates, p. 205, hemi-acetals, p. 206]:

$$Cl_3CC\overset{OH}{\underset{NH_2}{\diagup}}\overset{}{\diagdown}$$
$$(73)$$

With RNH_2 the products are also imines; these, too, are usually unstable unless one of the substituents on the carbonyl carbon atom is aromatic, e.g. ArCH=NR—the stable products are then known as *Schiff bases*. With R_2NH, the initial adduct (74) cannot lose water in the normal way; some such species have been isolated but they are not particularly stable. If, however, the adduct has any α-H atoms then a different dehydration can be made to take place yielding an *enamine* (75):

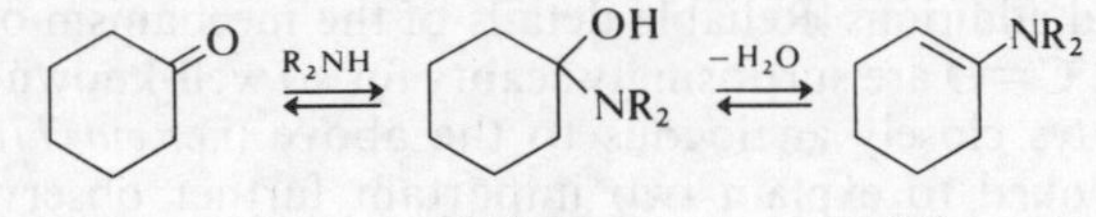

Enamines are of some importance as synthetic intermediates.

8.4 CARBON NUCLEOPHILE ADDITIONS

In discussing this group of reactions no formal distinction will be sought between those which are simple additions and those which are addition/eliminations. They are considered together, as a group on their own, because they result in the formation of carbon–carbon bonds, i.e. many of them are of great use and importance in synthetic organic chemistry. Before considering carbanion reactions in general, however, two other carbon nucleophile additions will first be mentioned.

8.4.1 Grignard, etc., reagents

The actual composition/structure of Grignard reagents—commonly written as RMgX—is still a matter of some dispute. It appears to depend on the nature of R, and also on the solvent in which the reagent is, or has been, dissolved. Thus the nuclear magnetic resonance (n.m.r.) spectrum of MeMgBr in Et_2O indicates that it is present largely as $MgMe_2 + MgBr_2$, while X-ray measurements on crystals of PhMgBr, isolated from EtO_2 solution, indicate that it has the composition $PhMgBr \cdot 2Et_2O$, with the four ligands arranged tetrahedrally round the Mg atom. Whatever the details may be, Grignard reagents may be regarded as acting as sources of negatively polarised carbon, i.e. as $^{\delta-}RMgX^{\delta+}$.

There is evidence of complexing of the Mg atom of the Grignard reagent with the carbonyl oxygen atom (76), and it is found that two molecules of RMgX are involved in the addition reaction, in some cases at least, possibly *via* a cyclic T.S. such as (77):

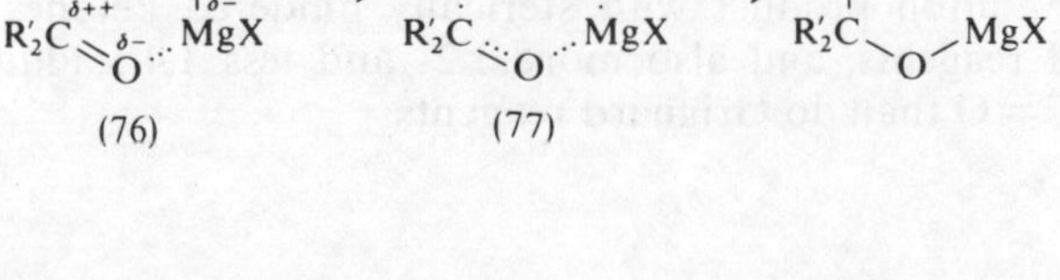

The second molecule of RMgX could be looked upon as a Lewis acid catalyst, increasing the positive polarisation of the carbonyl carbon atom through complexing with oxygen. It is indeed found in practice that the addition of Lewis acids e.g. $MgBr_2$, does speed up the rate of Grignard additions. Reliable details of the mechanism of Grignard addition to C=O are surprisingly scanty for so well-known a reaction, but pathways closely analogous to the above (i.e. *via* 77a and 77b) can be invoked to explain two important further observations: (*a*) that Grignard reagents having H atoms on their β-carbon atom (RCH_2CH_2MgX, 78) tend to *reduce* C=O $\rightarrow$ CHOH (79), being themselves converted to alkenes (80) in the process (transfer of H rather than RCH_2CH_2 taking place):

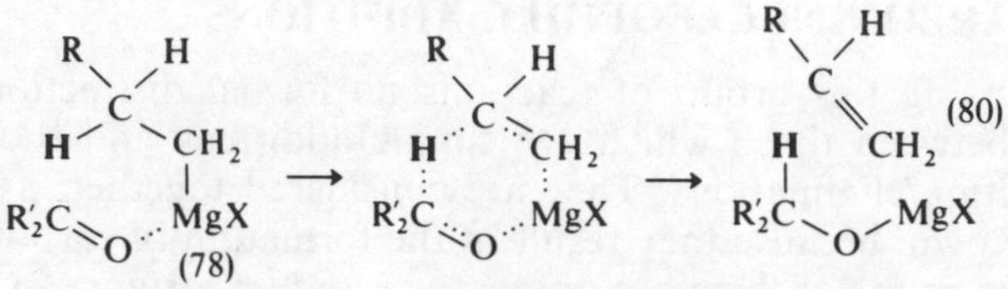

(*b*) that sterically hindered ketones having H atoms on their α-carbons, e.g. (81), tend to be converted to their enols (82), the Grignard reagent, RMgX being last as RH in the process:

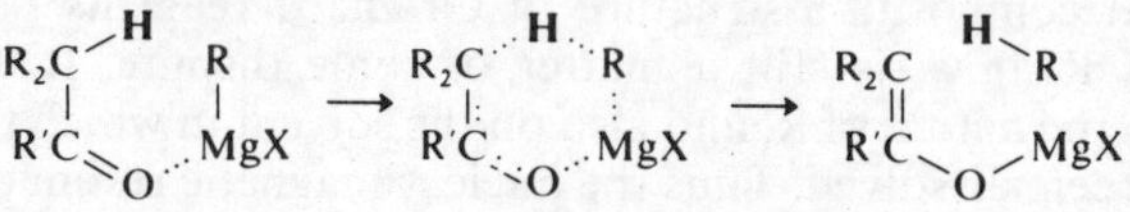

Grignard reagents act as strong nucleophiles and the addition reaction is often substantially irreversible (*cf.* conjugate addition to C=C—C=O that has already been referred to, p. 197). The initial products are alcohols but it is important to emphasise that the utility of Grignard, and similar, additions to C=O is as a general method of joining different groups of carbon atoms together, i.e. the original alcohol products can then be further modified in a wide variety of reactions. In the past organo-zinc compounds were used in a similar way, being largely displaced by Grignard reagents; in turn, Grignard reagents are tending to be displaced by lithium alkyls and aryls, RLi and ArLi, respectively. These latter reagents tend to give more of the normal addition product with sterically hindered ketones than do Grignard reagents, and also more 1,2- and less 1,4-, addition with C=C—C=O than do Grignard reagents.

8.4.2 Acetylide anions

Acetylenes, $RC\equiv CH$ and $HC\equiv CH$, are markedly acidic and may be converted by strong bases, e.g. $^{\ominus}NH_2$ in liquid ammonia, into the corresponding anions (*cf.* p. 266), which are somewhat more nucleophilic than $^{\ominus}CN$. Though these species, e.g. $RC\equiv C^{\ominus}$, are palpably carbanions, they are considered separately as, unlike the group we shall consider below, they require no stabilisation by electron-withdrawing groups such as $C=O$, $C\equiv N$, NO_2 etc. A useful group of carbon atoms may thus be added to $C=O$, and the reaction is especially useful synthetically in that the $C\equiv C$ linkage now present may be further modified in a variety of ways, e.g. reduced to the alkene (83) by H_2 with the Lindlar catalyst (*cf.* p. 188):

$$R_2C=\overset{\frown}{O} \atop \overset{\ominus}{C}\equiv CR' \quad \rightleftharpoons \quad R_2C\overset{O^{\ominus}}{\underset{C\equiv CR'}{\diagup}} \quad \overset{liq.\ NH_3}{\rightleftharpoons} \quad R_2C\overset{OH}{\underset{C\equiv CR'}{\diagup}} \quad \overset{H_2}{\underset{\substack{Lindlar \\ catalyst}}{\longrightarrow}} \quad R_2C\overset{OH}{\underset{CH=CHR'}{\diagup}}$$

$$(83)$$

8.4.3 Carbanions (general)

In general these reactions are base-catalysed in that it is necessary to remove a proton from HCXYZ in order to generate the carbanion, $^{\ominus}CXYZ$, the effective nucleophile; one or more of X, Y and Z are usually electron-withdrawing in order to stabilise it. The initial adduct (84) acquires a proton from the solvent (often H_2O or ROH) to yield the simple addition product (85). Whether or not this undergoes subsequent dehydration (86) depends on the availability of an H atom, either on an α-carbon or where X, Y or Z = H, and also on whether the $C=C$ so introduced would, or would not, be conjugated with other $C=C$ or $C=O$ linkages in the product:

$$R_2C=O \atop \overset{\ominus}{C}XYZ \quad \rightleftharpoons \quad R_2C\overset{O^{\ominus}}{\underset{CXYZ}{\diagup}} \quad \overset{R'OH}{\rightleftharpoons} \quad R_2C\overset{OH}{\underset{CXYZ}{\diagup}} \quad \rightleftharpoons \quad R_2C=CYZ$$

$$\qquad\qquad (84) \qquad\qquad\qquad (85) \qquad\qquad\qquad (86) \quad \text{(i.e. X = H)}$$

The initial carbon–carbon bond formation ($\rightarrow$ 84) is often reversible, and a subsequent step—such as dehydration—may be necessary to displace the equilibrium. The many different (often named) reactions really differ from each other only in the nature of the particular carbonyl compound (aldehyde, ketone, ester etc.), and in the type of carbanion, employed.

8.4.4   Aldol reactions

Here the carbanion (87), obtained from the action of base (usually $^{\ominus}OH$) on an α-H atom of one molecule of a carbonyl compound (88), adds to the carbonyl carbon of another (88) to yield a β-hydroxy-carbonyl compound. Thus with ethanal, CH_3CHO, the product is 3-hydroxybutanal (89)—aldol itself:

$$
\begin{array}{c}
\qquad\qquad\quad H \\
\qquad\qquad\quad | \\
(88)\qquad CH_2CHO \\
\\
\qquad (1) \updownarrow \,^{\ominus}OH
\end{array}
$$

$$
\underset{(88)}{\overset{\displaystyle C\overset{O}{\overset{\|}{}}}{\underset{\displaystyle H}{\overset{\displaystyle |}{Me C}}}}\quad \overset{\displaystyle \frown}{^{\ominus}CH_2} -\underset{\displaystyle H}{\overset{\displaystyle |}{C}}{=}O \overset{(2)}{\rightleftarrows} \overset{O^{\ominus}}{\overset{|}{Me CH}}{-}CH_2CHO \overset{H_2O}{\rightleftarrows} \underset{(89)}{MeCH(OH)CH_2CHO}
$$

$$
\updownarrow
$$

$$
\underset{(87)}{CH_2{=}\underset{H}{\overset{|}{C}}{-}O^{\ominus}}
$$

In the case of CH_3CHO the equilibrium is found to lie right over in favour of aldol. The forward reaction of step (2) and the reversal of step (1) are essentially competing with each other for the carbanion (87). Carrying out the reaction in D_2O fails to result in the incorporation of any deuterium into the CH_3 group as of yet unchanged ethanal, however, so that step (2) must be so much more rapid than the reverse of step (1) as to make the latter virtually irreversible.

For even simple ketones, e.g. propanone (90), the equilibrium is found to lie far over to the left ($\approx 2\%$ of 91)—reflecting the less ready attack of the carbanion (92) on a 'keto' (90), rather than on an 'aldehydo' (88, above), carbonyl carbon atom:

$$
\begin{array}{c}
\qquad\qquad\quad H \\
\qquad\qquad\quad | \\
(90)\qquad CH_2COMe \\
\\
\qquad (1) \updownarrow \,^{\ominus}OH
\end{array}
$$

$$
\underset{(90)}{\overset{\displaystyle C\overset{O}{\overset{\|}{}}}{Me_2C}}\quad \overset{\displaystyle \frown}{\underset{(92)}{^{\ominus}CH_2COMe}} \overset{(2)}{\rightleftarrows} \overset{^{\ominus}O}{\overset{|}{Me_2C}}{-}CH_2COMe \overset{H_2O}{\rightleftarrows} \underset{(91)}{Me_2C(OH)CH_2COMe}
$$

Thus it is found in the case of propanone (90) that carrying out the reaction in D_2O <u>does</u> result in the incorporation of deuterium into the CH_3 group of as yet unchanged propanone, i.e. step (2) is no longer rapid with respect to the reversal of step (1).

The reaction can, however, be made preparative for (91) by a continuous distillation/siphoning process in a Soxhlet apparatus: equilibrium is effected in hot propanone over solid $Ba(OH)_2$ (as base catalyst), the equilibrium mixture [containing $\approx 2\%$ (91)] is then siphoned off. This mixture is then distilled back on to the $Ba(OH)_2$, but only propanone (b.p. 56°) will distil out, the $\approx 2\%$ of 2-methyl-2-hydroxypentan-4-one ('diacetone alcohol', 91, b.p. 164°) being left behind. A second siphoning will add a further $\approx 2\%$ 'equilibrium's worth' of (91) to the first 2%, and more or less total conversion of (90) → (91) can thus ultimately be effected. These 'poor' aldol reactions can, however, be accomplished very much more readily under acid catalysis. The acid promotes the formation of an ambient concentration of the enol form (93) of, for example, propanone (90), and this undergoes attack by the protonated form of a second molecule of carbonyl compound, a carbonium ion (94):

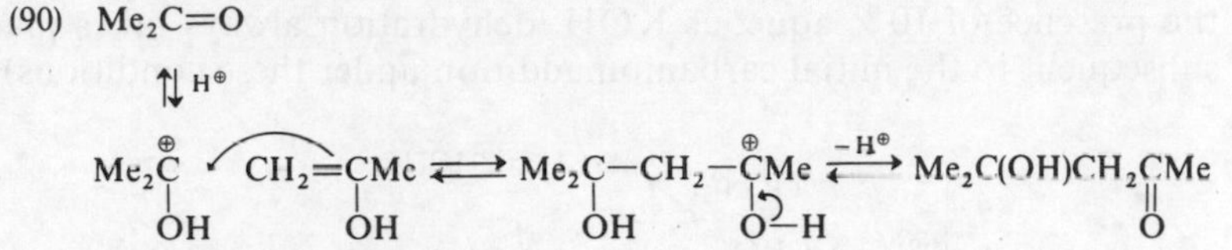

(90) $Me_2C{=}O$

$\updownarrow H^\oplus$

$\overset{\oplus}{Me_2C} \quad CH_2{=}CMe \rightleftharpoons Me_2C{-}CH_2{-}\overset{\oplus}{C}Me \underset{}{\overset{-H^\oplus}{\rightleftharpoons}} Me_2C(OH)CH_2CMe$
$\quad\ \ |\qquad\qquad\ |\qquad\qquad\quad\ |\qquad\quad |$
$\quad\ \ OH\qquad\quad\ OH\qquad\qquad\ OH\qquad\ O{-}H\qquad\qquad\qquad\qquad\ \ O$

$\quad\ (94)\qquad\quad\ (93)\qquad\qquad\qquad\qquad\qquad\qquad\qquad\qquad\qquad\ (91)$

Under acid conditions the tertiary alcohol (91) almost always undergoes acid-catalysed dehydration (*cf.* p. 241) to yield the $\alpha\beta$-unsaturated ketone, 2-methylpent-3-ene-2-one (mesityl oxide, 95):

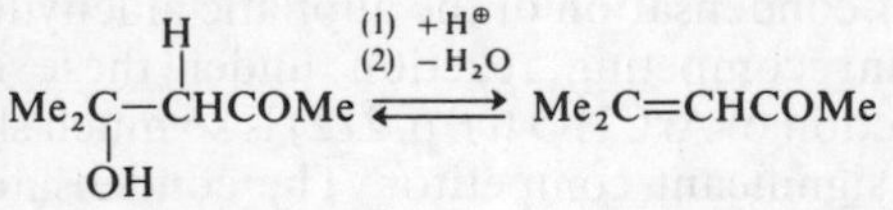

$\qquad\qquad\quad H \qquad\quad \overset{(1)\ +H^\oplus}{\underset{(2)\ -H_2O}{}}$
$\qquad\qquad\quad |$
$\qquad Me_2C{-}CHCOMe \rightleftharpoons Me_2C{=}CHCOMe$
$\qquad\qquad\quad |$
$\qquad\qquad\quad OH$

$\qquad\qquad\quad (91) \qquad\qquad\qquad\qquad\quad (95)$

Dehydration of aldols may also be effected under the influence of base, e.g. with aldol itself (89) to but-2-eneal (crotonaldehyde, 96):

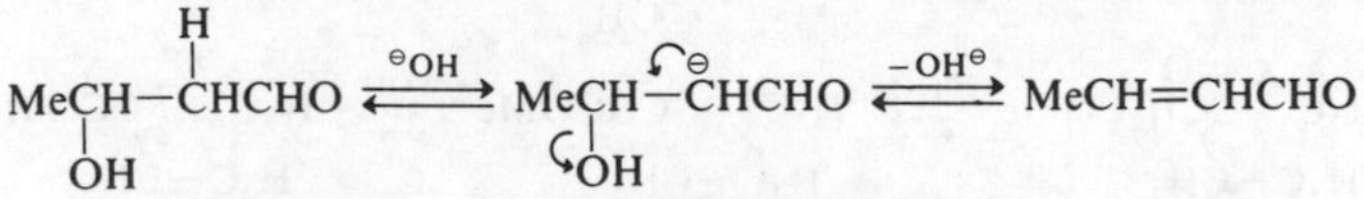

$\quad H$
$\quad |$
$MeCH{-}CHCHO \overset{\ominus OH}{\rightleftharpoons} MeCH{-}CHCHO \overset{-OH^\ominus}{\rightleftharpoons} MeCH{=}CHCHO$
$\quad |\qquad\qquad\qquad\qquad\ \ |$
$\quad OH\qquad\qquad\qquad\quad\ OH$

$\qquad (89)\qquad\qquad\qquad\quad\ (97)\qquad\qquad\qquad\qquad (96)$

Base-catalysed dehydrations are relatively unusual, and that one occurs here stems from the facts: (*a*) that (89) contains α-H atoms removable by base to yield an ambient concentration of the carbanion (97), and (*b*) that this carbanion possesses a goodish leaving group—$^\ominus OH$—on the adjacent (β-) carbon atom. The possibility of such an elimination may displace the equilibrium over to the right in a number of simple aldol additions, where it would otherwise lie far over to the left. It is important to remember, however, that the overall process aldol addition + dehydration is reversible, i.e. (88) $\rightleftarrows$ (96), and that $\alpha\beta$-unsaturated carbonyl compounds are thus cleaved by base under suitable conditions. It is also pertinent that (96) is still an aldehyde and can undergo further carbanion addition, followed by dehydration, and so on. This is how low molecular weight polymers are produced on heating simple aliphatic aldehydes with aqueous NaOH: to stop at the aldol stage something like dilute aqueous K_2CO_3 is required.

Crossed aldol condensations, where both aldehydes (or other suitable carbonyl compounds) have α-H atoms, are not normally of any preparative value as a mixture of four different products can result. Crossed aldol reactions can be of synthetic utility, where one aldehyde has no α-H, however, and can thus act only as a carbanion acceptor. An example is the *Claisen–Schmidt* condensation of aromatic aldehydes (98) with simple aliphatic aldehydes or (usually methyl) ketones in the presence of 10% aqueous KOH (dehydration always takes place subsequent to the initial carbanion addition under these conditions):

$$
\begin{array}{ccc}
 & \overset{\ominus}{C}H_2CHO \underset{\ominus OH}{\nearrow} & ArCH{=}CHCHO \\
(98) \quad ArCHO & & \\
 & \overset{\ominus}{C}H_2COMe \underset{\ominus OH}{\searrow} & ArCH{=}CHCOMe
\end{array}
$$

As would be expected, electron-donating groups in Ar slow the reaction, e.g. $p\text{-MeOC}_6\text{H}_4\text{CHO}$ reacts at only about one seventh the rate of C_6H_5CHO. Self-condensation of the aliphatic aldehyde can, of course, be an important competing reaction under these conditions, but Cannizzaro reaction of ArCHO (*cf.* p. 212) is so much slower a reaction as not to be a significant competitor. The condensation can also be effected under acid-catalysis (*cf.* p. 221).

Finally, aldol reactions can, with suitable dicarbonyl compounds e.g. (99), be intramolecular, i.e. cyclisations:

$$
\begin{array}{ccccc}
\text{Me} & & \text{Me} \quad \text{OH} & & \text{Me} \\
| & & \diagdown / & & | \\
C{=}O & & C & & C \\
H_2C \diagup \quad \diagdown CH_2COMe & \underset{}{\overset{\ominus OH}{\rightleftharpoons}} & H_2C \diagup \quad \diagdown CHCOMe & \underset{}{\overset{\ominus OH}{\rightleftharpoons}} & H_2C \diagup \quad \diagdown CCOMe \\
H_2C{-}CH_2 & & H_2C{-}CH_2 & & H_2C{-}CH_2 \\
\text{(99)} & & & &
\end{array}
$$

8.4.5   Nitroalkanes

Another synthetically useful reaction involves the addition to aldehydes and ketones of carbanions, e.g. (100), derived from aliphatic nitro compounds, e.g. nitromethane (101):

$$(101) \quad \overset{\text{H}}{\underset{|}{\text{CH}_2\text{NO}_2}}$$

$$\updownarrow \ominus\text{OH}$$

$$\underset{\text{R}_2\text{C}}{\overset{\text{O}}{\underset{||}{\text{C}}}} \curvearrowright \ominus\text{CH}_2-\text{N}\overset{\oplus}{\underset{\text{O}^\ominus}{\diagdown}}^{\diagup\text{O}} \rightleftarrows \text{R}_2\overset{\ominus\text{O}}{\underset{}{\text{C}}}-\text{CH}_2\text{NO}_2 \overset{\text{H}_2\text{O}}{\rightleftarrows} \text{R}_2\text{C(OH)CH}_2\text{NO}_2$$

$$(102)$$

$$\updownarrow$$

$$\text{CH}_2=\text{N}\overset{\oplus}{\underset{\text{O}^\ominus}{\diagdown}}^{\diagup\text{O}^\ominus}$$

$$(100)$$

Bases such as $\ominus$OH and $\ominus$OEt are used to obtain the carbanion, and whether or not the β-hydroxynitro compound (102) undergoes subsequent dehydration to $\text{R}_2\text{C}=\text{CHNO}_2$ depends on the conditions. Where the carbonyl compound is an aldehyde there is some danger of its undergoing an aldol reaction with itself, but the delocalised carbanion (100) usually forms more readily than does $\overset{\ominus}{\text{R}}\text{CHCHO}$, and the danger is thus relatively small. The product nitro compounds may be reduced to amines, and also modified in other ways.

8.4.6 Perkin reaction

In this reaction the carbanion (103) is obtained by removal of an α-H atom from a molecule of an acid anhydride (104), the anion of the corresponding acid acting as the necessary base; the carbonyl acceptor is pretty well confined to aromatic aldehydes. The products are $\alpha\beta$-unsaturated acids; a good example is the synthesis of 3-phenyl-propenoic acid (cinnamic acid, 105):

$$(104) \quad \underset{\text{MeCO}}{\overset{\overset{\text{H}}{\underset{|}{\text{CH}_2\text{CO}}}}{\diagdown}}{}^{\diagup\text{O}}$$

$$\updownarrow \text{MeCO}_2{}^\ominus$$

$$\underset{\underset{\text{H}}{|}}{\overset{\text{O}}{\underset{||}{\text{C}}}}\underset{\text{PhC}}{} \curvearrowright \underset{\text{MeCO}}{\overset{\ominus\text{CH}_2\text{CO}}{\diagdown}}{}^{\diagup\text{O}} \rightleftarrows \underset{\text{MeCO}}{\overset{\overset{\text{HO}}{\underset{|}{\text{PhCH}}}-\text{CH}_2\text{CO}}{\diagdown}}{}^{\diagup\text{O}} \overset{-\text{H}_2\text{O}}{\longrightarrow} \underset{\text{MeCO}}{\overset{\text{PhCH}=\text{CHCO}}{\diagdown}}{}^{\diagup\text{O}}$$

$$(103) \qquad\qquad (106) \qquad\qquad (107)$$

Dehydration of the original adduct (106) takes place under the conditions of the reaction (excess anhydride as solvent at $\approx 140°$), and the actual product in the reaction mixture is the mixed anhydride (107). Pouring the reaction mixture into water then effects hydrolysis to the mixed anhydride's constituent acids, $PhCH=CHCO_2H$ (105) + $MeCO_2H$. With some anhydrides of the form $(R_2CHCO)_2O$, where no dehydration subsequent to addition can take place, aldol type intermediates corresponding to (106) have been isolated.

8.4.7   Knoevenagel and Stobbe reactions

This addition involves carbanions from a wide variety of CH_2XY types but particularly where X and/or Y are CO_2R groups, e.g. $CH_2(CO_2Et)_2$; organic bases are often used as catalysts. In most cases the intermediate aldol is dehydrated to the $\alpha\beta$-unsaturated product (ester). An interesting example is with carbanions, e.g. (108), derived from esters of 1,4-butandioic (succinic) acid, e.g. $(CH_2CO_2Et)_2$, and aldehydes or ketones, employing alkoxide ions as base catalysts: the *Stobbe* condensation. These esters react a great deal more readily than might, *a priori*, have been expected, and one of the CO_2R groups is always, and unexpectedly, converted to $CO_2^{\ominus}$ in the course of the reaction; the product is always the $\alpha\beta$-unsaturated derivative (109), never the aldol. A pathway that will account for all these facts involves a cyclic (*lactone*) intermediate (110):

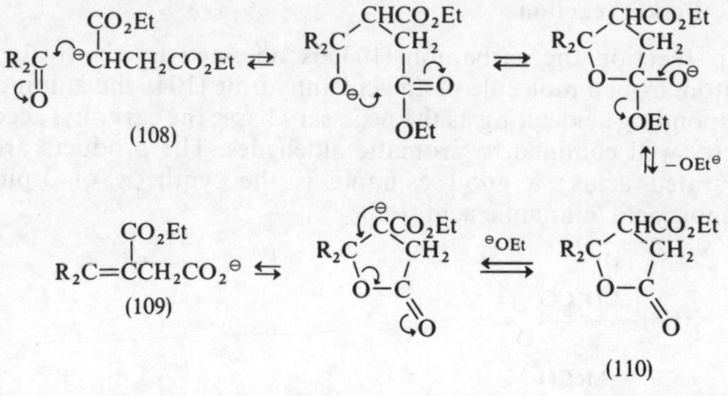

It has, in a few cases, proved possible actually to isolate cyclic intermediates such as (110).

8.4.8   Claisen ester condensation

This is another reaction that involves carbanions derived from esters, e.g. (111), but this time adding to the carbonyl carbon atom of another ester molecule. The reason for considering it here rather than under carboxylic derivatives (p. 233) is that it can, in its initiation, be regarded as something of an analogue, for esters, of the aldol condensation on aldehydes (*cf.* p. 220), e.g. with ethyl ethanoate (acetate, 112):

$$
\begin{array}{c}
\text{H} \\
| \\
(112) \quad \text{CH}_2\text{CO}_2\text{Et}
\end{array}
$$

(1) ⇅ $^\ominus$OEt

↕

$$
\begin{array}{ccc}
\text{CH}_2\text{=C—O}^\ominus & \left[\begin{array}{cc} {}^\ominus\text{O} & \text{O} \\ \| & \| \\ \text{MeC=CHCO}_2\text{Et} & \leftrightarrow & \text{MeC—}{}^\ominus\text{CHCO}_2\text{Et} \end{array} \right] \\
| & \\
\text{OEt} & (115)
\end{array}
$$

(111)

One significant difference from the simple aldol reaction, however, is that the original adduct (113) now possesses a good leaving group (OEt); thus instead of adding a proton, as in the aldol reaction proper (p. 220), $^\ominus$OEt is lost to yield a β-ketoester, ethyl 3-ketobutanoate (ethyl acetoacetate, 114). This is finally converted by base ($^\ominus$OEt) into its stabilised (delocalised) carbanion, (115).

Classically the base catalyst, $^\ominus$OEt, is introduced by adding just over one mole of sodium (as wire, or in other finely divided form) plus just a little EtOH to generate an initial small concentration of Na$^\oplus$$^\ominus$OEt. Further EtOH is generated in step (1), which yields further Na$^\oplus$$^\ominus$OEt with sodium, and the concentration of $^\ominus$OEt is thereby maintained. A whole mole is required as it is essential for the β-keto ester (114) to be converted (step 3) into its anion (115)—MeCOCH$_2$-CO$_2$Et is more acidic than EtOH (*cf.* p. 266)—if the overall succession of equilibria is to be displaced to the right. This is necessary because the carbanion-formation equilibrium—step (1)—lies even further over to the left than that with, for example, CH$_3$CHO; this reflects the less effective stabilisation through delocalisation in the ester carbanion (111) than in that from the aldehyde (116):

$$
\begin{array}{cccc}
{}^\ominus\text{CH}_2\text{—C=O} & \leftrightarrow & {}^\ominus\text{CH}_2\text{—C—O}^\ominus & \qquad & {}^\ominus\text{CH}_2\text{—C=O} & \leftrightarrow & \text{CH}_2\text{=C—O}^\ominus \\
| & & \| & & | & & | \\
\text{:OEt} & & {}^\oplus\text{OEt} & & \text{H} & & \text{H}
\end{array}
$$

(111)   (116)

This requirement to pull the equilibrium of step (1) over to the right is reflected in the fact that no reaction occurs with R$_2$CHCO$_2$Et in the presence of $^\ominus$OEt, despite the fact that a normal β-ketoester,

$R_2CHCOCR_2CO_2Et$, could be formed. It is significant, however, that this product β-ketoester has no α-H atom, and so cannot be converted into a carbanion corresponding to (115), i.e. step (3) cannot take place! Use of a base, e.g. $Ph_3C^\ominus Na^\oplus$, that is sufficiently strong to make step (1) virtually irreversible in the forward direction

$$R_2CHCO_2Et + Ph_3C^\ominus \rightleftarrows R\overset{\ominus}{C}CO_2Et + Ph_3CH$$

is found to induce a normal Claisen reaction in R_2CHCO_2Et, despite the fact that step (3) is still impossible.

It is important to emphasise the complete reversibility of normal Claisen reactions under suitable conditions, e.g. the so-called 'acid decomposition' (because both products—(117) and (118)—are derivatives of acids) of β-ketoesters (119):

$$\underset{(119)}{\underset{\overset{|}{\underset{\ominus}{OEt}}}{R\overset{O}{\overset{||}{C}}-CHR'CO_2Et}} \rightleftarrows \underset{\overset{|}{OEt}}{R\overset{\ominus O}{\overset{|}{C}}-CHR'CO_2Et} \rightleftarrows \underset{\underset{(117)}{\overset{|}{OEt}}}{R\overset{O}{\overset{||}{C}}} + \underset{(118)}{R'\overset{\ominus}{C}HCO_2Et}$$

1,3-(i.e. β-)diketones, e.g. (120), are also cleaved under these conditions to yield a derivative of an acid (121), and one of a simple ketone (122):

$$\underset{(120)}{\underset{\overset{|}{\underset{\ominus}{OEt}}}{R\overset{O}{\overset{||}{C}}-CH_2COR}} \rightleftarrows \underset{\overset{|}{OEt}}{R\overset{\ominus O}{\overset{|}{C}}-CH_2COR} \rightleftarrows \underset{\underset{(121)}{\overset{|}{OEt}}}{R\overset{O}{\overset{||}{C}}} + \underset{(122)}{\overset{\ominus}{C}H_2COR}$$

'Crossed' Claisen reactions with two different esters, each of which has α-H atoms, are seldom useful synthetically as there are, of course, four possible products. Crossed Claisen reactions are, however, often useful when one of the two esters has no α-H atoms, e.g. HCO_2Et, $ArCO_2Et$, $(CO_2Et)_2$ etc., as this can act only as a carbanion acceptor. Such species are in fact good acceptors, and the side reaction of the self-condensation of the other, e.g. RCH_2CO_2Et, ester is not normally a problem. Intramolecular Claisen reactions, where both CO_2Et groups are part of the same molecule [e.g. (123)], are referred to as *Dieckmann* cyclisations. These work best, under simple conditions, for the formation of the anions of 5-, 6- or 7-membered cyclic β-ketoesters [e.g. (124)], i.e. with $EtO_2C(CH_2)_nCO_2Et$ where $n = 4$–6:

$(x = 1$–3)   (123)

(124)

Big ring ketones (*cf.* the acyloin condensation, p. 214) may be obtained also by working at high dilution, i.e. the carbanion carbon atom then has a greater chance of reacting with the ester carbonyl carbon atom at the other end of its own chain than with one that is attached to a different molecule (intermolecular reaction).

8.4.9 Benzoin condensation

This reaction of aromatic aldehydes, ArCHO, resembles the Cannizzaro reaction in that the initial attack [rapid and reversible—step (1)] is by an anion—this time $^{\ominus}CN$—on the carbonyl carbon atom of one molecule, the 'donor', (125); but instead of hydride transfer (*cf.* Cannizzaro, p. 212) it is now carbanion addition by (127) to the carbonyl carbon atom of the second molecule of ArCHO, the 'acceptor' (128), that occurs. This, in common with cyanohydrin formation (p. 208) was one of the earliest reactions to have its pathway established— correctly!—in 1903. The rate law commonly observed is, as might be expected,

$$\text{Rate} = k[\text{ArCHO}]^2[^{\ominus}\text{CN}]$$

and the reaction is believed to follow the general pathway:

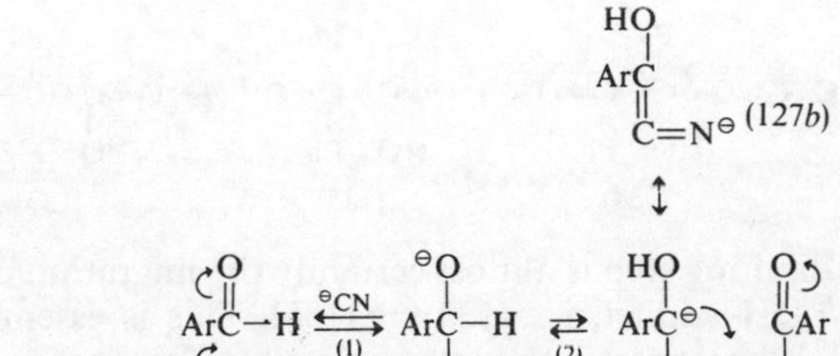

When the reaction is carried out in MeOH neither step (2), the formation of the carbanion (127), nor step (3), addition of this carbanion to the carbonyl carbon of the acceptor molecule (128), is completely rate-limiting in itself. These steps are followed by rapid proton transfer, (129) → (130), and, finally, by rapid loss of $^{\ominus}CN$—a good leaving group—i.e. reversal of cyanohydrin formation (*cf.* p. 208) on the product

2-hydroxyketone (131). Where Ar = Ph, the product is called benzoin. The reaction is completely reversible.

$^{\ominus}$CN is a very highly specific catalyst for the reaction; not so much because it is both a good nucleophile and a good leaving group, but because its electron-withdrawing ability increases the acidity of the C—H linkage in (126) and prompts the formation of the carbanion (127), which the CN group can stabilise by delocalisation (127a ↔ 127b). The catalyst in the Cannizzaro reaction—$^{\ominus}$OH—lacks any such ability, and the equivalent of (126) thus acts as a hydride donor to a second molecule of ArCHO.

8.4.10 Benzilic acid rearrangement

Oxidation of benzoin, PhCH(OH)COPh (above) yields benzil, PhCOCOPh (132), and this, in common with non-enolisable 1,2-diketones in general, undergoes base-catalysed rearrangement to yield the anion of an α-hydroxy acid, benzilate anion, $Ph_2C(OH)CO_2^{\ominus}$ (133). The rate law commonly observed is,

$$Rate = k[PhCOCOPh][^{\ominus}OH]$$

and the reaction is believed to follow the general pathway:

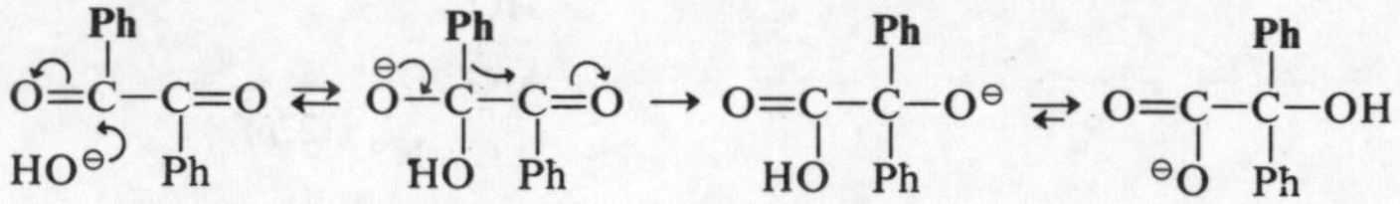

The slow, rate-limiting step is almost certainly the migration of phenyl in the initial $^{\ominus}$OH adduct, i.e. (134) → (135). This is essentially the analogue for 1,2-diketones of the intramolecular Cannizzaro reaction on the 1,2-dialdehyde glyoxal, OHCCHO (p. 213). In the latter it was an H atom that migrated with its electron pair, i.e. as hydride ion, to the adjacent C=O group, whereas in benzil (132) it is Ph that migrates with its electron pair, i.e. as a carbanion; hence the reason for considering this reaction as an (intramolecular) carbanion addition to C=O. There seem to be no examples of the equivalent of intermolecular Cannizzaro reactions on ketones, involving, as they necessarily must, migration of R with its electron pair from one molecule to another, i.e. $2R_2CO \rightsquigarrow RCO_2^{\ominus} + R_3COH$.

8.4.11 Wittig reaction

This is an extremely useful reaction for the synthesis of alkenes. It involves the addition of a phosphonium *ylid*, e.g. (136: ylids are species which, in their ground state, have charges of opposite sign on adjacent atoms), also known as a *phosphorane*, to an aldehyde or ketone: the ylid is indeed a carbanion involving also a hetero-atom. These species are generated by the reaction of an alkyl halide, RR'CHX (137), on a trialkyl- or triaryl-phosphine (138)—very often Ph_3P—to yield a phosphonium salt (139), followed by abstraction of a proton from it by a very strong base, e.g. PhLi:

$$Ph_3P + RR'CHX \longrightarrow \underset{X^\ominus \ (139)}{Ph_3\overset{\oplus}{P}-CHRR'} \xrightarrow{PhLi} Ph_3\overset{\oplus}{P}-\overset{\ominus}{C}RR'$$
$$(138) \quad (137)$$

$$\updownarrow$$

$$Ph_3P{=}CRR'$$
$$(136)$$

Addition of the Wittig reagent (136) to C=O, e.g. (140), is believed to follow the pathway:

$$(140) \ \underset{\underset{(136)}{RR'\overset{\ominus}{C}-\overset{\oplus}{P}Ph_3}}{R_2''C{=}O} \ \underset{(1)}{\rightleftharpoons} \ \underset{\underset{(141)}{RR'C-\overset{\oplus}{P}Ph_3}}{R_2''C-O^\ominus} \ \overset{(2)}{\rightarrow} \ \underset{RR'C{-}PPh_3}{R_2''C{-}O} \ \overset{(3)}{\rightarrow} \ \underset{\underset{(142)}{RR'C \quad PPh_3}}{R_2''C \quad O}$$

Step (1) may, or may not, be an equilibrium; step (1) or steps (2)/(3) may be rate-limiting, and step (2)/(3) may indeed be simultaneous; it has been possible, in the odd case, actually to isolate the zwitterion intermediate (141). Where the original phosphonium salt is chiral, e.g. $RR'R''P^\oplus{-}CH_2R$, its configuration is found to be retained in the product phosphine oxide, $RR'R''P{=}O$, e.g. (142).

Because of the variations possible in the R groups of the original halide (137) and in the carbonyl component (140), this is an extremely useful and versatile method for the synthesis of substituted alkenes. The presence of C=C or C≡C, even when conjugated with the C=O group, does not interfere. A CO_2R group, though it will react with the ylid (136) does so very much more slowly than with C=O, and thus does not interfere either. The reaction is particularly valuable for getting a double bond into positions that are difficult, e.g. exocyclic methylene groups (143),

or all but impossible, e.g. $\beta\gamma$-unsaturated acids (144),

$$R_2C{=}O + Ph_3\overset{\oplus}{P}-\overset{\ominus}{C}HCH_2CO_2^\ominus \rightarrow R_2C{=}CHCH_2CO_2^\ominus + Ph_3P{=}O$$
$$(144)$$

by other methods. In the case of (144), most methods tend to induce isomerisation to the thermodynamically more stable, conjugated ($\alpha\beta$-unsaturated) acid. The Wittig reaction has also been used intra-molecularly to prepare cyclic alkenes containing 5–16 carbon atoms.

8.5   STEREOSELECTIVITY IN CARBONYL ADDITION REACTIONS

Whether nucleophilic addition of HY to C=O proceeds stereo-selectively CIS or TRANS clearly has no meaning—unlike electro-philic addition of HY to C=C [(145) → (146) or (147)]—for with C=O [(148) → (149) or (150)] the alternative products are identical, because of free rotation about their C—O bonds:

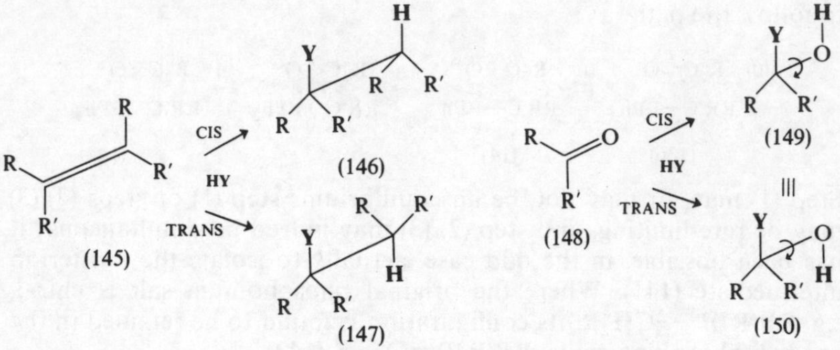

Addition of HY to RR′C=O introduces a chiral centre into the adduct (151), but the product will always be the ($\pm$) form—the racemate (151ab)—because initial nucleophilic attack from above (a), or below (b), on the planar carbonyl compound (148) will be statistically equally likely:

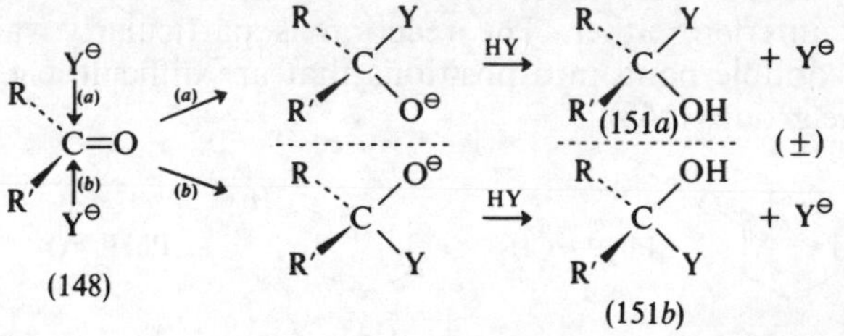

If, however, R or R′ is chiral—and particularly if this stems from the α-carbon atom—then the two faces of the carbonyl compound (148) are no longer equivalent, and addition from above and below not therefore statistically equally likely. Where the reaction is rever-sible it is likely that there will be a greater proportion of the thermodynamically more stable of the two alternative products in the reaction mixture (thermodynamic or equilibrium control, *cf.* p. 43). For essentially irreversible reactions, e.g. with RMgX, LiAlH$_4$ etc., the product that is formed the more rapidly is likely to preponderate (kinetic control); which this is likely to be can often be forecast from *Cram's rule*: a ketone will react in that conformation in which the O of the C=O group is *anti* to the largest of the three substitutions on the α-carbon atom (152). Preferential nucleophilic attack (e.g. by R′MgBr) will then take place from the least hindered side of the carbonyl carbon atom, i.e. (*a*). This is best seen using Newman projection formulae (*cf.* p. 7):

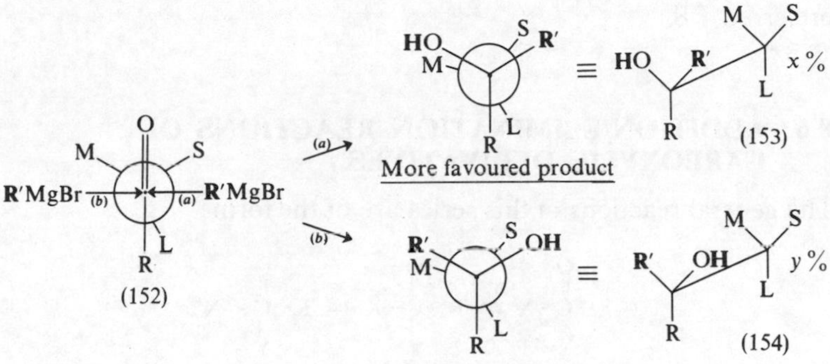

More favoured product (153)

Less favoured product (154)

This is saying, reasonably enough, that preferential reaction will take place *via* the less crowded (lower energy) T.S. We should thus expect the ratio x/y to increase: (i) as the difference in size between M and S increases, and (ii) as the size of R′ in R′MgBr increases. In practice, it is found that for attack of MeMgI on the aldehyde C$_6$H$_5$-(Me)CHCHO (152, L = C$_6$H$_5$, M = Me, S = H, R = H) x/y = 2:1, while replacing Me by the rather bulkier Et raises x/y to 2·5:1. Similarly, replacing MeMgI by the much bulkier C$_6$H$_5$MgBr for attack on the Me compound (152, M=Me) is found to raise the x/y ratio to >4:1.

The operation of Cram's rule has been investigated very largely for Grignard additions, and some hydride additions, to C=O; in general it works quite well at forecasting which will be the more favoured product, but there are a number of exceptions. This is hardly surprising for the rule assumes that product control depends only on steric interactions, whereas complex formation—between groups in the substrate, e.g. hydrogen bonding, or between substrate and attacking nucleophile, e.g. RMgX and carbonyl oxygen atom—and dipole/dipole interaction may also play a part. As an example of the latter effect α-chloro-aldehydes and -ketones are found to react (because of electrostatic repulsion) in that conformation, e.g. (155), in which Cl and carbonyl oxygen atom are *anti* to each other,

$$
\begin{array}{c}
\text{O} \\
\| \\
M-\!\!\!\overset{\displaystyle}{\underset{\underset{\displaystyle R}{|}}{\bigcirc}}\!\!\!-L \\
\text{Cl}
\end{array}
$$

(155)

irrespective of the size, relative to Cl, of the other groups attached to that α-carbon atom. Either of these types of effect may outweigh purely steric considerations in determining the geometry of the preferred T.S.

8.6 ADDITION/ELIMINATION REACTIONS OF CARBOXYLIC DERIVATIVES

The general reactions of this series are of the form:

$$
\underset{(156)}{\overset{\displaystyle \overset{\displaystyle C}{\overset{O}{\|}}}{R-\underset{\underset{Y^{\ominus}}{\displaystyle\frown}}{C}-X}} \; \overset{Y^{\ominus}}{\rightleftarrows} \; \underset{(157)}{\overset{\displaystyle \overset{\ominus O}{\overset{C}{|}}}{R-\underset{\underset{Y}{|}}{C}-X}} \; \overset{-X^{\ominus}}{\rightleftarrows} \; \overset{O}{\underset{\underset{Y}{|}}{R-\overset{\|}{C}}} + X^{\ominus}
$$

The reaction pathway is normally nucleophilic addition/elimination, *via* a so-called *tetrahedral intermediate* (157), leading to overall substitution. The difference between the reactions of carboxylic derivatives (156) and those of simple carbonyl compounds (aldehydes and ketones) stems from the fact that in the former there is a good leaving group —X— attached to the carbonyl carbon atom. The relative reactivity of the series (156, with differing X) towards a particular nucleophile $Y^{\ominus}$ (e.g. $^{\ominus}OH$) depends on: (*a*) the relative electron-donating or -withdrawing power of X towards the carbonyl carbon atom, and (*b*) the relative ability of X as a leaving group. The reactivity series is not necessarily exactly the same for every $Y^{\ominus}$, but in general it follows the order:

$$
\underset{}{\overset{O}{\overset{\|}{RC}}-Cl} > \underset{}{\overset{O}{\overset{\|}{RC}}-OCOR} > \underset{}{\overset{O}{\overset{\|}{RC}}-OR'} > \underset{}{\overset{O}{\overset{\|}{RC}}-NH_2} > \underset{}{\overset{O}{\overset{\|}{RC}}-NR'_2}
$$

Thus acid chlorides and anhydrides react readily with ROH and NH_3 to yield esters and amides, respectively, while esters react with NH_3 or amines to give amides, but the simple reversal of any of these reactions on an amide, though not impossible, is usually pretty difficult. The relative reactivity will also depend on both the electronic and, more particularly the steric, effect of R. A slightly unusual leaving group is $^\ominus CX_3$ (e.g. $^\ominus CI_3$) in the *haloform* (158) *reaction* (*cf.* p. 289):

$$R-\overset{\overset{O}{\|}}{\underset{\underset{OH}{|}}{C}}-CX_3 \xrightarrow{\overset{\ominus OH}{\rightleftharpoons}} R-\overset{\overset{\ominus O}{|}}{\underset{\underset{OH}{|}}{C}}-CX_3 \longrightarrow R-\overset{O}{\underset{\underset{OH}{|}}{C}} + {}^\ominus CX_3 \rightleftharpoons R-\overset{O}{\underset{\underset{O_\ominus}{|}}{C}} + HCX_3$$

$$(158)$$

The rate law followed by these reactions is generally of the form,

$$\text{Rate} = k[\text{RCOX}][Y^\ominus]$$

and the question arises whether they might perhaps proceed by a direct, one step (*cf.* S_N2) displacement on the carbonyl carbon atom. It is not normally possible to isolate tetrahedral intermediates such as (157), but it has proved possible to obtain evidence of the formation of one (159) spectroscopically, in a case where R carries strongly electron-withdrawing groups (*cf.* Cl_3CCHO, p. 205), i.e. $^\ominus OMe$ in dibutyl ether on CF_3CO_2Et (160):

$$CF_3-\overset{\overset{O}{\|}}{\underset{\underset{OMe}{|}}{C}}-OEt \underset{}{\overset{\ominus OMe}{\rightleftharpoons}} CF_3-\overset{\overset{\ominus O}{|}}{\underset{\underset{OMe}{|}}{C}}-OEt$$

$$(160) \qquad\qquad (159)$$

The $C{=}O$ absorption in the i.r. ($\nu_{max} 1790 \text{ cm}^{-1}$) disappears completely on adding $Na^\oplus OMe^\ominus$, and (160) may then be recovered unchanged on adding dry HCl gas to the solution. In proceeding from the original carboxylic derivative (156) to the tetrahedral intermediate (157), the carbonyl carbon atom changes its hybridisation from $sp^2 \rightarrow sp^3$ and, in so far as the T.S. of the rate-limiting stage of the overall reaction resembles (157), we might expect these reactions to be susceptible to steric effects: such is indeed the case (see below).

We have already discussed carbanion addition to RCO_2Et (Claisen ester condensation, p. 224), and also their reduction by $Li^\oplus AlH_4{}^\ominus$ (p. 211); some further examples of nucleophilic attack will now be considered.

8.6.1 Grignard, etc., reagents

Attack of Grignard reagents on esters, e.g. (161), follows the general pathway indicated above, so that the initial product of addition/elimination ($^\ominus OR'$ as leaving group) is a ketone (162):

$$R-\overset{\overset{O}{\|}}{\underset{\underset{R''{-}MgX}{}}{C}}-OR' \xrightarrow{R''MgX} R-\overset{\overset{\ominus O}{|}}{\underset{\underset{R''}{|}}{C}}\overset{\oplus MgX}{\underset{}{}}OR' \xrightarrow{-OR'^\ominus} R-\overset{\overset{O}{\|}}{\underset{\underset{R''}{|}}{C}} \xrightarrow{R''MgX} R-\overset{\overset{O^\ominus \oplus MgX}{|}}{\underset{\underset{R''}{|}}{C}}-R''$$

$$(161) \qquad\qquad\qquad\qquad\qquad\qquad (162) \qquad\qquad (163)$$

The carbonyl carbon atom of (162) is, however, more reactive towards nucleophiles than that of the original ester (161), because of the electron-donating mesomeric effect of the ester oxygen atom:

$$
\begin{array}{cc}
\overset{\displaystyle O^{\delta--}}{\underset{\delta++}{R-C-R''}} & \overset{\displaystyle O^{\delta-}}{R-C\overset{+}{\underset{\delta+}{}}OR'} \\
(162) & (161)
\end{array}
$$

As soon as it is formed, (162) thus competes preferentially with as yet unreacted ester (161) for Grignard reagent, R''MgX, and the actual end product of the reaction is the salt of a tertiary alcohol (163); two of its alkyl groups having been supplied by the Grignard reagent. Hardly surprisingly, acyl halides, e.g. RCOCl, yield the same products with Grignard reagents, but the reaction can in this case be stopped at the ketone stage by use of organo-cadmium compound, CdR''_2. The reaction also stops at the ketone stage with esters when R''Li is used at higher temperatures in place of R''MgX.

8.6.2   Some other nucleophiles

A reaction that has been much investigated is the hydrolysis of esters, e.g. (164), by aqueous base, i.e. $^{\ominus}OH$. It is found to be kinetically second order, and ^{18}O isotopic labelling experiments on (164) have established that this normally undergoes *acyl-oxygen* cleavage (*cf.* p. 47), i.e. compatible with the general tetrahedral intermediate, e.g. (165), pathway we have already discussed above:

$$
\underset{(164)}{\overset{\displaystyle O}{R-C-{}^{18}OEt}}\ \underset{^{\ominus}OH}{\rightleftharpoons}\ \underset{(165)}{\overset{\displaystyle \overset{\ominus}{O}}{\underset{OH}{R-C-{}^{18}OEt}}}\ \overset{-OEt^{\ominus}}{\longrightarrow}\ \underset{(166)}{\overset{\displaystyle O}{\underset{OH}{R-C}}} + {}^{18}OEt^{\ominus}\ \rightleftharpoons\ \underset{(167)}{\overset{\displaystyle O}{\underset{O_{\ominus}}{R-C}}} + H^{18}OEt
$$

The rate-limiting step is almost certainly attack of $^{\ominus}OH$ on the original ester, (164), and the overall reaction is essentially irreversible as $^{\ominus}OEt$ would remove a proton from (166) rather than attack its carbonyl carbon atom, while the carboxylate anion (167) will be insusceptible to nucleophilic attack by EtOH or EtO$^{\ominus}$. This mechanism is generally referred to as $B_{AC}2$ (*base*-catalysed, *acyl*-oxygen cleavage, *bimolecular*). Where nucleophilic attack is by $^{\ominus}OR$, rather than $^{\ominus}OH$, *transesterification* occurs, and an equilibrium mixture of both esters, (164) + (168), is obtained; the position of equilibrium depends on the relative concentrations and nucleophilic abilities of $^{\ominus}OEt$ and

$^\ominus$OR:

$$R-\overset{\overset{\displaystyle O}{\|}}{\underset{\underset{\displaystyle \ominus OR}{|}}{C}}-OEt \;\overset{\ominus OR}{\rightleftharpoons}\; R-\overset{\overset{\displaystyle \ominus O}{|}}{\underset{\underset{\displaystyle OR}{|}}{C}}-OEt \;\overset{-OEt^\ominus}{\rightleftharpoons}\; R-\overset{\overset{\displaystyle O}{\|}}{\underset{\underset{\displaystyle OR}{|}}{C}} + OEt^\ominus$$

(164)   (168)

Attack by $^\ominus$OH on amides, $RCONH_2$, follows an analogous course to that with esters (above) except that here $^\ominus NH_2$—rather than $^\ominus OEt$—is the leaving group. This removes a proton from (166) to form the stabler pair of carboxylate anion (167) + NH_3; loss of the latter from the hot, basic solution tends to drive the reaction over to the right. The attack of amines, $R\overset{..}{N}H_2$, on esters, e.g. (164), to form amides (169) follows very much the same general course as the examples above (it has been shown that $RNH^\ominus$, the conjugate base of RNH_2, is not involved in nucleophilic attack on the ester):

$$R-\overset{\overset{\displaystyle O}{\|}}{\underset{\underset{\displaystyle H_2NR'}{|}}{C}}-OEt \;\overset{R'NH_2}{\rightleftharpoons}\; R-\overset{\overset{\displaystyle \ominus O}{|}}{\underset{\underset{\displaystyle \overset{\oplus}{H_2NR'}}{|}}{C}}-OEt \;\overset{R'NH_2}{\rightleftharpoons}\; R-\overset{\overset{\displaystyle \ominus O}{|}}{\underset{\underset{\displaystyle HNR'}{|}}{C}}-OEt \;\overset{BH}{\underset{slow}{\longrightarrow}}\; R-\overset{\overset{\displaystyle O}{\|}}{\underset{\underset{\displaystyle HNR'}{|}}{C}} + HOEt + B^\ominus$$

(164)   (170)   (169)

The slow, rate-limiting step seems to be loss of the leaving group from (170), and this normally needs the assistance of a proton donor BH, e.g. H_2O.

Acid chlorides, RCOCl, undergo ready attack by weaker nucleophiles, e.g. H_2O, ROH ($^\ominus$OH and $^\ominus$OR are not required). The reaction may proceed *via* a tetrahedral intermediate (as above), or *via* an S_N1 type process (*cf.* p. 79) involving slow, rate-limiting loss of $Cl^\ominus$ from RCOCl to yield the acylcarbonium ion, $R\overset{\oplus}{C}=O$, followed by rapid reaction of this with the nucleophile. Which pathway is followed is, hardly surprising, found to depend on the polarity and ion-solvating ability of the solvent (*cf.* p. 80), as well as on the structure of the substrate.

Acid anhydrides, $(RCO)_2O$, will also often react with weaker nucleophiles, though more slowly than acid chlorides; the S_N1 type reaction pathway (above) does not normally occur. Anhydrides are essentially intermediate in reactivity—towards a particular nucleophile—between acid chlorides and esters, reflecting the leaving group ability sequence:

$$Cl^\ominus > RCO_2{}^\ominus > RO^\ominus$$

8.6.3 Acid-catalysed reactions

It is difficult to effect attack on the carbonyl carbon atom of RCO_2H, (171), with nucleophiles of the general type $Y^\ominus$, as they commonly remove proton instead, and the resultant $RCO_2^{\ominus}$ is then insusceptible to nucleophilic attack. Weaker nucleophiles of the form YH, e.g. ROH, do not suffer this inability, but their reactions with the relatively unreactive carbonyl carbon atom of RCO_2H are slow. The carbonyl character may be enhanced by protonation, however, i.e. by acid catalysis in, for example, esterification [(171) → (172)]:

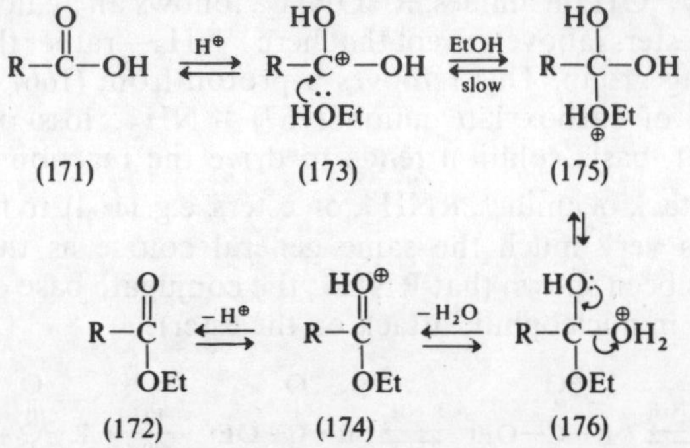

N.m.r. spectra support preferential protonation of the carbonyl oxygen of the acid (173) in the forward reaction (esterification), and of the ester (174) in the reverse reaction (hydrolysis). Acid catalysis also has the effect of promoting the loss of the leaving group, i.e. it is easier to lose H_2O from (176)—esterification—or EtOH from (175)—hydrolysis—than it is, for example, to lose $\ominus$OEt from (165) above. The equilibrium is normally displaced in the desired direction by using an excess of ROH (or of H_2O for hydrolysis). This mechanism is generally referred to as $A_{AC}2$ (*acid*-catalysed, *acyl*-oxygen cleavage, *bimolecular*). Reaction of R'OH with RCO_2R'', under these conditions, results in transesterification, the position of equilibrium being determined by the relative proportions of R'OH and R''OH. Acid anhydrides and amides undergo acid-catalysed hydrolysis in very much the same way as esters.

Esters, RCO_2R', where the alkyl group R' can form a relatively stable carbonium ion, e.g. (177) from (178), have been shown—by ^{18}O labelling experiments—to undergo *alkyl*-oxygen cleavage:

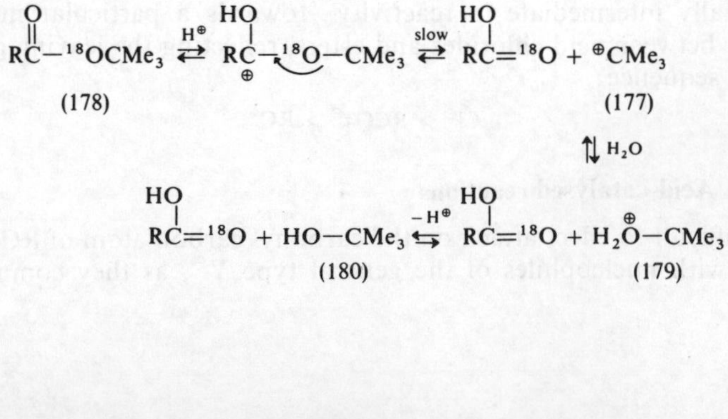

This mechanism is generally referred to as $A_{AL}1$ (*acid*-catalysed, *alkyl*-oxygen cleavage, *unimolecular*), it also occurs with ester alkyl groups such as Ph_2CH etc. When attempts are made to transesterify (178) with R'OH, the product is not now the expected ester, RCO_2R', but RCO_2H plus the ether $R'OCMe_3$; the latter arises from attack of R'OH on the carbonium ion intermediate (177), *cf*. the conversion of (177) to (179) by H_2O above.

Where the acid alkyl group, R in RCO_2R', is sufficiently bulky, e.g. R_3C (181), that bimolecular hydrolysis *via* a tetrahedral intermediate is inhibited (because of the degree of crowding there would be in the T.S.), a further, relatively rare, acid-catalysed mechanism is found to operate—$A_{AC}1$ (*acid*-catalysed, *acyl*-oxygen cleavage, *unimolecular*); it occurs only in powerful ionising solvents:

$$
\underset{(181)}{R_3C\overset{\overset{\displaystyle O}{\|}}{C}-\overset{..}{\underset{..}{O}}R'}
\;\underset{}{\overset{H^{\oplus}}{\rightleftharpoons}}\;
\underset{(183)}{R_3C\overset{\overset{\displaystyle O}{\|}}{C}\overset{\frown}{\underset{\oplus H}{O}}R'}
\;\underset{slow}{\overset{-R'OH}{\rightleftharpoons}}\;
\underset{(184)}{R_3C\overset{\overset{\displaystyle O}{\|}}{C}{}^{\oplus} + HOR'}
$$

$$\Big\updownarrow H_2O$$

$$
\underset{(182)}{R_3C\overset{\overset{\displaystyle O}{\|}}{C}-OH + R'OH}
\;\underset{}{\overset{-H^{\oplus}}{\rightleftharpoons}}\;
\underset{(185)}{R_3C\overset{\overset{\displaystyle O}{\|}}{C}-\overset{\oplus}{O}H_2 + HOR'}
$$

Exactly the same considerations apply to the esterification of hindered acids (182) in the reverse direction. It will be noticed that this mechanism requires protonation on the less favoured (*cf*. p. 236) hydroxyl oxygen atom (185) to allow the formation of the acyl carbonium ion intermediate (184). Apart from a number of R_3C types, a very well known example is 2,4,6-trimethylbenzoic (mesitoic) acid (186), which will not esterify under ordinary acid-catalysis conditions—and nor will its esters (187) hydrolyse. Dissolving acid or ester in conc. H_2SO_4 and pouring this solution into cold alcohol or water, respectively, is found to effect essentially quantitative esterification or hydrolysis as required; the reaction proceeds *via* the acyl carbonium ion (188):

(186)   (188)   (187)

Evidence for the formation of (188) is provided by the observation that while dissolution of unhindered benzoic acid itself (189) in conc. H_2SO_4 results in the expected *two*-fold freezing point depression;

$$\underset{(189)}{PhC\overset{O}{\|}-OH} + H_2SO_4 \rightleftarrows \underset{(190)}{Ph\overset{OH}{\underset{\oplus}{C}}-OH} + HSO_4{}^{\ominus}$$

dissolution of the hindered (186) results in *four*-fold depression:

$$\underset{(186)}{ArCO_2H} + 2H_2SO_4 \rightleftarrows \underset{(188)}{Ar\overset{\oplus}{C}=O} + H_3O^\oplus + 2HSO_4{}^\ominus$$

The two bulky *o*-Me groups block attack by MeOH on the carbonyl carbon atom of (186) when this is protonated normally (like 189) on the carbonyl oxygen atom [the two OH groups, *cf.* (190), block attack on it from the two remaining directions]. Abnormal protonation on the hydroxyl oxygen atom, *cf.* (185), allows formation of the planar species (188) in which ready attack on the $^\oplus C=O$ carbon atom by MeOH can now take place from directions at right angles to the plane of the molecule. That the reasons for this change of mechanism are steric is demonstrated by the fact that the acids (191) and (192), and their esters, undergo ready esterification/hydrolysis by the normal $A_{AC}2$ mode:

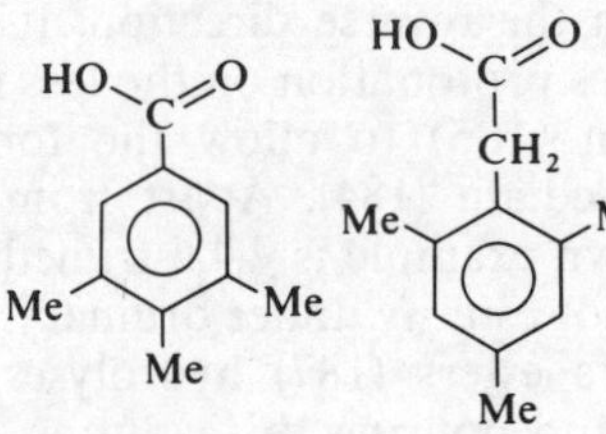

8.7   ADDITION TO C≡N

The C≡N linkage bears an obvious formal resemblance to C=O,

$$\underset{(193a)}{RC\equiv N} \leftrightarrow \underset{(193b)}{R\overset{\oplus}{C}=\overset{\ominus}{N}} \quad \text{i.e. } \underset{(193ab)}{R\overset{\delta+}{C}\overset{\delta-}{\equiv}N} \equiv RC\pm N$$

and might be expected to undergo a number of analogous nucleophilic addition reactions. Thus they add Grignard reagents to yield salts of ketimines (194), which may be hydrolysed to ketones (195):

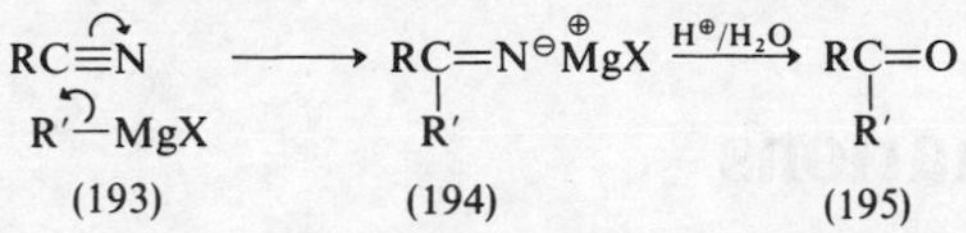

$$\underset{\substack{\overset{\displaystyle R'-MgX}{(193)}}}{RC\equiv N} \longrightarrow \underset{\substack{\overset{\displaystyle R'}{(194)}}}{RC=N^\ominus \overset{\oplus}{M}gX} \xrightarrow{H^\oplus/H_2O} \underset{\substack{\overset{\displaystyle R'}{(195)}}}{RC=O}$$

Reduction with $Li^\oplus AlH_4{}^\ominus$ (*cf.* p. 211) yields RCH_2NH_2, NH_3 adds to (193), in the presence of $NH_4{}^\oplus Cl^\ominus$ to yield salts of *amidines*, $RC(NH_2)=NH_2{}^\oplus Cl^\ominus$. Acid-catalysed addition of alcohols, e.g. EtOH, yields salts of *iminoethers* (196, *cf.* hemiacetals, p. 206):

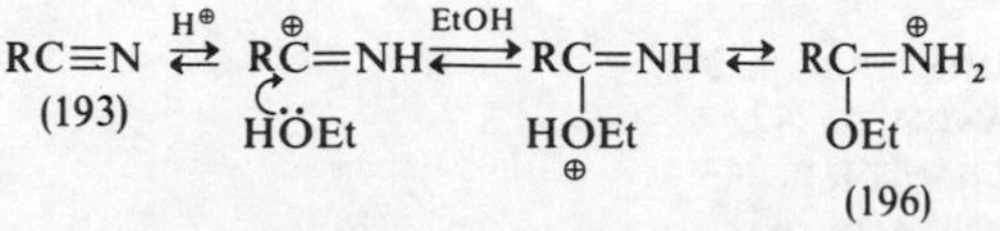

$$\underset{(193)}{RC\equiv N} \overset{H^\oplus}{\rightleftarrows} \underset{\substack{\overset{\displaystyle \ddot{C}}{HOEt}}}{R\overset{\oplus}{C}=NH} \overset{EtOH}{\rightleftarrows} \underset{\substack{\overset{\displaystyle \underset{\oplus}{HOEt}}}}{RC=NH} \rightleftarrows \underset{\substack{\overset{\displaystyle OEt}{(196)}}}{RC=\overset{\oplus}{N}H_2}$$

The addition of H_2O (hydrolysis) may be both acid- and base-catalysed:

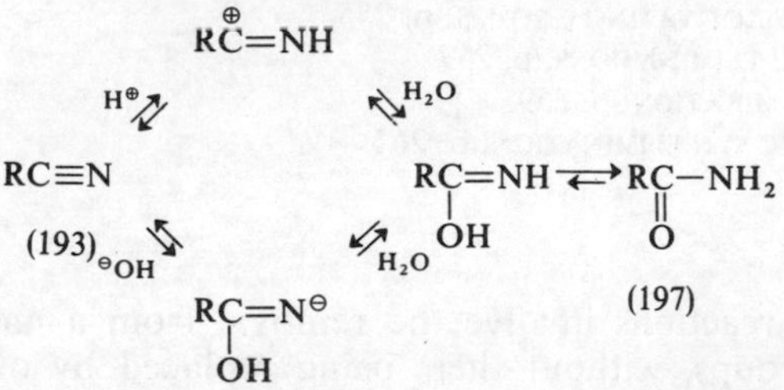

$$\overset{\oplus}{RC}=NH$$

$$\underset{(193)}{RC\equiv N} \qquad\qquad RC=NH \rightleftarrows RC-NH_2$$

with $H^\oplus$ and ${}_\ominus OH$ and H_2O paths, and:

$$\underset{\substack{\overset{\displaystyle OH}}}{RC=N^\ominus}$$

$$\underset{\substack{\overset{\displaystyle OH}}}{RC=NH} \qquad \underset{\substack{\overset{\displaystyle O}{(197)}}}{RC-NH_2}$$

The initial product is an amide (197), but this also undergoes ready acid- or base-catalysed hydrolysis (see above), and the actual reaction product is often the carboxylic acid, RCO_2H, or its anion.

9
Eliminations

Elimination reactions involve the removal from a molecule of two atoms or groups, without their being replaced by other atoms or groups. In the great majority of such reactions the atoms or groups are lost from adjacent carbon atoms, one of them very often being a proton and the other a nucleophile, $Y\!:$ or $Y^{\ominus}$, resulting in the formation of a multiple bond, a 1,2-(or $\alpha\beta$-) elimination:

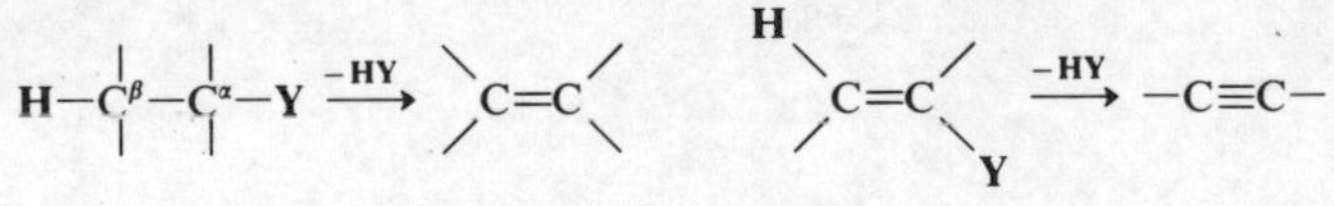

$$H-\overset{|}{\underset{|}{C^{\beta}}}-\overset{|}{\underset{|}{C^{\alpha}}}-Y \xrightarrow{-HY} \diagup C=C\diagdown \qquad \overset{H}{\diagdown}C=C\diagup \xrightarrow{-HY} -C\equiv C-$$

Eliminations from atoms other than carbon are also known;

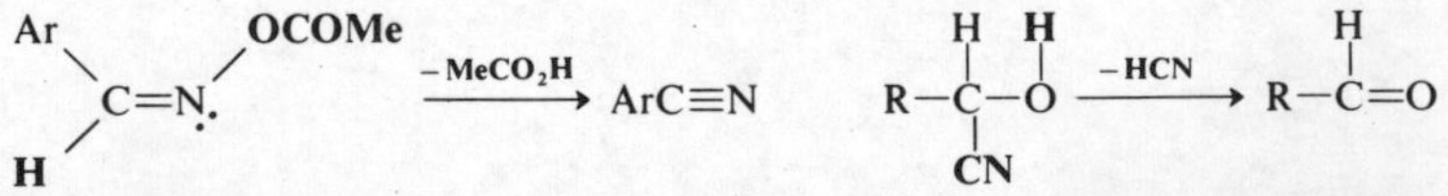

$$\overset{Ar}{\underset{H}{\diagdown}}C=\overset{OCOMe}{\underset{\cdot}{N}} \xrightarrow{-MeCO_2H} ArC\equiv N \qquad R-\overset{H}{\underset{CN}{C}}-O \xrightarrow{-HCN} R-\overset{H}{C}=O$$

as are eliminations both from the same atom, 1,1-(α-) eliminations (cf. p. 259), and from atoms further apart than 1,2- i.e. reversal of 1,4-addition (cf. p. 192), also 1,5- and 1,6-eliminations leading to cyclisation. 1,2-Eliminations are by far the most common and important, however, and most of our discussion will be concerned with them.

9.1   1,2-(β-)ELIMINATION

In 1,2-eliminations involving carbon atoms (i.e. most), the atom from which Y is lost is usually designated as the 1-(α-) carbon and that losing (usually) H as the 2-(β-) carbon; in the older αβ-terminology, the α- is commonly omitted, and the reactions are referred to as β-eliminations. Among the most familiar examples are base-induced elimination of hydrogen halide from alkyl halides—this almost certainly the most common elimination of all—particularly from bromides (1);

$$RCH_2CH_2Br \xrightarrow{\ominus OH} RCH=CH_2 + H_2O + Br^\ominus$$
(1)

acid-catalysed dehydration of alcohols (2);

$$RCH_2CR_2OH \xrightarrow{H^\oplus} RCH=CR_2 + H_3O^\oplus$$
(2)

and Hofmann degradation of quaternary alkylammonium hydroxides (3):

$$RCH_2CH_2\overset{\oplus}{N}Me_3{}^\ominus OH \rightarrow RCH=CH_2 + H_2O + NMe_3$$
(3)

Many other Y: or $Y^\ominus$ species are known as leaving groups, however, e.g. SR_2, SO_2R, OSO_2Ar etc.

Three different, simple mechanisms can be envisaged for 1,2-eliminations, differing from each other in the timing of H—C and C—Y bond-breaking. This could (a) be concerted,

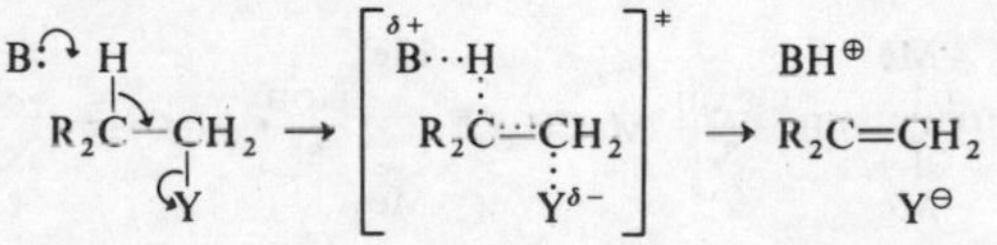

$$\overset{B \curvearrowright H}{\underset{\underset{Y}{|}}{R_2C - CH_2}} \rightarrow \left[\begin{array}{c} \overset{\delta+}{B \cdots H} \\ R_2\dot{C}{-}CH_2 \\ \overset{..}{Y^{\delta-}} \end{array} \right]^{\ddagger} \quad \begin{array}{c} BH^\oplus \\ \rightarrow R_2C=CH_2 \\ Y^\ominus \end{array}$$
(4)

i.e. a one-step process, passing through a single T.S. (4); this is referred to as the E2 mechanism (Elimination, bimolecular) and is somewhat reminiscent of S_N2 (cf. p. 78). Alternatively, the H—C and C—Y bonds can be broken separately in two-step processes. If the C—Y bond is broken first, (b), a carbonium ion intermediate (5) is involved;

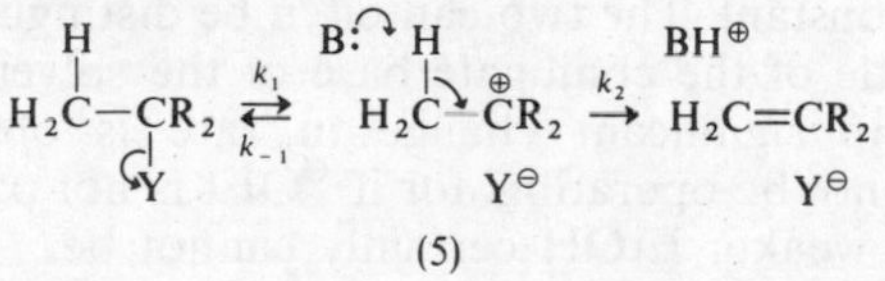

$$\underset{\underset{Y}{|}}{\overset{H}{\underset{|}{H_2C - CR_2}}} \underset{k_{-1}}{\overset{k_1}{\rightleftarrows}} \underset{Y^\ominus}{\overset{B \curvearrowright H}{\underset{|}{H_2C - \overset{\oplus}{C}R_2}}} \overset{k_2}{\rightarrow} \underset{Y^\ominus}{\overset{BH^\oplus}{H_2C=CR_2}}$$
(5)

this is referred to as the E1 mechanism (*E*limination, *unimolecular*). It is reminiscent of S_N1 (*cf.* p. 79), and the carbonium ion intermediates for S_N1 and E1 are, of course, identical. Finally, the H—C could be broken first, (*c*), in a two-step process involving a carbanion intermediate (6);

$$\underset{(6)}{\overset{\displaystyle B\colon\!\!\frown H}{\underset{Y}{\overset{\displaystyle |}{X_2C-CH_2}}} \underset{k_{-1}}{\overset{k_1}{\rightleftharpoons}} \overset{BH^\oplus}{\underset{\underset{Y}{\overset{\displaystyle |}{\frown}}}{X_2C^{\ominus}CH_2}} \overset{k_2}{\longrightarrow} \overset{BH^\oplus}{\underset{Y^\ominus}{X_2C{=}CH_2}}}$$

this is referred to as the E1cB mechanism [*E*limination, from *conjugate Base*, i.e. (6)]. Examples of reactions proceeding by all three mechanisms are known: E1cB is the least, and E2 probably the most, common. The three mechanisms will now be considered in turn, but it should be realised that they are only limiting cases (*cf.* S_N1/S_N2), and that in fact a continuous mechanistic spectrum, in the relative time of breaking of the two bonds, is available and is indeed observed in practice.

9.2 E1 MECHANISM

If, as is normally the case, carbonium ion, e.g. (5), formation is slow and rate-limiting, (i.e. $k_2 > k_{-1}$), then the rate law observed with, for example, the bromide $MeCH_2CMe_2Br$ is;

$$\text{Rate} = k[MeCH_2CMe_2Br]$$

the overall elimination is then completed (7) by rapid, non-rate-limiting removal of a proton from (8) usually by a solvent molecule, in this case EtOH:

$$\underset{(9)}{\overset{\displaystyle Me}{\underset{\displaystyle Me}{\overset{\displaystyle |}{MeCH_2C{-}OEt}}}} \xleftarrow[S_N1]{EtOH} \underset{(8)}{\overset{\displaystyle Me}{\underset{Me}{MeCH_2C^\oplus}}} \xrightarrow[E1]{EtOH} \underset{(7)}{\overset{\displaystyle Me}{\underset{Me}{MeCH{=}C}}}$$

It could be claimed that such an E1 solvolytic elimination would be indistinguishable kinetically from a bimolecular (E2) elimination, in which EtOH was acting as base, because the [EtOH] term in the E2 rate law,

$$\text{Rate} = k[MeCH_2CMe_2Br][EtOH]$$

would remain constant. The two can often be distinguished, however, by adding a little of the conjugate base of the solvent, i.e. $^\ominus OEt$ in this case. If no significant change in rate is observed, an E2 mechanism cannot be operating, for if $^\ominus OEt$ is not participating as a base the much weaker EtOH certainly cannot be.

The carbonium ion (8) is identical with that from S_N1 solvolysis (p. 78), and the latter reaction to yield the substitution product (9) is commonly a competitor with E1 elimination. Some evidence that the two processes do have a common intermediate is provided by the fact that the $E1/S_N1$ ratio is reasonably constant for a given alkyl group irrespective of the leaving group, $Y^\ominus$. The two processes do, however, proceed from (8) to products—(7) and (9), respectively—*via* different T.S.s, and the factors that influence elimination *v.* substitution are discussed subsequently (p. 253).

The factors that promote unimolecular, as opposed to bimolecular (E2), elimination are very much the same as those that promote S_N1 with respect to S_N2, namely: (*a*) an alkyl group in the substrate that can give rise to a relatively stable carbonium ion, and (*b*) a good ionising, ion-solvating medium. Thus (*a*) is reflected in the fact that with halides, increasing E1 elimination occurs along the series,

primary < secondary < tertiary

reflecting the relative stability of the resultant carbonium ions; primary halides hardly ever undergo E1 elimination. Branching at the β-carbon atom also favours E1 elimination; thus $MeCH_2CMe_2Cl$ yields only 34 % alkene, while Me_2CHCMe_2Cl yields 62 %. The reason may be partly steric: the greater the branching, the more the 'crowding strain' that is released on going from the sp^3 hybridised halide to the sp^2 hybridised carbonium ion intermediate; strain that is reintroduced on substitution (S_N1), but not on proton loss (E1) to yield an alkene. It is probably also related to the fact that the second chloride above will yield a more substituted, and hence thermodynamically more stable (*cf.* p. 26), alkene than the first. This is, with E1 reactions, also the major controlling factor (Saytzeff elimination, p. 250) in orientation of elimination, where more than one alkene can be derived by loss of different β-protons from a carbonium ion intermediate (8):

Thus in the above case the elimination product is found to contain 82 % of (7). Unexpected alkenes may arise, however, from rearrangement of the initial carbonium ion intermediate before loss of proton. E1 elimination reactions have been shown as involving a dissociated carbonium ion; they may in fact often involve ion pairs, of varying degrees of intimacy depending on the nature of the solvent (*cf.* S_N1, p. 90).

9.3 E1cB MECHANISM

If, as might be expected for this pathway, formation of the carbanion intermediate (6) is fast and reversible, while subsequent loss of the leaving group, $Y^\ominus$, is slow and rate-limiting, i.e. $k_{-1} > k_2$, then this reaction will follow the rate-law,

$$\text{Rate} = k[\text{RY}][\text{B}]$$

and will be kinetically indistinguishable from the concerted (E2) pathway. It should be possible to distinguish between them, however, by observing exchange of isotopic label, between as yet unchanged substrate and solvent, arising during fast, reversible carbanion (6) formation—something that clearly could not happen in the one-step, concerted (E2) pathway. A good example to test would be $PhCH_2$-CH_2Br (11), as the Ph group on the β-carbon would be expected to promote acidity in the β-H atoms, and also to stabilise the resultant carbanion (12) by delocalisation:

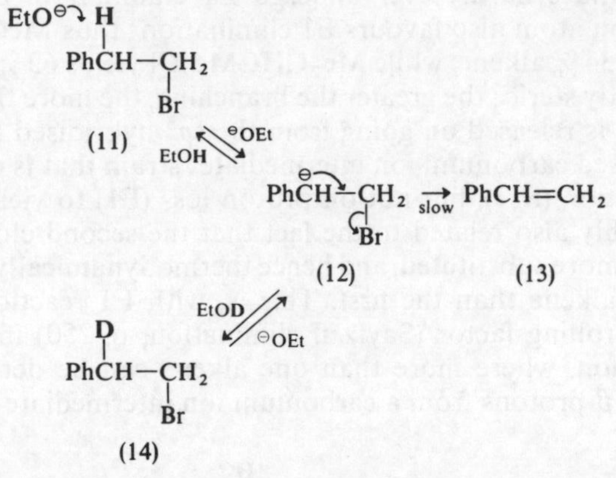

The reaction was carried out with $^\ominus OEt$ in EtOD, and (11) reisolated after $\approx$ half-conversion to (13): it was found to contain no deuterium, i.e. no (14); nor did the alkene (13) contain any deuterium, as might have been expected by elimination from any (14) formed. This potentially favourable case thus does not proceed by an E1cB pathway of the form described above; though we have not ruled out the case where $k_2 \gg k_{-1}$, i.e. essentially irreversible carbanion formation.

In fact reactions proceeding by this carbanion pathway are exceedingly rare; this is not altogether surprising as calculations suggest that the energy of activation for E2 is generally more favourable than that for E1cB, in most cases by $\approx$30–60 kJ (7–14 kcal) mol^{-1} (the reverse of step 2 would require addition of $Y^\ominus$ to C=C, which certainly doesn't happen at all easily). One example that almost certainly involves the latter pathway, however, is X_2CHCF_3 (15, X = Hal):

$$\underset{(15)}{X_2\overset{\displaystyle B:\curvearrowright H}{\underset{|}{C^{\beta}}}-\underset{|}{\overset{|}{C^{\alpha}}}F_2} \underset{fast}{\rightleftarrows} \underset{(16)}{X_2\overset{\ominus}{C}\overset{\curvearrowleft}{\underset{|}{C}}F_2} \underset{slow}{\rightarrow} \underset{(17)}{X_2C{=}CF_2}$$

This has all the right attributes in the substrate: (*a*) electronegative halogen atoms on the β-carbon to make the β-H more acid, (*b*) stabilisation of the carbanion (16) through electron-withdrawal by the halogen atoms on the carbanion carbon atom, and (*c*) a poor leaving group in F. Another structural feature favouring E1cB is (*d*) a positively charged substituent on the α-carbon atom also promoting the acidity of the β-H atoms, e.g. (18) which is found to undergo elimination to (19) in aqueous solution at room temperature—very much milder conditions than is common for such sulphonium hydroxides:

$$\underset{(18)}{\underset{\overset{|}{\underset{\oplus}{SMe_2}}}{PhSO_2CH-\overset{\displaystyle HO^{\ominus}\curvearrowright H}{\underset{|}{CH_2}}}} \rightleftarrows \underset{\underset{\oplus}{\overset{\curvearrowleft}{SMe_2}}}{PhSO_2\overset{\ominus}{CH}{-}CH_2} \rightarrow \underset{(19)}{PhSO_2CH{=}CH_2}$$

The powerfully electron-withdrawing $PhSO_2$ also fulfills criteria (*a*) and (*b*) above. Other examples of the E1cB pathway are benzyne formation (*cf.* p. 171), and reversal of simple addition to C=O, e.g. cyanohydrin (20) hydrolysis (*cf.* p. 208):

$$\underset{(20)}{\underset{\overset{|}{CN}}{O-\overset{\displaystyle B:\curvearrowright H}{\underset{|}{CR_2}}}} \rightleftarrows \underset{\overset{\curvearrowleft}{CN}}{{}^{\ominus}\overset{\curvearrowleft}{O}{-}CR_2} \rightarrow O{=}CR_2 + \overset{\ominus}{CN}$$

9.4 E2 MECHANISM

By far the commonest elimination mechanism is the one-step concerted (E2) pathway exhibiting, e.g. for the base-induced elimination of HBr from the halide RCH_2CH_2Br (21), the rate law:

$$Rate = k[RCH_2CH_2Br][B]$$

As B is often a nucleophile as well as a base, elimination is frequently accompanied by one-step, concerted (S_N2) nucleophilic substitution (*cf.* p. 78):

E2:

$$\text{B:} \curvearrowright \text{H} \\ \text{RCH} \overset{|}{\underset{\underset{Br}{|}}{-}} \text{CH}_2 \rightarrow \left[\begin{array}{c} \overset{\delta+}{B} \cdots H \\ R\dot{C}H \overset{\cdots}{=} \dot{C}H_2 \\ \ddot{B}r^{\delta-} \end{array} \right]^{\ddagger} \rightarrow \begin{array}{c} BH^{\oplus} \\ RCH = CH_2 \\ Br^{\ominus} \end{array}$$

(21)   (22)

S_N2:

$$\text{B:} \\ \text{RCH}_2\text{CH}_2 \rightarrow \left[\begin{array}{c} \overset{\delta+}{B} \\ RCH_2\dot{C}H_2 \\ \dot{B}r^{\delta-} \end{array} \right]^{\ddagger} \rightarrow \begin{array}{c} B^{\oplus} \\ | \\ RCH_2CH_2 \\ Br^{\ominus} \end{array}$$

(21)   (23)

The factors that influence elimination *v.* substitution are discussed subsequently (p. 253). Evidence for the involvement of C—H bond fission in the rate-limiting step—as a concerted pathway requires— is provided by the observation of a primary kinetic isotope effect (*cf.* p. 46) when H is replaced by D on the β-carbon.

One of the factors that affects the rate of E2 reactions is, hardly surprisingly, the strength of the base employed; thus we find:

$$^{\ominus}NH_2 > {}^{\ominus}OR > {}^{\ominus}OH$$

Some studies have been made with bases of the type $ArO^{\ominus}$, as this allows study of the effects of variation in basic strength (by introduction of *p*-substituents in $C_6H_5O^{\ominus}$) without concomitant change in the steric requirements of the base. With a given base, transfer from a hydroxylic solvent, e.g. H_2O or EtOH, to a bipolar aprotic one, e.g. $HCONMe_2$ (DMF) or $Me_2S^{\oplus}{-}O^{\ominus}$ (DMSO), can have a very pronounced effect as the strength of the base, e.g. $^{\ominus}OH$, $^{\ominus}OR$, is enormously increased thereby. This arises because the base has, in the latter solvents, no envelope of hydrogen-bonded solvent molecules that have to be stripped away before it can act as a base (*cf.* effect on nucleophilicity in S_N2, p. 81). Such change of solvent may result in a shift of mechanistic pathway from E1 to E2 for some substrate/base pairs.

The solvent may also exert an influence on the departure of the leaving group, Y: or $Y^{\ominus}$, *via* hydrogen bonding or other incipient solvation in the T.S. (22, above). For a substrate of given structure, but varying in Y, the better the leaving group the faster the reaction; the best leaving groups are the anions (conjugate bases) of strong acids, HY, and leaving group ability can be related in part to the relative strength of HY, e.g. $p\text{-MeC}_6H_4SO_3{}^{\ominus}$ ('tosylate') is an infinitely better leaving group than $^{\ominus}OH$. The strength of the C—Y bond also plays a part, and this may or may not follow the same sequence as the strength of HY. Relative C—Y bond strengths are borne out by the approximate relative rate sequence observed with $PhCH_2CH_2X$ with $^\ominus OEt$ in EtOH:

	$PhCH_2CH_2F$	$PhCH_2CH_2Cl$	$PhCH_2CH_2Br$	$PhCH_2CH_2I$
Rel. rate:	1	70	$4{\cdot}2 \times 10^3$	$2{\cdot}7 \times 10^4$

Incipient solvation in the T.S. can also play a part in determining relative leaving group ability, as we saw above, and this may or may not follow the same general sequence as acid strength of HY and/or C—Y bond strength. Change of solvent thus can, and does, change the sequence of relative leaving group ability in a series of different $Y^\ominus$s.

Finally, the major structural features in the substrate promoting E2 elimination are those that serve to stabilise the resultant alkene or, more particularly, the T.S. that precedes it. Such features include increasing alkyl substitution at both α- and β-carbon atoms (leading to alkenes of increasing thermodynamic stability), or introduction of a phenyl group that can become conjugated with the developing double bond.

9.4.1 Stereoselectivity in E2

With acylic molecules elimination could be envisaged as taking place from one or other of two limiting conformations—the *anti-periplanar* (24a) or the *syn-periplanar* (24b):

| (24a) | (25) | (24b) | (26) |

There is an obvious advantage in elimination taking place from a conformation in which H, C^β, C^α and Y are in the same plane as the *p* orbitals that are developing on C^β and C^α, as $H^\oplus$ and $Y^\ominus$ are departing, will then be parallel to each other, and thus capable of maximum overlap in the forming π bond. It will be energetically advantageous for the attacking atom of the base B to lie in this common plane also. Having established the desirability of elimination taking place from a planar conformation, there remains the question of whether either (24a) or (24b) is preferred over the other.

Two obvious grounds can be stated for preferring (24a), the anti-periplanar: (a) the electron pair liberated on C^β by loss of $H^\oplus$ will approach C^α from the side opposite to that from which an electron pair is departing with $Y^\ominus$, i.e. the incoming and outgoing electron pairs will be kept as far apart as possible—this 'backside' attack we have seen to be advantageous in the S_N2 pathway (p. 78); (b) elimination is taking place from the lower energy 'staggered' conformation (24a), rather than the higher energy 'eclipsed' conformation (24b, cf. p. 7). Though it should be emphasised that it is the electron distribution in, and the conformation of, the T.S.—rather than the ground state—that really matters. Nevertheless, we might well expect preferred elimination from conformation (24a)—ANTI ('opposite side') elimination—rather than from (24b)—SYN ('same side') elimination.

Where, as with (24) above, both C^β and C^α are chiral, elimination from the two conformations will lead to different products—the *trans*-alkene (25) from (24a) and the *cis*-alkene (26) from (24b). Thus knowing the configuration of the original diastereoisomer (e.g. 24), and establishing the configuration of the geometrical isomeride(s) that is formed, enables us to establish the degree of stereoselectivity of the elimination process. In most simple acyclic cases, ANTI elimination is found to be very much preferred, e.g. in about the simplest system, (26) and (27), that permits of stereochemical distinction:

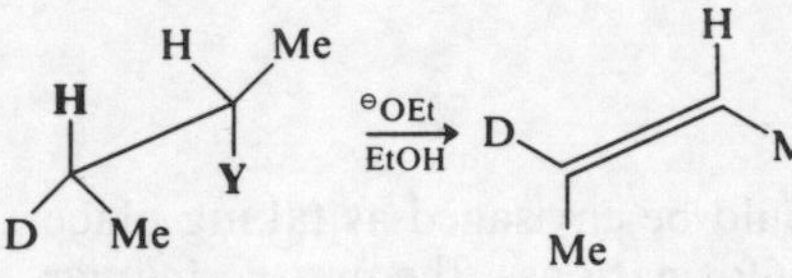

(26)   (28) 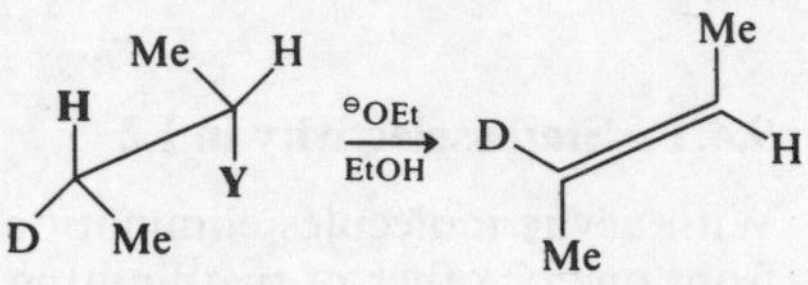   (27)   (29)

For Y = Br, Ts or $\overset{\oplus}{N}M3_3$, elimination was essentially 100% stereoselectively ANTI—only (28) was obtained from (26), and only (29) was obtained from (27). There are, however, numerous exceptions with longer chain $\overset{\oplus}{N}R_3$ compounds, perhaps because of some SYN elimination of the β-H *via* a cyclic transition state involving the ammonium hydroxide ion pair (30):

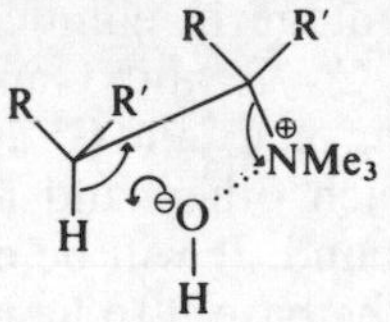

(30)

The degree of stereoselectivity may be influenced to some extent by the polarity and ion-solvating ability of the solvent.

In cyclic compounds the conformation from which elimination can take place may to a considerable extent be enforced by the relative rigidity of the ring structure. Thus for a series of eliminations from different sized rings, the following degrees of stereoselectivity were observed:

Ring size	%SYN elimination
Cyclobutyl	90
Cyclopentyl	46
Cyclohexyl	4
Cycloheptyl	37

The relative lack of stereoselectivity with cyclopentyl compounds is reflected in the behaviour of the *trans*- and *cis*-isomerides, (31) and (32). Each, if it eliminates by E2 at all, will be converted into the same alkene (33)—(31) *via* SYN elimination, and (32) *via* ANTI elimination:

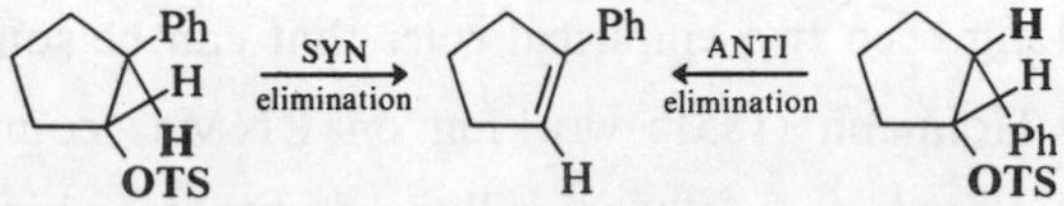

ANTI elimination [(32) → (33)] was found to proceed only 14 times faster than SYN elimination [(31) → (33)] reflecting the fact that the energy needed to distort the ring, so that (32) can assume an approximately anti-periplanar conformation, almost outweighs the normal energetic advantage of the staggered conformation over the, syn-periplanar, eclipsed one, i.e. (31).

The marked ANTI stereoselectivity observed with cyclohexyl systems (see above) reflects the ability to achieve, and the very marked preference to eliminate from, the so-called *trans*-diaxial conformation (34):

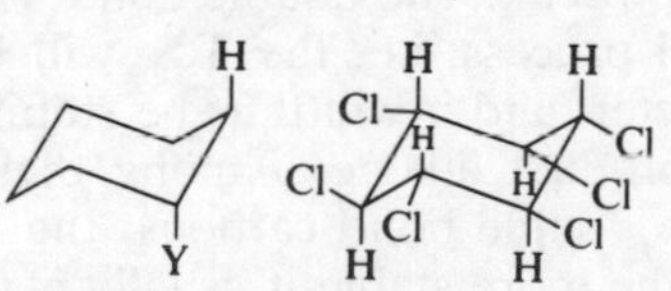

Thus of the geometrical isomers of hexachlorocyclohexane, $C_6H_6Cl_6$, one is found to undergo elimination of HCl at a rate slower, by a factor of $7-24 \times 10^3$, than any of the others: it is found to be the one (35) that cannot assume the above *trans*-diaxial conformation.

9.4.2   Orientation in E2: Saytzev v. Hofmann

In substrates which have alternative β-hydrogen atoms available, it is possible to obtain more than one alkene on elimination, e.g. (36)

where there are two possibilities:

	Y =	Br	$\overset{\oplus}{S}Me_2$	$\overset{\oplus}{N}Me_3$
MeCH$_2$CH=CH$_2$		19%	74%	95%
(37)				

$$\overset{②H}{|} \quad \overset{H①}{|} \quad ①\nearrow$$
$$MeCH-CH-CH_2 \quad ^\ominus OEt$$
$$\underset{Y}{|} \qquad ②\searrow$$

(36)

	Br	SMe$_2$	NMe$_3$
MeCH=CHCH$_3$	81%	26%	5%
(38)			

To help in forecasting which alkene is the more likely to be produced there have long been two empirical rules that can be summarised as follows: (*a*) Hofmann (1851; working on $R\overset{\oplus}{N}Me_3$ compounds, i.e. $Y = \overset{\oplus}{N}Me_3$) stated 'that alkene will predominate which has least alkyl substituents on the double bond carbons', i.e. (37) above; (*b*) Saytzev (1875; working on RBr compounds, i.e. Y = Br) stated 'that alkene will predominate which has most alkyl substituents on the double bond carbons', i.e. (38) above. Both generalisations are valid as the figures quoted above indicate. It is thus clear that the composition of the alkene mixture obtained on elimination is influenced by Y, the nature of the leaving group, and an explanation is required about how this influence is exerted; hardly surprisingly this involves both electronic and steric components.

If the C—Y bond in (36) is relatively readily broken then, as the attacking base begins to break the H—C$^\beta$ bond, the C—Y bond will also be beginning to break. The double bond will thus begin to form early in the overall process, i.e. the T.S. will have a good deal of double bond character, and it will thus be stabilised by any features that stabilise the product alkene. Among these is substitution by alkyl groups on the double bond carbons: the more heavily substituted an alkene is the more stable it usually is (*cf.* p. 26). What this means is that the attacking base will tend to remove preferentially that β-H which will lead to the more stable 'alkene-like' T.S. and then, in turn, to the more stable alkene product, i.e. a preference for ② above leading to (38): Saytzev elimination. This is what happens in (36) when Y is Br. The preference for Saytzev type elimination in the E1 pathway has already been mentioned (p. 243).

On this basis one might well expect Saytzeff elimination (→38) to be the norm. If, however, the C—Y bond in (36) is relatively less easily broken, then removal of a β-H by the attacking base will proceed further before the C—Y bond begins to break: a tendency that may well be promoted by a +ve charge on Y. The T.S. will now be less 'alkene-like' and more 'carbanion-like' in character; it will thus be stabilised by any features that stabilise carbanions. Among these is lack of alkyl substitution at the carbanion carbon atom, i.e. the relative stability of carbanions follows the sequence: primary > secondary > tertiary (*cf.* p. 267). The attacking base will thus tend to remove preferentially that β-H which will lead to the more stable 'carbanion-like' T.S., which will, in turn, result in the formation of the least substituted alkene, i.e. a preference for ① above leading to (37): Hofmann elimination.

Another manifestation of Hofmann elimination is that where β-H removal can take place from alternative alkyl substituents in 'onium' salts, e.g. (39), it is always the least substituted alkene that is eliminated preferentially i.e. (40), not (41):

$$
\underset{(39)}{\overset{\displaystyle ①H}{\underset{\displaystyle ②H}{\overset{|}{\underset{|}{\overset{\displaystyle CH_2-CH_2}{\underset{\displaystyle Me \rightarrow CH-CH_2}{\diagdown\,\diagup}}}}}}\quad\overset{\displaystyle NMe_2}{}\quad\underset{\ominus OEt}{\overset{①\nearrow}{\underset{②\searrow}{}}}\quad
\begin{array}{ll}
CH_2{=}CH_2 + MeCH_2CH_2NMe_2 & \underline{Hofmann} \\
\quad\quad\quad (40) & \\
MeCH{=}CH_2 + CH_3CH_2NMe_2 & \underline{Saytzev} \\
\quad\quad\quad (41) &
\end{array}
$$

This stems from the fact that the incipient carbanion developing by removal of the ② β-H will be destabilised, with respect to the one developing by removal of the ① β-H, by the electron-donating inductive effect of the Me group, i.e. primary carbanion ① > secondary carbanion ②. Although it has already arisen implicitly, it is perhaps as well to state specifically that for any given Y, increasing alkyl substitution at the β-carbon atom, provided one H is left, will discourage Hofmann and promote Saytzev elimination. Introduction of Ph, C=C etc. that can become conjugated with a forming double bond also promote Saytzev. β-Substituents that will help stabilise a developing -ve charge will, however, promote Hofmann elimination.

The effect of Y on the mode of elimination may also involve a steric element. Thus it is found that increase in the size of Y and, more particularly branching in it, leads to an increasing proportion of Hofmann elimination with the same alkyl group, e.g. with (42):

	$Y = Br$	$\overset{\oplus}{S}\diagup^{Me}_{\diagdown Me}$	$\overset{\oplus}{N}\diagup^{Me}_{\diagdown Me}{-Me}$	
	31	87	98	% Hofmann (i.e. ①)

$$
\underset{(42)}{\overset{\displaystyle ①H}{\underset{\displaystyle ②H}{\overset{|}{\underset{|}{\overset{\displaystyle CH_2}{\underset{\displaystyle MeCH_2CH}{\diagdown\,\diagup}}}}}}}\quad CH{-}Y
$$

The proportion of Hofmann elimination is also found to increase with increasing branching in the alkyl group of the substrate (constant

Y and base), and with increasing branching in the base, e.g. with (43), a bromide where preferential Saytzev elimination would normally be expected:

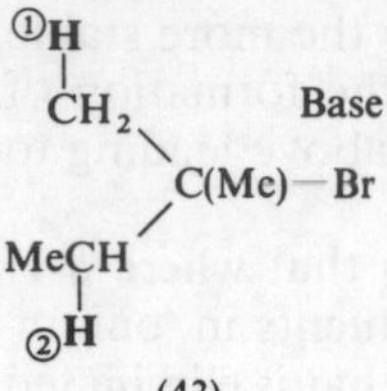

$$
\begin{array}{l}
\overset{①}{\text{H}} \\
| \\
\text{CH}_2 \\
\qquad\qquad\qquad \text{C(Me)—Br} \\
\text{MeCH} \\
| \\
\overset{②}{\text{H}} \\
(43)
\end{array}
$$

Base =	EtO$^\ominus$	Me$_3$CO$^\ominus$	Me$_2$EtCO$^\ominus$	Et$_3$CO$^\ominus$	
	30	72	77	78	% Hofmann (i.e. ①)

It may be mentioned in passing that the volume, and quantitative precision, of data available in this field owes much to the use of gas/liquid chromatography for the rapid, and accurate, quantitative analysis of alkene mixtures.

These several steric effects are explainable on the basis that <u>any</u> crowding, irrespective of its origin, will make the T.S. (44) that involves the removal of proton ② from (46*a*)—Saytzev elimination—relatively more crowded than the T.S. (45) that involves removal of proton ① from (46*b*)—Hofmann elimination. The differential will increase as the crowding increases (in R, Y or B), and Hofmann elimination will thus be progressively favoured over Saytzev:

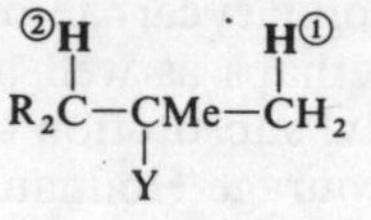

$$
\begin{array}{c}
\overset{②}{\text{H}} \qquad\quad \text{H}\overset{①}{} \\
| \qquad\qquad | \\
\text{R}_2\text{C—CMe—CH}_2 \\
\qquad\quad | \\
\qquad\quad \text{Y} \\
(46)
\end{array}
$$

Saytzev

(46*a*)   →   (44)   →

Hofmann

(46*b*)   →   (45)   →

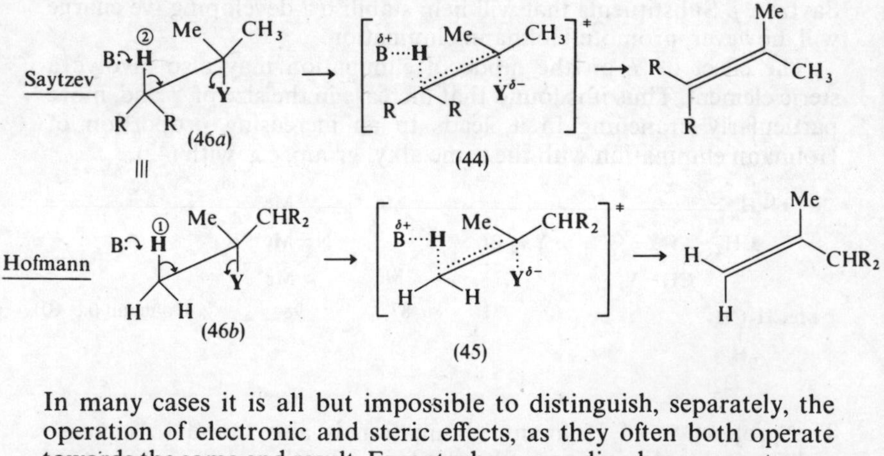

In many cases it is all but impossible to distinguish, separately, the operation of electronic and steric effects, as they often both operate towards the same end result. Except where crowding becomes extreme, however, it seems likely that the electronic effects are commonly in control.

In cyclic systems, the usual simple requirements of Saytzev or Hofmann rules may be overridden by other special requirements of the system, e.g. the preference for elimination from the *trans*-diaxial conformation in cyclohexane derivatives (*cf.* p. 249). Another such limitation is that it is not normally possible to effect an elimination so as to introduce a double bond on a bridgehead carbon atom in a fused ring system (Bredt's rule), e.g. (47) ↛ (48):

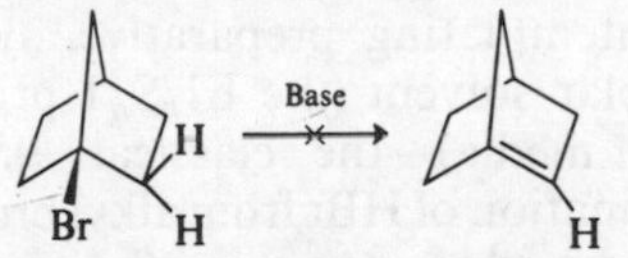

(47)   (48) 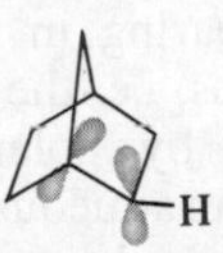   (49)

This is presumably the case because the developing *p* orbitals in an E2 reaction, far from being coplanar (*cf.* p. 247), would be virtually at right angles to each other (49), and so could not overlap significantly to allow development of a double bond. The relatively small ring system is rigid enough to make the distortion required for effective *p* orbital overlap energetically unattainable; there seems no reason why an E1 or E1cB pathway would be any more successful: the bicyclo-heptene (48) has, indeed, never been prepared. With bigger rings, e.g. the bicyclononene (50), or a more flexible system (51), sufficient distortion is now possible to allow the introduction of a double bond by an elimination reaction:

(50) 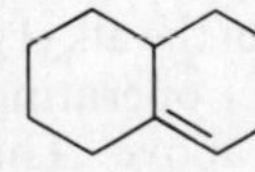   (51)

9.5 ELIMINATION *v.* SUBSTITUTION

E1 elimination reactions are normally accompanied by S_N1 substitution, as both have a common—carbonium ion—intermediate; though this is converted into *either* elimination *or* substitution products *via* different T.S.s in a fast, non rate-limiting step. Similarly, E2 elimination is often accompanied by S_N2 substitution, though in this case the parallel, concerted processes involve entirely separate pathways throughout. Thus considering elimination *v.* substitution there are really three main issues: (*a*) factors influencing $E1/S_N1$ product ratios, (*b*) factors influencing $E2/S_N2$ product ratios, and (*c*) factors influencing change of pathway, i.e. $E1/S_N1 \rightarrow E2/S_N2$ or vice versa.

The last of these, (*c*), may well be the most potent. Thus $E1/S_N1$ solvolysis of Me_3CBr, and of $EtMe_2CBr$, in EtOH (at 25°) was found to yield 19 % and 36 %, respectively, of alkene; while introduction of 2M $EtO^{\ominus}$—which shifts the mechanism in part at least to $E2/S_N2$—resulted in the alkene yields rising to 93 % and 99 %, respectively. It is indeed found generally, for a given substrate, that the $E2/S_N2$ ratio is substantially higher than the $E1/S_N1$ ratio. A point that is worth bearing in mind when contemplating preparative, synthetic operations, i.e. the use of a less polar solvent (the $E1/S_N1$ process is favoured by polar, ion-solvating media)—the classical alcoholic, rather than aqueous, potash for elimination of HBr from alkyl bromides. A shift in mechanism may also be induced by increasing the concentration of the base employed, e.g. $^{\ominus}OH$; hence the classical use of concentrated, rather than dilute, potash for elimination.

In either (*a*) or (*b*), the carbon structure of the substrate is of considerable importance, the proportion of elimination rising on going: primary < secondary < tertiary. In electronic terms this stems from increasing relative stabilisation of the T.S. for elimination as the number of alkyl groups on the carbon atoms of the developing double bond increases (*cf*. p. 250). Thus with $EtO^{\ominus}$ in EtOH on alkyl bromides, we find: primary $\rightarrow$ *ca*. 10 % alkene, secondary $\rightarrow$ *ca*. 60 %, and tertiary $\rightarrow$ > 90 %. This stems not only from an increasing rate of elimination, but also from a decreasing rate of substitution. Similarly, substituents such as C=C and Ar that can stabilise the developing double bond through conjugation (*cf*. p. 247) also strongly favour elimination: under comparable conditions, CH_3CH_2Br yielded ≈ 1 % alkene, while $PhCH_2CH_2Br$ yielded ≈ 99 %.

So far as the structure of the alkyl group of the substrate is concerned, there is also a steric effect operating essentially in the same direction as the electronic effect above. Thus in $E1/S_N1$ the sp^2 hybridised carbon atom in the carbonium ion intermediate (52) remains sp^2 hybridised ($\approx 120°$ bond angles) on elimination (53), but becomes sp^3 hybridised ($\approx 109°$ bond angles) on substitution (54):

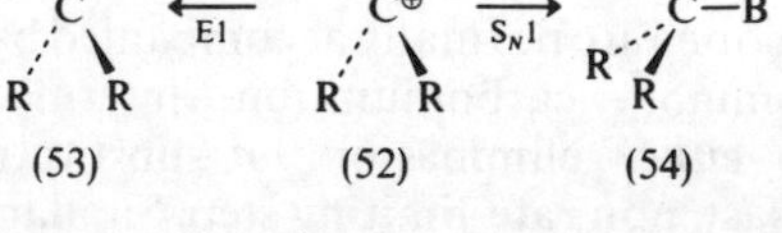

Crowding strain is thus reintroduced in the T.S. for substitution, but much less so, if at all, in the T.S. for elimination, and the differential between them will become greater—increasingly favouring elimination

—as the size and degree of branching in the R groups increases. A related, but slightly different, point is that the peripheral H will be much more accessible than the relatively hindered carbonium ion carbon; we should thus expect the proportion of elimination to rise as the size of the attacking base/nucleophile increases: as is indeed observed, i.e. $Me_3CO^\ominus$ is usually better than $EtO^\ominus$ for carrying out elimination reactions with. This discussion has tended to centre on the $E1/S_N1$ case, but essentially analogous steric effects are involved in the differential stabilisation of the T.S. for E2 with respect to the T.S. for S_N2.

The $E1/S_N1$ ratio is, of course, substantially independent of the leaving group Y, but this is not the case with $E2/S_N2$, where breaking of the C—Y bond is involved in each alternative T.S. The following rough sequence, in order of increasing promotion of elimination, is observed:

$$\text{Tosylate} < Br < \overset{\oplus}{S}Me_2 < \overset{\oplus}{N}Me_3$$

The attacking base/nucleophile is obviously of importance also; we require a species of high basicity and low nucleophicity, a 'hard' base (*cf.* p. 95). Thus, preparatively, tertiary amines, e.g. Et_3N, pyridine etc, are often used instead of $^\ominus OH$, $^\ominus OEt$ etc; the use of a base with a relatively high b.p. is also advantageous (see below).

Finally, elimination—whether E1 or E2—is favoured with respect to substitution by rise in temperature. This is probably due to elimination leading to an increase in the number of particles, whereas substitution does not. Elimination thus has the more favourable entropy term, and because this (ΔS^+) is multiplied by T in the relation for the free energy of activation, ΔG^+ ($\Delta G^+ = \Delta H^+ - T\Delta S^+$, *cf.* p. 38), it will increasingly outweigh a less favourable ΔH^+ term as the temperature rises.

9.6 EFFECT OF ACTIVATING GROUPS

We have to date considered the effect of alkyl substituents in promoting elimination reactions in suitable substrates, and also, in passing, that of Ar and C=C. Elimination is, in general, promoted by most electron-withdrawing substituents, e.g. CF_3, NO_2, $ArSO_2$, CN, C=O, CO_2Et etc. Their effect can be exerted: (*a*) through making the β-H atoms more acidic (55), and hence more easily removable by a base, (*b*) through stabilisation of a developing carbanion by electron-withdrawal (56), or in some cases, (*c*) through stabilisation of the developing double bond by conjugation with it (57):

$$\underset{(55)}{F_3C \overset{B:\curvearrowright H}{\overset{|}{\underset{|}{\overset{\downarrow}{CH}}}} - \overset{\delta+}{CH_2}}$$
$$\underset{Y}{}$$

$$\underset{(56)}{\overset{\ominus O}{\underset{\delta^- O}{}} \overset{\oplus}{N} \overset{B\cdots H}{\overset{|}{\underset{|}{CH}}} - CH_2}$$
$$\underset{Y}{}$$

$$\underset{(57)}{O \overset{B\cdots H}{\overset{|}{\underset{|}{\cdots C \cdots CH \cdots CH_2}}}}$$
$$\underset{OEt}{} \quad \overset{\cdot\cdot}{Y}^{\delta-}$$

The more powerfully electron-withdrawing the substituent the greater the chance that the T.S. in an E2 elimination will be 'carbanion-like' (*cf.* p. 251), or even that the reaction pathway may be shifted to the E1cB mode (*cf.* p. 244), e.g. possibly with NO_2 or $ArSO_2$, especially if the leaving group, Y, is a poor one.

A good example of elimination promotion is by the CHO group in aldol (58) making possible a base-catalysed dehydration to an $\alpha\beta$-unsaturated aldehyde (59, *cf.* p. 221):

$$\underset{(58)}{\overset{B:\curvearrowright H}{\underset{\overset{|}{H}}{O = C} - \overset{|}{\underset{OH}{CH}} - CHMe}} \rightleftarrows \left[\underset{(60)}{\overset{\overset{\delta+}{B\cdots H}}{\underset{\overset{|}{H}}{O \cdots C \cdots \overset{|}{\underset{\delta^- OH}{CH}} \cdots CHMe}}} \right]^{\ddagger} \rightleftarrows \underset{(59)}{\overset{BH^{\oplus}}{\underset{\overset{|}{H}}{O = C - CH = CHMe}}}$$
$$\qquad\qquad\qquad\qquad\qquad\qquad\qquad\qquad\qquad\qquad \ominus OH$$

Dehydrations are normally acid-catalysed (protonation of $\overset{\cdot\cdot}{O}H$ turning it into $^{\oplus}OH_2$, H_2O being a better leaving group than $^{\ominus}OH$), and a base-catalysed elimination is here made possible by the CHO group making the β-H atoms more acidic, and stabilising the resultant carbanion, i.e. (*a*) and (*b*) above. Stabilisation, by conjugation, of the developing double bond [(*c*) above] has been included in the T.S. (60) above, but how large a part this plays is not wholly clear. It is, however, significant that electron-withdrawing substituents are usually very much more effective in promoting elimination when they are on the β-, rather than the α-, carbon atom: they could conjugate with a developing double bond equally well from either position, but can only increase acidity of β-H, and stabilise a carbanion from the β-position. This is clearly seen in base-induced elimination of HBr from 1- and 2-bromoketones, (61) and (62), respectively,

$$\underset{(61)}{\overset{Me \quad H \curvearrowleft B}{\underset{\overset{|}{Br}}{O = C - CH - CH_2}}} \underset{slow}{\rightarrow} \left[\underset{(61)}{\overset{Me \quad H \cdots B^{\delta+}}{\underset{\overset{|}{Br^{\delta-}}}{O = C - CH \cdots CH_2}}} \right]^{\ddagger} \searrow$$
$$\underset{Me}{O = C - CH = CH_2}$$
$$(63)$$

$$\underset{(62)}{\overset{B:\curvearrowright H}{\underset{\overset{|}{\underset{Br}{Me}}}{O = C \overset{\downarrow}{\ll} CH - CH_2}}} \underset{fast}{\rightarrow} \left[\underset{(62)}{\overset{B^{\delta+} \cdots H}{\underset{\overset{|}{\underset{Br^{\delta-}}{Me}}}{O \cdots C \cdots CH \cdots CH_2}}} \right]^{\ddagger} \nearrow$$

where both give the same $\alpha\beta$-unsaturated (i.e. conjugated) ketone (63), but (62) is found to eliminate HBr very much faster than (61), under analogous conditions. Such β-substituents are often effective enough to promote loss of more unusual—and poor—leaving groups such as OR, NH_2 etc. (OH above).

9.7   OTHER 1,2-ELIMINATIONS

Attention has to-date been devoted almost entirely to eliminations in which it has been H that has been lost, as a proton, from the β-carbon atom. These are certainly the most important eliminations, but examples are known that involve the departure of an atom or group other than H from C^β, the commonest probably being 1,2-dehalogenations and, in particular, 1,2-debromination. This can be induced by a number of different species including iodide ion, $I^\ominus$, metals such as zinc, and some metal ions. e.g. $Fe^{2\oplus}$. The reaction with $I^\ominus$ in acetone is found to follow the rate law (after allowance has been made for the $I^\ominus$ complexed by the I_2 produced in the reaction),

$$\text{Rate} = k[\text{1,2-dibromide}][I^\ominus]$$

which would be compatible with a simple E2 pathway.

This is borne out by the high degree of ANTI stereoselectivity that is observed in acyclic examples (*cf.* p. 248), when either or both the bromine atoms are attached to secondary or tertiary carbon atoms, e.g. (64):

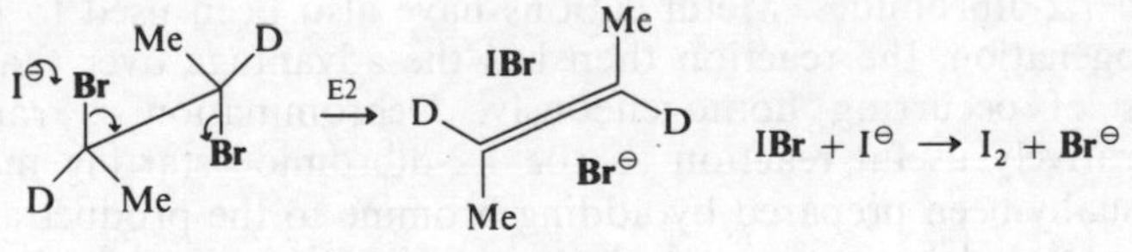

only the *trans*-alkene (65) is obtained. When either or both the bromine atoms are attached to primary carbon atoms, e.g. (66), however, the overall reaction is found to proceed stereoselectively SYN, i.e. the *cis*-alkene (67) is the only product. This somewhat surprising result is believed not to represent a stereochemical change in the elimination itself, but to result from a composite S_N2/E2 mechanism; in which S_N2 displacement of Br by $I^\ominus$, with inversion of configuration (68), is followed by a stereoselective ANTI elimination on the 1-iodo-2-bromide (68) to yield (67)—the overall reaction being an apparent

SYN elimination [(66) → (67)]:

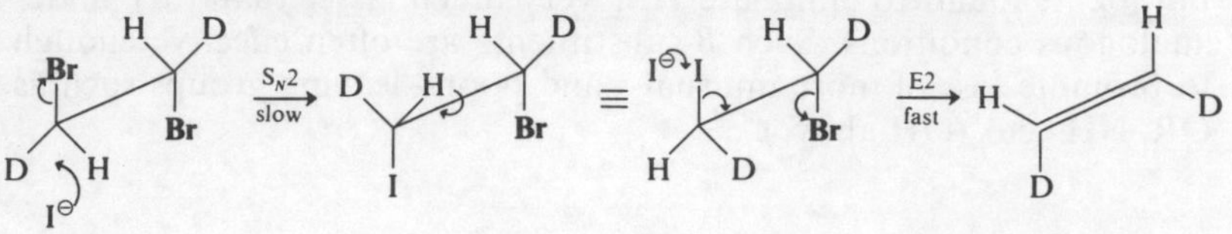

(66)   (68)   (68)   (67)

Support for the actual elimination step, in each case, being E2 is provided by the fact that changing the alkyl substituents on C^α and C^β results in reaction rates that, in general, increase with the relative thermodynamic stability of the product alkene.

$Br^\ominus$ and $Cl^\ominus$ are much less effective at inducing 1,2-dehalogenation than $I^\ominus$, but metals—particularly Zn—have long been used. Reaction takes place heterogeneously at the surface of the metal, the solvent renewing the active surface by removing the metal halide that is formed there. With simple examples, like those above, e.g. (69), there is a high degree of ANTI stereoselectivity, and the reaction pathway is probably simple E2:

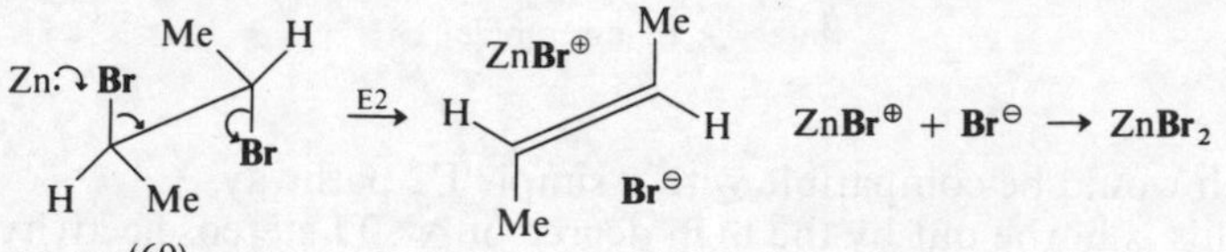

(69)

Strict ANTI stereoselectivity is, however, departed from with longer chain 1,2-dibromides, i.e. above C_4. The reaction may also be induced by Mg, hence the impossibility of making Grignard reagents from simple 1,2-dibromides. Metal cations have also been used to induce dehalogenation, the reaction then has the advantage over that with metals of occurring homogeneously. Debromination is rarely a preparatively useful reaction as the 1,2-dibromide starting material has usually been prepared by adding bromine to the product alkene! Bromination/debromination is, however, sometimes used for 'protecting' double bonds, e.g. in the oxidation of (70) → (71), which could not be carried out directly because the double bond would be attacked preferentially:

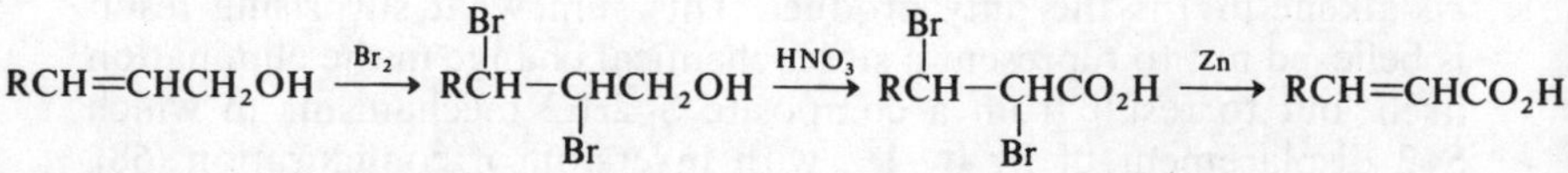

(70)   (71)

Eliminations have also been carried out on a number of compounds of the form $HalCH_2CH_2Y$, where Y = OH, OR, OCOR, NH_2 etc; these eliminations normally require conditions more drastic than for 1,2-dihalides, and metals or metal cations are found to be more effective than $I^{\ominus}$. The reactions tend to be somewhat indiscriminate in their stereochemistry. An interesting example is the elimination of $CO_2/Br^{\ominus}$ from the anion of 1,2-dibromo-3-phenylpropanoic acid (cinnamic acid dibromide, 72); this requires no external reagent to initiate elimination, and proceeds under extremely mild conditions:

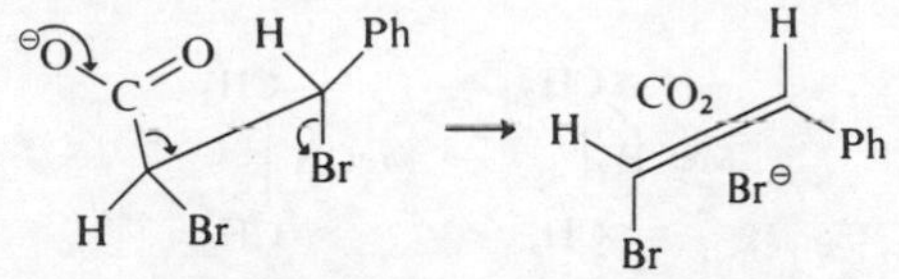

(72)

9.8 1,1-(α-)ELIMINATION

A relatively small number of examples are known of 1,1-eliminations in which both H and the leaving group, Y, are lost from the same (α-) carbon atom, e.g. (73) ⟶ (74). They tend to be favoured: (*a*) by powerfully electron-withdrawing Y groups—these increase the acidity of the α-H atoms, and stabilise a developing -ve charge on the α-carbon atom, (*b*) by using very strong bases, B, and (*c*) by the absence of β-H atoms—though this is not a requirement (*cf.* 73):

$$\underset{(73)}{MeCH_2CH_2CH(H)\text{—}Cl} \xrightarrow{B:} \underset{(75)}{MeCH_2CH_2\text{—}\ddot{C}H} \rightarrow \underset{(74)}{MeCH_2CH{=}CH_2}$$

In some, though not necessarily all, cases loss of $H^{\oplus}$ and $Cl^{\ominus}$ is thought to be concerted, leading directly to the *carbene* (*cf.* p. 49) intermediate (75); formation of the product alkene from (75) then requires migration of H, with its electron pair, from the β-carbon atom. A 1,1-elimination (Eα) will be indistinguishable kinetically from 1,2-(E2), and evidence for its occurrence rests on isotopic labelling, and on inferential evidence for the formation of carbenes, e.g. (75).

Thus introduction of 2D atoms at the α-position in (73) is found to result in one of them being lost in going to (74)—both would be retained in E2; while introduction of 2D at the β-position in (73) results in both being still present in (74), though one is now on the terminal (α- in 73) carbon atom—one would have been lost in E2. From such isotopic labelling data it is possible to determine how much of a given elimination proceeds by the 1,1-, and how much by the 1,2-pathway. Use of $C_6H_5^{\ominus}Na^{\oplus}$—an enormously strong base—in decane solution is found to result in 94 % 1,1-elimination from (73), while $Na^{\oplus}NH_2^{\ominus}$ caused much less, and $Na^{\oplus}OMe^{\ominus}$ hardly any at all, i.e. the operation of factor (*b*) above. It was also found that, for a given base, alkyl bromides and iodides underwent much less 1,1-elimination than the corresponding chlorides, i.e. operation of factor (*a*), above. Inferential evidence for the formation of the carbene intermediate (75) is provided by the isolation from the reaction mixture of the cyclo-propane (76),

$$\underset{(75)}{MeC\overset{CH_2}{\underset{\underset{\ddot{C}H}{|}}{H_2}}} \rightarrow \underset{(76)}{MeC\overset{CH_2}{\underset{\underset{CH_2}{|}}{H}}}$$

such intramolecular 'insertions' to form cyclopropanes being a common reaction of suitable carbenes; it is an example of 'internal trapping' (*cf.* p. 49). Only 4 % of (76) was isolated from the reaction of (73), but no less than 32 % of (76) was isolated from the 1,1-elimination of the isomeric chloride, $MeCH(Cl)CH_2CH_3$.

The most familiar, and most studied, example of 1,1-elimination occurs where no β-H atoms are available—the operation of factor (*c*) above—in the hydrolysis of haloforms, e.g. $CHCl_3$ (77), with strong bases. This involves an initial 1,1-elimination, probably *via* a two-step, i.e. 1,1-E1cB, pathway, to yield a dichlorocarbene intermediate (78);

$$\underset{(77)}{\overset{HO^{\curvearrowleft}}{\underset{CCl_3}{\overset{H}{|}}}} \underset{fast}{\rightleftarrows} \underset{(78)}{\overset{H_2O}{\underset{\overset{\ominus}{CCl_2}}{\underset{\overset{|}{Cl}}{}}}} \underset{slow}{\rightarrow} \ddot{C}Cl_2 \xrightarrow[fast]{\ominus OH/H_2O} CO + HCO_2^{\ominus}$$

The hydrolysis, as expected, follows the rate-law,

$$\text{Rate} = k[CHCl_3][^{\ominus}OH]$$

and the fast, reversible first step is supported by the fact that deuterated chloroform, $CDCl_3$, is found to undergo base-catalysed exchange with H_2O (loss of D) much faster than it undergoes hydrolysis. Support for the above mechanism is provided by the addition of species more nucleophilic than $^{\ominus}OH$, e.g. $PhS^{\ominus}$ which results in the formation of $HC(SPh)_3$, though $CHCl_3$ itself is inert to $PhS^{\ominus}$ in the absence of $^{\ominus}OH$. If generated in a non-protic solvent, the highly electron-deficient CCl_2 will add to electron-rich alkenes, e.g. *cis*-2-butene (79), to form cyclopropanes, e.g. (80),—a 'trapping' reaction (*cf*. p. 49):

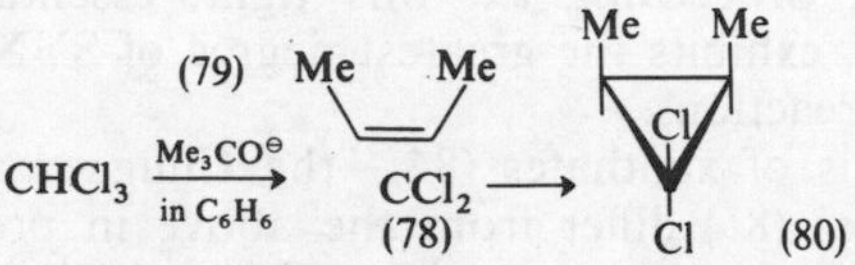

$$CHCl_3 \xrightarrow[\text{in } C_6H_6]{Me_3CO^{\ominus}} \underset{(78)}{CCl_2} \longrightarrow$$

(79)

(80)

Under suitable conditions, this can be a useful preparative method for cyclopropanes; another preparative 'trapping' reaction of CCl_2 is its electrophilic attack on phenols in the Reimer–Tiemann reaction (p. 284).

It should however, be emphasised that in protic solvents, with the common bases, and with substrates containing β-H atoms 1,1-elimination occurs to only a small extent if at all.

9.9 PYROLYTIC SYN ELIMINATION

There are a number of organic compounds including esters—especially acetates, xanthates (see below)—amine oxides, and halides that undergo pyrolytic elimination of HY, in the absence of added reagents, either in inert solvents or in the absence of solvent—in some cases in the gas phase. In general these eliminations follow the rate law,

$$\text{Rate} = k[\text{substrate}]$$

but are usually distinguishable from E1 eliminations (that follow the same rate law) by the degree of SYN stereoselectivity that they exhibit. They are sometimes referred to as E*i* eliminations (*elimination, intramolecular*), and the degree of SYN stereoselectivity reflects the extent to which they proceed *via* cyclic transition states, e.g. (81) below, that would dictate a SYN pathway.

The reaction that is perhaps of the greatest synthetic utility—because it proceeds at relatively low temperatures—is the Cope reaction of tertiary amine oxides, e.g. (82):

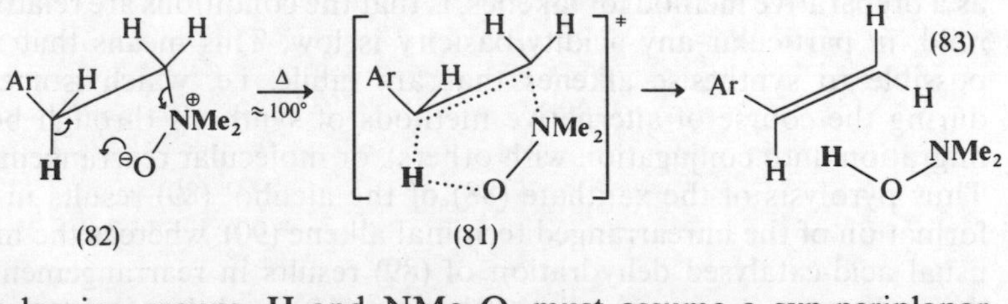

$$\xrightarrow[\approx 100°]{\Delta}$$

(82)   (81)   (83)

The leaving groups, H and NMe_2O, must assume a syn-periplanar conformation, with respect to each other, to be close enough together to permit the development of the O··H bond in the T.S. (81); the products are the alkene (83) and N,N-dimethylhydroxylamine. The Cope reaction, proceeding *via* this tight, essentially planar five-membered T.S., exhibits the greatest degree of SYN stereoselectivity of any of these reactions.

The pyrolysis of xanthates (84)—the Chugaev reaction—and of carboxylic esters (85) differ from the above in proceeding *via* six-membered, cyclic transition states, e.g. (86) and (87), respectively:

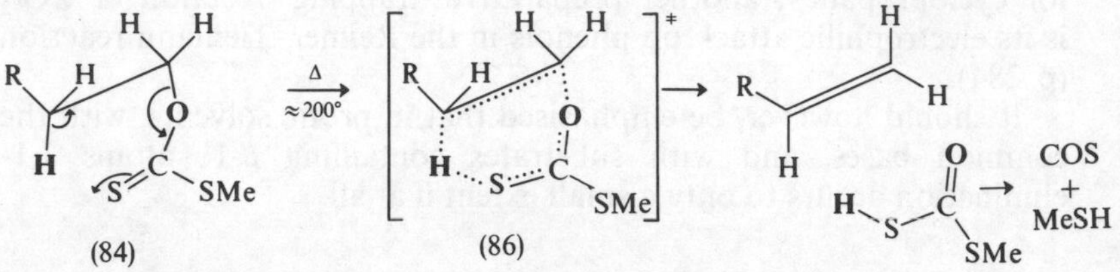

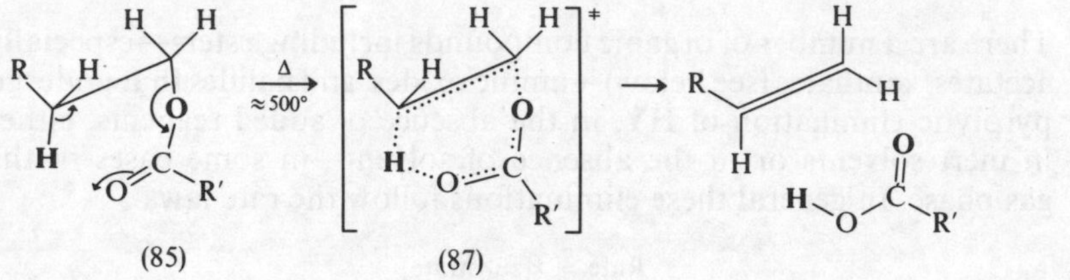

The six-membered rings in these T.S.s are more flexible than the five-membered T.S.—(81) above—and need not be planar (*cf.* cyclo-hexanes *v.* cyclopentanes). Elimination may thus proceed, in part at least, from conformations other than the syn-periplanar, with the result that the degree of SYN stereoselectivity in these eliminations may sometimes be lower than that observed in the Cope reaction. Both reactions require higher temperatures than for the Cope reaction, carboxylic esters particularly so.

One of the major advantages of this group of elimination reactions, as a preparative method for alkenes, is that the conditions are relatively mild, in particular any acidity/basicity is low. This means that it is possible to synthesise alkenes that are labile, i.e. which isomerise during the course of alternative methods of synthesis through bond migration (into conjugation with others), or molecular rearrangement. Thus pyrolysis of the xanthate (88) of the alcohol (89) results in the formation of the unrearranged terminal alkene (90), whereas the more usual acid-catalysed dehydration of (89) results in rearrangement in the carbonium ion intermediate (91, *cf.* p. 110), and thus in formation of the thermodynamically more stable, rearranged alkene (92):

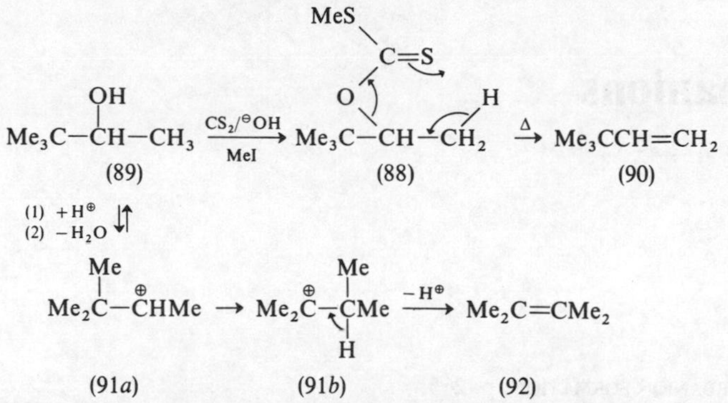

$$
\begin{array}{ccc}
& \overset{\displaystyle MeS}{\underset{\diagdown}{}} \\
& C{=}S \\
\underset{(89)}{\underset{|}{Me_3C{-}\overset{OH}{\overset{|}{CH}}{-}CH_3}} \xrightarrow[\text{MeI}]{CS_2/^{\ominus}OH} & \underset{(88)}{Me_3C{-}\overset{O}{\overset{|}{CH}}{-}\overset{H}{CH_2}} \xrightarrow{\Delta} & \underset{(90)}{Me_3CCH{=}CH_2}
\end{array}
$$

(1) $+H^{\oplus}$
(2) $-H_2O$

$$
\underset{(91a)}{Me_2\overset{Me}{\overset{|}{C}}{-}\overset{\oplus}{C}HMe} \rightarrow \underset{(91b)}{Me_2\overset{\oplus}{C}{-}\overset{Me}{\underset{\underset{H}{|}}{C}Me}} \xrightarrow{-H^{\oplus}} \underset{(92)}{Me_2C{=}CMe_2}
$$

Pyrolysis of alkyl chlorides and bromides (alkyl fluorides are too stable; alkyl iodides lead to some alkane, as well as alkene, through reduction by the eliminated HI) also results in the formation of alkenes, but temperatures up to 600° are required, and the elimination is seldom of preparative use; paradoxically it is the type that has received the most detailed study in this group. A cyclic T.S. would here involve a highly strained four-membered ring, and halide eliminations seem to be less concerted than those considered to-date. They probably involve a good deal of C—Hal bond-breaking in the T.S., normally heterolytically so that carbonium ion character is developed at the leaving group carbon atom. Not surprisingly halide eliminations are found to exhibit less SYN stereoselectivity than the others. Further mention will be made of Ei concerted eliminations, and of other reactions involving cyclic T.S.s, subsequently (p. 329).

10

Carbanions

In theory any organic compound such as (1) that contains a C—H bond, i.e. nearly all of them, can function as an acid in the classical sense by donating a proton to a suitable base, the resultant conjugate acid (2) being a *carbanion* (*cf.* p. 21):

$$R_3C\text{—}H + B: \rightleftarrows R_3C^{\ominus} + BH^{\oplus}$$
$$\quad (1) \qquad\qquad (2)$$

In considering relative acidity, it is normally only the thermodynamics of the situation that are of interest, in that the pK_a value for the acid (*cf.* p. 53) can be derived from the above equation; the kinetics of the situation is normally of little or no significance, as proton transfer from atoms such as O, N, etc., is extremely rapid in solution. With *carbon acids* such as (1), however, the *rate* at which proton is transferred to the base may well be sufficiently slow as to constitute the limiting factor: the acidity of (1) is then controlled kinetically rather than thermodynamically (*cf.* p. 274).

There are, however, other methods of generating carbanions than by proton removal as we shall see below. Carbanion formation is important—apart from the inherent interest of the species—because of their participation in a wide variety of reactions of synthetic utility: many of them of especial value in that they result in the formation of carbon–carbon bonds (*cf.* p. 217).

10.1 CARBANION FORMATION

The most general method of forming carbanions is by removal of an atom or group X from carbon, X leaving its bonding electron pair behind:

$$R_3C—X + Y \rightleftarrows R_3C^\ominus + XY^\oplus$$

By far the most common leaving group is X = H where, as above, it is a proton that is removed, (1) → (2), though other leaving groups are also known, e.g. Cl in triphenylmethyl chloride (3):

$$Ph_3C—Cl + 2Na \rightleftarrows Ph_3C^\ominus Na^\oplus + Na^\oplus Cl^\ominus$$
$$(3) \qquad\qquad\qquad (4)$$

Hardly surprisingly the tendency of alkanes to lose proton and form carbanions is not marked, as they possess no structural features that either promote acidity in their H atoms, or are calculated to stabilise the carbanion with respect to the undissociated alkane (*cf.* carboxylic acids, p. 54). Thus CH_4 has been estimated to have a pK_a value of ≈ 43, compared with 4·76 for $MeCO_2H$. The usual methods for determining pK_a do not, of course, work so far down the acidity scale as this, and these estimates have been obtained from measurements on the equilibrium:

$$RM + R'I \rightleftarrows RI + R'M$$

The assumption is made that the stronger an acid, RH, is the greater will be the proportion of it in the form RM (e.g. M = Li) rather than as RI. Determination of the equilibrium constant K allows a measure of the relative acidity of RH and R'H, and by suitable choice of pairs it is possible to ascend the pK_a scale until direct comparison can be made with an RH compound whose pK_a has been determined by other means.

Thus $Ph_3C—H$ (5) is found to have a pK_a value of 33, i.e. it is a very much stronger acid than CH_4, and the carbanion (4) may be obtained from it, preparatively, by the action of sodamide, i.e. $^\ominus NH_2$, in liquid ammonia:

$$Ph_3C—H + Na^\oplus NH_2^\ominus \underset{NH_3}{\overset{liq}{\rightleftarrows}} Ph_3C^\ominus Na^\oplus + NH_3$$
$$(5) \qquad\qquad\qquad\qquad (4)$$

Ph$_3$C$^\ominus$ may also be obtained, as we saw above, by the action of sodium on Ph$_3$C—Cl (3) in an inert solvent; the resulting solution of sodium triphenylmethyl is used as a very strong organic base (*cf.* p. 226) because of the proton-appropriating ability of the carbanion (4). Alkenes are slightly stronger acids than the alkanes—CH$_2$=CH$_2$ has a pK_a value of 37—but the alkynes are very much more strongly acidic, and HC≡CH itself has a pK_a value of 25. The carbanion HC≡C$^\ominus$ (or of course RC≡C$^\ominus$) may be generated from the hydrocarbon with $^\ominus$NH$_2$ in liquid ammonia: these acetylenic anions are of some synthetic importance (*cf.* p. 219).

Hardly surprisingly, the introduction of electron-withdrawing substituents also increases the acidity of hydrogen atoms on carbon. Thus we have already seen the formation of a somewhat unstable carbanion, $^\ominus$CCl$_3$, in the action of strong bases on chloroform (*cf.* p. 260), and the pK_a values of HCF$_3$ and HC(CF$_3$)$_3$ are found to be ≈ 28 and 11, respectively. The effect with substituents that can delocalise a -ve charge, as well as having an electron-withdrawing inductive effect, are even more marked; thus the pK_a values of CH$_3$CN, CH$_3$COCH$_3$ and CH$_3$NO$_2$ are found to be 25, 20 and 10·2, respectively. With CH$_3$NO$_2$, the corresponding carbanion, $^\ominus$CH$_2$NO$_2$, may be obtained by the action of $^\ominus$OEt in EtOH, or even of $^\ominus$OH in H$_2$O (*cf.* p. 223); but small concentrations of carbanion must be developed in aqueous solution even from the less acidic carbonyl compounds to enable the aldol reaction (*cf.* p. 220) to take place.

A table of some pK_a values for carbon acids is appended for convenience, before going on to discuss the factors that can contribute to the relative stabilisation of carbanions:

	pK_a		pK_a
CH$_4$	43	CH$_2$(CO$_2$Et)$_2$	13·3
CH$_2$=CH$_2$	37	CH$_2$(CN)$_2$	12
C$_6$H$_6$	37	HC(CF$_3$)$_3$	11
PhCH$_3$	37	MeCOCH$_2$CO$_2$Et	10·7
Ph$_3$CH	33	CH$_3$NO$_2$	10·2
CF$_3$H	28	(MeCO)$_2$CH$_2$	8·8
HC≡CH	25	(MeCO)$_3$CH	6
CH$_3$CN	25	CH$_2$(NO$_2$)$_2$	4
CH$_3$COCH$_3$	20	CH(NO$_2$)$_3$	0
C$_6$H$_5$COCH$_3$	19	CH(CN)$_3$	0

10.2 CARBANION STABILISATION

There are a number of structural features in R—H that promote the removal of H by bases through making it more acidic, and also features that serve to stabilise the resultant carbanion, R$^\ominus$; in some cases both effects are promoted by the same feature. The main features that serve to stabilise carbanions are (*cf.* factors that serve to stabilise carbonium ions, p.103): (*a*) increase in *s* character at the carbanion carbon, (*b*) electron-withdrawing inductive effects, (*c*) conjugation of the carbanion lone pair with a multiple bond, and (*d*) through aromatisation.

The operation of (*a*) is seen in the increasing acidity of the hydrogen atoms in the sequence: CH_3—CH_3 < CH_2=CH_2 < $HC\equiv CH$; the increase in acidity being particularly marked (see table above) on going from alkene to alkyne. This reflects the increasing *s* character of the hybrid orbital involved in the σ bond to H, i.e. $sp^3 < sp^2 < sp^1$. The *s* orbitals are closer to the nucleus than the corresponding *p* orbitals, and they are at a lower energy level; this difference is carried through into the hybrid orbitals resulting from their deployment. The electron pair in an sp^1 orbital is thus held closer to, and more tightly by, the carbon atom than an electron pair in an sp^2 or sp^3 orbital (effectively, the apparent electronegativity of the carbon atom increases). This serves not only to make the H atom more easily lost without its electron pair, i.e. more acidic, but also to stabilise the resultant carbanion.

The operation of (*b*) is seen in HCF_3 ($pk_a = 28$) and $HC(CF_3)_3$ ($pk_a = 11$), where the change from CH_4 ($pk_a \approx 43$) is brought about by the powerful electron-withdrawing inductive effect of the fluorine atoms making the H atom more acidic, and also stabilising the resultant carbanions, $^\ominus CF_3$ and $^\ominus C(CF_3)_3$ by electron-withdrawal. The effect is naturally more marked in $HC(CF_3)_3$ where nine F atoms are involved —compared with only three in HCF_3—despite the fact that they are not now operating directly on the carbanion carbon atom. We have already referred to the formation of $^\ominus CCl_3$ from $HCCl_3$ (p. 260), where a similar electron-withdrawing inductive effect must operate. This is likely to be less effective with Cl than with the more electro-negative F, but the deficiency may be overcome to some extent by the delocalisation of the carbanion electron pair into the vacant *d* orbitals of the second row element chlorine—this is, of course, not possible with the first row element fluorine.

The destabilising influence of the electron-donating inductive effect of alkyl groups is seen in the observed carbanion stability sequence:

$$CH_3^\ominus > RCH_2^\ominus > R_2CH^\ominus > R_3C^\ominus$$

Hardly surprisingly, it is the exact reverse of the stability sequence for carbonium ions (p. 83).

The operation of (*c*) is by far the most common stabilising feature, e.g. with CN (6), C=O (7), NO_2 (8), CO_2Et (9), etc.:

$$\underset{(6)}{\overset{\displaystyle B\!:\!\curvearrowright\overset{\textstyle H}{\underset{\displaystyle |}{}}}{CH_2\!\!\underset{\displaystyle }{\overset{\delta+}{\twoheadrightarrow}}\overset{\delta-}{C}\!\!\equiv\!\!N}} \quad \rightleftarrows \quad \underset{(10a)}{[^\ominus CH_2\!-\!C\!\equiv\!N} \quad \leftrightarrow \quad \underset{(10b)}{CH_2\!=\!C\!=\!N^\ominus]} + BH^\oplus \qquad pK_a = 25$$

$$\underset{(7)}{\overset{\text{B:}\curvearrowright\text{H}\quad\text{Me}}{\underset{|}{CH_2}\overset{|}{\twoheadrightarrow}\overset{|}{\underset{|}{C}}=O}} \;\rightleftharpoons\; \left[\underset{(11a)}{\overset{\text{Me}}{^\ominus CH_2-\overset{|}{C}=O}} \;\leftrightarrow\; \underset{(11b)}{\overset{\text{Me}}{CH_2=\overset{|}{C}-O^\ominus}}\right] + BH^\oplus \qquad pK_a = 20$$

$$\underset{(8)}{\overset{\text{B:}\curvearrowright\text{H}}{\underset{\underset{O_\ominus}{|}}{CH_2}\overset{\oplus}{\underset{|}{\twoheadrightarrow}\overset{\oplus}{N}}=O}} \;\rightleftharpoons\; \left[\underset{(12a)}{\overset{}{^\ominus CH_2-\underset{\underset{O_\ominus}{|}}{\overset{\oplus}{N}}=O}} \;\leftrightarrow\; \underset{(12b)}{\overset{}{CH_2=\underset{\underset{O_\ominus}{|}}{\overset{\oplus}{N}}-O^\ominus}}\right] + BH^\oplus \qquad pK_a = 10\cdot2$$

There is in each case an electron-withdrawing inductive effect increasing the acidity of the H atoms on the incipient carbanion atom, but the stabilisation of the resultant carbanion by delocalisation is likely to be of considerably greater significance. Overall, NO_2 is much the most powerful as might have been expected. The marked effect of introducing more than one such group on to a carbon atom may be seen from the table of pK_a values above (p. 266); thus $CH(CN)_3$ and $CH(NO_2)_3$ are as strong acids in water as HCl, HNO_3 etc. The question does arise however, about whether (10ab), (11ab) and (12ab) ought to be described as carbanions: O and N are more electronegative than C and (10b), (11b) and (12b) are likely to contribute markedly more to the hydrid anion structure than (10a), (11a) and (12a), respectively.

The carboxylate group, e.g. CO_2Et (9), is less effective in carbanion stabilisation than the $C=O$ group in simple aldehydes and ketones, as may be seen from the sequence of pK_a values: $CH_2(CO_2Et)_2$, 13·3; $MeCOCH_2CO_2Et$, 10·7; and $CH_2(COMe)_2$, 8·8. This is due to the electron-donating conjugative ability of the lone pair of electrons on the oxygen atom of the OEt group:

$$\underset{(9)}{\overset{\text{B:}\curvearrowright\text{H}}{\underset{\underset{\ddot{:}OEt}{|}}{CH}\overset{\delta+\;\delta-}{\underset{|}{\leftarrow}C}=O}} \;\rightleftharpoons\; \left[\underset{(13a)}{\overset{}{^\ominus CH_2-\underset{\underset{\ddot{:}OEt}{|}}{C}=O}} \;\leftrightarrow\; \underset{(13b)}{\overset{}{CH_2=\underset{\underset{\ddot{:}OEt}{|}}{C}-O^\ominus}}\right] + BH^\oplus \qquad pK_a = 24$$

With second row elements, as we saw above, any inductive effect they exert may be complemented by delocalisation, through use of their empty d orbitals to accommodate the carbanion carbon atom's lone pair of electrons; this happens with S in an $ArSO_2$ substituent and also with P in, for example $R_3P^\oplus$.

The operation of (d) is seen in cyclopentadiene (14) which is found to have a pK_a value of 16 compared with $\approx$ 37 for a simple alkene. This is due to the resultant carbanion, the cyclopentadienyl anion (15), being a 6p electron delocalised system, i.e. a $4n + 2$ Hückel system where $n = 1$ (*cf.* p. 17). The 6 electrons can be accommodated in three stabilised π molecular orbitals, like benzene, and the anion thus shows quasi-aromatic stabilisation; it is stabilised by *aromatisation*:

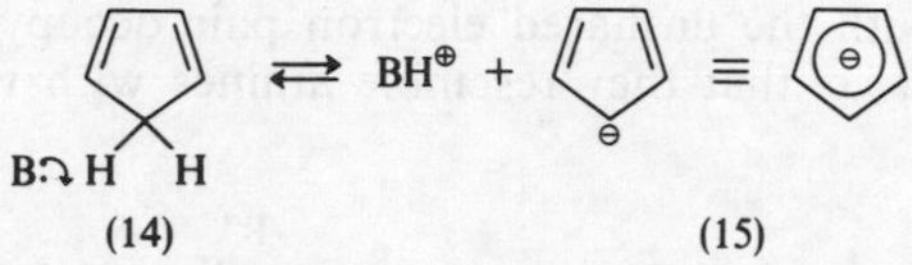

$$\text{(14)} \qquad \rightleftharpoons BH^\oplus + \text{(15)} \equiv$$

(14)   (15)

Its aromaticity cannot, of course, be tested by attempted electrophilic substitution, for attack by $X^\oplus$ would merely lead to direct combination with the anion. True aromatic character (e.g. a Friedel Crafts reaction) is, however, demonstrable in the remarkable series of extremely stable, neutral compounds obtainable from (15), and called metallocenes, e.g. *ferrocene* (16), in which the metal is held by π bonds in a kind of molecular 'sandwich' between the two cyclopentadienyl structures:

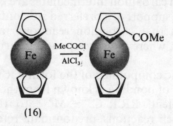

(16)

It is also possible to add two electrons to the non-planar, non-aromatic (*cf.* p. 17) cyclooctatetrane (17) by treating it with potassium, thereby converting it into the isolable, crystalline salt of the cyclooctatetraenyl dianion (18):

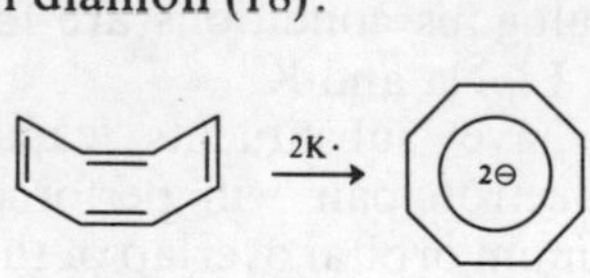

(17) $\xrightarrow{2K\cdot}$ (18)

This too is a Hückel $4n + 2$ p electron system ($n = 2$, this time) and shows quasi-aromatic stability; stabilisation by aromatisation has again taken place, remarkably this time in a doubly charged carbanion (18).

10.3 CARBANION STEREOCHEMISTRY

In theory a simple carbanion of the type $R_3C^\ominus$ could assume a pyramidal (sp^3) or a planar (sp^2) configuration, or possibly something in between these extremes, depending on R. The actual structure is not known with certainty, but what evidence there is suggests the pyramidal configuration with the unshared electron pair occupying the fourth sp^3 orbital (19), i.e. that they resemble amines with which they are isoelectronic:

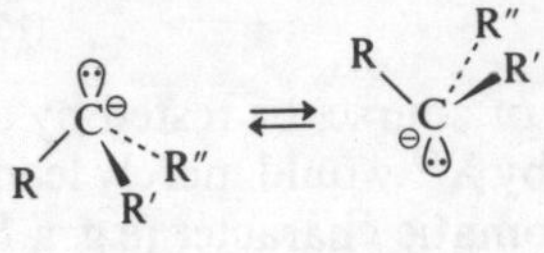

(19a)   (19b)

If such is the case, they should be capable of rapid interconversion with their mirror images, (19a $\rightleftarrows$ 19b) like chiral tertiary amines of the form RR'R''N:, which cannot be resolved into their enantiomers because of their too rapid interconversion. Thus it has proved impossible to establish substantial retention of configuration in RR'R''C groups in reactions in which carbanion intermediates are known to be involved. Positive evidence in support of a preferred sp^3 configuration is provided by the observation that carbanion reactions take place readily at bridgehead carbons, where the corresponding carbonium ion reactions fail (*cf*. p. 86).

In organometallic compounds of the form RR'R''C—M, pretty well the whole spectrum of bonding is known from the essentially covalent, *via* the polar-covalent, $RR'R''C^{\delta-}—M^{\delta+}$, to the essentially ionic, $RR'R''C^{\ominus}M^{\oplus}$. In their reactions, predominant retention, racemisation, and inversion of configuration have all been observed; the outcome in a particular case depending not only on the alkyl residue, but also on the metal, and particularly on the solvent. Even with the most ionic examples it seems unlikely that we are dealing with a simple carbanion; thus in the reaction of EtI with $[PhCOCHMe]^{\ominus}$ $M^{\oplus}$, the relative rates under analogous conditions are found to differ over a range of $\approx 10^4$ for M = Li, Na and K.

Carbanions which have substituents capable of conjugative delocalisation of the electron pair will perforce be planar (sp^2), in order to allow the maximum orbital overlap of the p orbital with those of the substituent, e.g. (4) and (10):

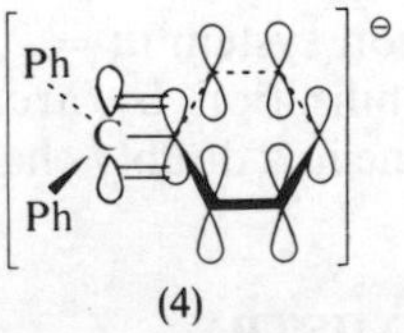

(4) 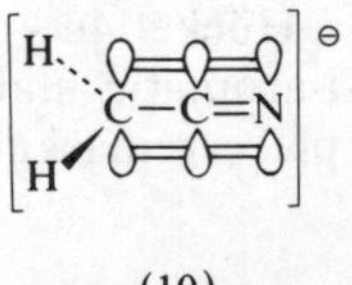   (10)

Where such alignment is prevented by structural or steric features, the expected stabilisation may not take place. Thus while pentan-2,4-

dione (20), with a pK_a value of 8·8, and cyclohexan-1,3-dione (21) are both readily soluble in aqueous NaOH (though not in water), and give a red colour with $FeCl_3$ solution (*cf.* phenol), the formally similar 1,3-diketone (22) does neither:

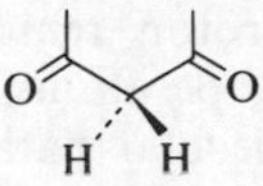

(20)

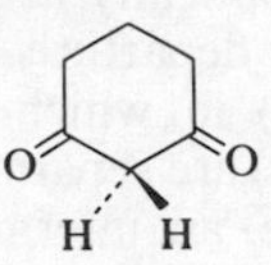

(21)

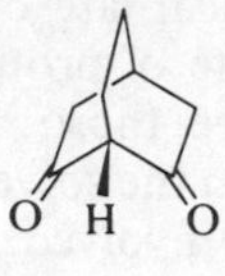

(22)

The H atom flanked by the two C=O groups in (22) exhibits hardly any more acidic character than the analogous one in the corresponding hydrocarbon. The different behaviour of (22) stems from the fact that the orbital containing the electron pair in the derived carbanion would be in a plane virtually at right angles to those embracing the *p* orbitals on the carbonyl carbon atoms (*cf.* p. 253): no stabilisation could occur in the carbanion, which thus does not form.

10.4 CARBANIONS AND TAUTOMERISM

Tautomerism, strictly defined, could be used to describe the reversible interconversion of isomers, in all cases and under all conditions. In practice, the term has increasingly been restricted to isomers that are fairly readily interconvertible, and that differ from each other only (*a*) in electron distribution, and (*b*) in the position of a relatively mobile atom or group. The mobile atom is, in the great majority of examples, hydrogen, and the phenomenon is then referred to as *prototropy*. Familiar examples are β-ketoesters, e.g. ethyl 2-ketobutanoate (ethyl acetoacetate, 23), and aliphatic nitrocompounds, e.g. nitromethane (24):

$$\mathrm{Me\overset{O}{\overset{\|}{C}}{-}\underset{H}{\underset{|}{C}}HCO_2Et \rightleftarrows Me\overset{OH}{\overset{|}{C}}{=}CHCO_2Et}$$

(23*a*) Keto (23*b*) Enol

$$\mathrm{\overset{H}{\overset{|}{C}}H_2{-}\underset{O_\ominus}{\underset{|}{\overset{\oplus}{N}}}{=}O \rightleftarrows CH_2{=}\underset{O_\ominus}{\underset{|}{\overset{\oplus}{N}}}{-}OH}$$

(24*a*) (24*b*)

Pseudo-acid Aci-form

Such interconversions are catalysed by both acids and bases.

10.4.1 Mechanism of interconversion

Prototropic interconversions have been the subject of much detailed study, as they lend themselves particularly well to investigation by deuterium labelling, both in solvent and substrate, and by charting the stereochemical fate of optically active substrates having a chiral centre at the site of proton departure. Possible limiting mechanisms (*cf.* S_N1/S_N2) are those: (*a*) in which proton removal and proton acceptance (from the solvent) are separate operations, and a carbanion intermediate is involved, i.e. an intermolecular pathway; and (*b*) in which one and the same proton is transferred intramolecularly:

(*a*)
$$R_2\overset{\overset{\displaystyle B\curvearrowright H}{|}}{C}-CH=Y \underset{R'OH}{\overset{B:}{\rightleftarrows}} \left[\begin{array}{c} R_2\overset{\ominus}{C}-CH=Y \\ \updownarrow \\ R_2C=CH-Y^{\ominus} \end{array}\right] \underset{B:}{\overset{R'OH}{\rightleftarrows}} R_2C=CH-\overset{\overset{\displaystyle H}{|}}{Y} \qquad \text{(Intermolecular)}$$

(25)
carbanion
intermediate

(*b*)
$$R_2\overset{\overset{\displaystyle H}{|}}{C}-CH=Y \overset{B:}{\rightleftarrows} \left[\begin{array}{c} B \\ \vdots \\ H \\ R_2C \overset{\diagdown}{} Y \\ \diagup \\ CH \end{array}\right]^{\ddagger} \overset{-B:}{\rightleftarrows} R_2C=CH-\overset{\overset{\displaystyle H}{|}}{Y} \qquad \text{(Intramolecular)}$$

(26)
T.S.

Many of the compounds that undergo ready base-catalysed keto $\rightleftarrows$ enol prototropic changes, e.g. β-keto esters, 1,3-(β-) diketones, aliphatic nitro compounds, etc., form relatively stable carbanions, e.g. (25), that can often be isolated. Thus it is possible to obtain carbanions from the 'keto' forms of the β-keto ester (23*a*) and nitromethane (24*a*) and, under suitable conditions, to protonate them so as to obtain the pure enol forms (23*b*) and (24*b*), respectively. It thus seems extremely probable that their interconversion follows the intermolecular pathway (*a*). The more acidic the substrate, i.e. the stabler the carbanion to which it gives rise, the greater the chance that prototropic interconversion will involve the carbanion as an intermediate.

The mechanism (*a*) nicely illustrates the difference between *tautomerism* and *mesomerism* that often gives rise to confusion. Thus taking ethyl 2-ketobutanoate (23) as an example,

$$\underset{\underset{\displaystyle \text{(23a) \underline{Keto}}}{H}}{\overset{\overset{\displaystyle O}{\|}}{MeC}}-CHCO_2Et \underset{R\,OH}{\overset{B:}{\rightleftarrows}} \left[\begin{array}{c} \overset{\overset{\displaystyle O}{\|}}{MeC}-\overset{\ominus}{C}HCO_2Et \\ O^{\ominus} \updownarrow \\ MeC=CHCO_2Et \end{array}\right] \underset{B:}{\overset{R'OH}{\rightleftarrows}} \overset{\overset{\displaystyle OH}{|}}{MeC}=CHCO_2Et$$

(23*a*) Keto   (27)   (23*b*) Enol
carbanion
intermediate

(23a) and (23b) are tautomers: quite distinct, chemically distinguishable and different species, readily interconverted but, in this case, actually separately isolable in a pure state. The two structures written for the carbanion intermediate (27) are mesomers: they have no real existence at all, they are merely somewhat inaccurate attempts to represent the electron distribution in the carbanion, which is a single individual only. It is perhaps better to represent (27) by a single structure of the form,

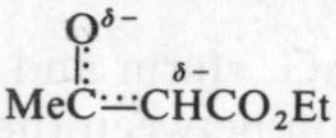

$$\overset{\displaystyle O^{\delta-}}{\underset{(27)}{MeC\cdots\overset{\delta-}{C}HCO_2Et}}$$

but this is still not wholly satisfactory in that it does not convey the important fact that more of the negative charge on the anion is located on the oxygen, rather than on the carbon, atom. It is very common to find a pair of tautomers, such as (23a) and (23b), 'underlain' as it were by a stabilised, delocalised carbanion such as (27).

The other extreme case, i.e. wholly intramolecular proton transfer—pathway (b), is seen in the Et_3N: catalysed conversion of the optically active substrate (28) into (29):

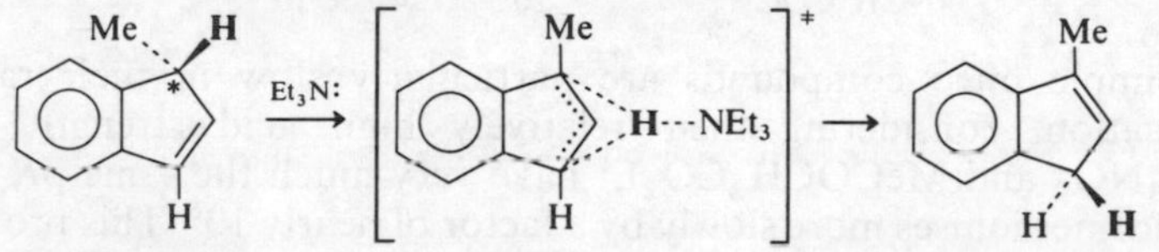

Here it is found that the rate of loss of optical activity and the rate of isomerisation are identical, and if the reaction is carried out in the presence of D_2O (five moles per mole of substrate) no deuterium is incorporated into the product. The reaction is thus wholly intramolecular under these conditions—no carbanion is involved—and is believed to proceed *via* a bridged T.S. such as (30). With a number of substrates features of both inter- and intra-molecular pathways are observed, the relative proportions being dependent not only on the substrate, but to a considerable extent on the base and solvent employed also.

10.4.2 Rate and structure

In virtually all the examples we have been talking about, the slow, rate-limiting stage is the breaking—or forming—of a C—H bond; this is one major respect in which carbon acids differ from those acids in which the incipient proton is attached to O, N etc. The rate of such

C—H bond-breaking can often be measured by determining the rate of hydrogen isotope exchange with suitable proton (deuteron) donors such as D_2O, EtOD etc. It is interesting, though hardly surprising, to find that this *kinetic acidity* scale (defined by k_1) does not correlate directly with the thermodynamic acidity scale (defined by K) that we have considered to date, i.e. pK_a values;

$$R_3C\text{—}H + B: \underset{k_{-1}}{\overset{k_1}{\rightleftarrows}} R_3C^\ominus + BH^\oplus \qquad (K = k_1/k_{-1})$$

for the former involves a $\Delta G^{\neq}$ term and the latter a $\Delta G^{\ominus}$ term, and there is no necessary relation between the two. It is very broadly true that structural changes in the substrate that lead to greater (thermodynamic) acidity also tend to lead to its more rapid conversion into the carbanion, but there are many exceptions as may be seen below:

	pK_a	$k_1 \, (\text{sec}^{-1})$
$CH_2(NO_2)_2$	4	8.3×10^{-1}
$CH_2(COMe)_2$	8.8	1.7×10^{-2}
CH_3NO_2	10.2	4.3×10^{-8}
$MeCOCH_2CO_2Et$	10.7	1.2×10^{-3}
$CH_2(CN)_2$	12	1.5×10^{-2}
$CH_2(CO_2Et)_2$	13.3	2.5×10^{-5}
CH_3COCH_3	20	4.7×10^{-10}

Simple nitro compounds are particularly slow in their rate of ionisation, considering their relatively high acid strength; thus CH_3NO_2 and $MeCOCH_2CO_2Et$ have very much the same pK_a, but the former ionises more slowly by a factor of nearly 10^5. This probably reflects a greater degree of delocalisation of charge in the carbanion derived from CH_3NO_2 than in that from $CH_3COCH_2CO_2Et$. In such cases both proton abstraction and donation tend to be slow, compared with those carbon acids in which the charge is more concentrated on carbon in their carbanions. This is borne out by the effect of $C\equiv N$ substituents on carbanions, where less charge delocalisation would be expected than with a $C=O$ substituent; thus $CH_2(CN)_2$ is found to have very much the same k_1 value as $CH_2(COMe)_2$, despite the fact that its pK_a value is larger (i.e. acidity lower) by 3.2 units. The relation between pK_a and k_1 can be much affected by the solvent, however.

10.4.3   Position of equilibrium and structure

In this context it is keto/enol systems that have been investigated by far the most closely, and most of our discussion will centre on them. The relative proportion of the two forms was commonly determined chemically, e.g. by titration of the enol form with bromine under conditions such that the rate of keto/enol interconversion was very low; it is, however, more accurate and more convenient to do this spectroscopically, e.g. in the i.r. for ethyl 3-ketobutanoate:

$$
\underset{\text{(23a) \underline{Keto}}}{\overset{\text{O}}{\underset{\text{OEt}}{\overset{\|}{\underset{|}{\text{MeC}}}}}\overset{\text{(1)}}{-}\text{CH}_2\overset{\text{(2)}}{-}\text{C}{=}\text{O}} \rightleftarrows \underset{\text{(23b) \underline{Enol}}}{\overset{\text{OH}}{\underset{\text{OEt}}{\overset{|}{\underset{|}{\text{MeC}}}}}{=}\text{CH}\overset{\text{(3)}}{-}\text{C}{=}\text{O}}
$$

(1) ν_{max} 1718 cm^{-1} (3) ν_{max} 1650 cm^{-1}
(2) ν_{max} 1742 cm^{-1}

In simple carbonyl compounds, e.g. MeCOMe, the proportion of enol at equilibrium is extremely small; the main structural features that result in its increase may be seen in the table below:

	% Enol in liquid
MeCOCH$_3$	$1 \cdot 5 \times 10^{-4}$
CH$_2$(CO$_2$Et)$_2$	$7 \cdot 7 \times 10^{-3}$
NCCH$_2$CO$_2$Et	$2 \cdot 5 \times 10^{-1}$
Cyclohexanone	$1 \cdot 2$
MeCOCH$_2$CO$_2$Et	$8 \cdot 0$
MeCOCHPhCO$_2$Et	$30 \cdot 0$
MeCOCH$_2$COMe	$76 \cdot 4$
PhCOCH$_2$COMe	$89 \cdot 2$

The major feature is a multiple bond, or a π orbital system such as Ph, which can become conjugated with the C=C double bond in the enol form. C=O is clearly effective in this respect, with an ordinary carbonyl C=O group being considerably more effective than the C=O in an ester group, *cf.* MeCOCH$_2$CO$_2$Et (8 %) and MeCOCH$_2$-COMe (76·4 %). The added effect of Ph may be seen in comparing MeCOCH$_2$CO$_2$Et (8 %) with MeCOCHPhCO$_2$Et (30 %), and MeCO-CH$_2$COMe (76·4 %) with PhCOCH$_2$COMe (89·2 %).

Another feature that will serve to stabilise the enol, with respect to the keto, form is the possibility of strong, intramolecular hydrogen bonding, e.g. in MeCOCH$_2$COMe (31) and MeCOCH$_2$CO$_2$Et (23):

$$
\underset{(31)}{\overset{\text{H}}{\underset{\text{Me}^{\diagup}\underset{\diagdown}{\text{C}}}{\overset{\text{O}^{\diagdown}\cdots\text{O}}{}}}\overset{}{\underset{\text{CH}}{\overset{\|}{\text{C}}}}\overset{}{\underset{\diagdown\text{Me}}{}}}
\qquad
\underset{(23)}{\overset{\text{H}}{\underset{\text{Me}^{\diagup}\underset{\diagdown}{\text{C}}}{\overset{\text{O}^{\diagdown}\cdots\text{O}}{}}}\overset{}{\underset{\text{CH}}{\overset{\|}{\text{C}}}}\overset{}{\underset{\diagdown\text{OEt}}{}}}
$$

Apart from any stabilisation effected with respect to the keto form, such intramolecular hydrogen bonding will lead to a decrease in the polar character of the enol, and to a more compact, 'folded-up' conformation of the molecule, compared with the more extended conformation of the keto form. This has the rather surprising result that where keto and enol forms can actually be separated, the latter usually has the lower b.p. despite its hydroxyl group. The effectiveness of intramolecular hydrogen bonding in stabilising the enol, with respect to the keto, form is seen on varying the solvent, and particularly on transfer to a hydroxylic solvent, e.g. with $MeCOCH_2COMe$ (31):

Solvent	% Enol
H_2O	15
MeCN	58
Liquid	80
C_6H_{14}	92
Gas phase	92

Thus the proportion of enol in the non-polar solvent hexane is the same as in the gas phase, and higher than in the liquid itself, the latter acting as a somewhat polar auto-solvent; the proportion drops again in the more polar MeCN, and more dramatically in water. What is happening is the increasing relative stabilisation of the keto form by solvation, this being particularly marked in water where *inter*molecular hydrogen bonding of the keto form's $C{=}O$ group can now take place as an alternative to its enolisation. The behaviour of $MeCOCH_2CO_2Et$ (23) is closely analogous; thus the 8 % enol present in the liquid rises to 46 % in hexane and to 50 % in the gas phase, but drops to 0·4 % in dilute aqueous solution. The percentage of enol present is also dependent on the temperature.

A particularly interesting—and extreme—example is provided by a comparison of MeCOCOMe (32) and the cylic 1,2-diketone, cyclopentan-1,2-dione (33):

(32*a*)	(32*b*)	(33*a*)	(33*b*)
	5.6×10^{-3}%		≈ 100%

With (32), despite the intramolecular hydrogen bonding possible in the enol form (32*b*), the equilibrium lies essentially completely over in favour of the keto form (32*a*), because this can take up an *anti*-conformation in which the two electronegative oxygen atoms are as far from each other as possible, and in which the carbonyl dipoles are opposed. With (33), the $C{=}O$ groups are 'locked' in the *syn*-conformation in both keto (33*a*) and enol (33*b*) forms, and the intramolecular hydrogen bonding open to (33*b*), but not to (33*a*), then decides the issue.

In the above examples the composition of the equilibrium mixture is, of course, governed by the relative thermodynamic stability of the two forms under the particular conditions being studied. An interesting situation arises with aliphatic nitrocompounds, e.g. phenylnitromethane (34), however. Here the normal nitro—pseudo-acid—form (34*a*) is the more stable of the two and, at equilibrium, predominates to the almost total exclusion of the aci-form (34*b*):

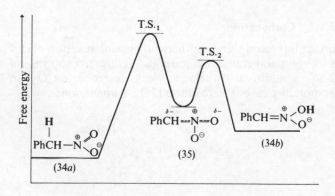

Despite this fact, acidification of the isolable sodium salt of the carbanion intermediate (35) is found to yield almost entirely the less stable aci-form (34*b*). This results from more rapid protonation taking place at the position of higher electron density, i.e. product formation under these conditions is kinetically controlled. The energy profile for the system has the form (Fig. 10.1),

Fig. 10.1

i.e. the transition state between (35) and (34*b*)—T.S.$_2$—is at a lower energy level than that between (35) and (34*a*)—T.S.$_1$, reflecting the greater ease of breaking an O—H than a C—H bond. Although the immediate result of the acidification of (35) is thus the formation of (34*b*), this can undergo spontaneous re-ionisation; equilibrium is thus gradually established, leading to the slow, but inexorable, formation of the more stable (34*a*): the ultimate composition of the product is thus thermodynamically controlled.

10.5   CARBANION REACTIONS

Carbanions can take part in most of the main reaction types, e.g. addition, elimination, displacement, rearrangement, etc. They are also involved in reactions, such as oxidation, that do not fit entirely satisfactorily into this classification, and as specific—*ad hoc*—intermediates in a number of other processes as well. A selection of the reactions in which they participate will now be considered; many are of particular synthetic utility, because they result in the formation of carbon–carbon bonds.

10.5.1   Addition

We have already discussed a large group of reactions in which carbanions add to the C=O group (*cf.* pp. 217–230), including examples of intramolecular carbanion addition, e.g. an aldol reaction (p. 222), Dieckmann reaction (p. 226), and the benzilic acid rearrangement (p. 228), and also to the C=C—C=O system, the Michael reaction (p. 196).

10.5.1.1   Carbonation

A further interesting, and synthetically useful, reaction of carbanions—and of organometallic compounds acting as sources of negative carbon—is addition to the very weak electrophile CO_2, to form the corresponding carboxylate anion (36)—*carbonation*:

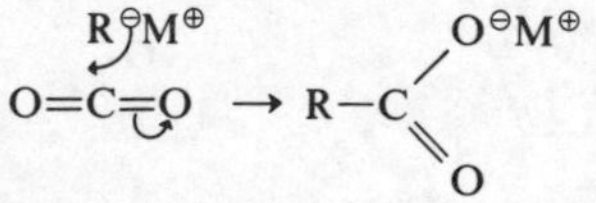

$$O=C=O \xrightarrow{R^{\ominus}M^{\oplus}} R-C\begin{matrix} O^{\ominus}M^{\oplus} \\ \diagdown \\ O \end{matrix}$$

(36)

It occurs with the alkyls, aryls or acetylides of metals more electropositive than magnesium, but including Grignard reagents, and is often carried out by adding a solution of the organometallic compound in an inert solvent to a large excess of powdered, solid CO_2; it is a particularly useful method for the preparation of acetylenic acids. The Kolbe–Schmidt reaction (p. 285) is another example of carbanion carbonation.

This reaction has been used a good deal in the study of carbanions, to detect their formation by converting them into stable, identifiable products. Thus substantial retention of configuration in an alkenyl carbanion (37) has been demonstrated, in the reaction of (38) with lithium, by converting it into (39):

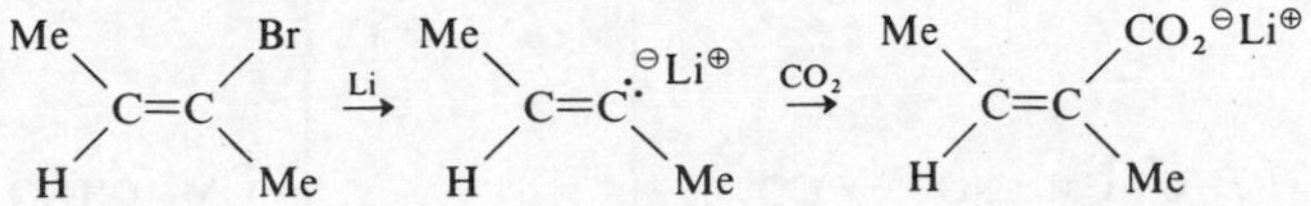

$$
\underset{(38)}{\overset{Me}{\underset{H}{\diagup}}C=C\overset{Br}{\underset{Me}{\diagdown}}} \xrightarrow{Li} \underset{(37)}{\overset{Me}{\underset{H}{\diagup}}C=C\overset{\ominus Li^{\oplus}}{\underset{Me}{\diagdown}}} \xrightarrow{CO_2} \underset{(39)}{\overset{Me}{\underset{H}{\diagup}}C=C\overset{CO_2{}^{\ominus}Li^{\oplus}}{\underset{Me}{\diagdown}}}
$$

The yield of (39) was $\approx 75\%$, while that of its geometrical isomeride was $<5\%$.

10.5.2 Elimination

We have already seen examples of carbanions involved as intermediates, e.g. (40), in elimination reactions, i.e. those that proceed by the E1cB pathway (p. 244), for example:

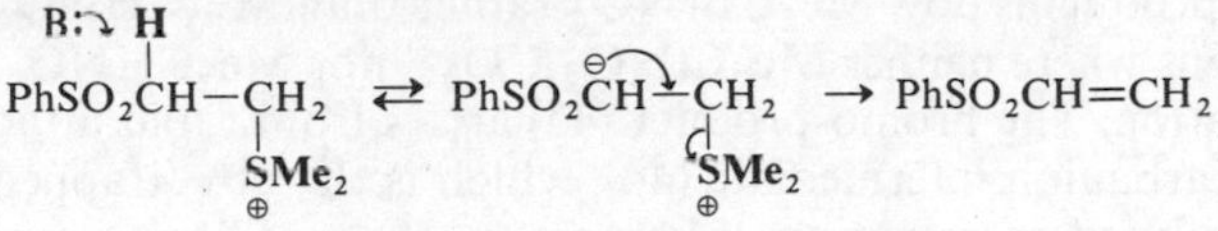

$$
\underset{\underset{\oplus}{SMe_2}}{\overset{B:\frown H}{PhSO_2CH-CH_2}} \rightleftarrows \underset{\underset{\oplus}{SMe_2}}{\overset{\ominus}{PhSO_2CH\frown CH_2}} \rightarrow PhSO_2CH=CH_2
$$

$$(40)$$

Another example is decarboxylation.

10.5.2.1 Decarboxylation

Loss of CO_2 from carboxylate anions (41) is believed to involve a carbanion intermediate (42) that subsequently acquires a proton from solvent, or other source:

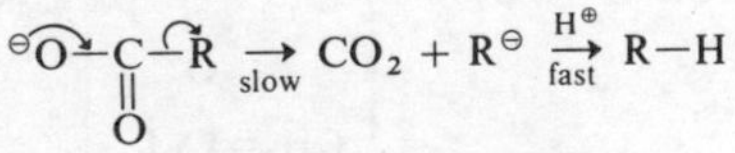

$$
\underset{(41)}{\overset{\ominus}{O}\overset{\frown}{\underset{\parallel}{\underset{O}{C}}}\overset{\frown}{R}} \underset{slow}{\rightarrow} CO_2 + R^{\ominus} \underset{(42)}{\overset{H^{\oplus}}{\underset{fast}{\rightarrow}}} R-H
$$

Loss of CO_2 is normally rate-limiting, i.e. the rate law is,

$$\text{Rate} = k[RCO_2{}^{\ominus}]$$

subsequent proton abstraction being rapid. Decarboxylation should thus be promoted by electron-withdrawing substituents in R that could stabilise the carbanion intermediate (42) by delocalisation of its negative charge. This is borne out by the very much readier de-carboxylation of the nitro-substuted carboxylate anion (43) than of

$Me_2CHCO_2^\ominus$ itself:

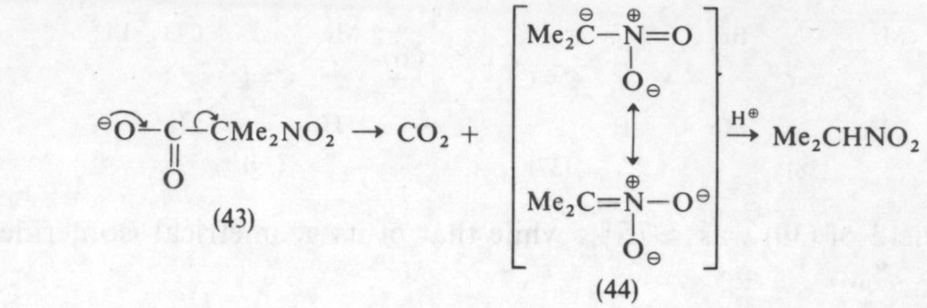

$$\ominus O-C-CMe_2NO_2 \rightarrow CO_2 + \left[\begin{array}{c} Me_2\overset{\ominus}{C}-\overset{\oplus}{N}=O \\ \underset{\ominus}{O} \\ \updownarrow \\ Me_2C=\overset{\oplus}{N}-O^\ominus \\ \underset{\ominus}{O} \end{array} \right] \xrightarrow{H^\oplus} Me_2CHNO_2$$

(43)   (44)

Similar ease of decarboxylation is seen in $Hal_3CCH_2CO_2^\ominus$, 2,4,6-$(NO_2)_3 C_6H_2CO_2^\ominus$ etc, but the reaction is not normally of preparative value with the anions of simple aliphatic acids other than $MeCO_2^\ominus$.

Evidence that carbanion intermediates, e.g. (44), are involved is provided by carrying out the decarboxylation in the presence of bromine. This is without effect on the overall rate of the reaction but the end product is now Me_2CBrNO_2 rather than Me_2CHNO_2—under conditions where neither $Me_2C(NO_2)CO_2^\ominus$ nor Me_2CHNO_2 undergo bromination. The bromo-product (45) arises from rapid attack of Br_2 on the carbanion intermediate (44), which is thereby 'trapped' (*cf.* the base-catalysed bromination of ketones, p. 288):

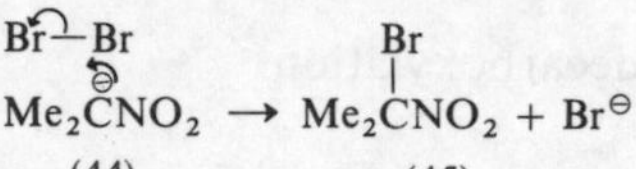

$$\begin{array}{cc} Br-Br & Br \\ \overset{\ominus}{} & | \\ Me_2\overset{\ominus}{C}NO_2 \rightarrow Me_2CNO_2 + Br^\ominus \\ (44) & (45) \end{array}$$

$C=O$ can also act like NO_2, and the anions of β-ketoacids (46) are decarboxylated very readily:

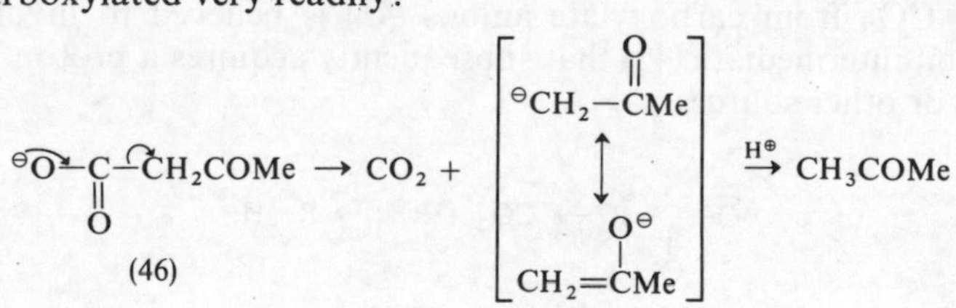

$$\ominus O-C-CH_2COMe \rightarrow CO_2 + \left[\begin{array}{c} O \\ \| \\ \ominus CH_2-CMe \\ \updownarrow \\ O^\ominus \\ | \\ CH_2=CMe \end{array} \right] \xrightarrow{H^\oplus} CH_3COMe$$

(46)

The overall rate law is, however, found to contain a term involving [keto acid] (47) as well as the term involving [keto acid anion]; this is believed to arise from incipient proton transfer to the keto group through hydrogen bonding:

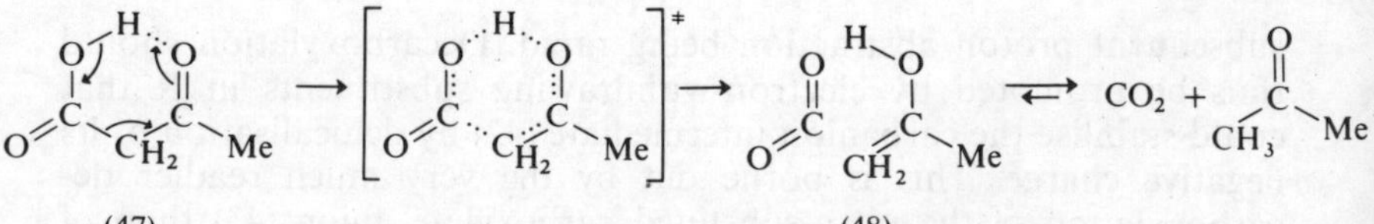

$$O=C\overset{H}{\underset{CH_2}{\diagdown}}C\diagup^O_{Me} \rightarrow \left[O=C\overset{H}{\underset{CH_2}{\cdots}}C\diagup^O_{Me} \right]^{\ddagger} \rightarrow O=C\overset{H}{\underset{CH_2}{\diagup}}O\diagdown C\diagdown_{Me} \rightleftharpoons CO_2 + C\diagup^O_{Me}_{CH_3}$$

(47)   (48)

Some evidence for this mode of decarboxylation of the free acid has been obtained by 'trapping' the enol intermediate (48). $\beta\gamma$-unsaturated acids (49) probably also decarboxylate by an analogous pathway:

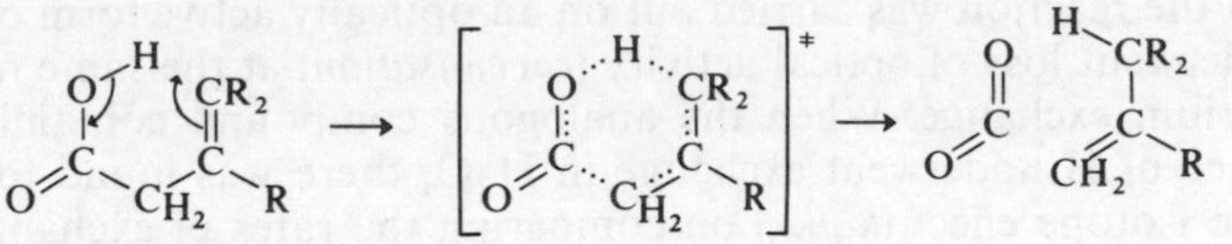

(49)

$\alpha\beta$-unsaturated acids, $R_2CHCR{=}CHCO_2H$, probably decarboxylate by this pathway also, as it has been shown that they isomerise to the corresponding $\beta\gamma$-unsaturated acid prior to decarboxylation.

Another example in which the free acid undergoes ready decarboxylation, but this time *via* a carbanion intermediate (50, actually an *ylid*), is pyridine-2-carboxylic acid (51), which is decarboxylated very much more readily that its 3- or 4-isomers:

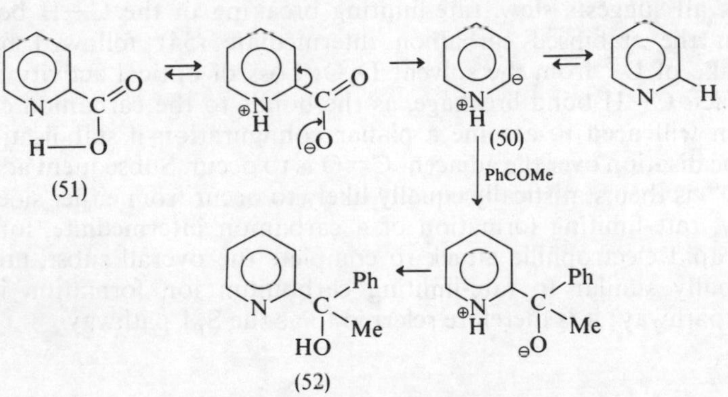

(51)   (50)

PhCOMe

(52)

The ylid intermediate (50) can be 'trapped' by carrying out the decarboxylation in the presence of carbonyl compounds, e.g. PhCOMe, to yield the carbanion addition product, e.g. (52); this process can indeed be used preparatively. The reason for the much easier decarboxylation of (51), than of its 3-, and 4-isomers, is the stabilisation that the $N^{\oplus}$ can effect on the adjacent carbanion carbon atom in the intermediate ylid (50).

10.5.3 Displacement

Carbanions, or similar species, are involved in a variety of displacement reactions, either as intermediates or as attacking nucleophiles.

10.5.3.1   Deuterium exchange

The ketone (53) is found to undergo exchange of its α-hydrogen atom for deuterium when treated with base ($^{\ominus}OD$) in solution in D_2O. When the reaction was carried out on an optically active form of (53), it underwent loss of optical activity (*racemisation*) at the same rate as deuterium exchange. When the analogous compound containing D in place of H underwent exchange in H_2O, there was found to be a kinetic isotope effect (k_H/k_D) on comparing the rates of exchange for the two compounds:

$$
\underset{\substack{\text{H} \ (+) \\ (53)}}{\overset{\substack{O \\ \| \\ MeEtC-CPh}}{\text{MeEtC}-\text{CPh}}} \xrightarrow[\text{slow}]{^{\ominus}OD}
\left[
\begin{array}{c}
\overset{O}{\overset{\|}{MeEtC-CPh}} \\
\overset{\ominus}{} \\
\updownarrow \\
\underset{MeEtC=CPh}{\overset{\ominus}{O}}
\end{array}
\right]
\xrightarrow[\text{fast}]{D_2O}
\underset{\substack{\text{D} \ (\pm)}}{\overset{\substack{O \\ \|}}{\text{MeEtC}-\text{CPh}}}
$$

$$(54)$$

This all suggests slow, rate-limiting breaking of the C—H bond to form the stabilised carbanion intermediate (54), followed by fast uptake of $D^{\oplus}$ from the solvent D_2O. Loss of optical activity occurs at each C—H bond breakage, as the bonds to the carbanion carbon atom will need to assume a planar configuration if stabilisation by delocalisation over the adjacent C=O is to occur. Subsequent addition of $D^{\oplus}$ is then statistically equally likely to occur from either side. This slow, rate-limiting formation of a carbanion intermediate, followed by rapid electrophilic attack to complete the overall substitution, is formally similar to rate-limiting carbonium ion formation in the S_N1 pathway; it is therefore referred to as the S_E1 pathway.

10.5.3.2   Carbanion nucleophiles

Both overt carbanions and organometallic compounds, such as Grignard reagents, are powerful nucleophiles as we have seen in their addition reactions with C=O (p. 217 *et seq.*); they tend therefore to promote an S_N2 pathway in their displacement reactions. Particularly useful carbanions, in preparative terms, are those derived from $CH_2(CO_2Et)_2$, β-ketoesters, 1,3-(β-)diketones, e.g. (55), etc.—the so-called 'reactive methylenes':

$$
\underset{(55)}{(MeCO)_2CH_2} \underset{}{\overset{^{\ominus}OEt}{\rightleftarrows}} (MeCO)_2\overset{\ominus}{CH} + R-Br \xrightarrow{RBr} \underset{(56)}{(MeCO)_2CH-R} + Br^{\ominus}
$$

The S_N2 character of the process has been confirmed kinetically, and in suitable cases inversion of configuration has been demonstrated at the carbon atom attacked in RBr above. The alkylated product (56) still contains an acidic hydrogen, and the process may be repeated to yield the dialkyl product, $(MeCO)_2CRR'$. Synthetically useful alkylations can also be effected on acetylide anion (57):

$$HC{\equiv}CH \overset{\ominus NH_2}{\rightleftarrows} HC{\equiv}C^{\ominus} + R{-}Br \overset{RBr}{\rightarrow} HC{\equiv}C{-}R + Br^{\ominus}$$

$$(57)$$

Here too, a second alkylation can be made to take place yielding $RC{\equiv}CR$ or $R'C{\equiv}CR$. It should, however, be remembered that the above carbanions—particularly the acetylide anion (57)—are the anions of very weak acids, and are thus themselves strong bases, as well as powerful nucleophiles. They can thus induce elimination (p. 253) as well as displacement, and reaction with tertiary halides is often found to result in alkene formation to the exclusion of alkylation.

Grignard reagents can also act as sources of negative carbon in displacement reactions, e.g. in the synthetically useful reaction with triethoxymethane (ethyl orthoformate, 58) to yield acetals (59) and, subsequently, their parent aldehydes (60):

$$\overset{\delta-}{R}{\frown}\overset{\delta+}{MgBr}$$
$$CH(OEt)_2 \rightarrow RCH(OEt)_2 \xrightarrow{H^{\oplus}/H_2O} RCHO$$
$$OEt$$

$$(58) \qquad\qquad (59) \qquad\qquad (60)$$

It is also possible, under suitable conditions, to generate the alkyls of more electropositive metals, e.g. sodium (61), and then subsequently to react these with alkyl halides:

$$RCH_2CH_2{-}Cl \xrightarrow{2Na\cdot} RCH_2CH_2{}^{\ominus}Na^{\oplus} \xrightarrow{R'Br} RCH_2CH_2R'$$

$$(61)$$

This is, of course, the Wurtz reaction, and support for such a mechanism involving carbanions (radicals may be involved under some conditions, however) is provided by the observation that in some cases it is possible, with optically active halides, to demonstrate inversion of configuration at the carbon atom undergoing nucleophilic attack. The carbanion, e.g. (61), can also act as a base and promote elimination:

$$RCH_2CH_2{}^{\ominus})$$
$$H$$
$$\quad RCH{\frown}CH_2 \qquad\rightarrow$$
$$\qquad Cl \quad Na^{\oplus}$$

$$\begin{array}{c} RCH_2CH_2 \\ | \\ (62)\ H \end{array}$$
$$RCH{=}CH_2 + Na^{\oplus}Cl^{\ominus}$$
$$(63)$$

Thus leading to the disproportionation—alkane (62) + alkene (63)— that is often observed as a side-reaction to the normal Wurtz coupling.

10.5.3.3   Reimer–Tiemann reaction

This involves a delocalised aryl carbanion (64), and also $^{\ominus}CCl_3$ derived from the action of strong bases on $HCCl_3$ (p. 260), though the latter has only a transient existence decomposing to CCl_2, a highly electron-deficient electrophile that attacks the aromatic nucleus:

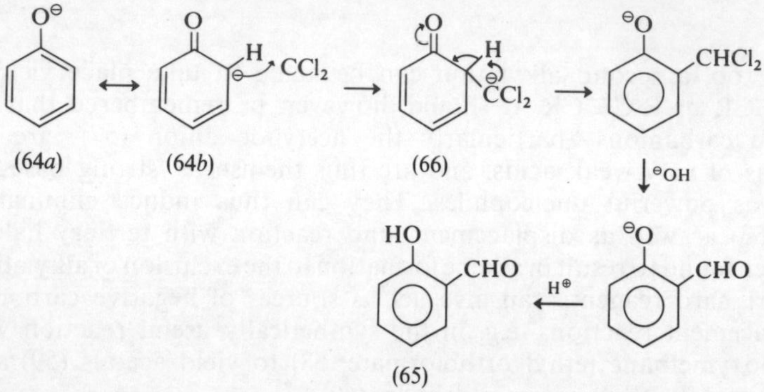

(64a)   (64b)   (66)

(65) $\xleftarrow{H^{\oplus}}$

The product from phenoxide ion (64) is, after acidification, very largely the *o*-aldehyde (salicaldehyde, 65) plus just a small amount of the *p*-isomer.

Some support for the reaction pathway suggested above is provided by what is observed when the analogous reaction is carried out on the anion of *p*-hydroxytoluene (*p*-cresol, 67):

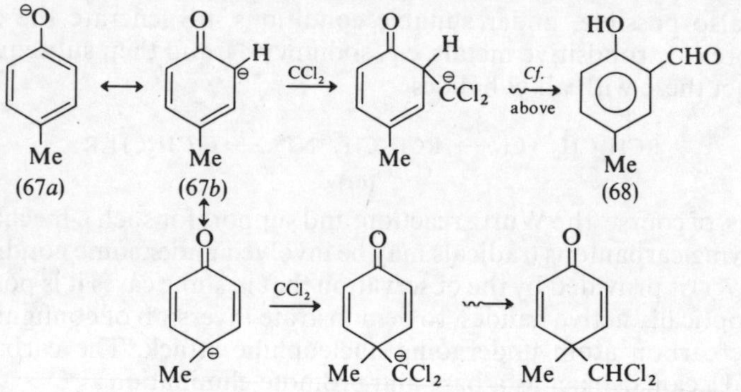

Me   Me   Me   Me
(67a)   (67b)   (68)

Me   Me CCl_2 Me $CHCl_2$
(67c)   (70)   (69)

In addition to the expected *o*-aldehyde (68), it is also possible to isolate the unhydrolysed dichloro-compound (69). Attack by CCl_2 at the *p*-position in (67c) yields the intermediate (70) which, unlike the *p*-analogue of (66) above, has no H-atom that can be lost as $H^{\oplus}$ to allow the ring to re-aromatise; (70) thus just acquires a proton, on final acidification, to yield (69). The dichloro-compound (69) owes its resistance to hydrolysis partly to its insolubility in the aqueous base medium, but also to the sterically hindered, neopentyl-type environment (*cf.* p. 86) of the chlorine atoms.

The somewhat analogous Kolbe–Schmidt reaction involves CO_2 as the electrophile in attack on powdered sodium phenoxide (64b):

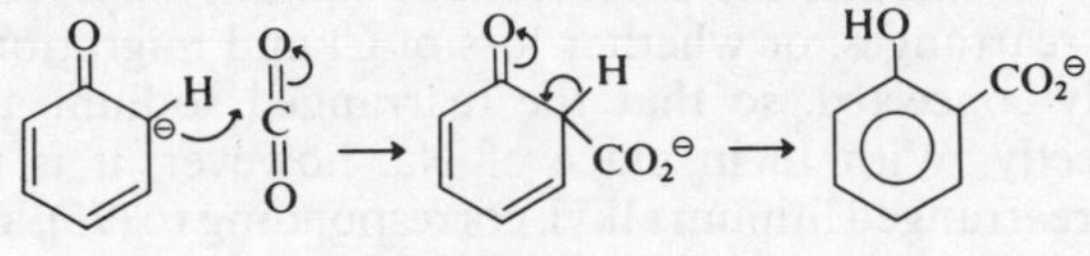

(64b)   (71)

The product is almost exclusively sodium *o*-hydroxybenzoate (salicylate, 71) only traces of the *p*-isomer being obtained; if, however, the reaction is carried out on <u>potassium</u> phenoxide the salt of the *p*-acid becomes the major product. It has been suggested that the preferential *o*-attack with sodium phenoxide may result from stabilisation of the T.S. (72) through chelation by $Na^{\oplus}$ in the ion pair:

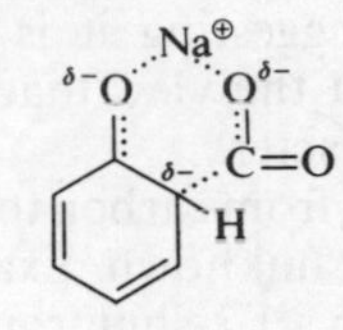

(72)

The $K^{\oplus}$ cation is larger and likely to be less effective in this role, so that attack on the *p*-position will therefore become more competitive.

10.5.4 Rearrangement

Rearrangements involving carbanions, in which the migrating group would move to the carbanion carbon atom without an electron pair, are very much less common than those involving carbonium ions (p. 108), in which the migrating group brings an electron pair with it. 1,2-Shifts of aryl groups from carbon are known, however, e.g. in the action of sodium on the chloride (73):

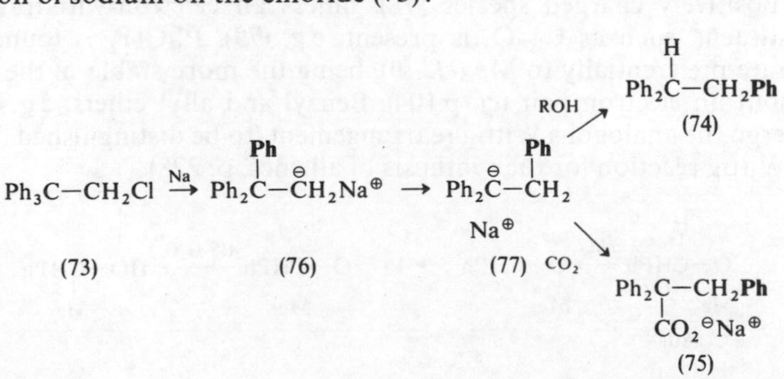

The product is a sodium alkyl, as expected, but protonation and carbonation yield the rearranged products (74) and (75), respectively. It is not known whether the unrearranged sodium alkyl (76) is formed which then rearranges, or whether loss of Cl and migration of Ph are substantially concerted, so that the rearranged sodium alkyl (77) is formed directly. With Li in place of Na, however, it is possible to form the unrearranged lithium alkyl, corresponding to (76), as witnessed by the products of its protonation and carbonation, and then rearrange it subsequently by raising the temperature. The tendency to rearrangement on reacting (73) with metals, or metal derivatives, is found to decrease in the order,

$$K \approx Na > Li > Mg$$

i.e. in the order of decreasing ionic character of the carbon–metal bond. This coupled with a study of the relative migratory aptitude of *p*-substituted Ar groups—suggesting it is $Ar^\oplus$ rather than Ar· that migrates—strongly support the view that the 1,2-shift is carbanionic, rather than radical, in character.

Simple 1,2-shifts of alkyl, from carbon to carbon, that are carbanionic in character are essentially unknown. Examples are known, however, in which alkyl is involved in a 1,2-shift from other atoms such as N and S to a carbanion atom—the Stevens rearrangement:

$$
\underset{\substack{|\\ \textbf{Me}\\ (78)}}{\overset{\textbf{H}}{\underset{}{\overset{|}{\underset{}{Me_2\overset{\oplus}{N}-CHPh}}}}}
\xrightarrow{PhLi}
\underset{\substack{|\\ \textbf{Me}}}{\overset{Li^\oplus}{\underset{}{Me_2\overset{\oplus}{N}-\overset{\ominus}{C}HPh}}}
\rightarrow
\underset{\substack{|\\ \textbf{Me}}}{Me_2N-CHPh}
$$

$$
\underset{\substack{|\\ \textbf{PhCH}_2\\ (79)}}{\overset{\textbf{H}}{\underset{}{\overset{|}{\underset{}{MeS-CHCOPh}}}}}
\underset{\ominus OH}{\rightleftharpoons}
\underset{\substack{|\\ \textbf{PhCH}_2}}{MeS-\overset{\ominus}{C}HCOPh}
\rightarrow
\underset{\substack{|\\ \textbf{PhCH}_2}}{MeS-CHCOPh}
$$

Very strong bases, e.g. PhLi, are required to remove a proton from the positively charged species (78), unless an electron-withdrawing substituent, such as C=O, is present, e.g. (79). PhCH$_2$ is found to migrate preferentially to Me (*cf.* 79), being the more stable of the two without an electron pair (*cf.* p.104). Benzyl and allyl ethers, e.g. (80), undergo the analogous Wittig rearrangement (to be distinguished from the Wittig reaction for the synthesis of alkenes, p. 228):

$$
\underset{\substack{|\\ \textbf{Me}\\ (80)}}{\overset{\textbf{H}}{\underset{}{\overset{|}{\underset{}{O-CHPh}}}}}
\xrightarrow{PhLi}
\underset{\substack{|\\ \textbf{Me}}}{\overset{Li^\oplus}{\underset{}{O-\overset{\ominus}{C}HPh}}}
\rightarrow
\underset{\substack{|\\ \textbf{Me}}}{Li^\oplus{}^\ominus O-CHPh}
\xrightarrow{H^\oplus/H_2O}
\underset{\substack{|\\ \textbf{Me}}}{HO-CHPh}
$$

Finally, there are base-induced carbanion rearrangements that involve a 1,3-elimination to form a cyclopropanone intermediate, e.g. (81)—the Favorskii rearrangement of 1-(α-)haloketones, e.g. (82):

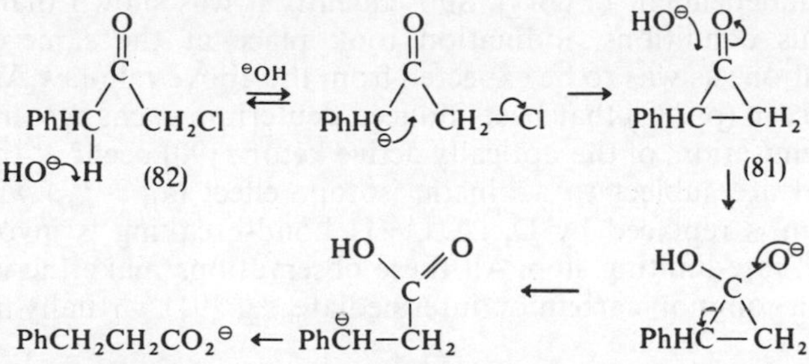

The cyclopropanone intermediate (81) undergoes subsequent addition of $^{\ominus}$OH, followed by ring-opening to yield the more stable of the two possible carbanions (83, benzyl > primary), followed by proton exchange to yield the rearranged carboxylate anion end-product (84).

10.5.5 Oxidation

Carbanions can, under suitable conditions, be oxidised; thus the triphenylmethyl anion (85) is oxidised, fairly slowly, by air:

$$Ph_3C^{\ominus}Na^{\oplus} \underset{Na/Hg}{\overset{O_2}{\rightleftarrows}} Ph_3C\cdot + NaO_2\cdot$$
$$\text{(85)} \qquad\qquad \text{(86)}$$

The resultant radical (86) can, in turn, be reduced back to the carbanion by shaking with sodium amalgam. In suitable cases, e.g. (87), the oxidation of carbanions with one-electron oxidising agents, usually iodine, can be useful synthetically for forming a carbon–carbon bond, through dimerisation (→ 88) of the resultant radical (89):

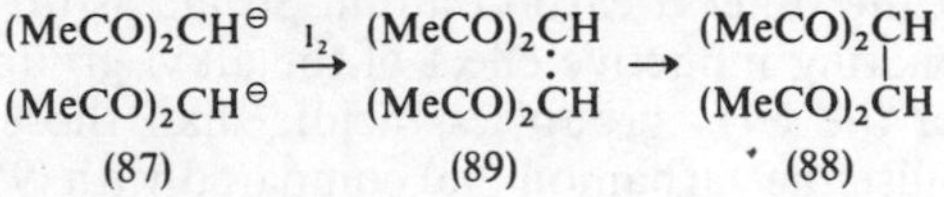

$$\begin{array}{ccc} (MeCO)_2CH^{\ominus} & (MeCO)_2\overset{\cdot}{C}H & (MeCO)_2CH \\ \xrightarrow{\text{I}_2} & & \rightarrow & | \\ (MeCO)_2CH^{\ominus} & (MeCO)_2\overset{\cdot}{C}H & (MeCO)_2CH \\ \text{(87)} & \text{(89)} & \text{(88)} \end{array}$$

10.5.6 Halogenation of ketones

One of the earliest observations relating to the possible occurrence of carbanions as reaction intermediates was that the bromination of acetone, in the presence of aqueous base, followed the rate-law,

$$\text{Rate} = k[\text{MeCOMe}][^{\ominus}\text{OH}]$$

i.e. was independent of $[\text{Br}_2]$. Subsequently it was shown that, under analogous conditions, iodination took place at the same rate as bromination; as was to be expected from the above rate-law. We have already seen (p. 282) that base-induced deuterium exchange (in D_2O), and racemisation, of the optically active ketone (90) occur at the same rate, and are subject to a kinetic isotope effect ($k_H > k_D$) when the α-H atom is replaced by D, i.e. C—H bond-breaking is involved in the slow, rate-limiting step. All these observations make the involvement of a common carbanion intermediate, e.g. (91), virtually inescapable:

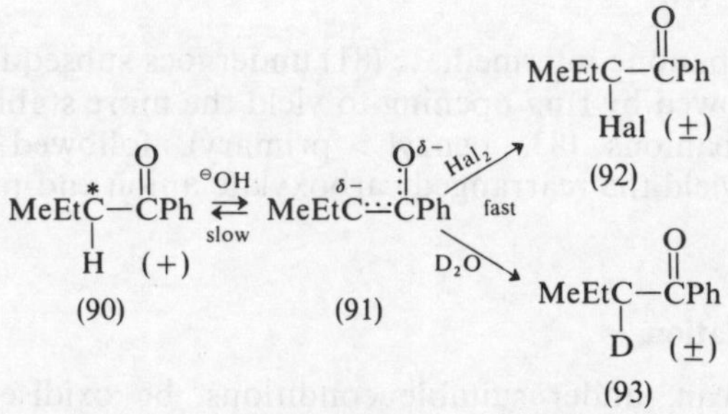

This intermediate is then attacked in a fast, non rate-limiting step by any one of the series of electrophiles—$Cl_2, Br_2, I_2, H_2O, D_2O$ etc—to yield end products such as (92), (93) etc; all of which will necessarily be produced at the same rate. This process has a formal resemblance to slow, rate-limiting formation of a carbonium ion intermediate, followed by rapid nucleophilic attack, in the S_N1 pathway; it is therefore referred to as an S_E1 process.

With ketones such as (94), that have alternative groups of α-H atoms to attack, two questions arise: (*a*) which group, the CH_2 or the CH_3, is attacked preferentially, and (*b*) when one H has been substituted by halogen, will a second halogen become attached to the same or to the other α-carbon atom. So far as (*a*) is concerned, the electron-donating inductive effect of the alkyl group R will make the H atoms of the CH_2 group less acidic than those of the CH_3, and will destabilise the carbanion (96) compared with (95):

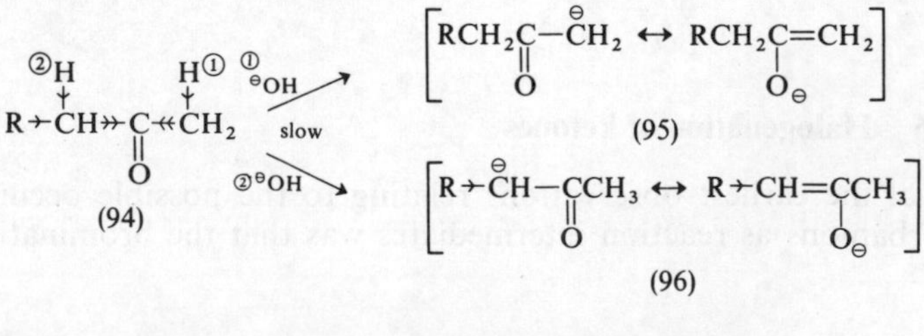

This effect is not very pronounced, however, as bromination of $MeCH_2COCH_3$ yields 1- and 3-bromobutanones in almost equal amount.

The answer to (b) also stems from a consideration of the alternative carbanions (98) and (99) obtainable from the mono-halogenated ketone (97):

$$\left[RCH_2\overset{\ominus}{C}-\overset{\ }{\underset{\underset{O}{\parallel}}{C}}H\rightarrow Br \leftrightarrow RCH_2C=CH\rightarrow Br \right]$$
(98)

$$\underset{(97)}{R\rightarrow \overset{\overset{\textcircled{2}H}{\mid}}{C}H\rightarrow \underset{\underset{O}{\parallel}}{C}\ll \overset{\overset{H\textcircled{1}}{\mid}}{C}H\rightarrow Br} \quad \begin{array}{c} \overset{\textcircled{1}\ \ominus OH}{\longrightarrow}\\ \text{slow} \\ \underset{\textcircled{2}\ \ominus OH}{\searrow} \end{array}$$

$$\left[R\rightarrow \overset{\ominus}{C}H-\underset{\underset{O}{\parallel}}{C}CH_2Br \leftrightarrow R\rightarrow CH=\underset{\underset{\ominus O}{\mid}}{C}CH_2Br \right]$$
(99)

The electron-withdrawing inductive effect of Br in (97) makes the α-H atoms on the CH_2Br α-carbon atom more acidic than those on the RCH_2 α-carbon, and will also serve to stabilise the carbanion (98) compared with (99); this differential will of course be reinforced by the effect of the R group. The former will thus be formed preferentially, and further halogenation is found to take place on the CH_2Br group, rather than on the RCH_2. Not only that, because of the electron-withdrawing inductive effect of the Br atom, (98) will be formed more readily than (95) was, i.e. the second halogenation will be easier than the first was. The third halogenation will be correspondingly easier still, and the normal product of base-catalysed halogenation is thus RCH_2COCX_3 (100). In CX_3 we now have a good leaving group, and addition of base to the C=O group of the ketone results in C—C bond fission (cf. p. 233):

$$\underset{(100)}{RCH_2\overset{\overset{O}{\parallel}}{\underset{\underset{\ominus OH}{}}{C}}\rightarrow CX_3} \rightleftarrows \underset{(101)}{RCH_2\overset{\overset{\ominus O}{\mid}}{\underset{\underset{OH}{\mid}}{C}}CX_3} \rightarrow \underset{(102)}{RCH_2\overset{\overset{O}{\parallel}}{\underset{\underset{OH}{\mid}}{C}} + \overset{\ominus}{C}X_3} \rightleftarrows \underset{(103)}{RCH_2\overset{\overset{O}{\parallel}}{\underset{\underset{O_\ominus}{}}{C}} + HCX_3}$$

CX_3 is a good leaving group because of the electron-withdrawing inductive effect of the three halogen atoms; this activates the carbonyl carbon atom in (100) to nucleophilic attack, and also stabilises the departing carbanion (101). The end-product, apart from the carboxylate anion (102), is the haloform (103), and the overall process— $RCH_2COCH_3 \rightarrow RCH_2CO_2^{\ominus} + HCX_3$—is known as the *haloform reaction*. It has been employed as a diagnostic test for methyl ketones, using I_2 and aqueous base as the resultant CHI_3 ('iodoform') is yellow, has a highly characteristic smell, and is insoluble in the reaction medium.

The halogenation of ketones is also catalysed by acids (general acid catalysis, *cf.* p. 73), the rate law observed is,

$$\text{Rate} = k[\text{ketone}][\text{acid}]$$

and, as with the base-catalysed reaction, the rates of bromination, iodination, deuterium exchange and racemisation are identical. This time the common intermediate, whose formation is slow and rate-limiting, is the enol (104):

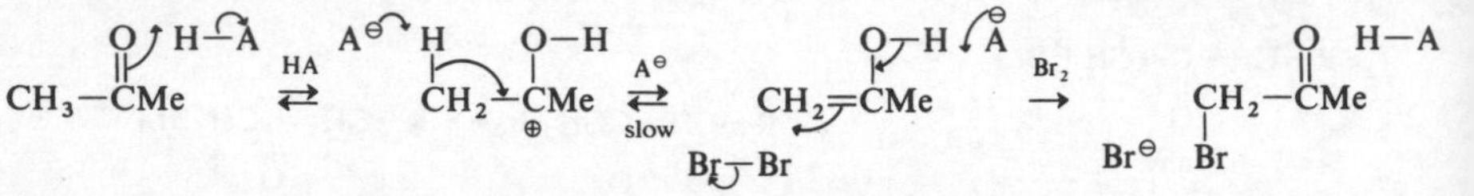

(104)

This then undergoes rapid, non rate-limiting attack by Br_2 or any other electrophile present.

To discover, (*a*) which of the groups of α-H atoms undergo preferential substitution in RCH_2COCH_3—and (*b*) which then undergoes further attack in an initial, mono-halogenated product—now requires a comparison of the formation of the relevant enols. For (*a*) this will be (105) and (106):

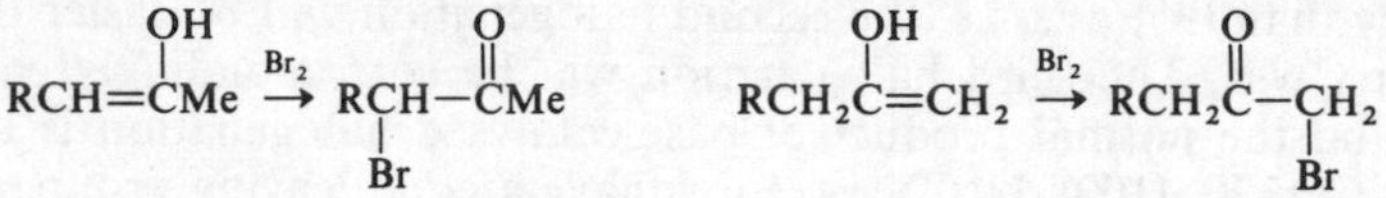

Of these (105) is likely to be more stable than (106) as it has the more heavily substituted double bond of the two (*cf.* p. 26); the favoured bromination product here is thus (107)—the reverse of what happened in the base-catalysed reaction—but again the preference is often not marked.

For (*b*) it is easier to understand what happens by considering propanone. The intermediate (and T.S.) involved in enol-formation (104) from CH_3COMe carries a +ve charge. With $BrCH_2COMe$ the analogous intermediate will be destabilised (relative to that from CH_3COMe) by the electron-withdrawing inductive effect of its bromine atom; as yet unchanged propanone will thus be converted to the enol, in preference to the mono-bromo product. It is thus possible to prepare and isolate the mono-bromo product, $MeCOCH_2Br$, from acid-catalysed bromination—this is not the case with base-catalysed bromination. Also unlike the base-catalysed reaction, further acid-catalysed bromination of CH_3COCH_2Br results in the formation of the 1,3-dibromoketone, $BrCH_2COCH_2Br$.

11
Radicals

11.1 INTRODUCTION

Most of the reactions that have been considered to-date have involved the participation of polar reactants and intermediates, i.e. carbonium ions and carbanions, or related highly polarised species, involving the *heterolytic* fission, and formation, of covalent bonds:

$$R_3C^{\ominus}: \quad X^{\oplus} \rightleftarrows R_3C-X \rightleftarrows R_3C^{\oplus} \quad :X^{\ominus}$$

But *homolytic* fission can also take place, thus generating species possessing an unpaired electron—radicals, e.g. (1) and (2):

$$R_3C-X \rightleftarrows R_3C\cdot \quad \cdot X$$
$$\text{(1)} \quad \text{(2)}$$

Reactions involving radicals occur widely in the gas phase: the combustion of any organic compound is nearly always a radical reaction, and the oxidative breakdown of alkanes in internal combustion engines is the largest scale, and most widespread, chemical reaction of all! Radical reactions also occur in solution, particularly if carried out in non-polar solvents, and if catalysed by light or the simultaneous decomposition of substances known to produce radicals themselves, e.g. organic peroxides. Radicals, once formed in solution, are generally found to be less selective in their attack on other species, or on alternative positions within the same species, than are carbonium ions or carbanions.

Another characteristic of many radical reactions is that, once initiated, they often proceed with great rapidity owing to the establishment of fast chain reactions of low energy requirement, e.g. in the halogenation of alkanes (3, *cf.* p. 314):

$$Br—Br$$
$$\downarrow hv$$
$$R—H + \cdot Br \rightarrow R\cdot + H—Br$$
$$(3) \quad \uparrow \quad \downarrow Br_2$$
$$\cdot Br + R—Br$$

In this case, the radical obtained photochemically, a bromine atom $Br\cdot$, generates another, $R\cdot$, on reaction with the neutral substrate, $R—H$ (3). This radical reacts in turn with a further neutral molecule, Br_2, generating $Br\cdot$ once again: the cycle thus proceeds without the need for further photochemical generation of $Br\cdot$, i.e. it is self-perpetuating. It is also characteristic of such radical reactions that they can be inhibited by the introduction of substances that themselves react particularly readily with radicals (*inhibitors*, or radical 'scavengers'), e.g. phenols, quinones, diphenylamine, iodine etc. These and similar substances can also be used to bring a radical reaction, already in progress, to a stop (*terminators*).

The first radicals to be studied were, hardly surprisingly, those that were somewhat less reactive, and thus capable of rather longer independent existence. The first such radical to be detected unequivocally was $Ph_3C\cdot$ (4), obtained in 1900 on reacting Ph_3CCl with finely divided silver (*cf.* p. 43). The radical reacted with halogens to reform the triphenylmethyl halide (5), or with oxygen from the air to form a peroxide (6):

$$Ph_3C\cdot + X—X \rightarrow Ph_3C—X + X\cdot \xrightarrow{Ph_3C\cdot} 2Ph_3C—X$$
$$(4) \qquad\qquad (5) \qquad\qquad (5)$$

$$Ph_3C\cdot + O_2 \rightarrow Ph_3COO\cdot \xrightarrow{Ph_3C\cdot} Ph_3COOCPh_3$$
$$(4) \qquad\qquad\qquad (6)$$

The yellow radical (4) was in equilibrium in solution in inert solvents with a colourless dimer, the proportion of radical increasing on dilution, and with rise of temperature. Thus a 3 % solution of dimer in benzene contained about 2 % of $Ph_3C\cdot$ at 20° and about 10 % at 80°; on removal of the solvent only the dimer was obtained. This was, not unnaturally, assumed to be hexaphenylethane, $Ph_3C—CPh_3$, and, as mentioned previously (p. 44), it was only 70 years later that the dimer was shown (by n.m.r. spectra) not to be this, but to have the structure (7):

$$Ph_3C—\text{〈 〉}=CPh_2$$
$$H$$

(7)

Hexaphenylethane has not, indeed, ever been prepared, and may well be not capable of existing under normal conditions due to the enormous steric crowding that would be present. The reasons for the relatively high stability of $Ph_3C\cdot$ are discussed below (p. 302).

Simple alkyl radicals are very much more reactive, and were first studied systematically only in 1929. The radicals were generated by the thermal decomposition of organo metallic compounds, such as $PbMe_4$,

$$PbMe_4 \rightleftarrows Pb + 4Me\cdot$$

and conveyed along a glass tube in a stream of an inert carrier gas, e.g. nitrogen. It was found that thin lead mirrors deposited at various distances along the inner wall of the tube were attacked by the stream of radicals. By measurements of how far along the tube mirrors continued to be attacked, coupled with a known rate of flow of carrier gas, it was possible to make accurate estimates of the half-life of alkyl radicals; for Me this was found to be 8×10^{-3} sec. The fate of such alkyl radicals, in the absence of metal mirrors to attack, is very largely dimerisation:

$$CH_3\cdot + \cdot CH_3 \rightarrow CH_3—CH_3$$

Once recognised in this way, alkyl radicals were invoked as intermediates in a number of reactions (see below).

Radicals, of varying degrees of stability, involving atoms other than carbon—*heteroradicals*—were also recognised. Thus it was discovered in 1911 that on warming N,N,N′,N′-tetraarylhydrazines, e.g. (8), in non-polar solvents resulted in the development of a green colour due to the radical (9):

$$2Ph_2NH \xrightarrow{MnO_4^{\ominus}} Ph_2N—NPh_2 \rightleftarrows Ph_2N\cdot + \cdot NPh_2$$
$$\quad\quad\quad\quad\quad\quad (8)\quad\quad\quad\quad\quad (9)$$

Another nitrogen radical, of considerable importance, is 1,1-diphenyl-2-picrylhydrazyl (10) obtained by PbO_2 oxidation of the triarylhydrazine (11):

$$Ph_2NNH_2 \xrightarrow[\text{chloride}]{\text{Picryl}} Ph_2NNH—\text{〈 〉}—NO_2 \xrightarrow{PbO_2} Ph_2N\overset{\cdot}{N}—\text{〈 〉}—NO_2$$

(11)   (10)

This is sufficiently stable (the reasons for its stability are discussed below, p. 303) to be recrystallised from various solvents, and obtained as violet prisms that may be kept more or less indefinitely. It is relatively unreactive towards other neutral molecules, but reacts readily with other radicals; it is indeed used as a 'trap', forming stable products, e.g. (12), with almost any other radical:

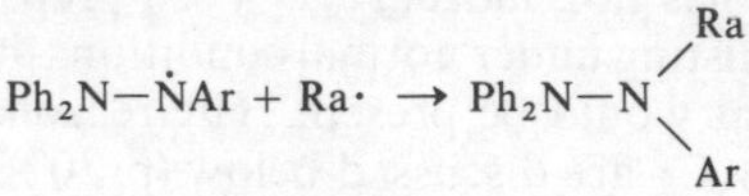

$$Ph_2N-\overset{\cdot}{N}Ar + Ra\cdot \;\rightarrow\; Ph_2N-N\overset{\displaystyle Ra}{\underset{\displaystyle Ar}{\Big\backslash}}$$

$$(10) \qquad\qquad\qquad (12)$$

As its solutions are highly coloured, its reaction with other radicals to form colourless products can be followed colorimetrically.

Solutions of diphenyl disulphide (13) are found to become yellow on heating, the colour disappearing again on cooling:

$$PhS-SPh \overset{\Delta}{\rightleftarrows} PhS\cdot + \cdot SPh$$

$$(13) \qquad\qquad (14)$$

The radicals (14) formed may be trapped with, for example, (10) above. Simple alkyl thiyl radicals such as $MeS\cdot$ have been detected as reaction intermediates; they are highly reactive. Relatively stable oxygen-containing radicals are also known. Thus the phenoxy radical (15),

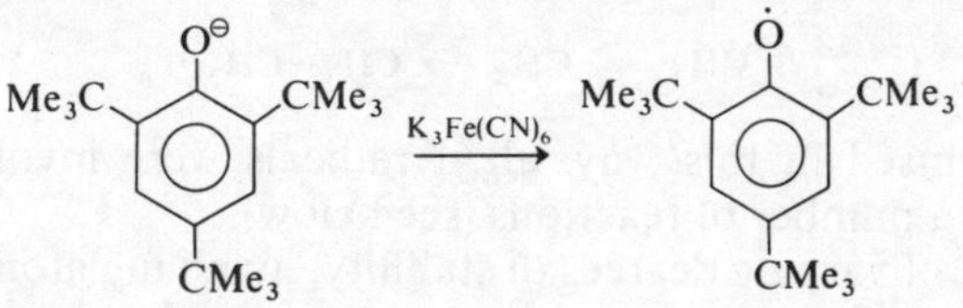

$$(15)$$

exists in this form, i.e. not as the dimer, both in solution and in the solid state; it is a dark blue solid (m.p. 97°). The reason for its relative unreactivity is almost certainly hindrance, by the bulky CMe_3 in both *o*-positions, to the approach of either another molecule of (15), or of other species, to the radical oxygen atom.

11.2   RADICAL FORMATION

There are a number of ways in which radicals may be generated from neutral molecules, several of which we have already seen; the most important are (*a*) photolysis, (*b*) thermolysis, and (*c*) redox reactions—by inorganic ions, metals or electrolysis—that involve one-electron transfers.

11.2.1 Photolysis

The prerequisite of this method is the ability of the molecule concerned to absorb radiation in the ultra-violet or visible range. Thus acetone in the vapour phase is decomposed by light having a wave-length of ≈ 320 nm (3200 Å);

$$\underset{(16)}{\overset{O}{\underset{\|}{Me-C-Me}}} \xrightarrow{h\nu} \underset{(16)}{Me\cdot} + \underset{(17)}{\overset{O}{\underset{\|}{\cdot C-Me}}} \rightarrow CO + \underset{(16)}{\cdot Me}$$

this happens because carbonyl compounds have an absorption band in this region. The photochemical decomposition yields the initial pair of radicals, (16) and (17), and the latter then breaks down spontaneously to yield another methyl radical and the stable species CO. Other species that undergo ready photolysis are alkyl hypochlorites (18) and nitrites (19), both of which can be used to generate alkoxyl radicals (20):

$$\underset{(18)}{RO-Cl} \xrightarrow{h\nu} \underset{(20)}{RO\cdot} + \cdot Cl$$

$$\underset{(19)}{RO-NO} \xrightarrow{h\nu} \underset{(20)}{RO\cdot} + \cdot NO$$

Another very useful photolytic homolysis is that of halogen molecules to yield atoms,

$$Cl-Cl \xrightarrow{h\nu} Cl\cdot + \cdot Cl$$

$$Br-Br \xrightarrow{h\nu} Br\cdot + \cdot Br$$

which can then initiate, for example, the halogenation of alkanes (p. 314), or addition to alkenes (p. 304).

The two major advantages of photolysis over thermolysis (see below) for the generation of radicals are: (*a*) it is possible to cleave strong bonds that do not break readily—or at all—at reasonable temperatures, e.g. azoalkanes (21),

$$\underset{(21)}{R-N=N-R} \xrightarrow{h\nu} R\cdot + N\equiv N + \cdot R$$

and (*b*) energy at only one particular level is transferred to a molecule, so that it is a more specific method of effecting homolysis than is pyrolysis. Thus the cleavage of diacyl peroxides, e.g. (22), occurs cleanly on photolysis,

$$RCO{-}OCR \xrightarrow{h\nu} 2R{-}\overset{O}{\overset{\|}{C}}{-}O\cdot \rightarrow 2R\cdot + 2CO_2$$

(22)

whereas in a number of cases thermolysis gives rise to other side reactions.

A very interesting technique for radical generation is *flash photolysis*, which employs a very intense pulse of radiation (visible or u.v.) of very short duration. This produces a very high immediate concentration of radicals, which may be detected—and whose fate may be followed—by spectroscopy through one or more subsequent pulses of lower intensity radiation of suitable wavelength. This is, of course, primarily a technique for the study of radicals rather than for their use in preparative procedures.

11.2.2   Thermolysis

Much of the early work on alkyl radicals of short life was, as we have seen (p. 293), carried out in the vapour phase through decomposition of metal alkyls, e.g. (23):

$$PbR_4 \rightleftarrows Pb + 4R\cdot$$

(23)

This stems from the weakness, i.e. ease of thermal fission, of the Pb—R bond, and radicals may be generated in solution in inert solvents, as well as in the vapour phase, through such thermolysis of weak enough bonds, e.g. those with a bond dissociation energy of $< \approx 165$ kJ (40 kcal) mol^{-1}. Such bonds very often involve elements other than carbon, and the major sources of radicals in solution are the thermolysis of suitable peroxides (O$\dotplus$O) and azo-compounds (C$\dotplus$N). Relatively vigorous conditions may, however, be necessary if the substrate does not contain substituents capable of stabilising the product radical, or promoting initial decomposition of the peroxide. Thus (Me$_3$CCOO)$_2$ has a half-life of ≈ 200 hr at 100°, while (PhCOO)$_2$ has one of only ≈ 0.5 hr at the same temperature. As was mentioned above, simple alkyl azo compounds, e.g. (21), are too stable to undergo thermolysis at reasonable temperatures, but can be made useful sources of radicals by the introduction of suitable substituents, e.g. (24):

$$\underset{\underset{CN}{|}}{Me_2C}{-}N{=}N{-}\underset{\underset{CN}{|}}{CMe_2} \xrightarrow{\Delta} 2[Me_2\dot{C}{-}C{\equiv}N \leftrightarrow Me_2C{=}C{=}\dot{N}] + N{\equiv}N$$

(24)

Thus $MeN=NMe$, despite the driving force supplied by $N\equiv N$ as among the best of all leaving groups, is stable up to $\approx 200°$, while (24) has a half-life of only ≈ 5 min at $100°$.

In the absence of other species with which a radical can react (e.g. abstraction of H from a suitable solvent), their life is terminated largely by dimerisation,

$$CH_3CH_2\cdot + \cdot CH_2CH_3 \rightarrow CH_3CH_2-CH_2CH_3$$

but also by disproportionation:

$$CH_3CH_2\cdot + H-CH_2CH_2\cdot \rightarrow CH_3CH_3 + CH_2=CH_2$$

The use of $PbEt_4$ as an anti-knock agent in petrol depends in part on the ability of the ethyl radicals, generated on its thermal decomposition, to combine with radicals produced in the over-rapid combustion of petroleum hydrocarbons; chain reactions which are building up to explosion (*knocking*) are thus terminated short of this. The complete details of how $PbEt_4$ operates are not known, but there is some evidence that minute PbO_2 particles derived from it can also act as 'chain-stoppers'.

Radical formation through carbon–carbon bond-fission is seen in the radical-induced 'cracking' at $\approx 600°$ of long-chain alkanes. The radicals introduced initially into the system probably act by abstracting a hydrogen atom from a CH_2 group of the chain; the resultant long chain, non-terminal radical (25) then undergoes fission β- to the radical carbon atom to yield a lower molecular weight alkene (26) plus a further radical (27) to maintain a chain reaction:

$$
\begin{array}{ccc}
Ra\cdot \quad H & Ra-H & \\
\overset{|}{RCH}-CH_2R' \rightarrow & R\dot{C}H-CH_2R' \rightarrow & RCH=CH_2 + \cdot R' \\
(25) & (26) & (27)
\end{array}
$$

Termination of the reaction by radical/radical interaction is unlikely to occur to any significant extent, until the concentration of long chain alkane has dropped to a very low level.

11.2.3 Redox reactions

These reactions all involve one-electron transfers in generating the radical, and it is therefore no surprise to find metal ions such as $Fe^{2\oplus}/Fe^{3\oplus}$ and $Cu^{\oplus}/Cu^{2\oplus}$ involved. Thus $Cu^{\oplus}$ ions are found to accelerate greatly the decomposition of acyl peroxides, e.g. (28):

$$
\left(\begin{matrix}O\\ \|\\ ArCO\end{matrix}\right)_2 + Cu^\oplus \rightarrow Ar\overset{\overset{O}{\|}}{C}-O\cdot + ArCO_2^\ominus + Cu^{2\oplus}
$$
$$
(28) \qquad\qquad\qquad (29)
$$

This constitutes a useful method for generating $ArCO_2\cdot$, as in the thermolysis of (28) there is a danger of the further decomposition of (29) to $Ar\cdot + CO_2$. $Cu^\oplus$ is also involved in the conversion of diazonium salts, $ArN_2{}^\oplus Cl^\ominus$, to $ArCl + N_2$ (Sandmeyer reaction), where $Ar\cdot$ is very probably formed transiently as an intermediate:

$$ArN_2{}^\oplus + Cu^\oplus \rightarrow Ar\cdot + N_2 + Cu^{2\oplus}$$

Both of these reactions are reductions, another is the use of $Fe^{2\oplus}$ to catalyse the oxidation reactions of aqueous hydrogen peroxide solution:

$$H_2O_2 + Fe^{2\oplus} \rightarrow HO\cdot + {}^\ominus OH + Fe^{3\oplus}$$

The mixture is known as Fenton's reagent, and the effective oxidising agent in the system is the hydroxyl radical, $HO\cdot$. This is particularly good as an abstractor of H, and can be used either to generate the resultant radical, e.g. (30), for further study, or, in some cases preparatively through the latter's dimerisation, e.g. (31):

$$HO\cdot + H{-}CH_2CMe_2OH \rightarrow H_2O + \cdot CH_2CMe_2OH \rightarrow HOCMe_2CH_2CH_2CMe_2OH$$
$$\qquad\qquad\qquad (30) \qquad\qquad\qquad\qquad\qquad (31)$$

Generation of a radical through an oxidative process probably occurs in the initiation of the autoxidation of benzaldehyde (p. 319), which is catalysed by a number of heavy metal ions capable of one-electron transfers, e.g. $Fe^{3\oplus}$:

$$\overset{O}{\overset{\|}{PhC}}{-}H + Fe^{3\oplus} \rightarrow \overset{O}{\overset{\|}{PhC}}\cdot + H^\oplus + Fe^{2\oplus}$$

We have already seen (p. 294) the generation of a stable phenoxy radical (32) through a one-electron oxidation by $Fe(CN)_6{}^{3\ominus}$,

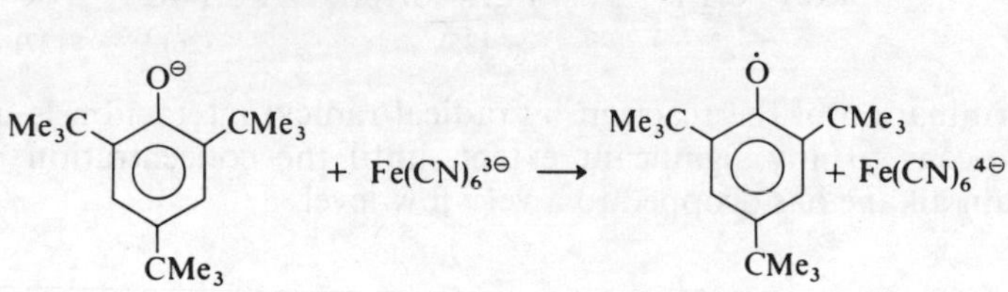

$$(32)$$

and also the dimeric oxidation of carbanions, e.g. (33), with iodine (p. 287):

$$2(MeCO)_2CH^\ominus \overset{I_2}{\rightarrow} 2(MeCO)_2CH\cdot \rightarrow (MeCO)_2CH{-}CH(COMe)_2$$
$$(33)$$

Radicals, (34), that subsequently dimerise, are also obtained through the anodic oxidation of carboxylate anions, $RCO_2{}^\ominus$, in the Kolbe electrolytic synthesis of hydrocarbons:

$$2RCO_2^{\ominus} \xrightarrow[\text{anode}]{-e^{\ominus}} 2RCO_2 \cdot \xrightarrow{-CO_2} 2R \cdot \rightarrow R-R$$

(34)

Conversely, electrolysis of ketones, (35), results in their cathodic reduction to radical anions (36), which dimerise to the dianions of pinacols (37):

$$2R_2C{=}O \xrightarrow[\text{cathode}]{+e^{\ominus}} 2R_2C{-}O^{\ominus} \rightarrow \begin{matrix} R_2C{-}O^{\ominus} \\ | \\ R_2C{-}O^{\ominus} \end{matrix} \xrightarrow{H^{\oplus}} \begin{matrix} R_2C{-}OH \\ | \\ R_2C{-}OH \end{matrix}$$

(35)　　　　　(36)　　　　　　(37)

We have seen similar radical anions generated from ketones in pinacol reduction with sodium or magnesium (p. 214), and also from esters with sodium in the acyloin condensation (p. 214).

11.3 RADICAL DETECTION

We have already seen how the high chemical reactivity of short-lived radicals can be enlisted to aid in their detection through their ability to etch metal mirrors (p. 293). The fact that the transition of an unpaired electron between the energy levels of a radical involves less energy than the transition of the paired electrons in the stable parent molecule means that the radical tends to absorb at longer wavelength. A number of radicals are thus coloured—where their precursors are not—and may readily be detected in this way, e.g. (11, p. 293) and (15, p. 294). Radicals may also be detected by their rapid discharge of the colour of solutions containing species such as 1,1-diphenyl-2-picrylhydrazyl (11).

Another useful, and quite sensitive, test is the initiation of polymerisation (*cf.* p. 311). Polymerisation can be initiated, in suitable substrates, by cations and anions as well as by radicals, but the effect of these several species can be differentiated by using a 50/50 mixture of phenylethene (styrene), $PhCH{=}CH_2$, and methyl 2-methyl-propenoate (methyl methacrylate), $CH_2{=}C(Me)CO_2Me$, as substrate: cationic initiators are found to produce polystyrene only, anions polymethyl methacrylate only, while radicals produce a copolymer containing equal amounts of the two monomers.

By far the most useful method for detecting radicals is, however, electron spin resonance (e.s.r.) spectroscopy, which utilises the permanent magnetic moment conferred on a radical by virtue of the spin of its unpaired electron (radicals are *paramagnetic*, species containing only electron pairs are *diamagnetic*). The electron spin can have one of two values ($+\frac{1}{2}$ or $-\frac{1}{2}$, *cf.* p. 2) and, in the presence of an applied magnetic field, these correspond to different energy levels;

transitions are possible between them resulting in a characteristic, and detectable, absorption spectrum. E.s.r. spectroscopy of unpaired electrons is thus the analogue of n.m.r. spectroscopy of nuclei that have a permanent magnetic moment, e.g. ^1H, ^{13}C etc; hardly surprisingly, they occur in different energy ranges (an unpaired electron has a much larger magnetic moment than a proton—^1H—and more energy is required to reverse its spin).

In e.s.r. spectroscopy, interaction ('*splitting*') occurs between the unpaired electron and neighbouring magnetic nuclei—especially ^1H—leading to quite complex patterns of lines; analysis of these can provide a great deal of detailed information about the structure and shape of a radical. Thus hydrogen abstraction from cycloheptatriene (38) by ·OH is found to lead to a radical having a very simple e.s.r. spectrum: eight equally spaced lines, indicating interaction of the unpaired electron with seven equivalent ^1H nuclei. The product radical thus cannot have the expected structure (39)—which would have a very much more complex e.s.r. spectrum—but must be the delocalised species (40, *cf.* p. 105):

(38)   (39)   (40)

Radicals have been detected by e.s.r. spectroscopy, under the most favourable conditions, in concentrations as low as 10^{-10} M. Sometimes this is achieved by generating the radical—from its precursors, or as a reaction intermediate—just outside the instrument, and then using a continuous flow technique to maintain a 'standing' concentration in the cavity of the spectrometer. A disadvantage of this method is that it requires relatively large volumes, and quantities of starting material. The longer the life of the radical the greater the chance of observing its spectrum; thus species such as $Ph_3C\cdot$ are easily observed, but species like $Ph\cdot$, $PhCH_2\cdot$, $C_2H_5\cdot$ etc, are a little more difficult. A technique that has been used to 'prolong' the life of short-lived species is to introduce a suitable diamagnetic substance, e.g. (41), which will react with the transient radical, and convert it into a longer-lived radical (42) that can be detected quite readily:

$$Ra\cdot + Me_3C-N=O \rightarrow Me_3C-\overset{\displaystyle Ra}{\underset{\displaystyle |}{N}}-O\cdot$$

(41)   (42)

This is known as *'spin trapping'*. Another technique, that has been used to study very short-lived radicals, is to generate them photo-lytically, from precursors, in a solid inert matrix, e.g. frozen argon. Their life is thus artificially prolonged because they are shielded from collision either with each other, or with other species that could terminate their existence.

11.4   RADICAL SHAPE AND STABILISATION

As with carbonium ions (p. 103) and carbanions (p. 269), the question arises of whether simple radicals—of the type $R_3C\cdot$—accommodate their unpaired electron in a p orbital (planar shape, 43) or an sp^3 hybrid orbital (pyramidal shape, 44),

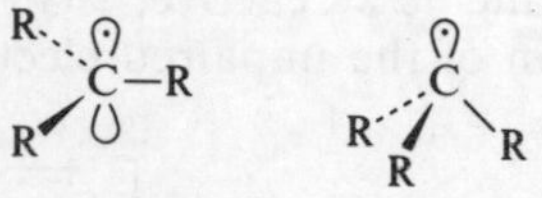

(43)   (44)

or whether the shape is somewhere between the two. Direct physical evidence for $CH_3\cdot$ comes from the e.s.r. spectrum of $^{13}CH_3\cdot$. Analysis of the lines, resulting from interaction between the unpaired electron and the paramagnetic ^{13}C nucleus, provides information about the degree of s character of the orbital in which the unpaired electron is accommodated. That in $^{13}CH_3\cdot$ is found to have little or none, and the radical is thus essentially (within $\approx 5\%$) planar, i.e. (43, R = H); a conclusion that is supported by evidence derived from u.v. and i.r. spectra. The s character of the half-filled orbital is found to increase across the series,

$$CH_3\cdot < CH_2F\cdot < CHF_2\cdot < CF_3\cdot$$

however, being essentially sp^3 in $CF_3\cdot$, i.e. the latter radical is thus pyramidal (44, R = F); the radicals $\cdot CH_2OH$ and $\cdot CMe_2OH$ are also substantially 'bent'. Comparison of the ease of formation, and reactivity once formed, of bridged radicals—such as (45) and (46)—

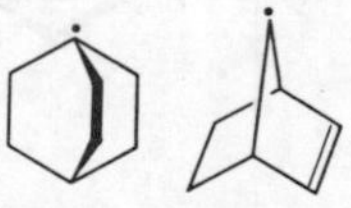

(45)   (46)

with their acylic equivalents would suggest that alkyl radicals do exhibit some preference for the planar state. This is nothing like so marked as with carbonium ions, however, and, unlike the latter (p. 86), there is little difficulty in generating radicals at bridge-head positions.

The relative stability of simple alkyl radicals is found to follow the sequence:

$$R_3C\cdot > R_2CH\cdot > RCH_2\cdot > CH_3\cdot$$

This reflects the relative ease with which the C—H bond in the alkane precursor will undergo homolytic fission, and more particularly, decreasing stabilisation, by hyperconjugation or other means, as the series is traversed. There will also be decreasing relief of strain (when R is large) on going from sp^3 hybridised precursor to essentially sp^2 hybridised radical, as the series is traversed. The relative difference in stability is, however, very much less than with the corresponding carbonium ions.

Radicals of allylic, $RCH=CHCH_2\cdot$ (47), and benzylic, $Ph\dot{C}HR$ (48), type are more stable, and less reactive, than simple alkyl radicals, because of delocalisation of the unpaired electron over the π orbital system in each case:

$$[RCH\!\cdots\!CH\!\cdots\!CH_2]\cdot$$
(47)

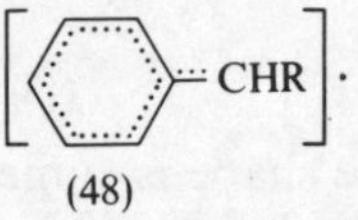

(48)

Both are essentially planar, i.e. sp^2 hybridised, at the radical carbon atom for only in this configuration is maximum p/π orbital overlap—with consequent stabilisation—possible. The stability of a radical increases as the extent of potential delocalisation increases; thus $Ph_2CH\cdot$ is more stable than $PhCH_2\cdot$, and $Ph_3C\cdot$ (*cf.* p. 292) is a pretty stable radical.

The shape of $Ph_3C\cdot$ (49) is a matter of some interest as it has a bearing on the extent to which delocalisation of the unpaired electron, with consequent stabilisation, can occur. The radical carbon atom is certainly sp^2 hybridised in (49), i.e. the bonds joining it to the three benzene nuclei all lie in the same plane; but maximum stabilisation will only occur if all three benzene nuclei can be simultaneously co-planar (49*a*),

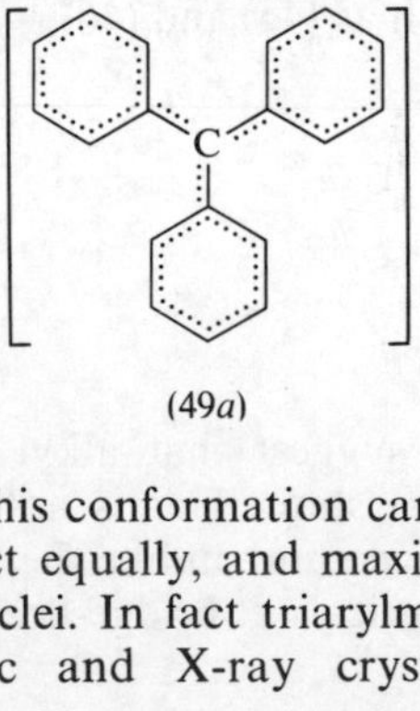

(49*a*) 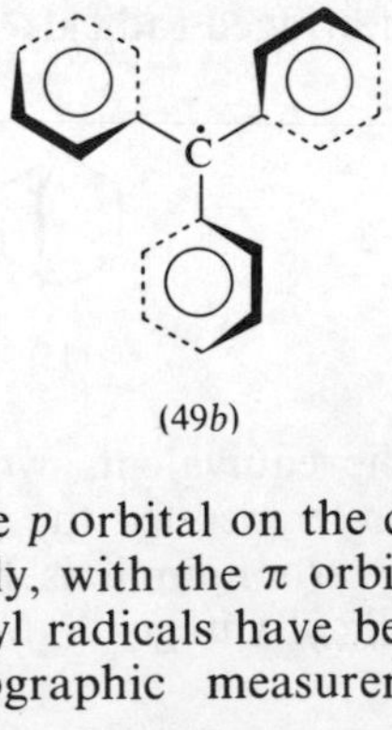   (49*b*)

for only in this conformation can the p orbital on the central carbon atom interact equally, and maximally, with the π orbital systems of the three nuclei. In fact triarylmethyl radicals have been shown, by spectroscopic and X-ray crystallographic measurements, to be propeller-shaped (49b), the benzene rings being angled at about 30° out of the common plane. Thus though delocalisation occurs in (49)— as indicated by its e.s.r. spectrum—it is not maximal, and its extent is not enormously greater in $Ph_3C\cdot$ than in $Ph_2CH\cdot$, or even in $PhCH_2\cdot$.

The benzene rings are forced out of the co-planar conformation by steric interaction of the o-H atoms of adjacent rings with each other; as would be expected, o-substituents bulkier than H are found to increase the out of plane dihedral angle of the aromatic nuclei—to 50° or more. Delocalisation must then be even further decreased, but such radicals with bulky o-substituents are nevertheless found to be more stable, i.e. more reluctant to form their dimers than is $Ph_3C\cdot$ itself. This must, of course, be due to a steric effect—the o-substituents are very close to the radical carbon atom and are thus capable of preventing its access to other species, or other species access to it [cf. (15), p. 294]. It is significant, in the light of what has been said above, that their effectiveness at 'masking' the radical carbon atom will increase the more the benzene rings are angled out of the coplanar conformation, i.e. the greater the dihedral angle.

If each aromatic nucleus in the radical has a bulky p-substituent, e.g. (50), then, irrespective of any substitution at the o-positions, dimerisation will be greatly inhibited, or even prevented [cf. (7), p. 293]:

$$(p\text{-}RC_6H_4)_3C\cdot \ + \ R\!-\!\langle\ \ \rangle\!= C(C_6H_4R\text{-}p)_2 \ \nrightarrow \ \text{Dimer}$$

(50a)   (50b)

The hetero radicals that have already been referred to—(9, p. 293), (10, p. 293), (14, p. 294) and (15, p. 294)—owe their relative stability [with respect to their dimers—apart from 1,1-diphenyl-2-picrylhydrazyl (10)] to a variety of factors: (a) the relative weakness of N—N, S—S and O—O bonds, (b) the delocalisation through the agency of aromatic nuclei, and (c) steric inhibition of access to the atom with the unpaired electron, or to an aryl p-position, cf. (50). The latter factor bulks large (in addition to the weakness of O—O bonds) in the great stability of (15, cf. p. 294); and all three factors operate to stabilise (51), which is wholly dissociated in solution:

(51)

This radical has been shown, from calculations based on e.s.r. spectra, to have the *p*-phenyl group co-planar with the central phenoxy nucleus, but the two *o*-phenyl groups angled at 46° to it. The *p*-group can thus effect maximum delocalisation—(*b*)—and also act as a bulky group to inhibit dimerisation—(*c*), *cf.* (50) above, while the two angled *o*-substituents inhibit access to the O atom, preventing formation of an O—O dimer [dimerisation does occur in the solid state, but it is then through one *p*-position, *cf.* (7, p. 293)].

11.5 RADICAL REACTIONS

It is possible, and logical, to classify the multifarious reactions of radicals from the point of view of the radical itself: (*a*) unimolecular reactions, e.g. fragmentation, rearrangement; (*b*) bimolecular reactions between radicals, e.g. dimerisation, disproportionation; and (*c*) bimolecular reactions between radicals and molecules, e.g. addition, displacement, atom (often H) abstraction. Such a grouping has, for our purpose, the disadvantage of fitting much less well into the general classification of reaction types that has been adopted throughout. We shall therefore discuss the reactions in which radicals are involved, either as reactants or intermediates, under the general heads of addition, displacement and rearrangement.

11.5.1 Addition

Additions to C=C are almost certainly the most important group of reactions involving radicals. This is due largely to the importance of addition (vinyl) polymerisation (p. 311), and the consequent extent to which its mechanism has been investigated; but addition of halogens and of halogen hydracids are also of significance.

11.5.1.1 Halogens

In addition to the polar mechanism already considered (p. 176), halogen addition to alkenes can proceed *via* radical intermediates. The former is favoured by polar solvents and by the presence of Lewis acid catalysts, the latter by non-polar solvents (or in the gas phase), by sunlight or u.v. irradiation, and by the addition of radical precursors (*initiators*) as catalysts. An example is the photochemically catalysed addition of chlorine to tetrachloroethene (52), which involves a chain reaction (*cf.* p. 292):

$$Cl-Cl$$
$$\downarrow h\nu$$
$$Cl_2C=CCl_2 + \cdot Cl \rightarrow Cl_2\overset{\cdot}{C}-CCl_3$$
$$(52) \qquad\qquad (53)$$

$$\cdot Cl + Cl_3C-CCl_3$$
$$(54)$$

Each molecule of chlorine, on photochemical fission, will give rise to two chlorine atoms, i.e. radicals, each of which is capable of initiating a continuing reaction chain. That each quantum of energy absorbed does indeed lead to the initiation of two reaction chains is confirmed by the observation that:

$$\text{Rate} \propto \sqrt{\text{Intensity of absorbed light}}$$

Chlorine atoms are electrophilic (the element is electronegative, and Cl· will readily take up an electron to complete its octet) and thus add readily to the double bond of (52) to yield the radical (53). This, in turn, can abstract a chlorine atom from a second molecule (the process can equally well be regarded as a radical displacement reaction on Cl—Cl) to yield the end-product of addition (54), plus a further atom of chlorine to continue the reaction chain, i.e. a very fast, continuing chain reaction is set up by each chlorine atom initiator generated photochemically. Each quantum of energy absorbed is found to lead to the conversion of several thousand molecules of (52) into (54); the reaction chains are, in this case, said to be long. Until the later stages of the reaction, when nearly all of (52) and Cl_2 have been used up, the concentrations of (53) and of Cl· will be very small compared with those of the starting materials; collision of a radical with a molecule will thus be very much more frequent than collision of a radical with another radical. Chain termination will ultimately take place through radical/radical collision, however, and this is generally found to involve (53) + (53) → (55):

$$Cl_3C\overset{Cl}{\underset{Cl}{\overset{|}{\underset{|}{C}}}}\cdot + \cdot\overset{Cl}{\underset{Cl}{\overset{|}{\underset{|}{C}}}}CCl_3 \rightarrow Cl_3C\overset{Cl}{\underset{Cl}{\overset{|}{\underset{|}{C}}}}-\overset{Cl}{\underset{Cl}{\overset{|}{\underset{|}{C}}}}CCl_3$$
$$(53) \qquad (53) \qquad\qquad (55)$$

The reaction is found to be inhibited by the presence of oxygen; this is because the molecule of oxygen has two unpaired electrons, and behaves as a diradical (*cf.* p. 325), $\cdot O\!-\!O\cdot$, albeit a not very reactive one. It can, however, combine with the highly reactive radical inter-mediates in the above addition, converting them into the very much less reactive peroxy radicals, $RaO\!-\!O\cdot$, which are unable to carry on the chain: it is thus a highly efficient inhibitor. That oxygen is reacting largely with the pentachloroethyl radicals (53) is shown by the forma-tion of (56),

$$\underset{(53)}{\overset{\displaystyle Cl}{\underset{\displaystyle Cl}{Cl_3C\!-\!\overset{\displaystyle |}{\underset{\displaystyle |}{C}}\cdot}}} \xrightarrow{\cdot O_2\cdot} \underset{(56)}{\overset{\displaystyle Cl}{Cl_3C\!-\!\overset{\displaystyle |}{C}\!=\!O}}$$

when the normal addition reaction is inhibited by oxygen.

The reactivity sequence for homolytic addition of the different halogens to alkenes is, hardly surprisingly, the same as that for electro-philic addition, i.e. $F_2 > Cl_2 > Br_2 > I_2$. The addition of fluorine—not requiring photochemical or other activation—is too vigorous to be of much use, and side reactions are common. Chlorination is generally rapid, with long reaction chains, and not readily reversible except at temperatures $>200°$; as the temperature rises, however, there is an increasing tendency to hydrogen abstraction leading to overall substitution by chlorine—rather than addition—in suitable cases (*cf.* p. 315). Bromination occurs readily, but with somewhat shorter reaction chains, and is usually reversible, while iodination takes place with difficulty, if at all, and is very readily reversible. The effect of increasing alkyl substitution at the double bond carbon atoms is found to have relatively little effect on the rate of halogen addition, certainly a good deal less than for addition by the polar mechanism (p. 179). Halogen substitution, e.g. by Cl, on the double bond carbon atoms results in a decreased reaction rate, e.g. $Cl_2C\!=\!CCl_2$ adds chlorine much more slowly than $CH_2\!=\!CH_2$.

The reversibility of addition of Br_2 and I_2—particularly the latter—has been made use of in the isomerisation (of the less to the more stable) of a pair of doubly bonded geometrical isomerides: in simple cases the *cis* to the *trans* e.g. (57) → (58). This may be carried out by u.v. irridation in the presence of catalytic quantities of Br_2 or I_2:

Normally, of course, an equilibrium mixture will be produced with the more stable form preponderating. That the interconversion does proceed, as above, *via* addition and elimination of Br· has been shown by using radioactive Br$_2$ as catalyst : both (58) and (57), in the equilibrium mixture obtained, are then found to contain radioactive Br.

The addition of chlorine or bromine to benzene—one of the few overall addition reactions of a simple benzene nucleus—has also been shown to proceed *via* a radical pathway, i.e. it is catalysed by light and by the addition of peroxides, and is slowed or prevented by the usual inhibitors. With chlorine this presumably proceeds ;

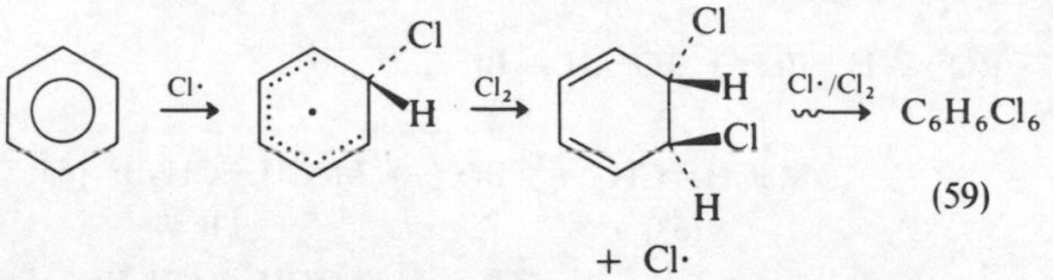

(59)

the product is a mixture of several of the eight possible geometrical isomerides of hexachlorocyclohexane (59). In the absence of light or peroxides no reaction takes place, while in the presence of Lewis acids overall electrophilic substitution takes place by an addition/elimination pathway (p. 137). With radicals other than Cl·, e.g. Ph·, overall homolytic substitution can be made to take place on benzene by an addition/elimination pathway too (p. 321).

Radical attack on methylbenzene (toluene, 60) results in preferential hydrogen abstraction by Cl· leading to overall substitution in the CH$_3$ group, rather than addition to the nucleus. This reflects the greater stability of the first formed (delocalised) benzyl radical, PhCH$_2$· (61), rather than the hexadienyl radical (62), in which the aromatic stabilisation of the starting material has been lost :

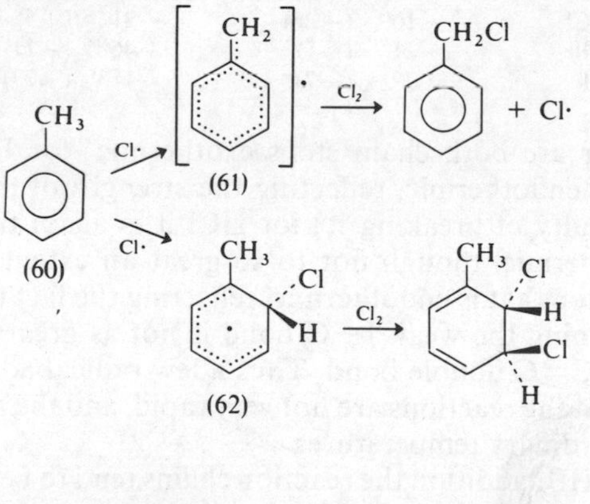

11.5.1.2 Hydrogen bromide

The addition of HBr to propene, MeCH=CH$_2$ (63), under polar conditions to yield 2-bromopropane, has already been referred to

(p. 181). In the presence of peroxides (or under other conditions that promote radical formation), however, the addition proceeds *via* a rapid chain reaction to yield 1-bromopropane (64); this is generally referred to as the *peroxide effect* leading to anti-Markownikov addition. This difference in orientation of HBr addition is due to the fact that in the first (polar) case it is initiated by $H^{\oplus}$ and proceeds *via* the more stable (secondary) carbonium ion, while in the second (radical) case it is initiated by Br· and proceeds *via* the more stable (secondary) radical (65):

$$RO\cdot + H—Br \rightarrow RO—H + Br\cdot$$
$$\downarrow$$

$$MeCH{=}CH_2 + \quad Br\cdot \rightarrow Me\overset{\cdot}{C}H—CH_2Br \quad (65)$$
$$(63) \qquad\qquad \uparrow \qquad\qquad\quad \downarrow_{H—Br}$$
$$Br\cdot \; + \; MeCH_2—CH_2Br$$
$$(64)$$

The initiation is by Br·, as hydrogen abstraction by RO· from HBr (as above) is energetically much more favourable than the alternative of bromine abstraction to form ROBr + H·. The alternative addition of Br· to (63) to form $MeCH(Br)CH_2\cdot$ (66) does not occur, as secondary radicals, e.g. (65), are more stable (*cf.* p. 302) than primary, e.g. (66).

HBr is the only one of the four hydrogen halides that will add readily to alkenes *via* a radical pathway. The reason for this is reflected in the ΔH values—in kJ (kcal) mol^{-1}—below for the two steps of the chain reaction for addition of HX to $CH_2{=}CH_2$, for example:

	(1) X· + $CH_2{=}CH_2$		(2) $XCH_2CH_2\cdot$ + HX	
H—F	-188	(-45)	$+155$	$(+37)$
H—Cl	-109	(-26)	$+21$	$(+5)$
H—Br	-21	(-5)	-46	(-11)
H—I	$+29$	$(+7)$	-113	(-27)

Only for HBr are both chain steps exothermic; for HF the second step is highly endothermic, reflecting the strength of the H—F bond and the difficulty of breaking it; for HCl it is again the second step that is endothermic, though not to so great an extent; while for HI it is the first step that is endothermic, reflecting the fact that the energy gained in forming the weak I—C bond is not as great as that lost in breaking the C=C double bond. Thus a few radical additions of HCl are known, but the reactions are not very rapid, and the reaction chains are short at ordinary temperatures.

Even with HBr addition the reaction chains tend to be rather short—much shorter than those in halogen addition—and more than a trace of peroxide is thus needed to provide sufficient initiator radicals: for preparative purposes up to 0·01 mol peroxide per mol of alkene. In practical terms, however, there may already be sufficient peroxide present in the alkene—through its autoxidation (p. 319) by the oxygen in the air—to auto-initiate the radical pathway of HBr addition, whether this is wanted or not. Once initiated, reaction by this pathway is very much faster than any competing addition *via* the polar pathway, and the anti-Markownikov product, e.g. (64), will thus predominate. If the Markownikov product, e.g. MeCH(Br)CH$_3$ from propene, is required it is necessary either to purify the alkene rigorously before use, or to add *inhibitors* (good radical acceptors such as phenols, quinones etc.) to mop up any radicals, or potential radicals, present: the latter is much the easier to do preparatively. Essentially complete control of orientation of HBr addition, in either direction, can thus be achieved, under preparative conditions, by incorporating either peroxides (radical initiators) or radical inhibitors in the reaction mixture. This is particularly useful as such control is not confined purely to alkenes themselves: CH$_2$=CHCH$_2$Br, for example, can be converted into 1,2- or 1,3-dibromopropane at will.

In any consideration of stereoselectivity in radical addition to acylic substrates, interpretation of the results is complicated by the knowledge that alkenes may be converted, at least in part, into their geometrical isomerides by traces of bromine (or of HBr, i.e. by Br·, *cf.* p. 306). This may, however, be minimised by working at low temperatures, and by using a high concentration of HBr. Thus addition of liquid HBr at −80° to *cis* 2-bromobut-2-ene (67) was found to proceed with high TRANS stereoselectivity, and to yield (68) almost exclusively:

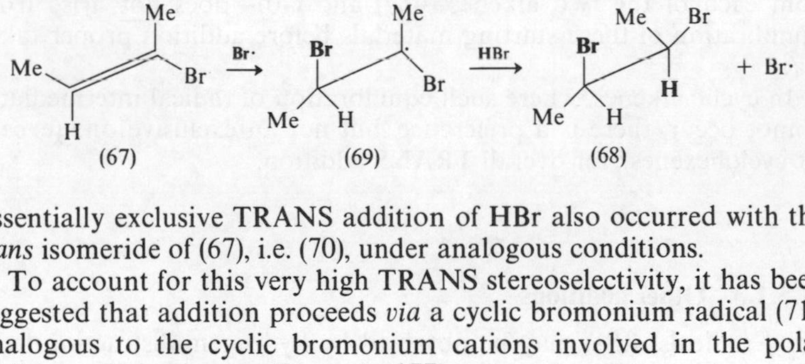

Essentially exclusive TRANS addition of HBr also occurred with the *trans* isomeride of (67), i.e. (70), under analogous conditions.

To account for this very high TRANS stereoselectivity, it has been suggested that addition proceeds *via* a cyclic bromonium radical (71), analogous to the cyclic bromonium cations involved in the polar addition of bromine to alkenes (p. 177):

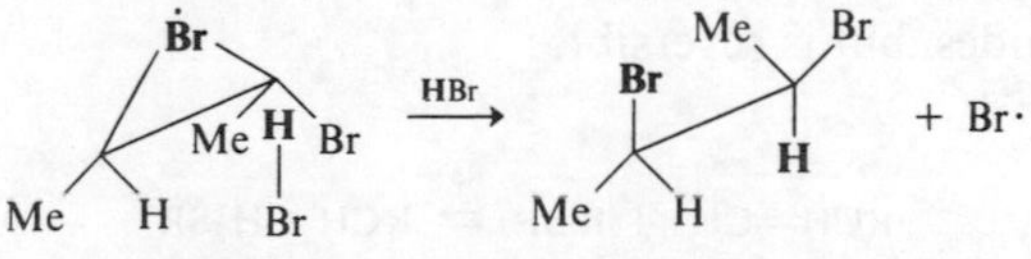

The overall addition would then be completed through attack by HBr from the less hindered side (away from bridging Br) to yield the product of overall TRANS addition (68). However, addition of HBr to (67) at room temperature, and using a lower concentration of HBr, is found to result in a 78 % : 22 % mixture of the products arising from TRANS and CIS stereoselective addition, (68) and (72), respectively. Significantly, the same mixture of products, in the same proportion, is obtained under these conditions from the *trans* isomeride of (67), namely (70). This strongly suggests that there is time for conformational equilibrium to be established between the respective radical inter-mediates—(69) and (73)—before H is abstracted from HBr to complete the overall addition:

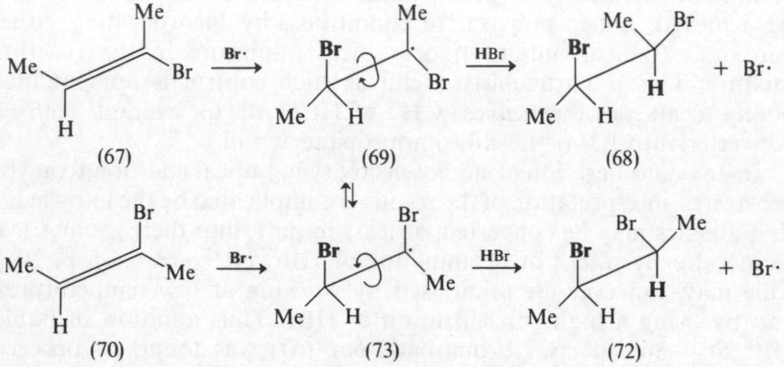

There are reasons for believing that this common product mixture from each of the two alkenes—(67) and (70)—does <u>not</u> arise from equilibration of these starting materials before addition proper takes place.

In cyclic alkenes, where such equilibration of radical intermediates cannot occur, there is a preference, but not an exclusive one (except for cyclohexenes), for overall TRANS addition.

11.5.1.3   Other additions

Thiyl radicals, RS·, may be obtained by hydrogen abstraction from RSH, and will then add readily to alkenes by a chain reaction analogous to that for HBr. The addition is of preparative value for making dialkyl sulphides, but is reversible:

$$RCH{=}CH_2 + R'SH \overset{Ra\cdot}{\rightleftarrows} RCH_2CH_2SR'$$

Sulphenyl chlorides, e.g. Cl_3CSCl, can also be used as sources of thiyl radicals, but here the addition is initiated by Cl· and the R'S will thus become attached to the other carbon atom of the double bond:

$$RCH{=}CH_2 \xrightarrow{Cl\cdot} R\dot{C}H{-}CH_2Cl \xrightarrow{Cl_3CSCl} R\underset{\underset{SCCl_3}{|}}{C}HCH_2Cl + Cl\cdot$$

Carbon–carbon bonds may be formed by the addition, among other things, of halomethyl radicals to alkenes. The $\cdot CX_3$ (X = Br, Cl) may be generated by the action of peroxides on, or by photolysis of, CX_4:

$$RCH{=}CH_2 \xrightarrow{\cdot CCl_3} R\dot{C}HCH_2CCl_3 \xrightarrow{CCl_4} R\underset{\underset{Cl}{|}}{C}HCH_2CCl_3 + \cdot CCl_3$$

$$\text{(74)} \qquad\qquad\qquad \text{(75)}$$

That the relatively inert CCl_4 adds in this way may seem a little surprising, but the ΔH values for both steps of the reaction chain are exothermic: $-75(-18)$, and $-17(-4)\,kJ\,(kcal)\,mol^{-1}$. The first formed radical (74) may, however, compete with $\cdot CCl_3$ in adding to $RCH{=}CH_2$, so that low molecular weight polymers are formed under some conditions, as well as the normal addition product (75).

11.5.1.4 Vinyl polymerisation

This reaction has been the subject of a great deal of theoretical and mechanistic study, largely because of the commercial importance of the polymers to which it can give rise. Like the other radical reactions we have discussed, it can be said to involve three stages—(a) initiation, (b) propagation, and (c) termination:

(a) *Initiation*:
  Formation of initiator from peroxides or azo compounds.
(b) *Propagation*:
  $$Ra\cdot + CH_2{=}CH_2 \rightarrow RaCH_2CH_2\cdot \xrightarrow[]{(n-1)CH_2{=}CH_2} Ra(CH_2)\cdot_{2n}$$
(c) *Termination*:
  (i) $Ra(CH_2)\cdot_{2n} + \cdot Ra \rightarrow Ra(CH_2)_{2n}Ra$
  (ii) $Ra(CH_2)\cdot_{2n} + \cdot(CH_2)_{2n}Ra \rightarrow Ra(CH_2)_{4n}Ra$
  (iii) $Ra(CH_2)_xCH_2\cdot + \cdot CH_2CH_2(CH_2)_yRa \rightarrow Ra(CH_2)_xCH_3 + CH_2{=}CH(CH_2)_yRa$

The propagation step is usually very rapid.

As the alkene monomers can absorb oxygen from the air, forming peroxides (*cf.* p. 319) whose ready decomposition can effect auto-initiation of polymerisation, it is usual to add a small quantity of inhibitor, e.g. quinone, to stabilise the monomer during storage. When subsequent polymerisation is carried out, sufficient radical initiator must therefore be added to 'saturate' the inhibitor before any polymerisation can be initiated; an induction period is thus often observed.

The radical initiators are not, strictly speaking, catalysts—though often referred to as such—for each radical that initiates a polymer chain becomes irreversibly attached to it and, if of suitable composition, may be detected in the molecules of product. The efficiency of some initiators may be so great that, after any induction period, every radical generated leads to a polymer chain.

Termination of a growing chain can result from collision with either an initiator radical (*c* i) or with another growing chain (*c* ii), but of these the latter is much the more frequent, as the initiator radicals will have been largely used up in starting the chains. Termination has been shown above only as dimerisation, but it can also involve disproportionation (*cf.* p. 297) between growing chains (*c* iii). H-abstraction can also occur by attack of a growing chain on 'dead' (no longer growing) polymer, leading to a new growing point and, hence, to *branching* (76):

$$Ra(CH_2)_{2n} \cdot H \qquad\qquad Ra(CH_2)_{2n-1}CH_3 \qquad\qquad (CH_2)_{2n} \cdot$$
$$Ra(CH_2)_x CH(CH_2)_y Ra \quad \rightarrow \quad Ra(CH_2)_x \dot{C}H(CH_2)_y Ra \xrightarrow{(CH_2=CH_2)_n} Ra(CH_2)_x CH(CH_2)_y Ra$$

$$(76)$$

The extent to which branching occurs can, hardly surprisingly, have a profound effect on the physical and mechanical properties of the resultant polymer.

Another major influence on the properties of the polymer is the average molecular weight, i.e. the average length of polymer molecules; this may vary from only a few monomer units to many thousand. Apart from the average length of polymer molecules, the actual spread of lengths among the polymer molecules also has a considerable influence; thus the properties of two polymers of approximately the same average m.w. will differ greatly if one is made up of molecules all of much the same length, while the other includes both very long and very short polymer molecules in its make-up. The length of molecules in a polymer may be controlled in a number of ways. Thus increase in the concentration of initiator, relative to that of alkene, will lead to shorter chain lengths: the number of growing chains is increased, and termination thus becomes more probable relative to continued propagation. Alternatively, actual terminators may be added or, more usually, *chain transfer agents*. These are compounds, usually of the form XH, that suffer H-abstraction by a growing polymer chain, thereby terminating the chain but generating a new radical, X·, in the process, that is capable of initiating a new chain (77) from monomer. Thiols, RSH, are often used:

$$Ra(CH_2)_n CH_2 \cdot + RSH \rightarrow Ra(CH_2)_n CH_3 + RS \cdot \xrightarrow{nCH_2=CH_2} RS(CH_2)_{2n} \cdot$$

$$(77)$$

A new growing chain is thus generated without slowing down the overall process of monomer conversion. In the case of terminators, XH is chosen so that X· is not reactive enough to initiate a new chain from monomer.

Radical-induced polymerisation of simple alkenes, e.g. ethene and propene, requires vigorous conditions including very high pressure, but many other alkene monomers carrying substituents polymerise readily. These include $CH_2{=}CHCl \rightarrow$ polyvinyl chloride (p.v.c.) for making pipes etc., $CH_2{=}CMeCO_2Me \rightarrow$ perspex, $PhCH{=}CH_2 \rightarrow$ polystyrene, the expanded form for insulation etc., and $CF_2{=}CF_2 \rightarrow$ teflon, which has an extremely low coefficient of friction, high chemical inertness and high m.p. (lining of frying pans etc.). The properties of a polymer may be varied even further—almost as required—by the *copolymerisation* of two different monomers so that both are incorporated, equally or in other proportions, in the polymer molecules. Reference has already been made (p. 299) to the analytical use of 50:50 copolymerisation of $PhCH{=}CH_2$ and $CH_2{=}CMeCO_2Me$ to distinguish radical-induced polymerisation from that initiated by anions or cations (*cf.* p. 185).

Radical-induced polymerisation has some drawbacks, however; thus branching induced by H-abstraction from the growing chain has already been referred to (p. 312). Another difficulty arises with monomers of the form $CH_2{=}CHX$ (i.e. with all the common monomers except $CH_2{=}CH_2$ and $CF_2{=}CF_2$) over the orientation of the substituent groups, X, with respect to the 'backbone' alkane chain of polymer molecules, whose conformation is 'frozen' in the final rigid solid. In radical polymerisation, the arrangement of the X groups is random, and such *atactic* polymers, e.g. atactic polypropene, are found to be non-crystalline, low density, low melting, and mechanically weak. It has been found, however, that use of a $TiCl_3 \cdot AlEt_3$ catalyst results not only in polymerisation occurring under very mild conditions, but with, for example, propene, the resultant polymer has all the Me groups oriented, regularly, in the same direction. This *isotactic* polypropene is found to be crystalline, high density (closer packing of chains), high melting, and mechanically strong—all desirable qualities —and branching has been largely avoided. This regular, *coordination* polymerisation is believed to result from groups of atoms in the surface of the heterogeneous catalyst acting as a template, so that each successive monomer molecule can be added to the growing polymer chain only through 'coordination', in one particular orientation, at the catalyst surface.

11.5.2 Displacement

Although most of the reactions to be considered under this head are nett, i.e. overall, displacements or substitutions, this is not commonly achieved directly, *cf.* S_N2. In some cases a radical is obtained from the substrate by abstraction (usually of H), and this radical then effects displacement on, or addition to, a further species. In some cases, however, the nett displacement is achieved by addition/abstraction.

11.5.2.1 Halogenation

The nett displacement occurring at carbon on chlorination, for example, of alkanes consists (after initial formation of Cl·) of H-abstraction from R—H by Cl·, followed by Cl-abstraction from Cl—Cl by R· (this step can also be regarded as direct displacement at Cl), the two steps alternating in a very rapid chain reaction:

$$
\begin{array}{c}
\text{Cl—Cl} \\
\downarrow {\scriptstyle h\nu} \\
\text{R—H} + \cdot\text{Cl} \longrightarrow \text{R·} + \text{H—Cl} \\
\uparrow \qquad \downarrow {\scriptstyle \text{Cl—Cl}} \\
\cdot\text{Cl} + \text{R—Cl}
\end{array}
$$

The chain length, i.e. number of RH $\longrightarrow$ RCl conversions per Cl· produced by photolysis, is $\approx 10^6$ for CH_4, and the reaction can be explosive in sunlight. Chlorination can also be initiated thermolytically, but considerably elevated temperatures are required to effect $Cl_2 \longrightarrow$ 2Cl·, and the rate of chlorination of C_2H_6 in the dark at 120° is virtually indetectable. It becomes extremely rapid on the introduction of traces of $PbEt_4$, however, as this decomposes to yield ethyl radicals, Et·, at this temperature, and these can act as initiators: Et· + Cl—Cl $\longrightarrow$ Et—Cl + Cl·. Chlorination of simple alkanes such as these is seldom useful for the preparation of mono-chloro derivatives, as this first product readily undergoes further attack by the highly reactive chlorine, and complex product mixtures are often obtained.

Ease of attack on differently situated hydrogen atoms in an alkane is found to increase in the sequence,

$$
\underset{\displaystyle 1}{\underset{\text{primary}}{\text{H—}\overset{\displaystyle H}{\underset{\displaystyle H}{C}}\text{—H}}} < \underset{\displaystyle 4\cdot4}{\underset{\text{secondary}}{\text{H—}\overset{\displaystyle H}{C}\text{—H}}} < \underset{\displaystyle 6\cdot7}{\underset{\text{tertiary}}{\overset{\displaystyle H}{C}\text{—H}}} < \overset{\displaystyle H}{C}\text{—H}
$$

i.e. in the order of weakening of the C—H bond, and of increasing stability of the product radical (*cf.* p. 302); the figures quoted are for the relative rates of abstraction of H by Cl· at 25°. This differential may often be opposed by a statistical effect, i.e. relative numbers of the different types of hydrogen atom available; thus in $(CH_3)_3CH$

there are 9 primary hydrogen atoms available to every one tertiary hydrogen atom. On chlorination $(CH_3)_3CH$ is found to yield mono-chloro products in the ratio of $\approx 65\%$ $(CH_3)_2CHCH_2Cl$ to 35% $(CH_3)_3CCl$—which is only roughly in accord with the rate ratios quoted above, after 'statistical' allowance has been made. If chlorination is carried out in solution, the product distribution is found to depend on the nature of the solvent, and particularly on its ability to complex with Cl·, thereby stabilising it and increasing its selectivity (see below).

Halogenation, and particularly chlorination, unlike most radical reactions, is markedly influenced by the presence in the substrate of polar substituents; this is because Cl·, owing to the electronegativity of chlorine, is markedly electrophilic (*cf.* p. 305), and will therefore attack preferentially at sites of higher electron density. Chlorination will thus tend to be inhibited by the presence of electron-withdrawing groups, as is seen in the relative amounts of substitution at the four different carbon atoms in 1-chlorobutane (78) on photochemically initiated chlorination at 35° :

$$CH_3-CH_2-CH_2-CH_2-Cl \quad (78)$$
$$25\% \quad 50\% \quad 17\% \quad 3\%$$

The variation over the three different CH_2 groups nicely demonstrates the falling off with distance of the electron withdrawing inductive effect of Cl. The γ-(3-)CH_2 group is behaving essentially analogously to that in $CH_3CH_2CH_2CH_3$, while the lower figure for the CH_3 group reflects the greater difficulty of breaking the C—H bond in CH_3 than in CH_2 (see above).

With propene, $CH_3CH=CH_2$ (79), there is the possibility of either addition of chlorine to the double bond, or of attack on the CH_3 group. It is found that at elevated temperatures, e.g. $\approx 450°$ (Cl· then being provided by thermolysis of Cl_2), substitution occurs to the total exclusion of addition. This is because the allyl radical (80) obtained by H-abstraction is stabilised by delocalisation, whereas the one (81) obtained on Cl· addition is not, and its formation is in any· case reversible at elevated temperatures, the equilibrium lying over to the left :

$$
\begin{array}{c}
[CH_2\cdots CH\cdots CH_2]\cdot + HCl \\
(80) \\
\end{array}
$$

H
|
$CH_2-CH=CH_2$
(79)

$$+Cl\cdot \quad CH_2-\dot{C}H-CH_2$$
$$|$$
$$Cl$$
$$(81)$$

Cyclohexene undergoes analogous 'allylic' chlorination for the same reasons.

So far as the other halogens are concerned, the ΔH values—in kJ (kcal) mol^{-1}—for the two steps of the halogenation chain reaction (p. 314) on CH_4 are as follows:

	(1) X· + H—CH$_3$		(2) CH$_3$· + X$_2$	
F$_2$	-134	(-32)	-292	(-70)
Cl$_2$	-4	(-1)	-96	(-23)
Br$_2$	$+63$	$(+15)$	-88	(-21)
I$_2$	$+138$	$(+33)$	-75	(-18)

The figures for fluorination reflect the weakness of the F—F [150 kJ (36 kcal) mol^{-1}], and the strength of the H—F [560 kJ (134 kcal) mol^{-1}], bonds. Fluorination normally requires no specific initiation (*cf.* p. 314), and is explosive unless carried out at high dilution. That fluorination does proceed by a radical pathway, despite not requiring specific initiation, is demonstrated by the fact that chlorination may be initiated in the dark, and at room temperature, by the addition of small traces of F$_2$. Bromination is a good deal slower than chlorination, under comparable conditions, as step (1)—H-abstraction by Br·—is commonly endothermic. This step is usually so endothermic for I· that direct iodination of alkanes does not normally take place.

The markedly lower reactivity of Br· than Cl· towards H-abstraction means that bromination is much more *selective* than chlorination (the figures refer to H-abstraction by Br· at 300°):

primary	secondary	tertiary
1	80	1600

A fact that can be put to preparative/synthetic use; thus bromination of $(CH_3)_3CH$ is found to yield only $(CH_3)_3CBr$ (*cf.* chlorination, p. 315). The effect is more pronounced when substituents are present that can stabilise the initial radical; thus across the series, CH_4, $PhCH_3$, Ph_2CH_2 and Ph_3CH the relative rates of bromination differ over a range of 10^9, but only over a range of 10^3 for chlorination. Selectivity decreases with rise of temperature, however.

Halogenation of an optically active form of a chiral alkane, RR′R″CH, is normally found to yield a racemic ($\pm$) halide—a result that tells us nothing about the preferred conformation of the intermediate radical, RR′R″C·, as racemisation would be observed with *either* a planar, *or* a rapidly inverting pyramidal, structure (*cf.* p. 301). However, bromination of (+)1-bromo-2-methylbutane (82) is found to yield an optically active bromide, (−)1,2-dibromo-2-methylbutane (83), i.e. the overall substitution occurs with <u>retention</u> of configuration. This is believed to result from the original (1-)bromo substituent interacting with one side of the intermediate radical (84)—the one opposite to that from which H has been abstracted—and so promoting attack by Br_2 on the other, thus leading to retention of configuration:

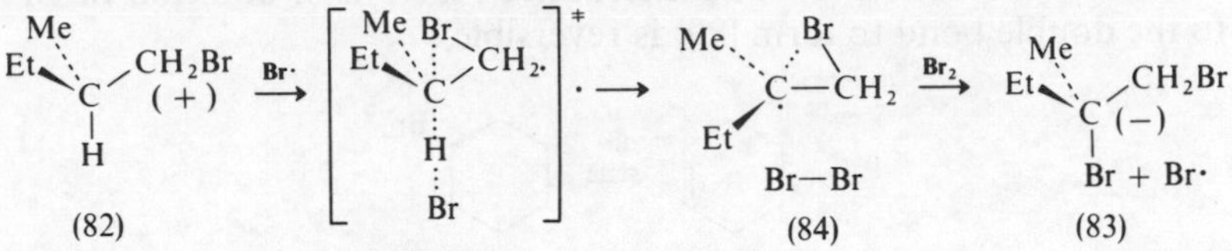

Bromination of an optically active form of the corresponding chloro-compound (1-chloro-2-methylbutane) also results in an optically active product, and retention of configuration. It may be that an actual bridged radical is formed, but a somewhat less concrete inter-action seems more likely, as halogenation with the more reactive chlorine is found to lead wholly to racemisation.

Radical halogenation, particularly chlorination, by other reagents is often of greater synthetic importance than that by the halogens themselves. Thus chlorination may be effected by Me_3COCl in the presence of a radical initiator, the latter abstracting Cl to leave $Me_3CO\cdot$ which has been shown to be the species that abstracts H from RH; this reagent is used particularly for allylic chlorination. Another useful reagent for preparative chlorination is SO_2Cl_2, the radical initiator again abstracts Cl to yield $\cdot SO_2Cl$, and both this species and the $Cl\cdot$ it yields by loss of SO_2 can act as H-abstractors from RH.

Another reagent that is extremely useful synthetically is N-bromo-succinimide (NBS, 85), which is highly selective in attacking only weak C—H bonds, i.e. at allylic, benzylic etc. positions. It requires the presence of radical initiators, and has been shown to effect bromina-tion through providing a constant, but very low, ambient concentration of Br_2—this is maintained through reaction of the HBr produced in the reaction with NBS. (c, below). There is usually a trace of Br_2 or HBr in the NBS that can react with the initiator to generate the initial $Br\cdot$ to start reaction (a, below):

(a) Br_2 or HBr + initiator $\rightarrow Br\cdot$

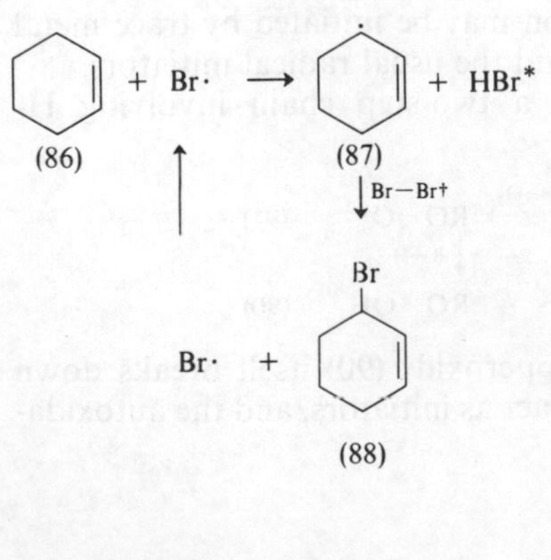

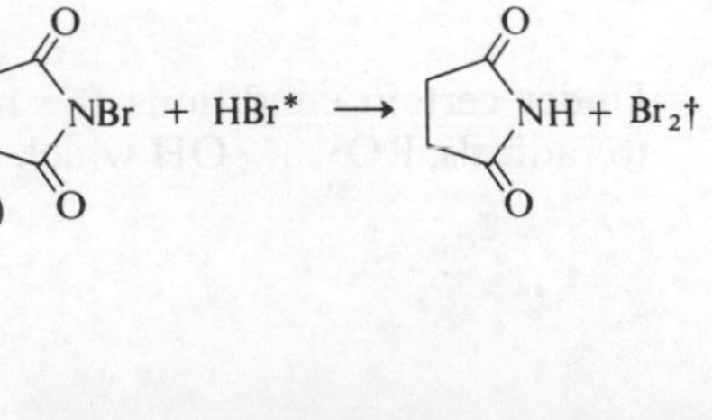

Control of the bromine concentration is maintained by reaction (*c*) which is fast, though ionic, but can be activated only by HBr produced in the chain reaction (*b*). The alternative reaction of addition of Br· to the double bond to form (89) is reversible,

while formation of (87) is not; overall substitution is thus favoured over addition so long as [Br$_2$] is kept low. The radical (87) is also stabilised by delocalisation, while (89) is not (*cf.* p. 302). Support for the above interpretation of the reaction of NBS is provided: (i) by the fact that NBS shows exactly the same selectivity ratios as does Br$_2$, and (ii) by the fact that cyclohexene (86) is found to undergo largely addition with high concentrations of bromine, but largely allylic substitution with low (it is necessary to remove the HBr produced—as happens with NBS).

11.5.2.2   Autoxidation

Autoxidation is the low temperature oxidation of organic compounds by O$_2$, involving a radical chain reaction. The initial stage is commonly the formation of hydroperoxides, RH $\rightarrow$ ROOH, so it is a nett, overall displacement, though the actual pathway involves H-abstraction and O$_2$ addition (see below). The first-formed hydroperoxides frequently undergo further reactions. Autoxidation is of importance in the hardening of paints, where unsaturated esters in the oils used form hydroperoxides, whose decomposition to RO· initiates polymerisation in further unsaturated molecules to form a protective, polymeric, surface film. But autoxidation is also responsible for deleterious changes, particularly in materials containing unsaturated linkages, e.g. rancidity in fats, and perishing of rubber. Indeed, the gradual decomposition of most organic compounds exposed to air and sunlight is due to photosensitised autoxidation. Autoxidation may be initiated by trace metal ions (*cf.* below), as well as by light and the usual radical initiators.

The main reaction pathway is a two-step chain involving H-abstraction:

$$\text{Ra·} + \text{H--R} \rightarrow \text{Ra--H} + \text{R·} \xrightarrow{\cdot\text{O}_2\cdot} \text{RO--O·} \quad (91)$$

$$\uparrow \qquad\qquad \downarrow \text{R--H}$$

$$\text{R·} \;+\; \text{RO--OH} \quad (90)$$

Under certain conditions the hydroperoxide (90) itself breaks down to radicals, RO· + ·OH which can act as initiators, and the autoxidation then becomes autocatalytic. The peroxy radicals (91) are usually of relatively low reactivity (*cf.* ·O—O· itself, p. 305), and are thus highly selective in the positions from which they will abstract H. Thus allylic and benzylic C—H are relatively readily attacked, because the C—H bonds are slightly weaker and the resultant radicals stabilised by delocalisation, while in simple alkanes it is generally only tertiary C—H that is affected, e.g. the allylic position to form (92), and the tertiary (N.B. bridgehead) position to form (93):

(92)   (93)

Relative reactivities towards H-abstraction by RO_2· at 30° are observed as follows: $PhCH_3(1)$, $Ph_2CH_2(30)$ and $PhCH_2CH=CH_2(63)$.

With alkenes, rather than alkanes, autoxidation can involve addition of RO_2· to the double bond as well as, or in place of, H-abstraction, particularly where there are no allylic, benzylic or tertiary C—H linkages available. The effect of the presence of such peroxides in alkenes on the orientation of HBr addition to the latter has already been referred to (p. 308). Ethers are particularly prone to autoxidation, initial attack taking place at a C—H linkage α- to the oxygen atom to yield a stabilised radical; the first-formed hydroperoxide reacts further to yield dialkyl peroxides that are highly explosive on heating— not to be forgotten on evaporating ethereal solutions to dryness! Accumulated peroxides, in ether that has been standing, may be safely decomposed before its use by washing with a solution of a reducing agent, e.g. $FeSO_4$.

Autoxidation may in some cases be of preparative use; thus reference has already been made to the large scale production of phenol + acetone by the acid-catalysed rearrangement of the hydroperoxide from 2-phenylpropane (cumene, p. 127). Another example involves the hydroperoxide (94) obtained by the air oxidation at 70° of tetrahydro-naphthalene (tetralin); the action of base then yields the ketone (α-tetralone, 95), and reductive fission of the O—O linkage the alcohol (α-tetralol, 96):

(96)   (94)   (95)

Aldehydes, and particularly aromatic ones, are highly susceptible to autoxidation; thus benzaldehyde (97) is rapidly converted into benzoic acid (98) in air at room temperature. This reaction is catalysed by light and the usual radical initiators, but is also highly susceptible to the presence of traces of metal ions that can act as one-electron oxidising agents (*cf.* p. 297), e.g. $Fe^{3\oplus}$, $Co^{3\oplus}$ etc:

$$(a) \quad Fe^{3\oplus} + \underset{(97)}{PhC-H} \rightarrow Fe^{2\oplus} + H^{\oplus} + \underset{(99)}{PhC\cdot} \xrightarrow{\cdot O_2} \underset{(100)}{PhC-O-O\cdot}$$

$$(b) \uparrow \qquad \downarrow \underset{PhC-H}{\overset{O}{\parallel}} (97)$$

$$\underset{(99)}{PhC\cdot} + PhC-O-OH \qquad (101)$$

$$(c) \quad \underset{(101)}{PhC-O-OH} + \underset{(97)}{PhC-H} \xrightarrow{H^{\oplus}} \underset{(98)}{2PhC-OH}$$

The oxidation is initiated (*a*) by $Fe^{3\oplus}$ to yield the benzoyl radical (99) which adds on a molecule of oxygen to form the perbenzoate radical (100), this reacts with benzaldehyde (97) to yield perbenzoic acid (101) and another benzoyl radical (99)—these two steps constituting the chain reaction (*b*). The actual end-product is not perbenzoic acid (101), however, as this undergoes a rapid acid-catalysed, non-radical reaction (*c*) with more benzaldehyde (97) to yield benzoic acid (98). This latter reaction (*c*), being acid-catalysed, speeds up as the concentration of product benzoic acid (98) builds up, i.e. it is *autocatalytic*. That benzoyl radicals (99) are involved is borne out by the observation that carrying out the reaction at higher temperatures ($\approx 100°$), and at low oxygen concentrations, results in the formation of CO, i.e. by $PhCO \rightarrow Ph\cdot + CO$.

The autoxidation of aldehydes, and of other organic compounds, may be lessened considerably by very careful purification—removal of existing peroxides, trace metal ions etc.—but much more readily and effectively by the addition of suitable radical inhibitors, referred to in this context as *anti-oxidants*. The best of these are phenols and aromatic amines which have a readily abstractable H atom, the resultant radical is of relatively low reactivity, being able to act as a good chain terminator (by reaction with another radical) but only as a poor initiator (by reaction with a new substrate molecule).

An interesting, and slightly different, autoxidation is photo-oxidation of hydrocarbons such as 9,10-diphenylanthracene (102) in solvents such as CS_2. The light absorbed converts the hydrocarbon into the stabilised diradical (103, *cf.* p. 325), or something rather like it, stabilisation occurring through delocalisation of its unpaired electrons and also by conversion of a partially aromatic state in (102) to a completely aromatic one in (103). The diradical then adds on a molecule of oxygen to yield the trans-annular peroxide (104) in a non-chain step:

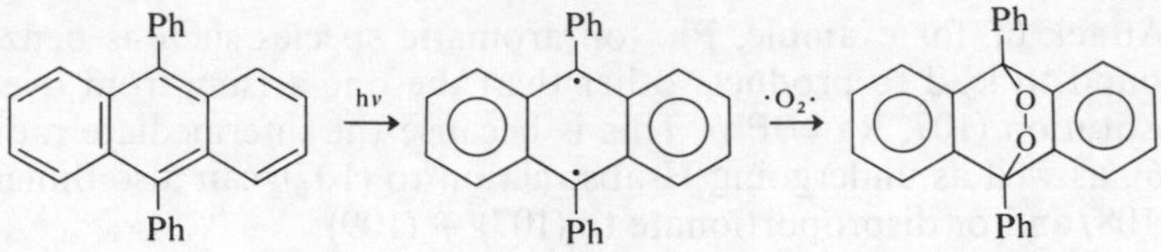

(102)   (103)   (104)

Similar photo-oxidation occurs with increasing readiness as the number of benzene rings in the *lin* (rings joined successively in the same line) hydrocarbon increases, i.e. as its overall aromatic character decreases; this occurs so readily with, for example, the very dark green hydrocarbon hexacene (105),

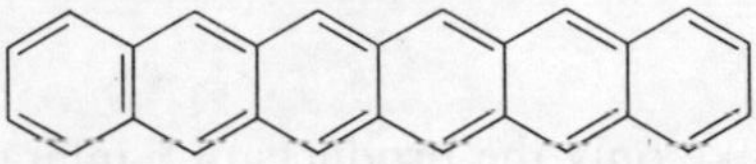

(105)

as to make it impossible to work with in the presence of sunlight and air (*cf.* p. 325).

11.5.2.3   Aromatic substitution

Attack on aromatic species can occur with radicals, as well as with the electrophiles (p. 130) and nucleophiles (p. 164) that we have already considered; as with these polar species, homolytic aromatic substitution proceeds by an addition/elimination pathway:

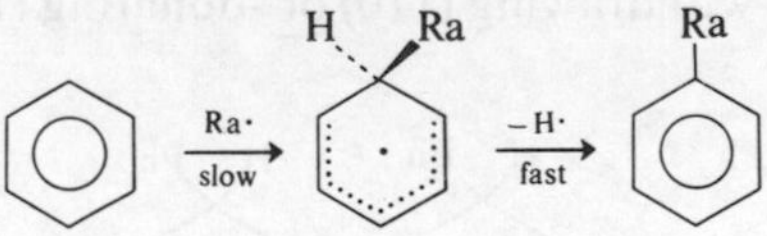

(106)   (107)

Loss of a hydrogen atom from the delocalised cyclohexadienyl radical intermediate (106) to yield the substituted end-product (107) does not proceed spontaneously, but requires the intervention of another radical (or of a radical precursor—in which case a continuous chain reaction may be set up). Reaction between two radicals—(106) and the H-abstractor—is likely to be fast, i.e. non rate-limiting, and no significant k_H/k_D kinetic isotope effect is observed, i.e. attack of Ra· on the original aromatic substrate is rate-limiting. Overall substitution reactions have been investigated in which Ra· is Ar· (especially Ph·), PhCO$_2$· (and some RCO$_2$·), R·, and ·OH; but of these, the first—phenylation or arylation—has been studied in by far the greatest detail.

Attack of, for example, Ph· on aromatic species such as benzene is found to lead to products other than the one arising from overall substitution (107, Ra = Ph). This is because the intermediate radical (106), as well as undergoing H-abstraction to (107), can also dimerise to (108) and/or disproportionate to (107) + (109):

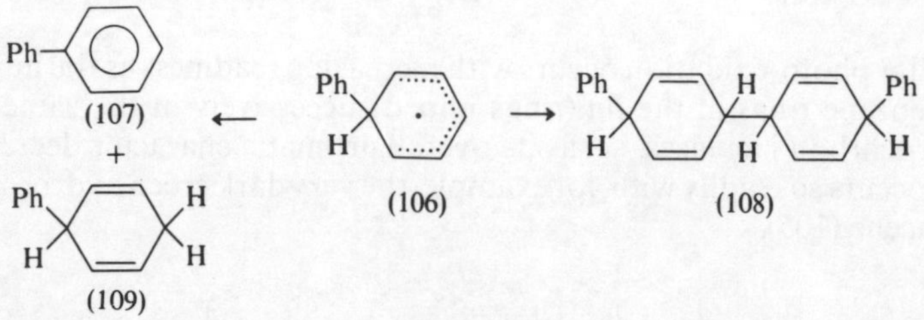

For simplicity's sake only the products of *p*-interaction in (106) have been shown above: *o*-interaction can also lead to an *o*-dihydro isomer of (109), and to both *o*-/*o*- and *o*-/*p*-coupled isomers of (108). Product mixtures from arylation of aromatic species can thus be quite complex.

So far as the overall substitution reaction (→ 107) is concerned, marked differences from electrophilic and nucleophilic attack become apparent as soon as the behaviour of substituted benzene derivatives (C$_6$H$_5$Y) is considered. Thus homolytic attack on C$_6$H$_5$Y is found to be faster than on C$_6$H$_6$, no matter whether Y is electron-attracting or -withdrawing. Further, preferential attack always takes place *o*-/*p*- to Y, no matter what its nature. This is understandable in that the electron brought by the attacking radical, e.g. Ph·, to the intermediate (106) can be delocalised—and the intermediate thereby stabilised—by either an electron-withdrawing (110) or -donating (111) substituent:

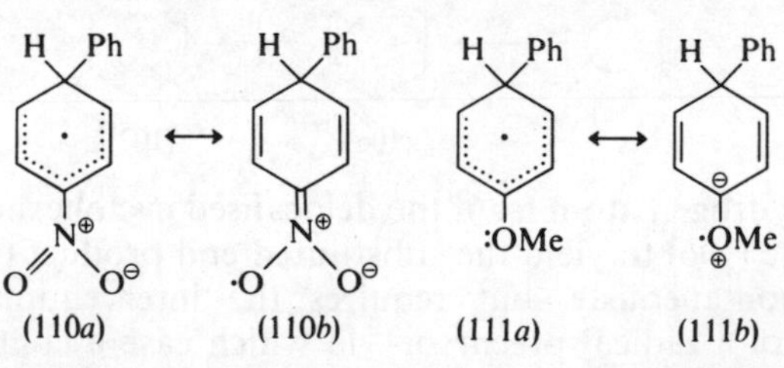

There is, however, no entirely satisfactory explanation of two facts illustrated by the partial rate (*cf.* p. 156) data below: (*a*) that attack at the position *m*- to Y is also normally faster than attack on a position in benzene, or (*b*) that attack is always fastest at the position *o*- to Y, unless Y is so large as to impede approach of the reagent, e.g. CMe_3 :

PhY	$f_{o\text{-}}$	$f_{m\text{-}}$	$f_{p\text{-}}$
PhOMe	5·60	1·23	2·31
$PhNO_2$	5·50	0·86	4·90
PhMe	4·70	1·24	3·55
PhCl	3·90	1·65	2·12
PhBr	3·05	1·70	1·92
$PhCMe_3$	0·70	1·64	1·81

The above figures all refer to phenylation by Ph· generated from $(PhCO_2)_2$. The very small spread in these figures is in marked contrast to the very large one observed for electrophilic substitution (p. 156). This is not altogether surprising seeing that attack is here by a non-polar species, but it does mean—in contrast to electrophilic substitution —that some of all three isomers is normally obtained.

The attacking aryl radicals may be generated from a variety of different precursors, but none are entirely satisfactory. Thus diacyl-peroxides such as $(PhCO_2)_2$ have been much used, but as their decomposition involves the step, $PhCO_2· \rightarrow Ph· + CO_2$, they always lead to some ester formation (acyloxylation) either by Ph· + $PhCO_2·$, or by attack of $PhCO_2·$ on the aromatic substrate. Aromatic azo and diazo compounds have also been employed, e.g. N-nitrosoacetanilide PhN(NO)COMe, and diazonium salts + base, $ArN_2^{\oplus} + {}^{\ominus}OH$ (or $MeCO_2{}^{\ominus}$)—the Gomberg reaction for the synthesis of unsymmetrical diaryls, Ar—Ar'. In each case the radical precursor is decomposed in the presence of an excess of the aromatic substrate, which is often in fact used as the solvent. Although they work reasonably well for individual cases, such arylation reactions are seldom of any general applicability. Some intramolecular radical arylations work quite well, however, e.g. the Pschorr reaction; this involves the thermal decomposi-tion of diazonium salts, e.g. (112), in the presence of copper powder as catalyst, and is used in the synthesis of phenathrenes such as (113):

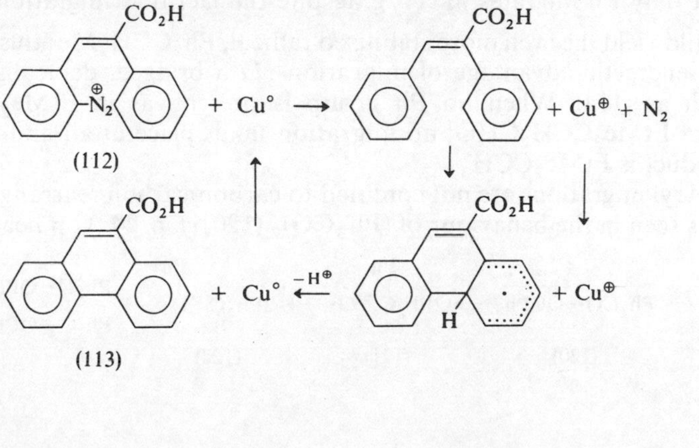

This is essentially a chain reaction involving $Cu° \rightleftarrows Cu^\oplus$ inter-conversions. Similar results have been achieved by photolysis of the aryl iodide corresponding to the diazonium salt (112).

11.5.3   Rearrangement

So far as rearrangement is concerned, radicals resemble carbanions (*cf.* p. 285) very much more closely than they do carbonium ions (*cf.* p. 108). Thus, as with carbanions, 1,2-shifts of alkyl from carbon to carbon are unknown, as are similar 1,2-shifts of hydrogen, but a number of 1,2-shifts of aryl are known which almost certainly involve a stabilised, bridged intermediate or, more likely, transition state, e.g. (114). A good example is with the aldehyde (115), which undergoes H-abstraction from the CHO group by $Me_3CO\cdot$ (ex. $Me_3COOCMe_3$) to yield the acyl radical (116), which readily loses CO to form (117). This can, in turn, abstract H from RCHO (115) to form a hydrocarbon, but the only hydrocarbon actually obtained is not the one derivable from (117), but the one (118) from the rearranged radical (119):

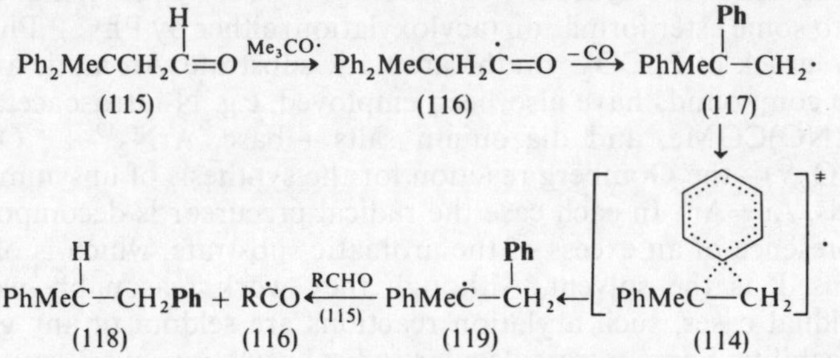

The rearranged radical (119) is more stable than the original one (117) not only because the former is tertiary and the latter primary, but also because (119) is stabilised by delocalisation of the unpaired electron over the π orbital system of a benzene nucleus. It is significant that only Ph migrates in (117), despite the fact that migration of Me would yield the even more stabilised radical, $Ph_2\overset{\cdot}{C}CH_2Me$; this reflects the energetic advantage of migration *via* a bridged, delocalised T.S. such as (114). When no Ph group is present, as in $EtMe_2CCH_2\cdot$ from $EtMe_2CCH_2CHO$, no migration takes place at all and the end product is $EtMe_2CCH_3$.

Aryl migrations are not confined to carbon/carbon rearrangements, as is seen in the behaviour of $(Ph_3CO)_2$ (120, *cf.* p. 292) on heating:

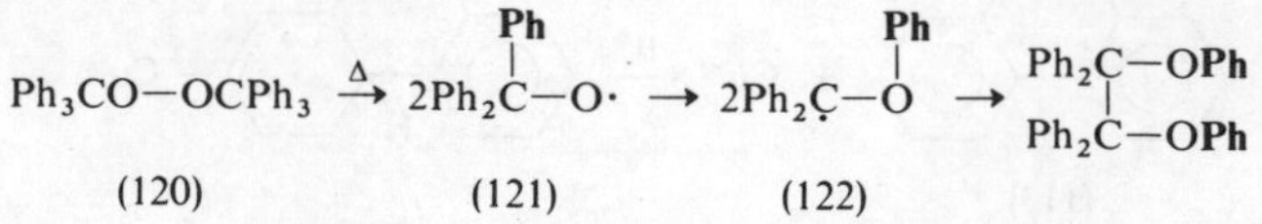

This too proceeds *via* a bridged T.S.; again the driving force of the rearrangement is the much greater stability of (122) than (121). As well as 1,2-aryl shifts, similar migrations of vinyl, acyl and acyloxy groups are known, occurring *via* bridged transition states or intermediates, and also 1,2-chlorine shifts in which an empty *d* orbital on the halogen atom is used to accommodate the unpaired electron in a bridged intermediate, e.g. (123). Thus photo-catalysed addition of HBr to $CCl_3CH=CH_2$ (124) yields none of the expected $CCl_3CH_2CH_2Br$ (125), but 100% of $CHCl_2CHClCH_2Br$ (126):

$$Cl_3C-CH=CH_2 \xrightarrow{Br^{\cdot}} Cl_2\overset{Cl}{\underset{|}{C}}-\overset{\cdot}{C}H-CH_2 \xrightarrow[\times]{HBr} Cl_2\overset{Cl}{\underset{|}{C}}-\overset{H}{\underset{|}{C}}H-CH_2 + Br^{\cdot}$$

(124) (127) $\downarrow Br$ (125) Br

$$Cl_2C\overset{Cl}{\diagup}\underset{|}{CH}-CH_2 \rightarrow Cl_2\overset{Cl}{\underset{|}{\dot{C}}}-CH-CH_2 \xrightarrow{HBr} Cl_2\overset{H}{\underset{|}{C}}-\overset{Cl}{\underset{|}{C}}H-CH_2 + Br^{\cdot}$$

(123) Br (128) Br (126) Br

The driving force of the reaction is the formation of a more stable radical, i.e. the unpaired electron is delocalised more effectively by Cl in (128) than by H in (127). Migration of fluorine does not occur as its *d* orbitals are not accessible, and migration of Br only rarely as the intermediate radicals undergo elimination (to alkene) more readily than rearrangement.

Radical migration of hydrogen is also known, though only over longer distances than 1,2-shifts, e.g. a 1,5-shift to oxygen *via* a 6-membered cyclic T.S. in the photolysis of the nitrite ester (129)—an example of the Barton reaction:

$$Ph(CH_2)_5ONO \xrightarrow{hv} PhCH_2\overset{H}{\underset{}{C}H}\overset{\cdot O}{\diagup}CH_2 \longrightarrow \left[PhCH_2\overset{H\cdots O}{\underset{}{C}H}\diagdown CH_2 \right]^{\cdot} \rightarrow PhCH_2\dot{C}H(CH_2)_3OH$$

(129) H_2C-CH_2 H_2C-CH_2

11.6 BIRADICALS

The oxygen molecule, a paramagnetic species with an unpaired electron on each atom, has already been referred to as biradical, albeit an unreactive one. The photochemical excitation of an anthracene to a biradical, or to something rather like one, has also been mentioned (p. 320); if this excitation is carried out in the absence of air or oxygen, instead of the trans-annular peroxide—(104)—a photo-dimer (130) is obtained:

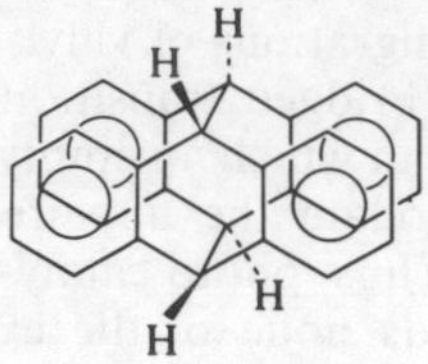

(130)

Biradicals have also been encountered as intermediates in the Mg reduction of ketones to pinacols (p. 214) and, as radical anions, in the acyloin condensation of esters (p. 214). The thermolysis of cyclopropane (131) to propene (132) at $\approx 500°$ is also believed to involve biradical intermediates, e.g. (133) and (134):

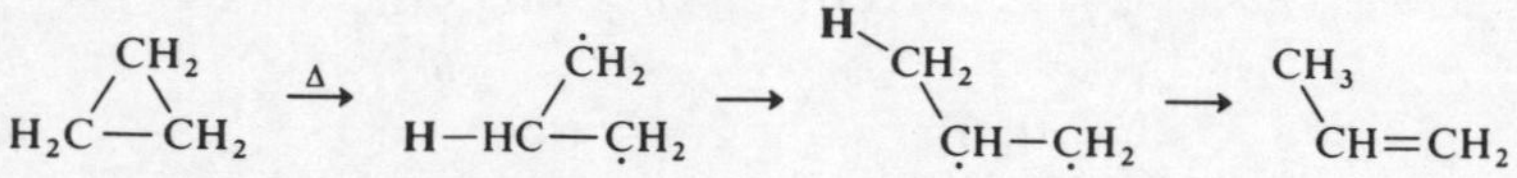

In order to form the biradical (133), the cyclopropane molecule becomes vibrationally excited by collision with another molecule; the C—C bond may then break provided the extra energy is not lost too rapidly by further collision. There is driving force here for a 1,2-shift of hydrogen—unlike in mono-radicals (p. 324)—because of the opportunity of electron-pairing to form a π bond (with evolution of energy) in (134). There is evidence that this H-migration is commonly the rate-limiting step of the reaction.

The above biradicals, with the exception of the oxygen molecule, are all highly unstable; there are, however, a number of much more stable species that show evidence of biradical character. Thus the hydrocarbon (135) exists, in part, in solution as a biradical:

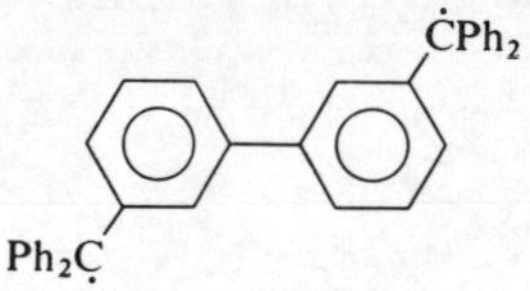

(135)

It behaves, hardly surprisingly, very like $Ph_3C·$ (*cf.* p. 292), existing out of solution as a colourless solid, but this latter is probably a polymer rather than a dimer as with $Ph_3C·$. The solid is dissociated in solution to about the same extent as the $Ph_3C·$ dimer. The unpaired electrons in the biradical form (135) cannot interact with each other to form a wholly paired, diamagnetic species, as such interaction across both central benzene nuclei would necessitate *m*-quinoid forms that cannot exist; the electrons are thus 'internally insulated' from each other. Such internal insulation in biradicals may also arise through steric rather than electronic causes. Thus the species (136) exists in solution as a biradical to the extent of $\approx 17\%$, being in equilibrium with a polymer (like 135):

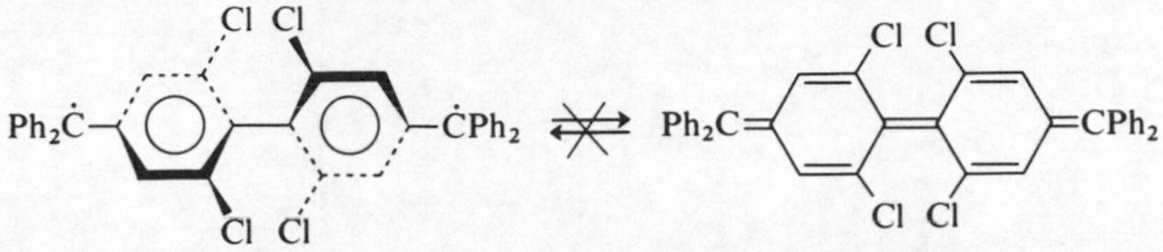

Ph₂Ċ—(ring: Cl Cl / Cl Cl)—(ring)—ĊPh₂ ⇄̸ Ph₂C=(ring: Cl Cl / Cl Cl)=(ring)=CPh₂

(136)   (137)

Here there is no formal electronic bar to interaction between the electrons, i.e. pairing to form the diamagnetic species (137); but this does not in fact happen, because the bulky chlorine atoms in the *o*-positions prevent the benzene rings from attaining a coplanar conformation—the only position in which significant *p* orbital overlap, required for electron-pairing, can occur.

Interestingly enough, some biradical character has been observed in systems similar to (136) even when there are no bulky chlorine atoms present to inhibit delocalisation sterically. Thus with the system (138) $\rightleftarrows$ (139),

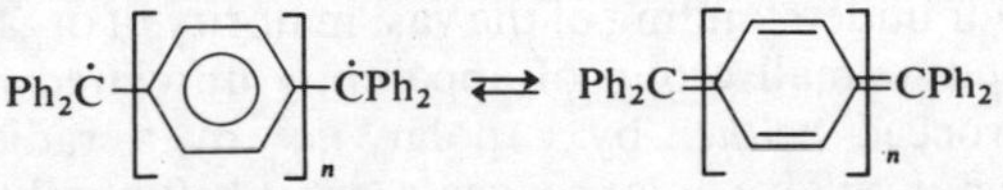

Ph₂Ċ—[(ring)]ₙ—ĊPh₂ ⇌ Ph₂C=[(ring)]ₙ=CPh₂

(138)   (139)

both $n = 3$ and $n = 4$ species are paramagnetic in the solid state, corresponding to $\approx 8\%$ biradical character for $n = 3$ and 15% for $n = 4$ at 20°.

12

Symmetry controlled reactions

12.1 INTRODUCTION

At a time when general mechanistic considerations had brought order and light to our understanding of the vast majority of organic reactions, there remained a small group of apparently unrelated reactions that appear to proceed *neither* by a polar, *nor* by a radical, pathway. Thus they do not involve polar reagents, are substantially uninfluenced by changes in solvent polarity, by the presence of radical initiators (or inhibitors) or other catalysts, and all attempts to isolate, detect, or trap intermediates were unsuccessful. Examples of such reactions that we have already encountered, are the Diels–Alder reaction (p. 193), involving the 1,4-addition of (usually) substituted alkenes to conjugated dienes;

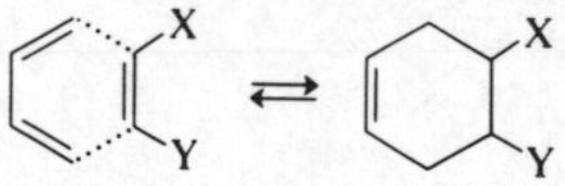

and the pyrolytic elimination reactions of carboxylic esters (1), xanthates (2), etc. (p. 262), to yield alkenes (3):

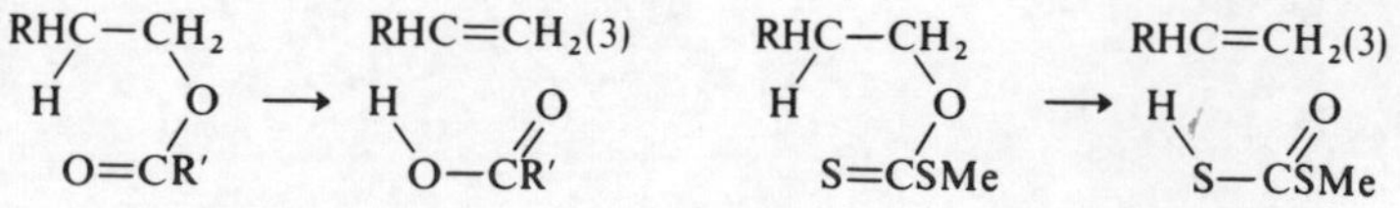

Such reactions are apparently concerted, i.e. the electronic rearrangements involved in bond-making/bond-breaking proceed simultaneously in a one-step process; though each bond undergoing change need not necessarily have been made or broken to the same extent by the time the T.S. has been reached. The transition states are cyclic, ones involving six *p* electrons (*cf.* aromaticity, p. 17) being preferred though not essential, and the reactions are normally attended by a high degree of stereoselectivity (*cf.* p. 261). Many of the reactions are reversible, e.g. Diels–Alder reactions, though the equilibrium often lies well over to one side or the other. The term *pericyclic* has been coined to describe such concerted reactions that proceed *via* cyclic transition states.

As pericyclic reactions are largely unaffected by polar reagents, solvent changes, radical initiators, etc., the only means of influencing them is thermally or photochemically. It is a significant feature of pericyclic reactions that these two influences often effect markedly different results, either in terms of whether a reaction can be induced to proceed readily (or at all), or in terms of the stereochemical course that it then follows. Thus the Diels–Alder reaction (*cf.* above), an example of a concerted *cycloaddition* process, can normally be induced thermally but not photochemically, while the concerted cycloaddition of two molecules of alkene, e.g. (4) to form a cyclobutane (5),

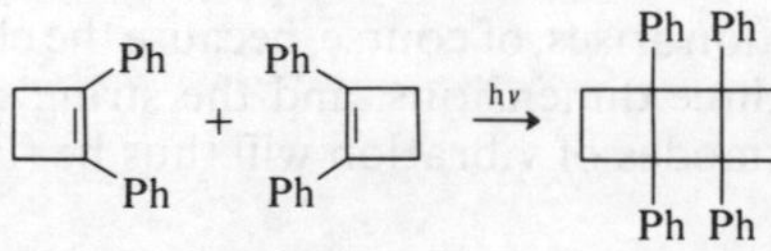

(4)   (4)   (5)

can be induced photochemically but not thermally. The differential stereochemical effect is clearly seen with *trans, cis, trans* 2,4,6-octatriene (6); this found to cyclise on heating to give the *cis* 1,2-dimethylcyclohexa-3,5-diene (7) only, while on photochemical irradiation it cyclises to give the *trans* 1,2-dimethylcyclohexa-3,5-diene (8) only:

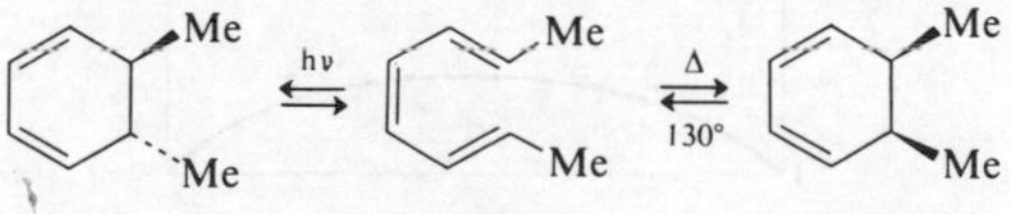

(8)   (6)   (7)

This type of reaction, whether it involves the concerted cyclisation of a polyene, as here, or the concerted ring-opening of a cyclic compound to form a polyene, is known as an *electrocyclic* reaction.

In each case above the reaction will proceed *via* the pathway of lowest energetic demand (the one with the lowest $\Delta G^{\ddagger}$), a concerted pathway having an obvious advantage in that part of the energy required for bond-breaking is being supplied by that produced from the simultaneous bond-formation. It seems reasonable, therefore, to suppose that the pathway of lowest energetic demand, in any particular case, will be the one that involves maintaining the greatest degree of residual bonding throughout the reaction, but particularly in the transition state. The maintenance of bonding implies the maintenance of orbital overlap, and we have therefore to establish the conditions necessary to ensure such maintenance of orbital overlap. This requires us to consider one property of atomic and molecular orbitals that we have not yet referred to, namely *phase*.

12.2 PHASE AND SYMMETRY OF ORBITALS

We have already seen (p. 2) that the individual electrons of an atom can be symbolised by wave functions, ψ, and some physical analogy can be drawn between the behaviour of such a 'wave-like' electron and the standing waves that can be generated in a string fastened at both ends—the 'electron in a one-dimensional box' analogy. The one-dimensional limitation arises, of course, because the electron's 'standing waves' will be in three dimensions, and the string's in only one; the first three possible modes of vibration will thus be (Fig. 12.1):

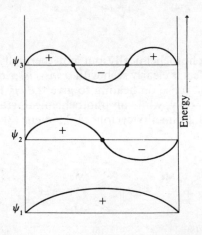

Fig. 12.1

In the first mode, ψ_1, the amplitude of the wave increases from zero to a maximum, and then decreases to zero again; in the second, ψ_2, the amplitude increases to a maximum, decreases through zero (a *node*, marked ● above) to a minimum, and then back to zero again, i.e. the *phase* of the wave changes once; while in the third mode, ψ_3, the amplitude changes from zero to a maximum, through zero to a minimum, through zero to a maximum again, and then finally back to zero, i.e. there are two nodes (marked ● above), and the phase of the wave changes twice. Displacements above the nodal plane are conventionally designated $+$, and those below $-$. The lobes of, for example, a $2p$ atomic orbital, which has one nodal plane, thus differ in phase, and are conventionally designated as $+$ and $-$, i.e. (9); this can, however, lead to confusion because of the usual association of $+$ and $-$ with charge[†], and phase differences, which are purely relative, will therefore be designated here by shading and no shading, i.e. (10):

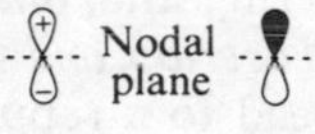

Nodal plane

(9)   (10)

Molecular orbitals are obtained by the linear combination of atomic orbitals, and the question of phase will, of course, arise with them too. Thus we can write the two MOs (π and π^*, *cf.* p. 9) arising from the two p atomic orbitals in ethene,

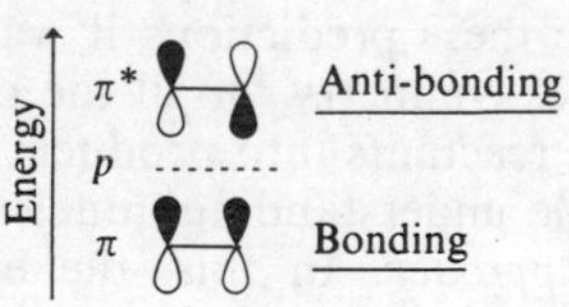

[†] It is important to emphasise that ψ^2, which represents the probability of finding an electron in a particular element of space, will always be positive, no matter whether ψ is positive or negative.

and the four MOs (ψ_1, ψ_2, ψ_3 and ψ_4) arising from the four p AOs in butadiene (*cf*. Fig. 1.2, p. 12) in the *cisoid* conformation (p. 194):

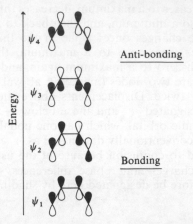

The importance of considering the phase of orbitals is that: *only orbitals of the same phase will overlap, and so result in a bonding situation*; orbitals of different phase lead to a repulsive, anti-bonding situation.

By a consideration of the relative phases, and hence overall symmetry, of the orbitals involved, Woodward and Hoffmann were able in 1965 to formulate a set of rules; these not only explained the behaviour of the pericyclic reactions that were known to date, but also made precise predictions about the behaviour to be expected of many others, that had not yet been carried out. These predictions included whether reactions would be induced thermally or photochemically, and the detailed stereochemistry that would then be followed. The achievement is all the greater in that a number of the predictions—since proved correct— appeared at the time to be highly implausible. To make these predictions it was necessary to consider the relative phases, i.e. symmetry, of all the orbitals involved during the transformation of reactants into products. It is, however, possible to obtain a reasonable understanding, much more simply, by use of the *frontier orbital approach*. In this, the electrons in the *highest occupied molecular orbital* (HOMO) of one reactant are looked upon as being analogous to the outer (valence) electrons of an atom, and reaction is then envisaged as involving the overlap of this (HOMO) orbital with the *lowest unoccupied molecular orbital* (LUMO) of the other reactant. Where, as in electrocyclic reactions, only one species is involved only the HOMO need be considered. A variety of pericyclic reactions will now be reviewed, using this approach.

12.3   ELECTROCYCLIC REACTIONS

We have already seen (p. 329) that the cyclisation of *trans,cis,trans* 2,4,6-octatriene (6) proceeds thermally to yield *cis* 1,2-dimethylcyclohexa-3,5-diene (7) only, and photochemically to yield the corresponding *trans* isomer (8) only; in either case the equilibrium lies almost completely over towards the cyclic form. The degree of stereoselectivity is in fact so great that the thermal cyclisation yields <0·1 % of the *trans* isomer (8), despite the latter being thermodynamically more stable than the *cis* form (7). The six MOs of (6)—ψ_1, ψ_2, ψ_3, ψ_4, ψ_5 and ψ_6, arising from the six p AOs—may be written (*cf.* butadiene, p. 332):

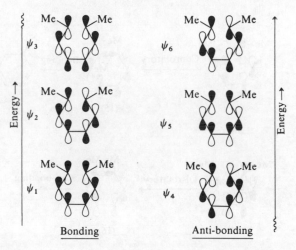

As there are 6 p electrons to accommodate—two per orbital—the HOMO will be ψ_3 (11). To form the C—C σ bond on cyclisation, the p orbitals on the terminal carbon atoms of the conjugated system (C$_2$ and C$_7$—the C atoms carrying the Me substituents) will have to rehybridise to sp^3 orbitals, and each rotate through 90° to allow of their potential overlap. This rotation could happen either (*a*) in the same direction—*conrotatory* (12), or (*b*) in opposite directions—*disrotatory* (13):

Conrotatory movement results in the apposition of sp^3 orbitals of opposite phase—an anti-bonding situation, while disrotatory movement results in the apposition of sp^3 orbitals of the same phase—a bonding situation, leading to formation of the cyclohexadiene (7) in which the two Me groups are *cis*.

On photochemical ring-closure, irradiation results in the promotion of an electron into the orbital of next higher energy level, i.e. $\psi_3 \xrightarrow{h\nu} \psi_4$ and the ground state LUMO (ψ_4) thus now becomes the HOMO (14):

Me Me→←Me   →Conrotatory→   Me   →Bonding→   Me   Me

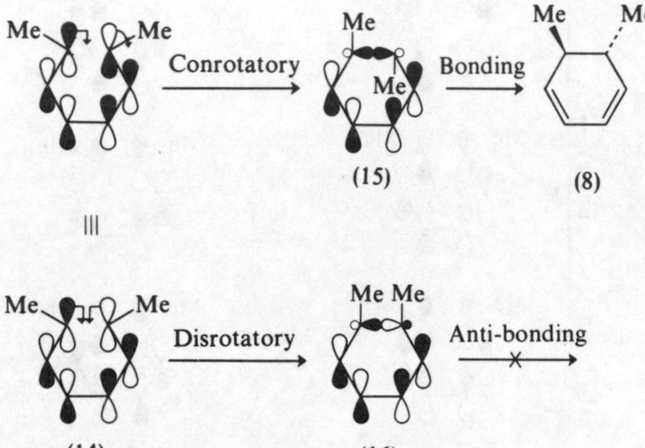

(15)   (8)

HOMO(ψ_4):

Me Me↓↓Me   →Disrotatory→   Me Me   →Anti-bonding→×

(14)   (16)

It is now conrotatory movement that results in the apposition of sp^3 orbitals of the same phase—the bonding situation, leading to formation of the *trans* isomer (8).

It is interesting to contrast the above with the hexa-2,4-diene $\rightleftarrows$ 3,4-dimethylcyclobutene situation. Here exactly the opposite stereochemical inter-relationships are observed, i.e. *trans, trans* hexa-2,4-diene (17) is associated thermally with *trans* 3,4-dimethylcyclobutene (18), and photochemically with the *cis* isomer (19):

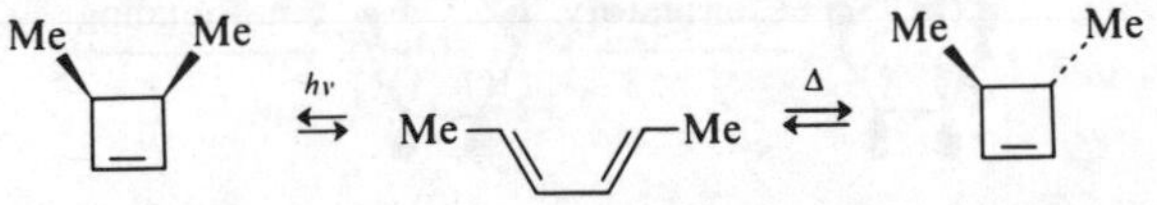

Me Me   Me   Me
$\overset{h\nu}{\leftrightarrows}$ Me⎯⎯⎯Me $\overset{\Delta}{\rightleftarrows}$

(19)   (17)   (18)

For the thermal interconversion (the equilibrium tends to lie over towards the diene), the HOMO for the diene (17) will be ψ_2 (20) as there are four *p* electrons to accommodate (p. 332):

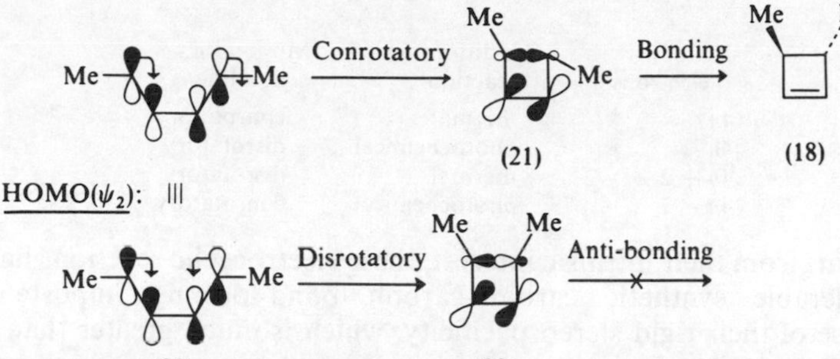

Me—[orbital diagram] Me $\xrightarrow{\text{Conrotatory}}$ Me [orbital diagram] Me $\xrightarrow{\text{Bonding}}$ Me Me [cyclobutene structure]

(21)   (18)

HOMO(ψ_2): |||

Me—[orbital diagram]—Me $\xrightarrow{\text{Disrotatory}}$ Me Me [orbital diagram] $\xrightarrow{\text{Anti-bonding}}{\times}$

(20)   (22)

This time it is controtatory movement that results in a bonding situation, and formation of the *trans* dimethylcyclobutene (18). For the photochemical interconversion (which tends to lie over in favour of the cyclobutene), irradiation of the diene will result in the promotion of an electron into the orbital of next higher energy level, i.e. $\psi_2 \rightarrow \psi_3$, and the ground state LUMO (ψ_3) thus now becomes the HOMO(23):

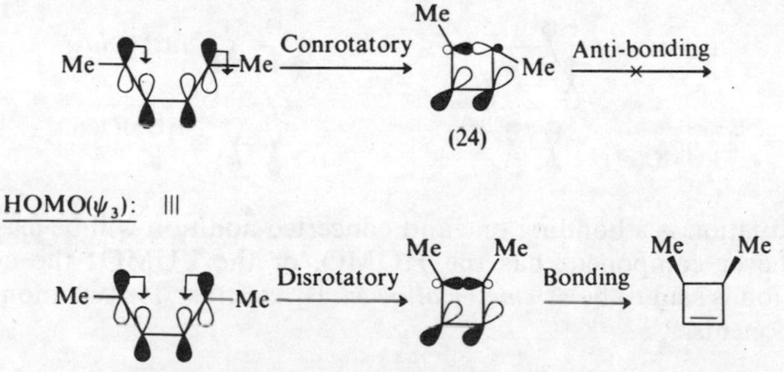

Me—[orbital diagram]—Me $\xrightarrow{\text{Conrotatory}}$ Me [orbital diagram] Me $\xrightarrow{\text{Anti-bonding}}{\times}$

(24)

HOMO(ψ_3): |||

Me—[orbital diagram]—Me $\xrightarrow{\text{Disrotatory}}$ Me Me [orbital diagram] $\xrightarrow{\text{Bonding}}$ Me Me [cyclobutene structure]

(23)   (25)   (19)

Thus disrotatory movement now results in a bonding situation, and formation of the *cis* dimethylcyclobutene (19).

This difference in behaviour obviously stems from the way in which the relative phase, at the terminal carbons, of the MOs of the two dienic systems are arranged, i.e. their symmetry: the phase being the same for the HOMO (ψ_3) of the 6π electron system, and for the LUMO (ψ_3) of the 4π electron system; the phase being different for the HOMO (ψ_2) of the 4π electron system, and for the LUMO (ψ_4) of the 6π electron system. The *same phase* leads to *disrotatory* movement for bond-making/bond-breaking to occur, the *opposite phase* to *conrotatory* movement for bond-making/bond-breaking to occur.

Alternatively, the thermal/photochemical antithesis is summarised in the generalisations:

No. of π electrons	Conditions for reaction	Motion for bonding
$4n$	thermal	conrotatory
$4n$	photochemical	disrotatory
$4n + 2$	thermal	disrotatory
$4n + 2$	photochemical	conrotatory

Apart from their intrinsic interest, these electrocyclic reactions have considerable synthetic carbon–carbon bond-forming importance because of their rigid stereospecificity, which is much greater than in the vast majority of other, non-concerted, reactions.

12.4 CYCLOADDITIONS

In cycloadditions two components are commonly involved, and the feasibility of a particular process will be determined by whether overlap can take place between the HOMO of one component and the LUMO of the other. Thus for a diene plus a monoene,

HOMO(ψ_2)   LUMO(π^*)   LUMO(ψ_3)   HOMO(π)

the situation is a bonding one and concerted addition will be feasible, whichever component has the HOMO, or the LUMO: the cyclo-addition is said to be *symmetry allowed*. By contrast, for two monoene components,

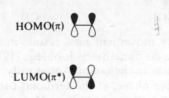

HOMO(π)

LUMO(π^*)

the situation is a non-bonding one and concerted addition will not be feasible: the cycloaddition is said to be *symmetry forbidden*.

This is a general situation for thermal, concerted reactions: those involving $4\pi e + 2\pi e$ systems proceed readily, e.g. the Diels–Alder reaction, whereas those involving $2\pi e + 2\pi e$ systems, e.g. the cyclo-dimerisation of alkenes, do not. We might, however, expect that photochemical cyclo-dimerisation of alkenes would be symmetry allowed, as irradiation will promote an electron, of one component, into the orbital of next higher energy level, i.e. $\pi \xrightarrow{hv} \pi^*$, and its ground state LUMO (π^*) thus now becomes the HOMO:

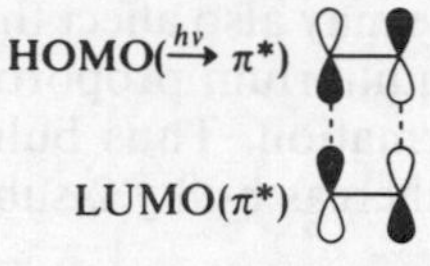

HOMO($\xrightarrow{hv} \pi^*$)

LUMO(π^*)

Many such reactions may indeed be carried out preparatively under photochemical conditions, though, for reasons that cannot be gone into here (the detailed mechanism of photochemical changes), they are often not concerted but proceed *via* biradical intermediates. One photochemical ($2\pi + 2\pi$) cycloaddition that does, however, proceed *via* a concerted process is the example we have already referred to:

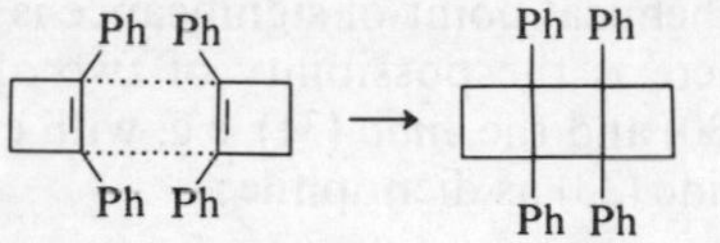

Ph Ph   Ph Ph

Ph Ph   Ph Ph

(4)   (4)   (5)

The importance of ($4\pi + 2\pi$) thermal, concerted cycloadditions are great enough to warrant their separate consideration.

12.4.1 Diels–Alder reaction

By far the best known ($4\pi + 2\pi$) cycloaddition is the Diels–Alder reaction. This has been discussed to some extent already (p. 193), including the fact that it proceeds rigorously, stereospecifically SYN, with respect to both diene (26) and dienophile (27):

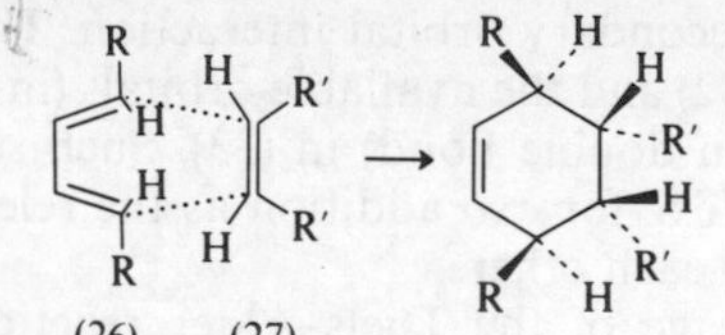

R   H   R'   R   H   H

(26)   (27)

This is confirmatory evidence of a concerted pathway, implying as it does the simultaneous formation of *both* new σ bonds in the T.S. That both new bonds are not necessarily formed to the same extent in the T.S. is, however, suggested by the fact that the reaction is markedly influenced by the electronic effect of substituents. It is found to be promoted by electron-donating substituents in the diene, and by electron-withdrawing substituents in the dienophile; the reaction does indeed proceed only poorly, if at all, in the absence of the latter. The presence of substituents, and even of hetero atoms, in the system appears not to affect the symmetry of the orbitals involved, however.

Substituents in the diene may also affect the cycloaddition sterically, through influencing the equilibrium proportion of the diene that is in the required *cisoid* conformation. Thus bulky 1-*cis* substituents (28) slow the reaction down, whereas bulky 2-substituents (29) speed it up, through this agency:

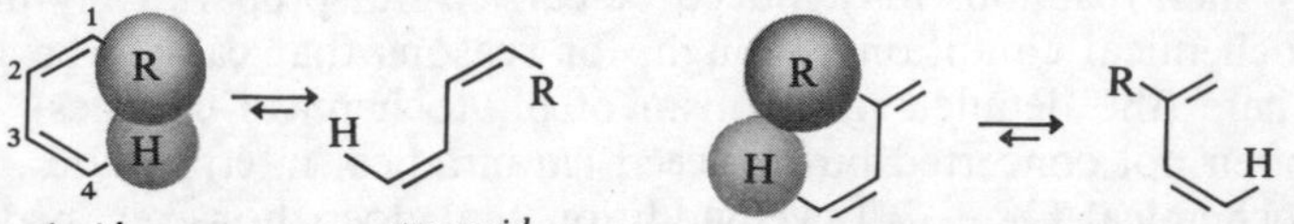

Another stereochemical point of significance is that in some Diels–Alder reactions there is the possibility of two alternative modes of addition, the *exo* (30) and the *endo* (31), e.g. with cyclopentadiene (32), and maleic anhydride (33) as dienophile:

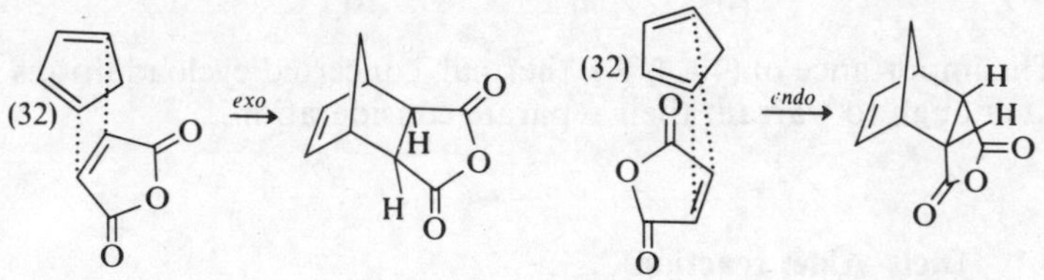

In this case (but by no means in all such cases) the *endo* adduct is found to predominate. It has been suggested that this occurs because in the T.S. for *endo* addition, but not in that for *exo*, further stabilisation (offering a less energy-demanding, and hence faster, reaction pathway) can occur through secondary orbital interaction. This involves other suitable orbitals in (32) and the available orbitals (in addition to those of the carbon–carbon double bond) in (33). Such secondary overlap cannot occur in the T.S. for *exo* addition as the relevant orbitals will be too far away from each other.

The great advantage of the Diels–Alder reaction (as a carbon–carbon bond-forming process) is its generality; the variety of different dienophiles that can be used preparatively is very wide indeed (possible variations in the diene are somewhat less wide), and conditions can usually be found to make the great majority of such cycloadditions go in satisfactory yield. Like other cycloadditions, Diels–Alder reactions are potentially reversible and in some cases the reverse process, the *retro* Diels–Alder reaction, can be made preparatively useful. Thus cyclopentadiene (32) will readily undergo an *auto* Diels–Alder reaction to form a tricyclic dimer; it is commonly stored in this, relatively stable, form and reconverted to (32) on heating, i.e. by distillation, as required. The thermal cracking of cyclohexene (the Diels–Alder adduct of butadiene and ethene—though not prepared that way!) has been used as a useful method for the laboratory preparation of butadiene. The pyrolytic SYN eliminations of carboxylic esters and xanthates, that have already been referred to (p. 262), can also be considered as close analogues of retro (4 + 2) cycloaddition reactions.

12.4.2 1,3-Dipolar additions

The $4\pi e$ component in a $(4\pi + 2\pi)$ cycloaddition need be neither a four-atom system (as in 1,3-dienes), nor involve carbon atoms only, so long as the HOMO/LUMO symmetry requirements for a concerted pathway can be fulfilled. The most common of these non-dienic $4\pi e$ systems involve three atoms, and have one or more dipolar canonical structures, e.g. (34a), hence the term—1,3-dipolar addition. They need not, however, possess a large permanent, i.e. residual, dipole, *cf.* (34a ↔ 34b):

$$H_2\overset{\ominus}{C}\diagup\overset{N}{}\diagdown\overset{\oplus}{N} \quad\leftrightarrow\quad H_2\overset{\oplus}{C}\diagup\overset{N}{}\diagdown\overset{\ominus}{N}$$

(34a)　　　　　　(34b)

The initial addition of ozone to alkenes to form molozonides (p. 189) can be regarded as a 1,3-dipolar addition, and many other such additions are of considerable synthetic utility in the preparation of heterocyclic systems. Thus we have already seen the preparation of a 1,2,3-triazole from $Ph\overset{\ominus}{N}-N\overset{\oplus}{=}N$ (p. 190), and another example involves preparation of the dihydropyrazole (35) from diazomethane (34):

(34)

$$H_2\overset{\oplus}{C}\diagup\overset{N}{}\diagdown\overset{\ominus}{N}$$
$$H_2C=CH$$
$$\diagdown$$
$$CO_2Et$$

→

(35) with CO_2Et

The only significant difference between these $(4\pi + 2\pi)$ cyclo-additions and the Diels–Alder reaction is that the former involve less symmetrical, somewhat dipolar, transition states.

12.5 SIGMATROPIC REARRANGEMENTS

The third major category of pericyclic reactions can be looked upon as involving the migration of a σ bond—hence the name—within a

π-electron framework. The simplest examples involve the migration of a σ bond that carries a hydrogen atom.

12.5.1   Hydrogen shifts

Such reactions, in acyclic polyenes, can be generalised in the form:

$$\underset{(36)}{R_2\overset{\overset{\displaystyle H}{|}}{C}(CH=CH)_xCH=CR_2'} \rightarrow \underset{(37)}{R_2C=CH(CH=CH)_x\overset{\overset{\displaystyle H}{|}}{C}R_2'}$$

Consideration of the feasibility of these shifts as concerted processes, i.e. *via* cyclic transition states, requires as usual a consideration of the symmetry of the orbitals involved. A model related to the transition state can be constructed by the device of assuming that the C—H σ bond that is migrating can be broken down into a hydrogen 1s orbital and a carbon 2p orbital. For the case where $x = 1$ in (36), the T.S. can then be considered as being made up from a pentadienyl radical (38), with a hydrogen atom (one electron in a 1s orbital) migrating between the terminal carbon atoms of its $5\pi e$ system (i.e. a 6e system overall is involved):

(38)

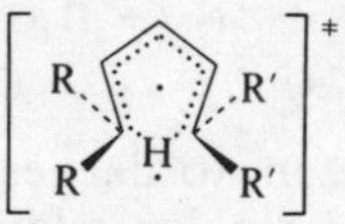

By analogy with the categories of pericyclic reactions we have already considered, the feasibility of the migration will then be decided by the relative phase of the terminal lobes, i.e. the symmetry, of the HOMO of the pentadienyl radical (38). As this is a $5\pi e$ system, its electron configuration will be $\psi_1^2\psi_2^2\psi_3^1$, and its HOMO is therefore ψ_3. This MO can be shown to have terminal lobes of the same phase, so that overlap between the hydrogen atom's 1s orbital and <u>both</u> the terminal lobes of (38)'s MO can be maintained in the T.S. (39):

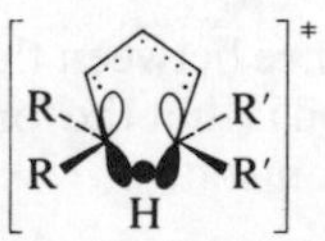

(39)

Thermal 1,5-hydrogen shifts are thus allowed and, because of the symmetry of the T.S. (39), the H atom in the product (37, $x = 1$) will be on the same side of the common plane of the polyene's carbon atoms as it was in the starting material (36, $x = 1$); this is described as a *suprafacial* shift. This latter point would not be experimentally verifiable in the above example, but that thermal 1,5-shifts (which are quite common) do involve strictly suprafacial migration has been demonstrated in the compound (40). This is found, on heating, to yield a mixture of (41) and (42), which are produced by suprafacial shifts in the alternative conformations (40*a*) and (40*b*), respectively:

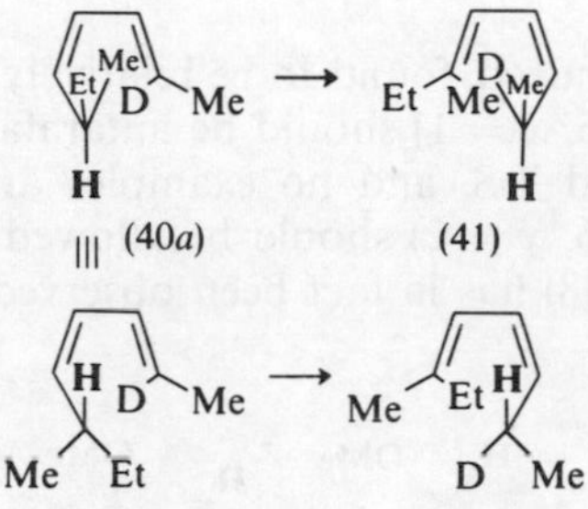

The terminal lobes of the HOMO will be of the same phase in a nonatetraenyl radical also, i.e. for (36, $x = 3$), and 1,9-shifts (in a 10*e* system overall) should thus be allowed, and suprafacial. Formation of the required 10-membered T.S. could present some geometrical difficulty, however, and it is somewhat doubtful whether any such concerted 1,9-shifts have actually been observed. Suprafacial thermal shifts have not been observed in other 'allowed', i.e. $(4n + 2)e$ overall —(36, $x = 3,5 \ldots$), systems either.

It is conceivable that the spherically symmetrical 1*s* hydrogen orbital could, alternatively, overlap across the plane of the polyene's carbon atoms, when the terminal lobes of the latter's HOMO were opposite in phase—*antarafacial* overlap. The terminal lobes of the HOMO will be opposite in phase for (36, $x = 0,2,4 \ldots$), leading to a T.S. such as (43) when $x = 0$ (an overall 4*e* system):

Such a transition state is likely to be highly strained, however, and no such 1,3-antarafacial shifts have actually been observed. A 1,7-thermal antarafacial shift in (36, $x = 2$), where the T.S. is likely to be much less strained has, however, been observed in the vitamin D series.

1,3-Photochemical shifts should, however, be allowed and suprafacial $(44 \rightarrow 45)$ as the HOMO of the T.S. (ψ_3, due to $\psi_1^2\psi_2^1 \rightarrow \psi_1^2\psi_3^1$) now has terminal lobes which are of the same phase (46):

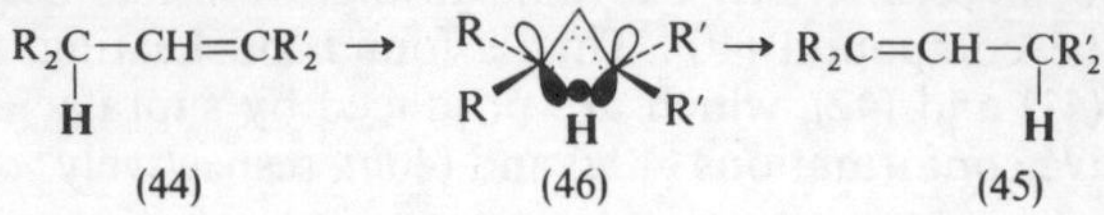

$$R_2\overset{\underset{|}{H}}{C}-CH=CR_2' \quad \rightarrow \quad R\text{-}{\underset{R \quad \underset{H}{} \quad R'}{\overset{\overset{R'}{}}{}}} \quad \rightarrow \quad R_2C=CH-\overset{\underset{|}{H}}{C}R_2'$$

$$(44) \qquad\qquad (46) \qquad\qquad (45)$$

Such 1,3-shifts are, indeed, found to be relatively common. 1,5-Photochemical shifts in $(36, x = 1)$ should be antarafacial, but this is likely to involve a strained T.S. and no examples are known. 1,7-Photochemical shifts in $(36, x = 2)$ should be allowed and suprafacial, and the example $(47 \rightarrow 48)$ has in fact been observed:

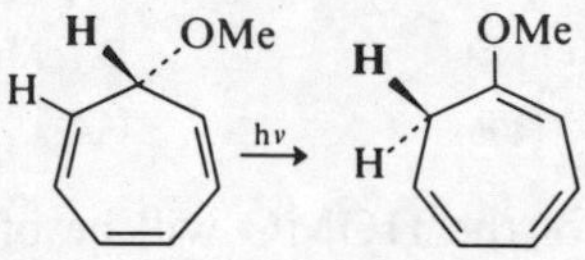

$$(47) \qquad\qquad\qquad (48)$$

The observation above does not, of course, of itself demonstrate that the shift is suprafacial, but the relatively rigid cyclic structure would in any case rule out an antarafacial pathway.

12.5.2 Carbon shifts

The most common examples involve a shift from carbon to carbon in the Cope rearrangement of 1,5-dienes $(49 \rightarrow 50$, not to be confused with Cope elimination, p. 261),

$$(49) \qquad\qquad\qquad\qquad\qquad (50)$$

and a shift from oxygen to carbon in the Claisen rearrangement of alkyl aryl ethers $(51 \rightarrow 52)$:

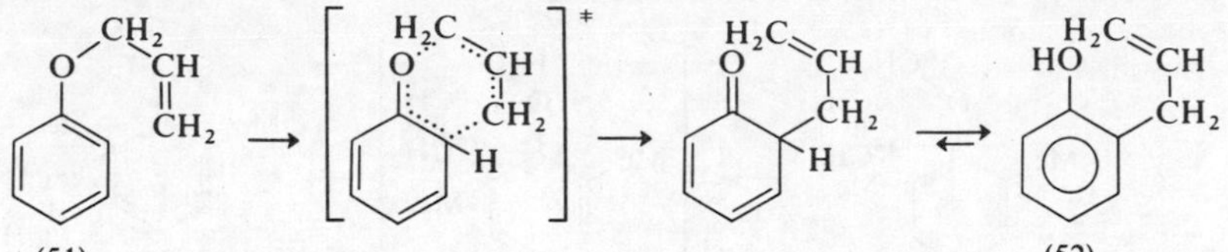

(51)   (52)

So far as thermal reactions are concerned, those that can proceed *via* 6-membered transition states go most readily, and are by far the commonest. That a 6-membered cyclic T.S. in the *chair* conformation is commonly preferred is shown by the fact that the *meso* form of (49) yields *only* (99·7%) the *cis,trans* form (50a), out of the three possible geometrical isomerides (*cis,cis*; *cis,trans*; and *trans,trans*) of (50):

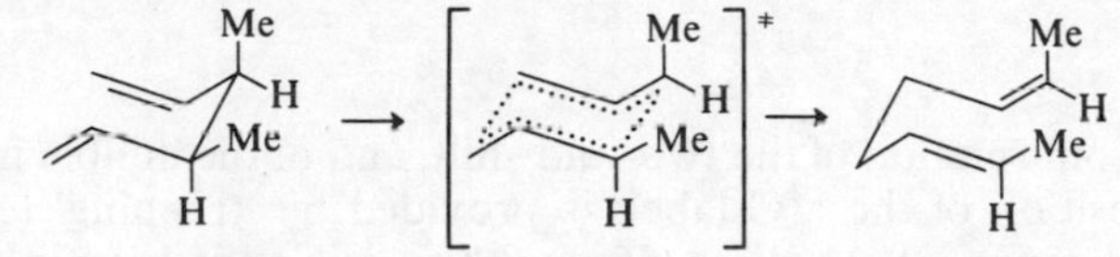

meso   *cis, trans*
(49a)   (50a)

This corresponds to a shift which is suprafacial at both 'ends' of the migrating system.

The Claisen rearrangement is strictly intramolecular, and shows the large negative value of ΔS^{+} characteristic of the degree of ordering required by a cyclic T.S. This latter requirement is also borne out by ^{14}C labelling, which indicates that the position of the ^{14}C atom in the allyl group is 'inverted' during migration $(51a \rightarrow 52a)$:

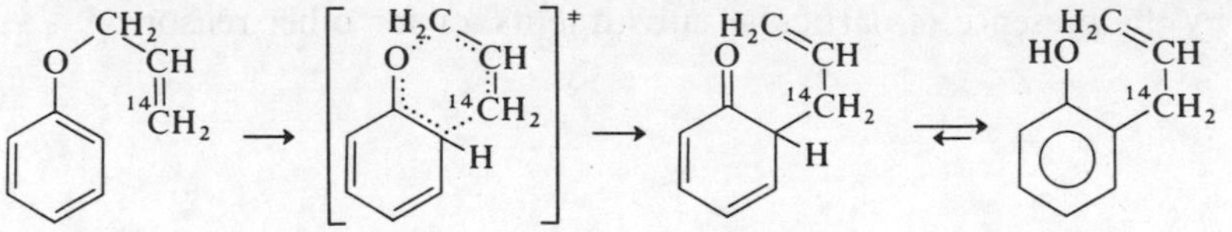

(51a)   (53a)   (52a)

The dienone intermediate (53a), as well as enolising to the phenol (52a), is itself capable of undergoing a Cope rearrangement to yield a second dienone (*cf.* 56a), whose enol is the *p*-substituted phenol (*cf.* 57a). Enolisation normally predominates, but where (51) has *o*-substituents, i.e. (54a), '*o*-enolisation' cannot take place, and only the *p*-phenol (57a) is then obtained. That this product is indeed formed not by <u>direct</u> migration of the allyl group, but by two successive shifts, is suggested by the 'double inversion' of the position of the ^{14}C label in the allyl group that is found to occur:

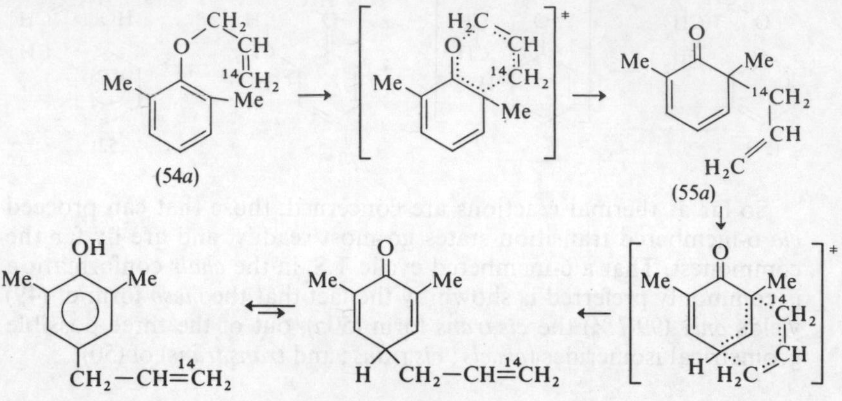

(54a)   (55a)

(57a)   (56a)

Further confirmation of the two-fold shift, and of the double inversion of the position of the ^{14}C label, is provided by 'trapping' (*cf.* p. 49) the first dienone intermediate (55a) with maleic anhydride in a Diels–Alder reaction.

Finally, it must be emphasised that where, in any of the electrocyclic reactions, cycloadditions or sigmatropic rearrangements considered above, a reaction has been described as symmetry forbidden, this applies to the concerted pathway only: it could well be that an energetically feasible, non-concerted pathway is still available, involving zwitterionic or biradical intermediates. Equally, the statement that a reaction is symmetry allowed does not necessarily guarantee that it will proceed readily in practice: the attainment of the required geometry in the T.S. could well be inhibited by the size of ring required, by the presence of particular substituents, or for other reasons.

13

Linear free energy relationships

13.1 INTRODUCTION

In previous chapters we have considered the relative reactivity of numerous series of compounds in specific reactions—such as nucleophilic displacement by $EtO^\ominus$ in the series of bromoalkanes below (*cf.* p. 85)—

$$CH_3CH_2Br > MeCH_2CH_2Br > Me_2CHCH_2Br > Me_3CCH_2Br$$

and have sought to account for the reactivity sequences observed in terms of the operation of electronic and steric effects. This has proved a useful and rewarding exercise, but a major disadvantage of such studies, and explanations, is that they remain *qualitative*: what is still needed is a method for relating structure and reactivity on a *quantitative* basis.

13.2   FIRST HAMMETT PLOTS

The first such relationship, on a thoroughly established basis, was observed by Hammett as long ago as 1933. He showed that for the reaction of a series of methyl esters (1) with NMe_3,

$$RCO_2Me + NMe_3 \xrightarrow{k} RCO_2^{\ominus} + {}^{\oplus}NMe_4$$

(1)

the rates of reaction were directly related to the ionisation constants, in water, of the corresponding carboxylic acids (2):

$$RCO_2H + H_2O \overset{K}{\rightleftarrows} RCO_2^{\ominus} + H_3O^{\oplus}$$

(2)

Thus on plotting log k for reaction of the esters (1) against log K for ionisation of the acids (2) (he actually plotted the $-$log values so as to have more easily handled numbers) a reasonable straight line resulted (Fig. 13.1):

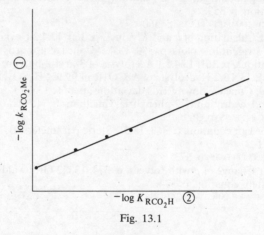

Fig. 13.1

Equilibrium constants, K, and rate constants, k, are each related to free energy changes (pp. 34, 37) in the relevant reactions in the following way:

$$\Delta G^{\ominus} = -2 \cdot 303 \, RT \log K$$
equilibrium constant

$$\Delta G^{\ddagger} = -2 \cdot 303 \, RT \log k \left(\frac{h}{k'T} \right)$$

rate constant

$$\left[\begin{array}{l} h = \text{Planck's constant} \\ k' = \text{Boltzmann's constant} \end{array} \right]$$

The fact that there is in Fig. 13.1 a straight line relationship between $-$log k for reaction of the esters (1), and $-$log K, for ionisation in water of the corresponding carboxylic acids (2), implies that there is also a straight line relationship between $\Delta G^{\ddagger}$, the free energy of activation for the ester reaction, and $\Delta G^{\ominus}$, the standard free energy change for ionisation in water of the acids. Because of this straight line relationship between the free energy terms for these two different reaction series, straight line plots like the one in Fig. 13.1 are generally referred to as *linear free energy relationships*.

Another early example of Hammett's is shown in Fig. 13.2, which represents a plot of log k for base-catalysed hydrolysis of a group of ethyl esters (3) against log K for ionisation in water of the corresponding carboxylic acids (2). Judged

$$RCO_2Et + {}^{\ominus}OH \xrightarrow{k} RCO_2^{\ominus} + EtOH$$

(3)

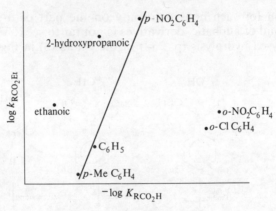

Fig. 13.2

by the standards of Fig. 13.1, the plot in Fig. 13.2 is pretty disappointing: there is a straight line relationship for benzoic acid and its p-Me and p-NO$_2$ derivatives, but the o-NO$_2$ and o-Cl benzoic acid derivatives then lie far off to one side of this straight line, while the aliphatic derivatives, ethanoic and 2-hydroxypropanoic acids, lie far off to the other side. Hammett found indeed that straight lines were not generally obtained if reaction data for either o-substituted benzene derivatives, or aliphatic species, were included in the plot. He did, however, find that if consideration was restricted to reactions of m- and p-substituted benzene derivatives, then—as shown in Fig. 13.3 (p. 348)—excellent linearity resulted, and this held for a very wide range of different reactions of such derivatives.

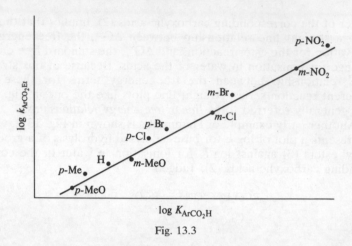

Fig. 13.3

A reason for such non-conformity on the part of *o*-substituted benzene, and of aliphatic, derivatives is not far to seek. Thus for the base catalysed hydrolysis (p. 234) of the esters (3) in Figs 13.2 and

13.3, the *m*- or *p*-substituent in (3*a*) is far removed from the reaction centre and, in this rigid molecule, can exert no steric effect upon it. By contrast, the *o*-substituent in (3*b*) is close at hand (*cf.* p. 237), and leads to increasing crowding in the transition state leading to the tetrahedral intermediate produced in slow, rate-limiting attack on ester (3*b*) by ⊖OH; very much the same is true also for the more flexible molecules of the aliphatic esters (3*c*).

13.3 THE HAMMETT EQUATION

Despite establishing such linear relationships for a wide range of reactions of *m*- and *p*-substituted benzene derivatives, we still lack any simple form of this quantitative relationship that can actually be used to investigate new situations: here again, it was Hammett who supplied the answer.

13.3.1 Derivation of Hammett equation

The general equation for a straight line is $y = mx + c$, and this can be applied to the straight line in Fig. 13.3 to give,

$$\log k_X = \rho \log K_X + c \qquad [1]$$

where ρ is the slope of this straight line, c the intercept, and X is the particular *m*- or *p*-substituent in the benzene ring of the species concerned. It is also possible to write an exactly analogous equation that is restricted to the *unsubstituted* ester and acid, i.e. where X = H:

$$\log k_H = \rho \log K_H + c \qquad [2]$$

Subtracting [2] from [1], we obtain,

$$\log k_X - \log k_H = \rho(\log K_X - \log K_H) \qquad [3]$$

which may also be written in the form:

$$\log \frac{k_X}{k_H} = \rho \log \frac{K_X}{K_H} \qquad [4]$$

13.3.2 Substituent constant, σ_X

Hammett then designated the ionisation, in water at 25°, of *m*- and *p*-substituted benzoic acids as his *standard reference reaction*. He chose this reaction because reasonably precise aqueous ionisation constant, K_X, data were already available in the literature for quite a range of differently *m*- and *p*-substituted benzoic acids. Knowing K_H and K_X for a variety of differently X-substituted benzoic acids, it is then possible to define a quantity, σ_X, as

$$\sigma_X = \log \frac{K_X}{K_H} \qquad [5]^*$$

where σ_X is a *substituent constant*, differing from one substituent to another but remaining constant for a particular substituent no matter what the reaction in which a compound carrying this substituent is involved.

* [5] may, of course, also be written in the form, $\sigma_X = pK_{a(H)} - pK_{a(X)}$; so that the numerical value of σ_X for a particular substituent is obtained by simple subtraction of the pK_a value for the substituted acid (where this is known) from the pK_a value for benzoic acid itself.

Substituting [5] into [4] we then get,

$$\log \frac{k_X}{k_H} = \rho\sigma_X \qquad [6]$$

which is the usual form of what has come to be called the Hammett equation.

By using known values of K_X (or pK_a) for aqueous ionisation of m- and p-substituted benzoic acids (or measuring K_X [pK_a] where the value is not already available for a particular m- or p-substituent) it is possible to calculate σ_X as required, and a selection of values obtained in this way is shown below:

Substituent, X	$\sigma_{m\text{-}X}$	$\sigma_{p\text{-}X}$	
Me$_3$C	−0·10	−0·20	
Me	−0·07	−0·17	
H	0	0	(by definition)
MeO	+0·12	−0·27	
HO	+0·12	−0·37	
F	+0·34	+0·06	
Cl	+0·37	+0·23	
MeCO	+0·38	+0·50	
Br	+0·39	+0·23	
CN	+0·56	+0·66	
NO$_2$	+0·71	+0·78	

Hardly surprisingly, the value of σ_X for a particular substituent is found to depend on the location of the substituent, having a different value in the m-position from that in the p-.

13.3.3   Reaction constant, ρ

Having thus obtained a range of substituent constant, σ_X, values it is now possible to use them to calculate the value of ρ, the *reaction constant*, in [6] for any further reactions in which we may be interested: this is often done graphically. Thus to evaluate ρ for, say, the base-catalysed hydrolysis of m- and p-substituted ethyl 2-arylethanoates (4) we would, from kinetic measurements (or from

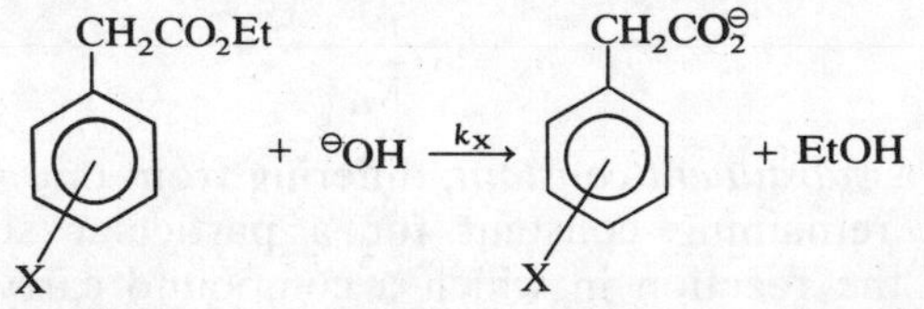

(4)

the literature if we're lucky!), obtain k_H for the unsubstituted ester, and k_X for at least three different substituted esters. Knowing the value of σ_X for each of these substituents, we can then plot log (k_X/k_H) against σ_X and, from [6], the slope of the resulting straight line will be the value of ρ for this reaction: it turns out to be $+0 \cdot 82$ for this particular hydrolysis, when carried out in aqueous ethanol at 30°. The ρ values for quite a wide range of different reactions of m- and p-substituted benzene derivatives are shown below:

Reaction	Type	ρ	
(1) ArNH$_2$ with 2,4-(NO$_2$)$_2$C$_6$H$_3$Cl in EtOH(25°)	k	$-3 \cdot 19$	
(2) ArNH$_2$ with PhCOCl in C$_6$H$_6$(25°)	k	$-2 \cdot 69$	
(3) ArCH$_2$Cl solvolysis in aq. Me$_2$CO(69·8°)	k	$-1 \cdot 88$	
(4) ArO$^\ominus$ with EtI in EtOH(25°)	k	$-0 \cdot 99$	
(5) ArCO$_2$H with MeOH (acid-catalyscd, 25°)	k	$-0 \cdot 09$	
(6) ArCO$_2$Me hydrolysis (acid) in aq. MeOH(25°)	k	$+0 \cdot 03$	
(7) ArCH$_2$CO$_2$H ionisation in H$_2$O(25°)	K	$+0 \cdot 47$	
(8) ArCH$_2$Cl with I$^\ominus$ in Me$_2$CO(20°)	k	$+0 \cdot 79$	
(9) ArCH$_2$CO$_2$Et hydrolysis (base) in aq. Et(OH(30°)	k	$+0 \cdot 82$	
(10) ArCO$_2$H ionisation in H$_2$O(25°)	K	$+1 \cdot 00$	(standard reaction)
(11) ArOH ionisation in H$_2$O(25°)	K	$+2 \cdot 01$	
(12) ArCN with H$_2$S (base) in EtOH(60·6°)	k	$+2 \cdot 14$	
(13) ArCO$_2$Et hydrolysis (base) in aq. EtOH(25°)	k	$+2 \cdot 51$	
(14) ArNH$_3^\oplus$ ionisation in H$_2$O(25°)	K	$+2 \cdot 73$	

The standard reaction, the aqueous ionisation of m- and p-substituted benzoic acids at 25°, will have a ρ value of $1 \cdot 00$ as a necessary concomitant of the definition of σ_X in [5], and its use in [6]. The value of the reaction constant, ρ, for a particular reaction, carried out under specified conditions, remains constant no matter what the m- or p-substituents present in the compounds involved.

13.3.4 Physical significance of σ_X and ρ

Before we can go on to consider the actual use that may be made of Hammett plots, it is necessary to provide some physical justification for σ_X and ρ in terms of the more familiar factors that we have already seen influencing reaction rates and equilibria.

If we consider σ_X, the substituent constant, first and look at the list of σ_{m-X} values (p. 350), we can see that m-Me$_3$C and m-Me each have a small $-$ve value, H has the value—by definition—of zero, while all the other m-substituents have (increasing) $+$ve values. The change in sign ($-$ve $\rightarrow$ $+$ve) does, of course, parallel the change in direction (electron-donating $\rightarrow$ electron-withdrawing) of the inductive effect exerted by these substituents. The substituents may also exert a field effect (p. 151), operating through the medium, but this will act in the same direction as the inductive effect. It would thus seem that σ_{m-X} represents, both in direction and magnitude, a measure of the *total polar effect* exerted by the substituent X on the reaction centre.

This is borne out by a comparison of the rates of base-catalysed hydrolysis (*cf.* p. 234) of *m*-NO$_2$ (5), and of *m*-Me (6), substituted ethyl benzoates with that of the unsubstituted ester: a reaction in which the slow, and hence rate-limiting, step is initial attack on the ester by $^\ominus$OH (p. 234):

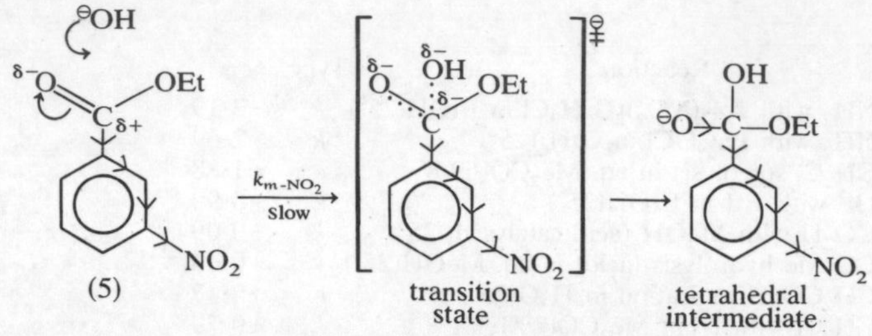

$$\sigma_{m\text{-NO}_2} = +0\cdot71 \qquad \frac{k_{m\text{-NO}_2}}{k_H} = 63.5$$

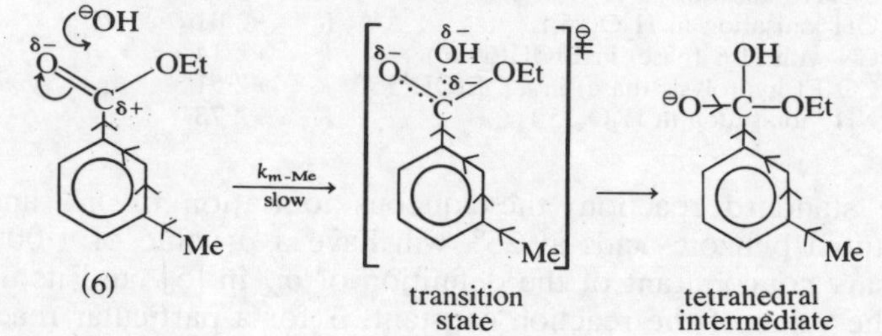

$$\sigma_{m\text{-Me}} = -0\cdot07 \qquad \frac{k_{m\text{-Me}}}{k_H} = 0\cdot66$$

The *m*-nitro ester (5), with $\sigma_{m\text{-NO}_2} = +0\cdot71$, is hydrolysed 63·5 times as fast as the unsubstituted ester (powerful electron-withdrawal markedly assisting $^\ominus$OH attack on the carbonyl carbon atom, and stabilising the transition state leading to the negatively charged tetrahedral intermediate); while the *m*-Me ester (6), with $\sigma_{m\text{-Me}} = -0\cdot07$, is hydrolysed 0·66 times as fast as the unsubstituted ester (very weak electron-donation slightly inhibiting $^\ominus$OH attack, etc).

If we now look at the list of $\sigma_{p\text{-X}}$ values (p. 350), it is apparent that not only does the $\sigma_{p\text{-X}}$ value for a particular substituent, X, vary in *magnitude* from the $\sigma_{m\text{-X}}$ value for the same substituent, it may differ in *sign* too: as is the case with *m*- and *p*-MeO. An examination of the effect of a *m*-MeO (7) and a *p*-MeO (8) substituent on the same reaction as above (base-catalysed ester hydrolysis) makes plain the reason for this change in sign (p. 353).

In the *m*-position, the electronegative oxygen atom of the MeO group exerts an electron-withdrawing inductive effect ($\sigma_{m\text{-MeO}} = +0\cdot12$) and hydrolysis is faster than with the unsubstituted ester [*cf.*

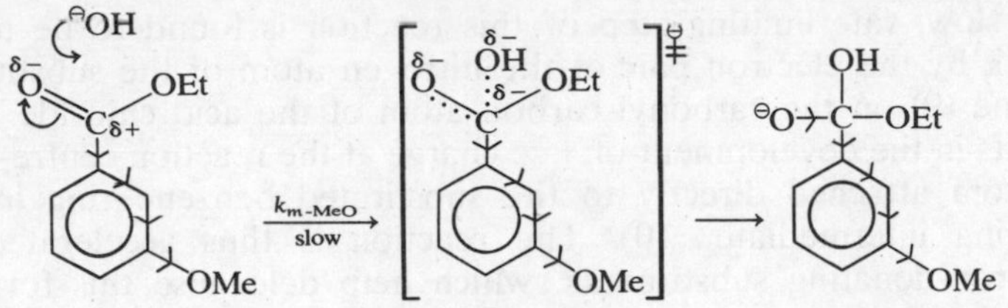

(7)   transition state   tetrahedral intermediate

$$\sigma_{m\text{-MeO}} = +0\cdot12 \qquad k_{m\text{-MeO}} > k_H$$

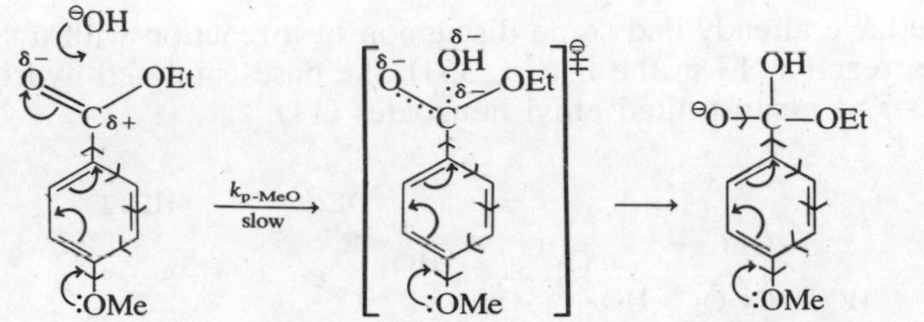

(8)   transition state   tetrahedral intermediate

$$\sigma_{m\text{-MeO}} = -0\cdot27 \qquad k_H > k_{p\text{-MeO}}$$

the m-NO_2 ester (5)]. In the p-position, MeO will still exert an electron-withdrawing inductive effect, but in addition it can, through its electron pairs, exert an electron-donating mesomeric effect on the ring carbon atom to which the CO_2Et group is attached. The latter effect, because it involves the more readily polarisable π electron system, is the greater of the two, and the overall result is therefore *net* electron-donation ($\sigma_{p\text{-MeO}} = -0\cdot27$); as is required by the observation that the p-MeO ester is hydrolysed markedly more slowly than the unsubstituted compound (*cf.* p. 153).

Now let us consider ρ, the reaction constant. Looking at the list of ρ values (p. 351), we can select first a reaction with a sizeable $-$ve ρ value, say reaction 2—the benzoylation of m- and p-substituted anilines (9)—with $\rho = -2\cdot69$, and look at this reaction rather more closely:

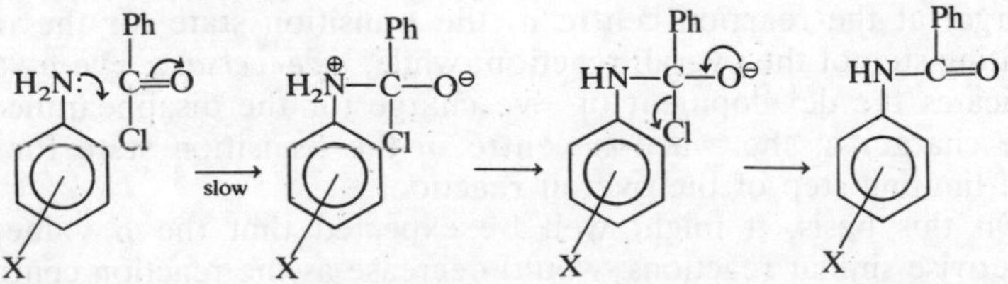

(9)   (10)

The slow, rate-limiting step of this reaction is found to be initial attack by the electron pair of the nitrogen atom of the substituted aniline (9) on the carbonyl carbon atom of the acid chloride. This results in the development of +ve charge at the reaction centre—the N atom attached directly to the substituted benzene ring in the forming intermediate (10). The reaction is thus accelerated by electron-donating substituents, which help delocalise this forming +ve charge in the transition state leading to the intermediate (10), and correspondingly retarded by electron-withdrawing substituents; this behaviour is found to hold in general for reactions with −ve ρ values.

We have already had some discussion of a reaction with a +ve ρ value, reaction 13 in the list (p. 351), the base-catalysed hydrolysis of *m*- and *p*-substituted ethyl benzoates (11):

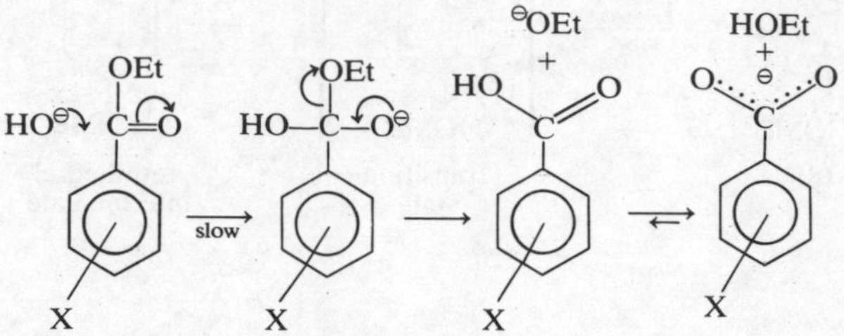

This has ρ value of +2·51, the known slow, rate-limiting step in this reaction is attended by the development of −ve charge adjacent to the reaction centre in the transition state leading to the intermediate (12), and the overall reaction is, as we have already seen (p. 352), accelerated by electron-withdrawing, and retarded by electron-donating, substituents.

The *magnitude* of the reaction constant, ρ, is essentially a measure of the susceptibility of a reaction to the polar effect of substituents in the adjacent benzene ring—the larger the value of ρ, the greater the susceptibility of the reaction. The *sign* of ρ is also of diagnostic use, as we have seen, in that a −ve ρ value indicates the development of +ve charge (or, of course, the disappearance of −ve charge) at the reaction centre in the transition state for the rate-limiting step of the overall reaction; while, *vice-versa*, a +ve ρ value indicates the development of −ve charge (or the disappearance of +ve charge) at the reaction centre in the transition state for the rate-limiting step of the overall reaction.

On this basis, it might well be expected that the ρ value, of otherwise similar reactions, would decrease as the reaction centre is moved further away from the substituents that are exerting a polar, electronic effect upon it. This is borne out by the ρ values for the aqueous ionisation of the acids (13)–(16):

Acid ionisation (H₂O)	ρ
(13) $XC_6H_4CO_2H$	1·00 (standard reaction)
(14) $XC_6H_4CH_2CO_2H$	0·49
(15) $XC_6H_4CH_2CH_2CO_2H$	0·21
(16) $XC_6H_4CH{=}CHCO_2H$	0·47

Introduction of first one, and then two, CH_2 groups between the benzene ring and CO_2H progressively reduces the susceptibility of the acid's ionisation to the polar effect of the substituent X in the benzene ring. The susceptibility, as revealed by the value of ρ, rises again for (16), however, as $CH{=}CH$ is a markedly better transmitter of electronic effects than is $CH_2{-}CH_2$.

13.3.5 Through-conjugation—σ_x^- and σ_x^+

Before we go on to consider the major uses of σ_x and ρ, it is first necessary to take a little closer look at just how constant the σ_x value for a particular substituent really is. If we plot data for the aqueous ionisation of m- and p-substituted benzoic acids (13)—the standard reaction—against that for ionisation of the corresponding substituted phenols (17), a very reasonable straight line (Fig. 13.4) is obtained for a wide range of different substituents:

① $XC_6H_4CO_2H + H_2O \rightleftharpoons XC_6H_4CO_2^\ominus + H_3O^\oplus$
   (13)
② $XC_6H_4OH + H_2O \rightleftharpoons XC_6H_4O^\ominus + H_3O^\oplus$
   (17)

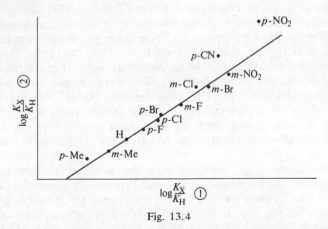

Fig. 13.4

Two substituents however, the powerfully electron-withdrawing p-NO_2 and p-CN, lie *above* this straight line: indicating that p-NO_2

phenol and *p*-CN phenol are in fact *stronger* acids than we would have expected them to be. Why this is so becomes apparent if we write out the structures of the species involved in both ionisation equilibria for, say, the *p*-NO$_2$ compounds, (18 and 19) and examine the polar, electronic effects that can operate in them:

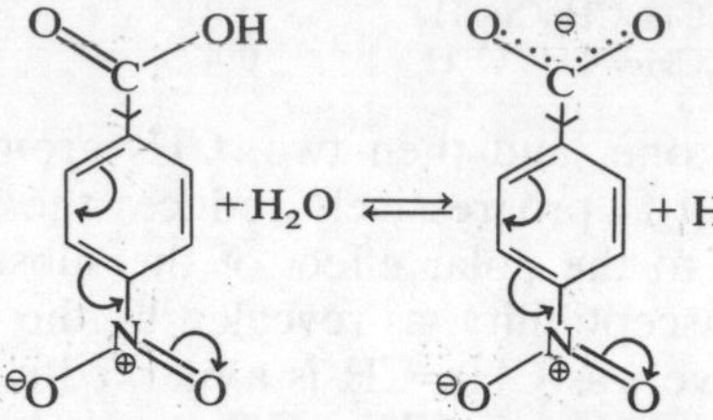

(18a)   +H$_2$O ⇌   (18b)   + H$_3$O$^\oplus$

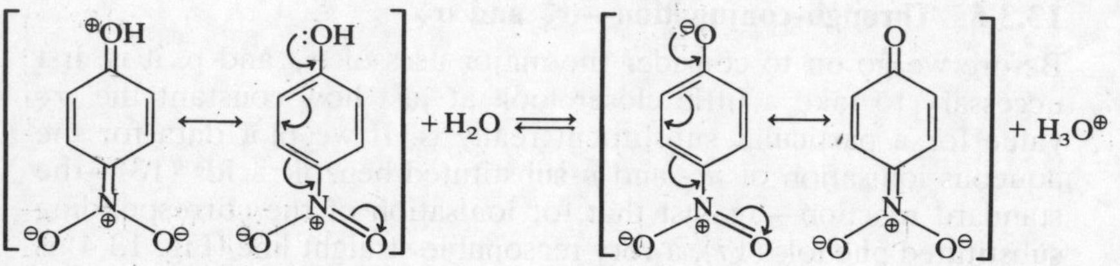

(19a)   +H$_2$O ⇌   (19b)   + H$_3$O$^\oplus$

For each species, the inductive effect of the *p*-NO$_2$ substituent—which will be essentially similar in each of the sets of species—has been omitted, but the mesomeric or conjugative effect has been included. In (18a) ⇌ (18b), the standard reaction that was used to evaluate $\sigma_{p\text{-NO}_2}$, the conjugative effect of the *p*-NO$_2$ substituent is transmitted ultimately to the reaction centre only through an inductive effect—operating from the ring carbon atom carrying the CO$_2$H, or CO$_2^\ominus$, on that group itself. In (19a) ⇌ (19b), however, the conjugative effect can be transmitted right through from the *p*-NO$_2$ substituent to the electron pairs on the oxygen atom which is now the reaction centre. This effect will be particularly marked in (19b), where the anion will be stabilised substantially by delocalisation of its −ve charge, and the ionisation equilibrium for *p*-NO$_2$ phenol thereby displaced over towards the right in the anion's favour.

The value for $\sigma_{p\text{-NO}_2}$ obtained from the standard reaction, (18a ⇌ 18b), clearly does not take into account the heightened effect of this '*through-conjugation*', which is why the point for *p*-NO$_2$—and for *p*-CN—is off the line in Fig. 13.4. Such through-conjugation can, however, be allowed for by using the aqueous ionisation of phenols to establish a set of new, alternative, σ values, for *p*-NO$_2$ and other comparable electron-*withdrawing* substituents: these new values may then be used for reactions in which through-conjugation can occur.

This can be achieved by first plotting $\log K_X/K_H$ against σ_X for *m*-substituted phenols only (which cannot be involved in through-conjugation), then the slope of the resulting straight line will give the value of ρ, the reaction constant, for this reaction. Using this value in the normal Hammett equation ([6], p. 350), enables us to calculate the new, revised, σ_{p-} value for *p*-NO$_2$, and for similar substituents capable of through-conjugation. These revised figures are generally referred to as σ_{p-}^- values, and a number are compared with the normal σ_{p-} values below:

Substituent, X	$\sigma_{p\text{-}X}^-$	$\sigma_{p\text{-}X}$
CO$_2$Et	0·68	0·45
COMe	0·84	0·50
CN	0·88	0·66
CHO	1·03	0·43
NO$_2$	1·27	0·78

An exactly analogous situation will arise where there is the possibility of direct through-conjugation between a suitable electron-*donating* p-substituent and a reaction centre at which +ve charge is developing. A good example is solvolysis (S$_N$1) of the tertiary halides, 2-aryl-2-chloropropanes (20), shown in Fig. 13.5:

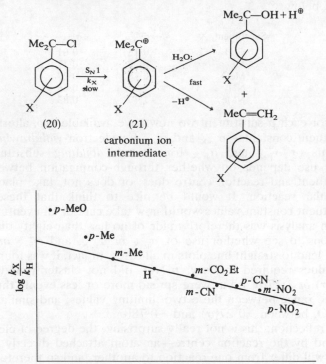

Fig. 13.5

Solvolysis of the p-MeO and p-Me chlorides is found to be faster than expected because of stabilisation, by through-conjugation, of the carbonium ion intermediates (21a and 21b) developing in the slow, rate-limiting step of the solvolysis:

p-MeO (21a)   p-Me(21b)

The fact that development of +ve charge, in the transition state for this slow step, is substantial is borne out by the large −ve ρ value, −4·54, for the reaction. By using this solvolysis as a new standard reaction, it is possible, as with σ_p^-, to obtain in an exactly analogous manner a set of σ_p^+ values that make allowance for through-conjugation by powerful electron-*donating* p-substituents; a number of these revised figures are compared with the corresponding σ_p-values below:

Substituent, X	$\sigma_{p\text{-}X}^+$	$\sigma_{p\text{-}X}$
C_6H_5	−0·18	−0·01
Me	−0·31	−0·17
MeO	−0·78	−0·27
NH_2	−1·30	−0·66
NMe_2	−1·70	−0·83

So for each p-substituent we now have available *two*, alternative, substituent constants—$\sigma_{p\text{-}X}^-$ and $\sigma_{p\text{-}X}$ for electron-*withdrawing* substituents or $\sigma_{p\text{-}X}^+$ and $\sigma_{p\text{-}X}$ for electron-*donating* substituents—whose use depends on whether through-conjugation between p-substituent and reaction centre does, or does not, take place in a particular reaction. It would be nice to think that these dual substituent constant values would now take care of all eventualities, and an analysis was therefore made of no less than eighty different reactions to see whether use of $\sigma_{p\text{-}X}^-$ or $\sigma_{p\text{-}X}$, and $\sigma_{p\text{-}X}^+$ or $\sigma_{p\text{-}X}$, would lead to straight line plots in all cases. In fact, it was found that the values required for, say, p-NO_2 did not cluster round *either* 0·78(σ) or 1·27(σ^-), but were spread more or less evenly throughout the range between these two, limiting values; and similarly for p-MeO, between −0·27(σ) and −0·78(σ^+).

On reflection, this is not really surprising: the degree of electron-demand by the reaction centre—an atom attached directly to the ring—will differ from one reaction to another, and so therefore will the degree of response, *via* through-conjugation, by a particular *p*-substituent; hence the need for a whole range of different $\sigma_{p\text{-}X}$ values for such a substituent.

13.3.6 Yukawa–Tsuno equation

There have been a number of attempts, by the introduction of a further parameter into the Hammett equation, to quantify this graded response—*via* through-conjugation—on the part of a *p*-substituent. Among the best known of these is the Yukawa–Tsuno equation, [7], which, in the form shown here, is

$$\log \frac{k_X}{k_H} = \rho[\sigma_X + r(\sigma_X^+ - \sigma_X)]\qquad [7]$$

applicable to electron-*donating* *p*-substituents; for electron-*withdrawing* *p*-substituents σ_X^+ would, of course, be replaced by σ_X^-. The new parameter, *r*, intended as a measure of the through-conjugation operating in a particular reaction, is given the value of 1·00 for solvolysis of the tertiary halides, 2-aryl-2-chloropropanes (20). For this reaction [7] does, of course, then simplify to [8],

$$\log \frac{k_X}{k_H} = \rho\sigma_X^+\qquad [8]$$

which is reasonable enough as it was this reaction that we used (p. 357) to define σ_X^+ in the first place, for electron-donating *p*-substituents capable of through-conjugation!

To evaluate *r* for other reactions, we can obtain ρ for the reaction by measuring k_X values for *m*-substituted compounds only, and then measure k_X for *p*-substituted compounds where the values of $\sigma_{p\text{-}X}$ and $\sigma_{p\text{-}X}^+$, or $\sigma_{p\text{-}X}^-$, are already known. Using [7], *r* can then be evaluated by calculation, or by graphical methods. Thus for the base-catalysed hydrolysis of *p*-substituted phenoxytriethylsilanes

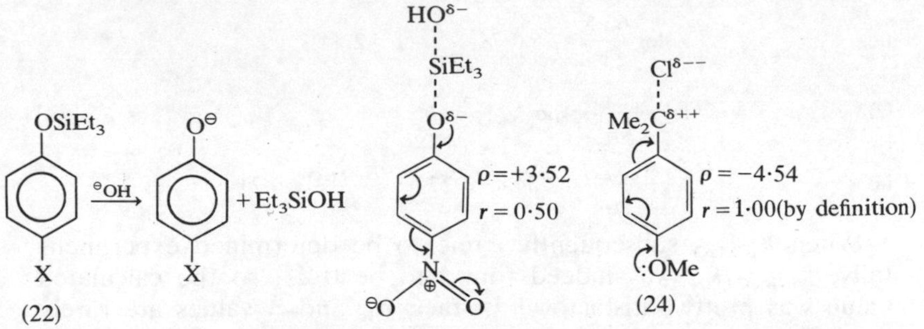

(22), the value of *r* is found to be 0·50. This extent of through-conjugation—by a substituent such as *p*-NO$_2$—suggests the development of substantial $-$ve charge ($\rho = +3\cdot52$) in the transition

state (23) for the rate-limiting step. This will not, however, be so far advanced as the development of +ve charge ($\rho = -4\cdot54$) in the transition state (24) for the standard reaction, halide solvolysis, where $r = 1\cdot00$. As, in each case, the development of charge in the transition state goes hand-in-hand with bond-breaking between the reaction centre and the leaving group, the magnitude of r can perhaps be construed as some indication of the extent of such bond-breaking by the time the transition state has been reached.

13.4 USES OF HAMMETT PLOTS

Having now given some consideration to the significance that can be attached to σ_X and ρ in more familiar physical terms, it is possible to go on and discuss the actual uses that can be made of them in providing information about reactions and the pathways by which they take place.

13.4.1 Calculation of *k* and *K* values

The simplest possible use that can be made of the Hammett equation is to calculate k or K for a specific reaction of a specific compound, where this information is not available in the literature, or indeed where the actual compound has not even been prepared yet. Thus it is known that the base-catalysed hydrolysis of ethyl *m*-nitrobenzoate is 63·5 times as fast as the hydrolysis of the corresponding unsubstituted ester under parallel conditions; what then will be the comparable rate for base-catalysed hydrolysis of ethyl *p*-methoxybenzoate under the same conditions? Looking at the table of σ_X values (p. 350), we find that $\sigma_{m\text{-}NO_2} = 0\cdot71$, while $\sigma_{p\text{-}MeO} = -0\cdot27$. Then from the Hammett equation [6] (p. 350):

① $$\log \frac{k_{m\text{-}NO_2}}{k_H} = \rho\sigma_{m\text{-}NO_2} \qquad [6a]$$

i.e. $$\log \frac{63\cdot5}{1} = \rho \times 0\cdot71 \quad \therefore \ \rho = 2\cdot54$$

② $$\log \frac{k_{p\text{-}MeO}}{k_H} = \rho\sigma_{p\text{-}MeO} \qquad [6b]$$

i.e. $$\log \frac{k_{p\text{-}MeO}}{k_H} = 2\cdot54 \times -0\cdot27 \quad \therefore \ \frac{k_{p\text{-}MeO}}{k_H} = 0\cdot21$$

When $k_{p\text{-}MeO}$ subsequently came to be determined experimentally, $k_{p\text{-}MeO}/k_H$ was indeed found to be 0·21, so the calculated value was pretty satisfactory! In fact, σ_X and ρ values are rarely used for such a purpose, they are employed much more often in providing salient data about reaction pathways.

13.4.2 Deviations from straight line plots

We have already seen (p. 354) how the sign and magnitude of ρ, the reaction constant, can provide useful information about the development (or dissipation) of charge (+ve or −ve) on going from starting materials to the transition state for the rate-limiting step of a reaction. We have also seen (p. 355) how deviations from straight line plots using normal σ_X, substituent constant, values led to the definition of σ_X^+ or σ_X^- values to take into account through-conjugation between certain p-substituents and the reaction centre. The need to use other than the normal σ_X values indicates the occurrence of such through-conjugation in a particular reaction, and the Yukawa–Tsuno parameter, r, then provides a measure of its extent.

Paradoxically, Hammett plots are usually most informative at the very point at which they depart from linearity, but the major inference that can be drawn from this departure is found to differ depending on whether the deviation is concave 'upwards' or concave 'downwards'.

13.4.3 Concave upwards deviations

13.4.3.1 Acetolysis of 3-aryl-2-butyl brosylates

An interesting case in point is the acetolysis of 3-aryl-2-butyl p-bromobenzenesulphonates or brosylates (25), for which the Hammett plot is shown in Fig. 13.6. The lower right-hand side of the

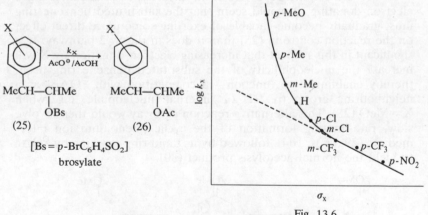

Fig. 13.6

plot—where the substituents are powerfully electron-withdrawing—is a straight line whose slope indicates a ρ value for the reaction of −1·46. On moving across to the left—as the substituents become less electron-withdrawing—the plot now curves upwards, indicating that the rate of acetolysis of these species is *faster* than we would have expected it to be on the basis of the σ_X values for their substituents.

What we might expect as a pathway for this reaction would be simple S_N2 displacement (p. 97) of the good leaving group— brosylate anion—by acetate anion:

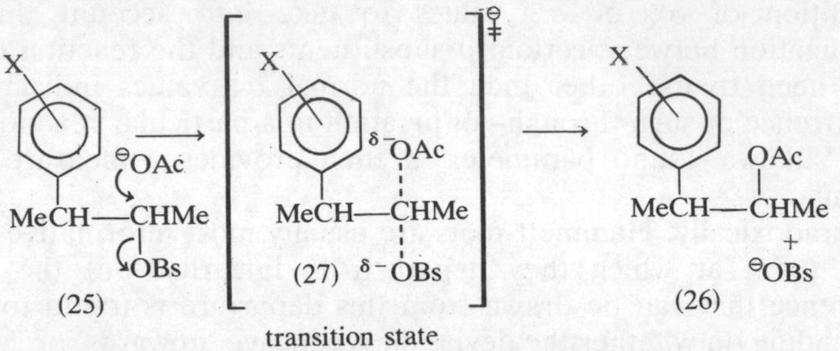

(25)   (27) $^{\delta-}$OBs   (26)

transition state

The smallish $-ve$ ρ value ($-1\cdot46$) is compatible with such a path-way, given that in the transition state (27) breaking of the C—OBs bond, is somewhat more fully advanced than formation of the AcO—C bond, resulting in the transient development of a small amount of $+ve$ charge at the reaction centre. This is in no sense unreasonable with (a) a secondary carbon atom as reaction centre (cf. p. 82), and (b) so good a leaving group (cf. p. 97); this pathway would be increasingly aided, albeit weakly, as the substituent X becomes less electron-withdrawing, i.e. the reaction would slowly increase in rate.

To account for the departure from linearity, as X becomes more electron-donating, it would seem that the substituted benzene ring must gradually become capable of exerting some more direct effect on the reaction centre in (25) than it does in the S_N2 pathway. It is significant in this respect that increasing electron-donation by X will increase the nucleophilicity of the substituted benzene ring itself, thereby enabling it to function—in competition with $^{\ominus}$OAc—as a neighbouring group (p. 92) or 'internal' nucleophile, e.g. when X = MeO (28). This alternative reaction pathway would then involve slow, rate-limiting formation of the cyclic phenonium ion inter-mediate (29, cf. p. 104), followed by its rapid ring-opening by $^{\ominus}$OAc to yield the normal acetolysis product (30):

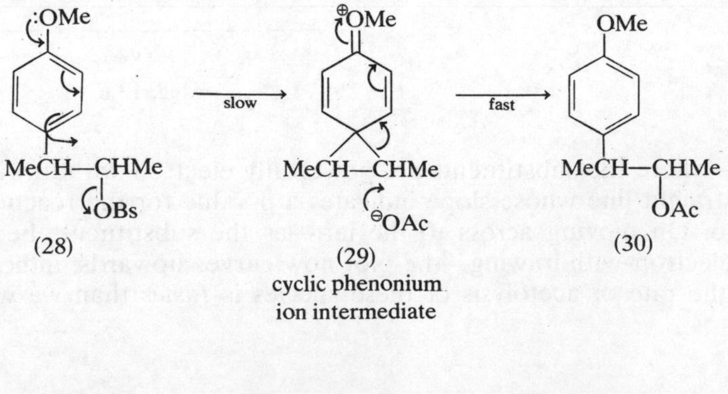

(28)   (29)   (30)

cyclic phenonium
ion intermediate

Support for the suggestion that Fig. 13.6 involves a change in actual reaction pathway is provided by acetolysis of the *threo* isomer of the brosylate (31). Acetolysis will then lead to two different, distinguishable, diastereoisomers whose relative proportion will depend on how much of the total reaction proceeds by *external* nucleophilic attack *via* the S_N2 pathway (*erythro* product, 32), and how much by *internal* nucleophilic attack *via* a cyclic phenonium ion intermediate (*threo* product, 33):

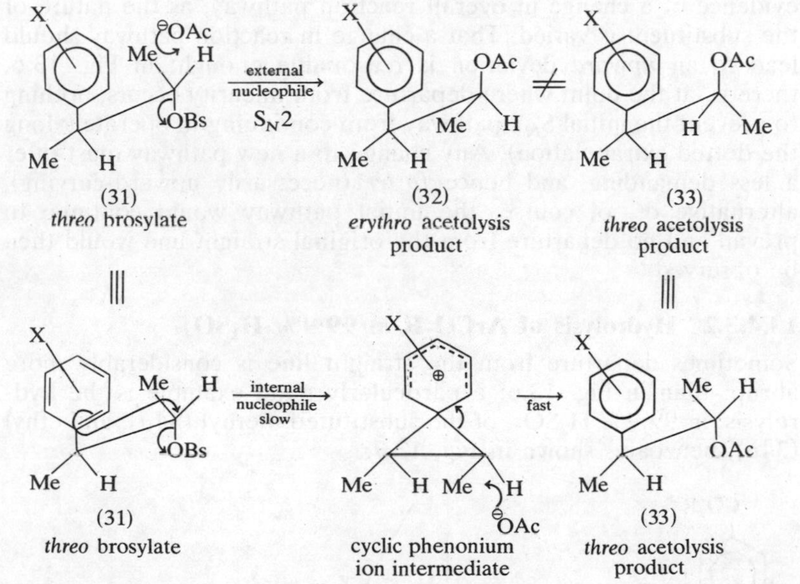

(31)
threo brosylate

(32)
erythro acetolysis
product

(33)
threo acetolysis
product

(31)
threo brosylate

cyclic phenonium
ion intermediate

(33)
threo acetolysis
product

The two, alternative, acetolysis products (32 and 33), being diastereoisomers not mirror images, may then be separated, or their relative yields estimated by spectroscopic methods. It is found that the yield of *threo* product (33) varies considerably as the nature of X, the substituent in the benzene ring, is changed:

Substituent, X	Yield of *threo* product* (33)	
p-MeO	100	
p-Me	88	
m-Me	68	
H	59	* = percentage of reaction
p-Cl	39	proceeding *via*
m-Cl	12	*internal*
m-CF$_3$	6	nucleophilic
p-NO$_2$	1	attack

When X = *p*-MeO, the most electron-donating substituent at the top left-hand corner of Fig. 13.6, 100% of acetolysis is proceeding *via* internal nucleophilic attack by *p*-MeOC₆H₄; when X = *m*-Cl, just coming on to the straight line part of the plot in Fig. 13.6, only 12% of the total reaction is proceeding *via* the internal route; while when X = *p*-NO₂, the most electron-withdrawing substituent, only 1% of the total reaction is now proceeding by this route.

When a simple Hammett plot exhibits an upward deviation, i.e. is concave upwards as in Fig. 13.6, then this can usually be taken as evidence of a change in overall reaction pathway, as the nature of the substituent is varied. That a change in reaction pathway should lead to an *upward* deviation is reasonable enough: in Fig. 13.6, there is, at the point where departure from linearity occurs, nothing to prevent the initial S_N2 pathway from continuing to operate (along the dotted extrapolation). Any change to a new pathway must offer a less demanding, and hence *faster* (necessarily upward-curving), alternative or, of course, the initial pathway would continue to prevail and no departure from the original straight line would then be observed.

13.4.3.2 Hydrolysis of ArCO₂R in 99·9% H₂SO₄

Sometimes departure from the straight line is considerably more abrupt than in Fig. 13.6; a particularly good example is the hydrolysis, in 99·9% H₂SO₄, of the substituted methyl (34*a*), and ethyl (34*b*), benzoates shown in Fig. 13.7:

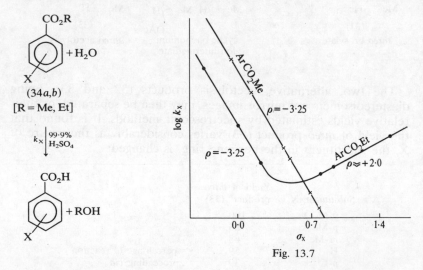

Fig. 13.7

Considering first the simpler of the two cases, the straight line for the methyl esters (34*a*) which has a ρ value of −3·25. From this ρ

value it is apparent that this reaction cannot be proceeding *via* the normal ($A_{AC}2$) pathway (p. 236) for acid-catalysed ester hydrolysis which, as we know (reaction 6, p. 351), has a ρ value of +0·03. That value refers, however, to hydrolysis being carried out with dilute sulphuric acid, while here 99·9% sulphuric acid is being used: one result of which is to make the concentration of water available for hydrolysis very low.

We have, however, already seen an alternative acid hydrolysis pathway ($A_{AC}1$, p. 237) in which a water molecule is not involved in the slow, rate-limiting step. In addition, this step is one in which considerable +ve charge is developed at the reaction centre as the protonated ester (35a) is converted into the acyl cation intermediate (36a): a necessary requirement for a reaction with a large −ve (−3·25) ρ value:

$$
\underset{(35a)}{\underset{\overset{|}{H}}{\underset{\overset{\oplus}{O}Me}{Ar\!-\!C\!=\!O}}} \underset{slow}{\rightleftharpoons} \underset{(36a)}{\underset{\oplus}{Ar\!-\!C\!=\!O}} \overset{H_2O:}{\rightleftharpoons} \underset{\overset{|}{H}}{\overset{H_2O^{\oplus}}{Ar\!-\!C\!=\!O}} \rightleftharpoons \overset{HO}{Ar\!-\!C\!=\!O}
$$

The same $A_{AC}1$ pathway must also be operating initially for the ethyl esters (34b), on the left-hand side of Fig. 13.7, as the ρ value (−3·25) for this reaction is the same as that for the methyl esters (34a). As the substituent in the benzene ring becomes more strongly electron-withdrawing, however, a sharp change in curvature is observed with the ethyl esters to a new straight line with a ρ value of +2.0. This now +ve ρ value requires a slow, rate-limiting step for hydrolysis in which +ve charge is <u>de</u>creased at the reaction centre— the overall reaction being increasingly accelerated as the substituent in the ring becomes more electron-withdrawing.

There is indeed yet another pathway for acid-catalysed ester hydrolysis ($A_{AL}1$, p. 236) that would fulfil this requirement:

$$
\underset{(35b)}{\underset{\overset{|}{H}}{\underset{\overset{\oplus}{O}\!-\!CH_2Me}{Ar\!-\!C\!=\!O}}} \underset{slow}{\rightleftharpoons} \underset{(37b)}{\underset{\overset{|}{H}}{\underset{O}{Ar\!-\!C\!=\!O}}} + {}^{\oplus}CH_2Me \overset{H_2O:}{\rightleftharpoons} HO\!-\!CH_2Me + H^{\oplus}
$$

Loss of MeCH₂$^{\oplus}$ the ethyl cation (37b), leads to a marked decrease in +ve charge adjacent to the reaction centre (had it actually been from the reaction centre itself the +ve value of ρ would have been larger); this carbonium ion intermediate (37b) reacts rapidly with any available water to yield ethanol.

The question does then arise, given the observed shift in reaction pathway for the ethyl esters (34b), why does a similar shift not occur with the corresponding methyl esters $(34a)$? Such a shift would, of course, necessitate the formation of a methyl, $CH_3^\oplus$, rather than an ethyl, $MeCH_2^\oplus$ $(37b)$, cation in the slow, rate-limiting step. $CH_3^\oplus$ is known to be considerably more difficult to form than is $MeCH_2^\oplus$ and this difference is apparently great enough to prevent, on energetic grounds, such an $A_{AC}1 \rightarrow A_{AL}1$ shift with the methyl esters, despite potential assistance (to $A_{AL}1$) from increasingly electron-withdrawing substituents.

13.4.4 Concave downwards deviations

There are, however, also examples of deviations from simple Hammett plots in which the curvature is in the opposite direction, concave downwards, and these deviations have a rather different significance.

13.4.4.1 Cyclodehydration of 2-phenyltriarylmethanols

A good example is the cyclodehydration of some substituted 2-phenyltriarylmethanols (38), in 80% aqueous ethanoic acid containing 4% H_2SO_4 at 25°, to yield the corresponding tetraarylmethanes (39), as shown in Fig. 13.8:

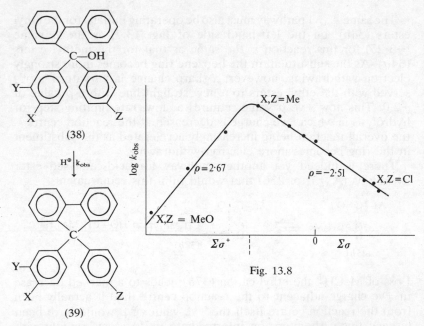

Fig. 13.8

Two of the benzene rings in (38) carry substituents, involving a total of three possible positions of substitution, but none of the compounds studied contains more than two substituents. Where there are two substituents, the value actually plotted is the sum of the separate σ_X value for each substituent. The plot—of k_{obs} for the overall reaction against σ_X—is clearly a composite of two straight lines, one on the left with $\rho = +2.67$, and one on the right with $\rho = -2.51$.

There seems little doubt that the overall reaction follows a four-step pathway, the first two steps constituting an E1 (p. 241) elimination of water to yield a carbonium ion intermediate (40), which then, in the last two steps, effects internal electrophilic substitution on the 2-phenyl nucleus to yield the product tetraaryl-methane (39):

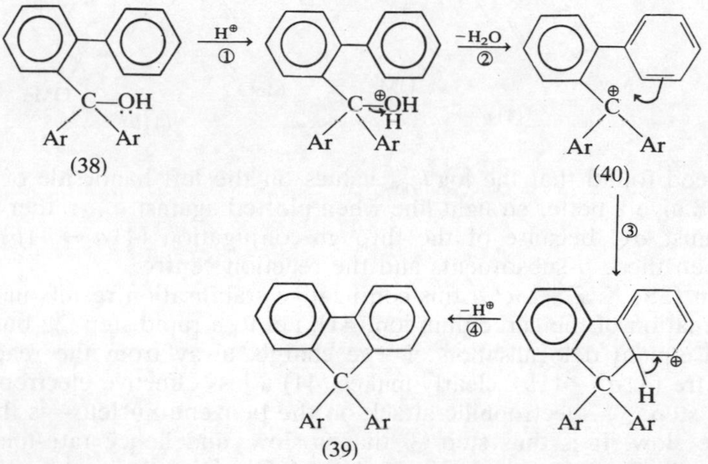

The question then arises—which step in the overall reaction is likely to be the slow, and hence rate-limiting, one? It's unlikely to be step ①: initial protonation in acid-catalysed dehydration is generally rapid; or step ④: final loss of proton in aromatic electrophilic substitution is also generally rapid. This leaves steps ② and ③ as possible candidates for the slow step overall, and fortunately a clear distinction can be made between them. In step ②, +ve charge is *increasing* at the reaction centre (the carbon atom carrying the two substituted Ar groups), while in step ③, +ve charge is *decreasing* at the reaction centre. How does this match up with the requirements of Fig. 13.8?

The right-hand side of the plot in Fig. 13.8 has a −ve ρ value (−2.51) indicating the development of substantial +ve charge at the reaction centre during the overall, rate-limiting step. This would, of course, be compatible with step ② being rate-limiting, but not with step ③. For the left-hand side of the plot in Fig. 13.8, exactly the reverse is true; here a +ve ρ value (+2·67) indicates a substantial decrease of +ve charge at the reaction centre, which would be compatible with step ③ being rate-limiting, but not with step ②.

It is significant that the substituents involved at the far left-hand side of the plot (38; X, Z = MeO) are powerfully electron-donating, and thus capable of stabilising the carbonium ion (41a ↔ 41b), developing in step ②, by delocalisation of its +ve charge. It is

(41a)   (41b)

indeed found that the log k_{obs} values on the left hand side of Fig. 13.8 give a better straight line when plotted against σ_X^+, rather than against σ_X, because of the through-conjugation (41a ↔ 41b) between these *p*-substituents and the reaction centre.

In (38; X, Z = MeO) this conjugative stabilisation results in easy formation of the carbonium ion (41), i.e. to a rapid step ②; but the consequent delocalisation of +ve charge, away from the reaction centre (41a ↔ 41b), clearly makes (41) a less effective electrophile, i.e. step ③—electrophilic attack on the benzene nucleus—is therefore slow. It is thus step ③ that is slow, and hence rate-limiting overall, for compound (38; X, Z = MeO). On moving across Fig. 13.8, from left to right, the substituents become less electron-donating, delocalisation of +ve charge thereby becomes less pronounced, and the reaction centre progressively more electrophilic.

Rate-limiting step ③ is thus speeded-up, and the overall reaction rate therefore increases, i.e. the slope of the plot is upwards from left to right ($\rho = +2·67$). Also on moving from left to right, decreasing through-conjugation, as the substituents become less electron-donating, makes carbonium ion formation more difficult; thus step ② is being slowed down as step ③ is being speeded-up. There must, therefore, come a point at which speeding-up step ③ catches up with the slowing-down step ②; any further decrease in electron-donation by the substituents must result in step ② becoming slower than step ③, thereby making it now rate-limiting for the overall reaction. This shift in rate-limiting step from step ③ → step ② occurs, in Fig. 13.8, with the compound (38; X, Z = Me).

Still further decrease in electron-donation by the substituents, beyond this point, will result in still further slowing-down of step

②—now the rate-limiting step—and hence slowing-down of the overall reaction, i.e. the slope on the right-hand side of the plot is now downwards from left to right ($\rho = -2.51$). For a reaction in which such a shift of rate-limiting step is observed (as the electron-donating/-withdrawing ability of the substituent is changed) there will be one substituent, or narrow range of substituents, for which the balance between the rates of step ② and step ③ is such as to make the overall reaction rate a maximum.

This happens in Fig. 13.8, as we have seen, with the compound (38; X, Z = Me). On each side of this maximum the, different, rate-limiting step will be slowing down progressively, and so therefore will the overall reaction rate. Shifts in rate-limiting step, within the same overall reaction pathway, are thus distinguished by concave *downwards* deviations in Hammett plots; this in contrast to the concave *upwards* deviations which, as we have already seen (p. 364), are characteristic of a change in actual reaction pathway.

13.5 STERIC EFFECTS

Quite early on (p. 348) in this discussion of linear free energy relationships consideration was restricted to the side-chain reactions of *m*- and *p*-substituted benzene derivatives. The reactions of *o*-substituted benzene derivatives, and indeed of aliphatic compounds, were excluded because of the operation of steric and other effects, which led to non-linear, or even to apparently random, plots.

The success and utility of Hammett plots, and the realisation that they are often of most value diagnostically when they do indeed diverge from linearity, has emboldened a number of workers to seek, with suitable modifications, to extend their scope to a much wider range of compounds. The most general and successful of these extensions was proposed by Taft.

13.5.1 The Taft equation

Acting on a suggestion originally made by Ingold, Taft began by comparing the relative susceptibility to polar substituent effects (the ρ value) of the hydrolysis—under acid-catalysed ($A_{AC}2$, p. 236) and under base-catalysed ($B_{AC}2$, p. 234) conditions—of *m*- and *p*-substituted benzoate esters (42) (p. 370).

The ρ value for base-catalysed hydrolysis (+2·51) is +ve and quite large, reflecting the development of not inconsiderable −ve charge at the reaction centre in the rate-limiting step—attack on this centre by $^{\ominus}OH$ (step ① in the $B_{AC}2$ pathway). By contrast, the ρ value for acid-catalysed hydrolysis (+0·03) is very nearly zero; which means, of course, that the rate of this hydrolysis does not vary significantly from one ester to another, no matter what the *m*- or

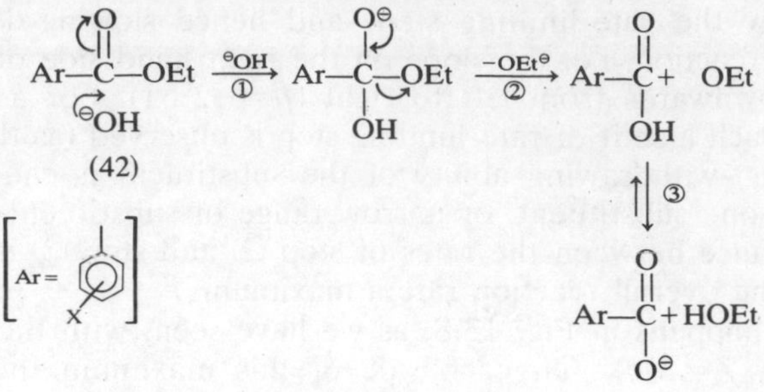

Base-catalysed hydrolysis $(B_{AC}2)$: $\rho = 2 \cdot 51$

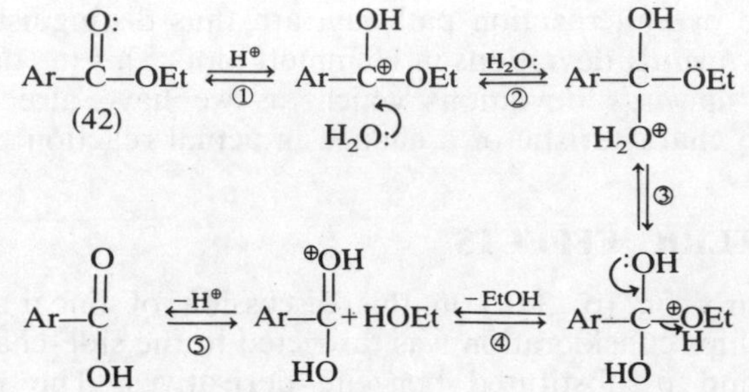

Acid catalysed hydrolysis $(A_{AC}2)$: $\rho = 0 \cdot 03$

p-substituent present. The ρ value for this hydrolysis is so small not because there is no significant redistribution of charge in the slow step (step ②)), but because the overall reaction rate, i.e. k_{obs} (which is plotted to evaluate ρ), is determined not solely by k_2 for this slow step, but involves also K_1 for the preceding, reversible, step ①. These two terms all but cancel each other out, in so far as susceptibility of the two steps to electron-donation/-withdrawal by polar substituents is concerned, and the overall ρ value for the reaction is thus virtually zero.

If we now extend our consideration of base-catalysed ($B_{AC}2$), and acid-catalysed ($A_{AC}2$), hydrolysis to esters in general, including aliphatic ones (RCO$_2$Et), we see that there is a close similarity between the transition states (42*b* or 42*a*) for the rate-limiting step in each of the two pathways: they are both tetrahedral; and differ

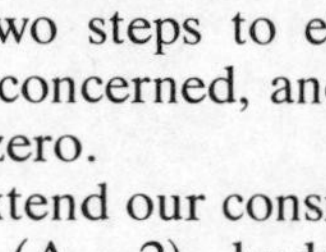

T.S. for base-catalysed   T.S. for acid-catalysed
hydrolysis ($B_{AC}2$)   hydrolysis ($A_{AC}2$)

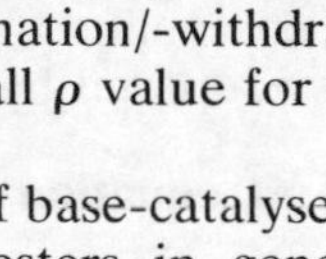

only in the second of them having two protons more than the first. Protons, being very small, exert comparatively little steric influence; it is therefore a not unreasonable assumption that any steric effect stemming from the group R is, because of the close spatial similarity of the two transition states, substantially the same in both acid- and base-catalysed hydrolysis. It then becomes possible to write a Hammett type equation, [9], to represent the operation of the *polar* effect only of substituent R in ester hydrolysis:

$$\log\left[\frac{k_R}{k_0}\right]_{base} - \log\left[\frac{k_R}{k_0}\right]_{acid} = \rho^* \sigma_R^* \qquad [9]$$

As the steric effect exerted by R is essentially the same in both modes of hydrolysis, the two steric terms will cancel each other out, and will thus not appear in equation [9].

Taft then gave ρ^* in [9], the value 2·48, derived by subtracting the ρ value for acid-catalysed hydrolysis of benzoate esters (0·03) from the ρ value for base-catalysed hydrolysis of the same esters (2·51). He took as his reference substituent R = Me, rather than R = H, so that k_0 in [9] refers to $\underline{Me}CO_2Et$ rather than $\underline{H}CO_2Et$. Then by kinetic measurements on the acid- and base-catalysed hydrolysis of a series of esters containing R groups other than Me, it is possible—using [9]—to evaluate σ_R^* for each of these different R groups with respect to Me, for which by definition $\sigma_{Me}^* = 0$ (*cf.* $\underline{H}$ with $\sigma_H = 0$ for benzoic acid ionisation, p. 350). By giving ρ^* here the value 2·48, the resulting σ_R^* values—which are a measure of the polar effect only exerted by R—do not differ too greatly in magnitude from the values of σ_X, σ_X^+ and σ_X^- with which we are already familiar (p. 350).

Then, employing the more general equation [10], it is possible to use these σ_R^* values, in conjunction with suitable kinetic measurements of k_R and k_{Me}, to evaluate ρ^* for other

$$\log\frac{k_R}{k_{Me}} = \rho^* \sigma_R^* \qquad [10]$$

reactions of a whole range of aliphatic compounds in addition to esters. Using [10] in this way, straight line plots were obtained for a number of different reactions of aliphatic compounds.

13.5.2 Steric parameters, E_S and δ

After all the emphasis we placed earlier (p. 348) on steric effects, obtaining a straight line plot may at first sight seem a rather surprising result; especially, given that the relation [10] takes into account only the *polar* effect exerted by R. However, obtaining a straight line plot, using [10], does not necessarily mean that *no* steric effects are operating in a reaction. It means only that there is no substantial *change* in the operation of such effects on going from starting materials to the transition state for the rate-limiting step of the overall reaction (or on going from starting materials to products for an equilibrium).

It is not necessary to look very far to find aliphatic reactions that do *not* yield straight line plots with [10], however; and, as with previous deviations from linearity (p. 361), these departures are commonly much more informative about the details of reaction pathways than are neat straight lines. Where such departures from linear (polar effects only) plots are observed, suggesting the operation of significant—and changing—steric effects, it is possible to incorporate a *steric substituent parameter*, E_S, whose evaluation is based on an earlier observation.

Thus we have already seen (p. 369) that the acid-catalysed hydrolysis of *m*- and *p*-substituted benzoate esters (42) is (with a ρ value of 0·03) essentially uninfluenced by any polar effect exerted by the substituent, X; and this substituent is sufficiently far removed from the reaction centre to be clearly incapable of exerting any

(42)

steric effect on it either. These esters thus all undergo acid-catalysed hydrolysis at essentially the same rate. There is no reason to believe that acid-catalysed hydrolysis of aliphatic esters, RCO_2Et, will be any more susceptible to polar effects than was the corresponding hydrolysis of benzoate esters. If then different hydrolysis rates *are* observed with aliphatic esters as R is varied, these must reflect differing steric effects exerted by the different R groups. Such aliphatic esters are indeed found to undergo hydrolysis at markedly different rates, so it is possible, taking Me as the standard substituent once again, to use equation [11]

$$\log \left[\frac{k_{RCO_2Et}}{k_{MeCO_2Et}} \right]_{acid} = E_S \qquad [11]$$

to evaluate E_S, the steric substituent parameter, for R. E_S values, obtained in this way for a number of different substituents, are listed below:

R in RCO_2Et	E_S		R in RCO_2Et	E_S
H	+1·24		$Me(CH_2)_3$	−0·39
Me	0	(by definition)	Me_2CHCH_2	−1·13
Et	−0·07		Me_3C	−1·54
$ClCH_2$	−0·24		Me_3CCH_2	−1·74
ICH_2	−0·37		Ph_2CH	−1·76
$PhCH_2$	−0·38		Et_3C	−3·81

From the form of equation [11], the E_S value for Me, the reference substituent, will of course be 0. All substituents other than H have $-ve$ E_S values because all substituents other than H are larger than Me, and the rate of hydrolysis of any ester RCO_2Et ($R \neq H$) will thus be slower than that of $MeCO_2Et$, in a reaction whose rate is governed solely by the steric effect of R.

It is found in practice that the value of the steric parameter, E_S, for a particular group, R, differs to some extent from one reaction to another. This is not altogether surprising as both the local environment of R and the size of the attacking reagent will vary from one reaction to another. It means, however, that on incorporating E_S into the Hammett type equation, [12], it is necessary to introduce a yet further parameter, δ, as a measure of a particular reaction's

$$\log \frac{k_R}{k_{Me}} = \rho^*\sigma_R^* + \delta E_S \qquad [12]$$

susceptibility towards steric effects. In that sense δ is the steric parallel to ρ^*—which measures the reaction's susceptibility towards polar effects. The δ parameter is given the value $1 \cdot 00$ for acid-catalysed ester hydrolysis, as the standard reaction, and its value for other reactions can then be determined experimentally in the usual way.

Now that steric parameters have been introduced in this way, the treatment can be extended to include the reactions of *o*-substituted benzene derivatives as well. Thus for the acid-catalysed hydrolysis of *o*-substituted benzamides (43), the value of δ is found to be $0 \cdot 81$; so this reaction is apparently slightly less susceptible to the steric

CONH₂ ... CO₂H structures with $\xrightarrow[H_2O]{H^\oplus}$

(43)

effect of substituents than is the earlier standard reaction, the acid-catalysed hydrolysis of aliphatic esters, RCO_2Et.

13.6   SOLVENT EFFECTS

One of the things our discussion of linear free energy relationship has not yet made any endeavour to take into account is the role played in reactions by the solvent. This despite the fact that the very great majority of organic reactions do take place in solution, with the solvent often playing a crucial role.

13.6.1   Change of ρ with solvent

It is, of course, true that some implicit consideration is given to the solvent in that the ρ value for a particular reaction is found to change when the solvent in which the reaction is carried out is changed:

Reaction		ρ
$ArCO_2H(44) + H_2O \rightleftarrows ArCO_2^{\ominus}(45) + H_3O^{\oplus}$ (H_2O)		1·00 (by definition)
" + " $\rightleftarrows$ " + " (50% aq. EtOH)		1·60
" + " $\rightleftarrows$ " + " (EtOH)		1·96
$ArCO_2Et + {}^{\ominus}OH \rightarrow ArCO_2^{\ominus} + EtOH$ (70% aq. dioxan)		1·83
" + " $\rightarrow$ " + " (85% aq. EtOH)		2·54

For ionisation of m- and p-substituted benzoic acids (44), the hydroxylic solvent is capable of solvating both the undissociated acid (44) and the carboxylate anion (45) obtained from its ionisation. The relative effectiveness of such solvation—of negatively charged anion (45) with respect to neutral, undissociated acid (44)—is a major factor in determining the position of equilibrium, i.e. K_X. As the solvent is changed from water, with a dielectric constant of 79, to ethanol, with a dielectric constant of only 24, there will be a marked decrease in advantageous solvation of the charged anion (45) with respect to the uncharged acid (44). The relative importance of the polar effect exerted by electron-withdrawing substituents, in overall stabilisation of the carboxylate anion (i.e. in acid-strengthening: increasing K_X), will therefore *increase* as the dielectric constant of the solvent *decreases*. The value of ρ, the susceptibility of the reaction to the polar effect of a substituent, will also increase, therefore, on changing the solvent from water to ethanol.

13.6.2   Grunwald–Winstein equation

Attempts to correlate the differing rate of a particular reaction, when carried out in a range of different solvents, with the dielectric constant values for these solvents have not proved very rewarding. Attempts have therefore been made to establish empirical reactivity/solvent correlations along general Hammett lines. Among the more significant of these attempts has been that of Grunwald and Winstein on the solvolysis of halides. They sought to establish a solvent parameter, designated **Y**, which would correlate with the different rate constants found for solvolysis of the same halide in a range of different solvents.

They took as their standard reaction the S_N1 solvolysis of the tertiary halide, 2-chloro-2-methylpropane (46), and selected as their standard solvent 80% aqueous ethanol (80% EtOH/20% H_2O):

$$Me_3C{-}Cl \xrightarrow[\text{slow}]{S_N1} Me_3C^{\oplus}\ Cl^{\ominus} \xrightarrow[\text{fast}]{S} Me_3C{-}S$$
$$(46) \qquad\qquad (47) \qquad\qquad [S = \text{solvent}]$$
$$\text{ion pair}$$
$$\text{intermediate}$$

It is then possible to set up the Hammett-like relation, [13],

$$\log k_A - \log k_0 = Y_A - Y_0 \qquad\qquad [13]$$

in which the rate constants, k_A and k_0, refer to solvolysis of the tertiary halide (46) in a solvent A and in the standard solvent (80% aq. EtOH), respectively; while Y_A and Y_0 are the empirical solvent parameters for solvent A and for this standard solvent. By setting the value of Y_0 at zero and measuring k_A for the solvolysis of (46) in a range of different solvents, it is then possible, using [13], to derive a Y_A value for each of them:

Solvent, A	Y_A	
H_2O	+3.49	
aq. MeOH (50% H_2O)	+1·97	
HCONH$_2$	+0·60	
aq. EtOH (30% H_2O)	+0·59	
aq. EtOH (20% H_2O)	0	(by definition)
aq. Me$_2$CO (20% H_2O)	−0·67	
MeOH	−1·09	
EtOH	−2·03	
Me$_2$CHOH	−2·73	
Me$_3$COH	−3·26	

These Y_A values are found not to run in parallel with the dielectric constant values for the solvents concerned. Obviously the dielectric constant value for the solvent must be involved in some way in Y_A, as separation of opposite charges is a crucial feature of the rate-limiting step in an S_N1 reaction: formation of the ion-pair intermediate (47; *cf.* p. 78). But major, specific solvation effects must also be involved, and the parameter Y_A will reflect these too, and quite possibly other properties of the solvent as well. It is common to describe Y_A as representing a measure of the 'ionising power' of the solvent A.

It is now possible to go a stage further, and write a not unfamiliar relation, [14], that now covers the solvoylsis of halides in general, and not merely that of the

$$\log \frac{k_A}{k_0} = m Y_A \qquad\qquad [14]$$

standard halide, 2-chloro-2-methylpropane (46). Here k_A and k_0 are the rate constants for solvolysis of any halide, in solvent A and in the standard solvent, respectively. Y_A has already been defined as a *solvent* parameter representing the ionising power of solvent A, while m is a *compound* parameter characteristic of the particular halide: it is given the value 1·00 for the standard halide, 2-chloro-2-methylpropane (46). The actual value of m can be taken as a measure of the susceptibility of the solvolysis of a particular halide towards the ionising power, Y_A, of that solvent:

Halide	m
PhCH(Me)Br(48)	1·20
Me₃CCl(46)	1·00 (by definition)
Me₃CBr	0·94
EtMe₂CBr	0·90
CH₂=CHCH(Me)Cl	0·89
EtBr(49)	0·34
Me(CH₂)₃Br(50)	0·33

An alternative interpretation of m is that it provides some measure of the extent of ion-pair formation in the transition state for the rate-limiting step of the overall solvolysis reaction: it can then be put to some diagnostic use. Thus, ion-pair formation is known to be well advanced in the transition state for S_N1 solvolysis of 2-chloro-2-methylpropane (46), the standard halide, for which $m = 1·00$. Not altogether surprisingly the value of m for 1-bromo-1-phenylethane (48), in which the developing benzyl type cation, $[PhCHMe]^{\oplus}$, is stabilised by delocalisation of its +ve charge over the π-system of the attached benzene nucleus (*cf.* p. 84), is even larger—at 1·20. By contrast, the m values for the primary halides, bromoethane (49) and 1-bromobutane (50), are much lower—0·34 and 0·33, respectively. These values, indicating low susceptibility towards the ionising power of the solvent, are characteristic of halides whose solvolysis is known to proceed *via* the S_N2 pathway. In general, an m value of 0·5 can be taken as an approximate indicator of an S_N1/S_N2 mechanistic borderline in solvolysis reactions of this kind.

The major defect of the Grunwald–Winstein treatment is that it is limited in its scope. It has been applied to reactions other than halide solvolysis, but is in general restricted to those reactions for which the major contribution to the rate-limiting step is of the form:

$$A{-}B \xrightarrow[\text{slow}]{k} A^{\oplus} B^{\ominus}$$

There have been several other attempts to define quantitative solvent parameters, including measuring the extent to which a particular band in the spectrum of an absorbing molecule is shifted as a result of its interaction with molecules of the different solvents. None of these approaches has been particularly successful in terms of general utility, however.

13.7 THERMODYNAMIC IMPLICATIONS

It is perhaps interesting, in view of the very considerable success of Hammett plots, to say a word finally about the thermodynamic implications of linear free energy relationships in general. We have already mentioned (p. 346) the relationship between free energy change, ΔG, and $\log k$ or $\log K$; and each ΔG term is, of course, made up of an enthalpy, ΔH, and an entropy, ΔS, component:

Equilibrium constant:
$$\underline{\Delta G^{\ominus}} = -2 \cdot 303 \, RT \log K$$
$$\underline{\Delta G^{\ominus}} = \Delta H^{\ominus} - T\Delta S^{\ominus}$$

Rate constant:
$$\underline{\Delta G^{\ddagger}} = -2 \cdot 303 \, RT \log k \frac{h}{k'T}$$
$$\underline{\Delta G^{\ddagger}} = \Delta H^{\ddagger} - T\Delta S^{\ddagger}$$

$$\left[\begin{matrix} k' = \text{Boltzmann's constant} \\ h = \text{Planck's constant} \end{matrix} \right]$$

Looking back at one of our earliest examples—Fig. 13.3 (p. 348) in which $\log K$ for the ionisation of $ArCO_2H$ is plotted against $\log k$ for the base-catalysed hydrolysis of $ArCO_2Et$—the straight line implies that there is also a linear relationship between the $\Delta G^{\ominus}$ values for the former reaction and the $\Delta G^{\ddagger}$ values for the latter. Such a straight line relationship between these two series of ΔG terms is to be expected only if, for *each* series, one or other of the following conditions is satisfied:

(1) ΔH is linearly related to ΔS for the series
(2) ΔH is constant for the series
(3) ΔS is constant for the series

Any of these conditions constitutes an extremely stringent limitation, and there has always been some doubt expressed over the extent to which any one of them is indeed satisfied in reactions which nevertheless give quite good straight line Hammett plots: thereby making the linear relationships that are observed all the more mysterious! Interestingly enough, doubt has also been expressed whether even the standard, reference reaction—the ionisation of m- and p-substituted benzoic acids in water at 25°—itself satisfied any one of these apparently necessary conditions.

A major obstacle to deciding the truth or otherwise of this assertion about benzoic acid ionisation has been the experimental difficulties involved in making the necessary measurements. The solubility of the acids in water is pretty low, and their $\Delta H^{\ominus}$ values are very small, with consequent imprecision in, and unreliability of, the results so obtained. Relatively recently, however, $\Delta G^{\ominus}$, $\Delta H^{\ominus}$ and $\Delta S^{\ominus}$ have been redetermined, with great precision, for a series of ten m- and p-substituted benzoic acids. Using this data, stringently linear plots were obtained for $\Delta H^{\ominus}$ against $\Delta S^{\ominus}$, for $\Delta G^{\ominus}$ against $\Delta H^{\ominus}$, and for $\Delta G^{\ominus}$ against $\Delta S^{\ominus}$. So it looks as though Hammett was on to a good thing after all when he made his choice of standard, reference reaction in the first place!

It is important, however, to remember that what theoretical interpretation there has been of the Hammett equation has come from circumstantial evidence rather than by rigorous proof. It remains an empirical relationship and, to that extent, there is no point in even trying to evaluate σ_X and ρ to several places of decimals. The sort of information we need, as an aid to the elucidation of reaction pathways, is of an 'order of magnitude' kind: such things as whether ρ is +ve or −ve, whether its value is large or small, whether there are noticeable deviations from linearity in plots of σ_X against $\log k_X$ and, if so, of what kind.

Having said all that, it is equally important to remember that the number and variety of useful correlations to which Hammett plots have given rise is quite astonishing, particularly when we consider the simplicity and convenience of the approach. Indeed, linear free energy relationships in general constitute a testament to the theoretical utility of concepts that are purely empirical in their genesis!

Select bibliography

Structural theory and spectroscopy

DEWAR, M. J. S. and DOUGHERTY, R. C. *The PMO Theory of Organic Chemistry* (Plenum, 1975).

FLEMING, I. *Frontier Orbitals and Organic Chemical Reactions* (Wiley, 1976).

MCWEENY, R. *Coulson's Valence* (O.U.P., 3rd Edition, 1979).

MORTIMER, C. T. *Reaction Heats and Bond Strengths* (Pergamon, 1962).

STREITWIESER, A. *Molecular Orbital Theory for Organic Chemists* (Wiley, 1961).

WILLIAMS, D. H. and FLEMING, I. *Spectroscopic Methods in Organic Chemistry* (McGraw-Hill, 3rd Edition, 1980).

Structure and reaction mechanism

ALDER, R. W., BAKER, R. and BROWN, J. M. *Mechanism in Organic Chemistry* (Wiley-Interscience, 1971).

AMIS, E. S. *Solvent Effects on Reaction Rates and Mechanisms* (Academic Press, 1966).

BANTHORPE, D. V. *Elimination Reactions* (Elsevier, 1963).

BUNCEL, E. *Carbanions: Mechanistic and Isotopic Aspects* (Elsevier, 1975).

COLLINS, C. J. and BOWMAN, N. S. (Eds.). *Isotope Effects in Chemical Reactions* (Van Nostrand Reinhold, 1970).

CRAM, D. J. *Fundamentals of Carbanion Chemistry* (Academic Press, 1965).

DE LA MARE, P. B. D. *Electrophilic Halogenation* (C.U.P., 1976).

DE LA MARE, P. B. D. and BOLTON, J. *Electrophilic Addition to Unsaturated Systems* (Elsevier, 1966).

ELIEL, E. L. *Stereochemistry of Carbon Compounds* (McGraw-Hill, 1962).

FORRESTER, A. R., HAY, J. M. and THOMSON, R. H. *Organic Chemistry of Stable Free Radicals* (Academic Press, 1968).

GARRATT, P. J. *Aromaticity* (McGraw-Hill, 1971).

GILCHRIST, T. L. and REES, C. W. *Carbenes, Nitrenes and Arynes* (Nelson, 1969).

GILCHRIST, T. L. and STORR, R. C. *Organic Reactions and Orbital Symmetry* (C.U.P., 2nd Edition, 1979).

GILLIOM, R. D. *Introduction to Physical Organic Chemistry* (Addison-Wesley, 1970).

GOULD, E. S. *Mechanism and Structure in Organic Chemistry* (Holt-Dryden, 1959).

HARTSHORN, S. R. *Aliphatic Nucleophilic Substitution* (C.U.P., 1973).

HINE, J. *Physical Organic Chemistry* (McGraw-Hill, 2nd Edition, 1962).

HINE, J. *Structural Effects on Equilibria in Organic Chemistry* (Wiley, 1975).

HOFFMANN, R. W. *Dehydrobenzene and Cycloalkynes* (Academic Press, 1967).

HOGGETT, J. G., MOODIE, R. B., PENTON, J. R. and SCHOFIELD, K. *Nitration and Aromatic Reactivity* (C.U.P., 1971).

INGOLD, C. K. *Structure and Mechanism in Organic Chemistry* (Bell, 2nd Edition, 1969).

ISAACS, N. S. *Reactive Intermediates in Organic Chemistry* (Wiley, 1974).

JENCKS, W. P. *Catalysis in Chemistry and Enzymology* (McGraw-Hill, 1969).

JOHNSON, C. D. *The Hammett Equation* (C.U.P., 1973).

JONES, R. A. Y. *Physical and Mechanistic Organic Chemistry* (C.U.P., 1979).

KIRMSE, W. *Carbene Chemistry* (Academic Press, 2nd Edition, 1971).

KOCHI, J. K. (Ed.). *Free Radicals* (Wiley-Interscience, Vols. I and II, 1973).

KOSOWER, E. M. *An Introduction to Physical Organic Chemistry* (Wiley, 1968).

LEFFLER, J. E. and GRUNWALD, E. *Rates and Equilibria of Organic Reactions* (Wiley, 1963).

LWOWSKI, W. (Ed.) *Nitrenes* (Interscience, 1970).

MILLER, J. *Aromatic Nucleophilic Substitution* (Elsevier, 1968).

NEWMAN, M. S. (Ed.). *Steric Effects in Organic Chemistry* (Wiley, 1956).

NONHEBEL, D. C. and WALTON, J. C. *Free-radical Chemistry* (C.U.P., 1974).

NORMAN, R. O. C. and TAYLOR, R. *Electrophilic Substitution in Benzenoid Compounds* (Elsevier, 1965).

OLAH, G. A. and SCHLEYER, P. VON R. (Eds.). *Carbonium Ions* (Interscience, Vols. I–V, 1968–76).

SAUNDERS, W. H. and COCKERILL, A. F. *Mechanisms of Elimination Reactions* (Wiley-Interscience, 1973).

SHORTER, J. *Correlation Analysis in Organic Chemistry* (O.U.P., 1973).

STEVENS, T. S. and WATTS, N. E. *Selected Molecular Rearrangements* (Van Nostrand Reinhold, 1973).

SYKES, P. *The Search for Organic Reaction Pathways* (Longman, 1972).

THYAGARAJAN, B. S. (Ed.). *Mechanisms of Molecular Migrations* (Interscience, Vol. I, 1968; Vol. II, 1969; Vol. III, 1971).

Review series

Advances in Physical Organic Chemistry. GOLD, V. and BETHELL, D.
(Eds.) (Academic Press, Vol. I, 1963–   ).
Organic Reaction Mechanisms. KNIPE, A. C. and WATTS, W. E. (Eds.)
(Interscience, Vol. I, 1966–   ).
Progress in Physical Organic Chemistry. TAFT, R. W. (Ed.) (Inter-
science, Vol. I, 1963–   ).
Reactive Intermediates—A Serial Publication. JONES, M. and MOSS, R. A.
(Eds.) (Wiley-Interscience, Vol. I, 1978–   ).

Audio cassettes

The following titles in the 'Chemistry Cassette' series are published
by, and are available from, The Chemical Society, Blackhorse
Road, Letchworth, Herts., SG6 1HN, England:
CC3 SYKES, P. *Some Organic Reaction Pathways: A). Elimination;
B). Aromatic Substitution* (1975).
CC6 SYKES, P. *Some Reaction Pathways of Double Bonds:
A). C = C; B). C = O* (1977).
CC7 SYKES, P. *Some Reaction Pathways of Carboxylic Acid Deriva-
tives* (1979)
CC9 SYKES, P. *Radicals and their Reaction Pathways* (1979).
CC11 SYKES, P. *Linear Free Energy Relationships* (1980).

Index